API Design

API Design for C++

Second Edition

Martin Reddy

Fellow of the IEEE

Fellow of the AAIA

Distinguished Member of the ACM

Morgan Kaufmann is an imprint of Elsevier
50 Hampshire Street, 5th Floor, Cambridge, MA 02139, United States

ISBN: 978-0-443-22219-1

For information on all Morgan Kaufmann publications visit our website at
https://www.elsevier.com/books-and-journals

Publisher: Mara Conner
Acquisitions Editor: Chris Katsaropoulos
Editorial Project Manager: Palak Gupta
Production Project Manager: Prasanna Kalyanaraman
Cover Designer: Greg Harris

Typeset by TNQ Technologies

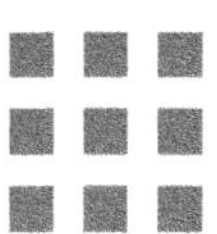

Contents

Author biography xi
Foreword xiii
Preface xv
Acknowledgments xxi

1. Introduction 1
   - What are APIs? 1
   - What's different about API design? 4
   - Why should you use APIs? 7
   - When should you avoid APIs? 11
   - API examples 13
   - Libraries, frameworks, and software development kits 17
   - File formats and network protocols 19
   - About this book 21

2. Qualities 25
   - Model the problem domain 25
   - Hide implementation details 31
   - Minimally complete 42
   - Easy to use 50
   - Loosely coupled 65
   - Stable, documented, and tested 80

3. Patterns 81
   - Pimpl idiom 83
   - Singleton 93

Factory Methods 101
API wrapping patterns 107
Observer pattern 115

4. Design 123
A case for good design 123
Gathering functional requirements 129
Creating use cases 132
Elements of API design 139
Architecture design 142
Class design 154
Function design 167

5. Styles 179
Flat C APIs 180
Object-oriented C++ APIs 186
Template-based APIs 189
Functional APIs 193
Data-driven APIs 197

6. C++ usage 209
Namespaces 209
Constructors and assignment 211
Const correctness 216
Templates 220
Operator overloading 227
Function parameters 236
Avoid #define for constants 239
Avoid using friends 242

Exporting symbols 244
Coding conventions 247

7. C++ revisions 251
Which C++ revision to use 251
C++11 API features 252
C++14 API features 283
C++17 API features 289
C++20 API features 307
C++23 API features 322

8. Performance 329
Pass input arguments by const reference 331
Minimize #include dependencies 333
Declaring constants 338
Initialization lists 341
Memory optimization 343
Don't inline functions until you need to 348
Copy on write 351
Iterating over elements 356
Performance analysis 360

9. Concurrency 367
Multithreading with C++ 367
Terminology 369
Accessing shared data 373
Concurrent API design 378

10. Versioning 383

Version numbers 383

Software branching strategies 388

Life cycle of an API 392

Levels of compatibility 393

How to maintain backward compatibility 401

API reviews 408

11. Documentation 415

Reasons to write documentation 415

Types of documentation 423

Documentation usability 430

Using Doxygen 434

12. Testing 443

Reasons to write tests 443

Types of API testing 446

Writing good tests 454

Writing testable code 459

Automated testing tools 476

13. Objective-C and Swift 487

Interface design in C++ and Objective-C 487

Data hiding in Objective-C 489

Objective-C behind a C++ API 491

C++ behind an Objective-C API 494

C++ behind a Swift API 497

14. Scripting 501

Adding script bindings 501

Script binding technologies 506

Adding Python bindings with Boost Python 510

Adding Ruby bindings with SWIG 521

15. Extensibility 533

Extending via plugins 533

Extending via inheritance 549

Extending via templates 560

Appendix A: Libraries 565

References 583

Index 589

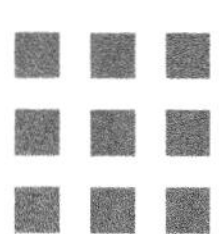

Author biography

Dr. Martin Reddy is a Fellow of the Institute of Electrical and Electronics Engineers (IEEE), a Fellow of the Asia-Pacific Artificial Intelligence Association (AAIA), and an Association for Computing Machinery (ACM) Distinguished Engineer. He holds a PhD in computer science from the University of Edinburgh and has over 30 years of experience in the software industry. He has published over 40 professional articles, 10 patents, and 2 books. And he was honored as Alumnus of the Year by his alma mater, the University of Strathclyde.

Martin was cofounder and Chief Technology Officer (CTO) of the artificial intelligence (AI) company PullString, where he oversaw the development of the company's technology until it was acquired by Apple in 2019. At Apple, Martin was a software architect within the artificial intelligence and machine learning (AI/ML) division and was responsible for the architecture and APIs of major components of the Siri virtual assistant, software that runs on over one billion devices worldwide.

Before that, Dr. Reddy worked for 6 years at Pixar Animation Studios where he was a lead software engineer working on APIs and tooling for the studio's in-house animation system, Presto. While at Pixar, Martin worked on several Academy Award–winning and nominated films, including *Finding Nemo, The Incredibles, Cars, Ratatouille,* and *Wall-E.* He was also the hair model for Mr. Incredible.

Martin began his career at SRI International, where he worked on a distributed 3D terrain visualization system called TerraVision and coauthored the geospatial features of the Virtual Reality Modeling Language (VRML) and Extensible 3D (X3D) ISO (International Organization for Standardization) international standards. He was elected a director of the Web3D Consortium for two consecutive years. His PhD was in the area of perceptually modulated level of detail for real-time 3D graphics systems.

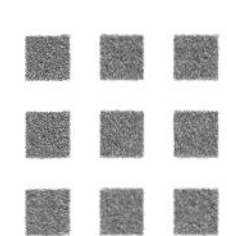

Foreword

I should begin by confessing that I do not consider myself a world-class application programming interface (API) designer or software engineer. I do, however, consider myself an expert researcher in the areas of computer graphics and geometric modeling. It was in this line of work that I first met Martin at Pixar Animation Studios.

As a graphics researcher I was accustomed to writing mathematically sophisticated papers. I was also formally trained as a computer scientist at a major university and had written my share of code. Armed with this background, when I was presented with the opportunity to lead a group of software engineers working on a new generation of animation software for Pixar, I figured that it couldn't be any more difficult than research. After all, research is, by definition, the creation of the unknown, whereas engineering is the implementation of well-understood subjects. I could not have been more wrong.

I came to realize that software engineering was without a doubt the most difficult challenge I had ever confronted. After more years than I care to admit, I eventually gave up and went back to graphics research.

I can't tell you how much I would have benefitted from a book like API Design for C++. Many of the lessons we learned the hard way have been captured by Martin in this insightful, easy to use book. Martin approaches the subject not from the perspective of an academic software researcher (although he draws heavily from results and insights gained there), but from the perspective of an in-the-trenches software engineer and manager. He has experienced firsthand the importance of good software design and has emerged as an articulate voice of what good means. In this book he presents effective strategies for achieving that goal.

I particularly like that Martin is not focusing just on API design, but more broadly on software life cycles, allowing him to cover topics such as versioning, strategies for backward compatibility, and branching methodologies.

In short, this book should be of great value to those creating or managing software activities. It is a comprehensive collection of best practices that have proven themselves time and time again.

Tony DeRose
Senior Scientist and Research Group Lead
Pixar Animation Studios

Preface

Preface to the second edition

The first edition of this book was written in 2011, right before the C++11 standard was published (it was referred to as C++0x at the time). Since then, C++11 has become well-established and further revisions have appeared at a regular cadence, including C++14, C++17, C++20, and C++23. Given all these changes to the language, I wanted to update this book to be more relevant to today's C++ engineers. So, in this second edition, I've included an entirely new chapter dedicated to the new application programming interface (API) design features that have been added to the standard over the past decade and more. I've also gone through the entire text to modernize it for these updates to the standard. This includes updating all of the code samples to use C++11 syntax where possible, because I feel that's the minimum standard today.

While working on such a large update, I decided to add some new chapters to cover topics I felt were missing from the first edition. This includes a new chapter on concurrency and multithreading, a topic that's more relevant now since C++11 introduced direct support for this important issue. Also, because I've since spent several years working at Apple, I wanted to include some of the things I learned about integrating C++ code with Objective-C and Swift, because these have become important languages for writing software on the iOS and macOS platforms.

Finally, I added several new sections and sidebars throughout the book to augment the original text, from discussing the SOLID principles and the Thread Safe Interface design pattern to covering functional programming styles and how to incorporate inclusive language in API names and documentation. Overall, this second edition represents a significant update and modernization of the book, from front cover to back cover. I hope that this new edition proves to be an even more invaluable resources for your software projects.

What is application programming interface design?

Writing large applications in C++ is a complex and tricky business. However, designing reusable C++ interfaces that are robust, stable, easy to use, and durable is even more difficult. The best way to succeed in this endeavor is to adhere to the tenets of good API design.

An API presents a logical interface to a software component and hides the internal details required to implement that component. It offers a high-level abstraction for a module and promotes code reuse by allowing multiple applications to share the same functionality.

Modern software development has become highly dependent upon APIs, from low-level application frameworks to data format APIs and graphical user interface (GUI) frameworks. In fact, common software engineering terms such as modular development, code reuse, componentization, dynamic link library, software frameworks, distributed computing, and service-oriented architecture all imply the need for strong API design skills.

Some popular C and C++ APIs that you may already recognize include the C++ Standard Library, Boost, the Microsoft Windows API (Win32), Microsoft Foundation Classes, libtiff, libpng, zlib, libxml++, OpenGL, MySQL++, Qt, wxWidgets, GTK+, KDE, SkypeKit, Intel's Threading Building Blocks, the Netscape Plugin API, and the Apache module API. In addition, many of Google's open-source projects are C++, as is a lot of the code on the GitHub website.

APIs such as these are used in all facets of software development, from desktop applications to mobile computing, and from embedded systems to Web development. For example, the Mozilla Firefox Web browser is built on top of more than 80 dynamic libraries, each of which provides the implementation for one or more APIs.

Elegant and robust API design is therefore a critical aspect of contemporary software development. One important way in which this differs from standard application development is the far greater need for change management. As we all know, change is an inevitable factor in software development; new requirements, feature requests, and bug fixes cause software to evolve in ways that were never anticipated when it was first devised. However, changes to an API that's shared by hundreds of end-user applications can cause major upheaval and ultimately may cause clients to abandon an API. The primary goal of good API design is therefore to provide your clients with the functionality they need while causing an minimal impact to their code (ideally zero impact) when you release a new version.

Why should you read this book?

If you write C++ code that another engineer relies upon, you're an API designer and this book has been written for you.

Interfaces are the most important code that you write because a problem with your interface is far more costly to fix than a bug in your implementation. For instance, an interface change may require all of the applications based upon your code to be updated, whereas an implementation-only change can be integrated transparently and effortlessly into client applications when they adopt the new API version. Put in more economic terms, a poorly designed interface can seriously reduce the long-term survival of your code. Learning how to create high-quality interfaces is therefore an essential engineering skill, and the central focus of this book.

As Michi Henning noted, API design is more important today than it was 20 years ago. This is because many more APIs have been designed in recent years. These also provide richer and more complex functionality and are shared by more end-user applications (Henning, 2009). Nevertheless, there are no other books currently on the market that concentrate on the topic of API design for C++.

It's worth noting that this book is not meant to be a general C++ programming guide. There are already many good examples of these on the market. I will certainly cover lots of object-oriented design material and present many handy C++ tips and tricks. However, I will focus on techniques for representing clean modular interfaces in C++. By corollary, I will not dive as deeply into the question of how to implement the code behind these interfaces, such as specific algorithm choices or best practices limited to the code within the curly braces of your function bodies.

On the other hand, this book will cover the full breadth of API development, from initial design through implementation, testing, documentation, release, versioning, maintenance, and deprecation. I will even cover specialized API topics such as creating scripting and plugin APIs. Although many of these topics are also relevant to software development in general, our focus here will be on the implications for API design. For example, when discussing testing strategies, I will concentrate on automated API testing techniques rather than attempting to include end-user application testing techniques such as GUI testing, system testing, or manual testing.

In terms of my own credentials to write this book, I have led the development of APIs for research code shared by several collaborating institutions, in-house animation system APIs that have been used to make Academy Award–winning movies, open source client/server APIs that have been used by millions of people worldwide, and APIs for the Siri virtual assistant that have been deployed to over a billion devices. Throughout all of these disparate experiences, I have consistently witnessed the need for high-quality API design. This book therefore presents a practical distillation of the techniques and strategies of industrial-strength API design that have been drawn from a range of real-world experiences.

Who is the target audience?

While this book is not a beginner's guide to C++, I have made every effort to make the text easy to read and to explain all terminology and jargon clearly. The book should therefore be valuable to new programmers who have grasped the fundamentals of C++ and want to advance their design skills, as well as senior engineers and software architects who are seeking to gain new expertise to complement their existing talents.

There are three specific groups of readers that I have borne in mind while writing this book:

1. **Practicing software engineers and architects.** Junior and senior developers who are working on a specific API project and need pragmatic advice on how to produce the most elegant and enduring design.
2. **Technical managers.** Program and product managers who are responsible for producing an API product and who want to gain greater insight into the technical issues and development processes of API design.
3. **Students and educators.** Computer science and software engineering students who are learning how to program and are seeking a thorough resource on software design that is informed by practical experience on large-scale projects.

Focusing on C++

Although there are many generic API design methodologies that can be taught (skills that apply equally well to any programming language or environment), ultimately an API must be expressed in a particular programming language. It's therefore important to understand the language-specific features that contribute to good API design. This book is therefore focused on the issues of designing APIs for a single language (C++) rather than diluting the content to make it applicable for all languages. Readers who wish to develop APIs for other languages, such as Java or C#, may still gain a lot of general insight from this text, but the book is directly targeted at C++ engineers who must write and maintain APIs for other engineers to consume.

C++ is still one of the most widely used programming languages for large software projects and tends to be the most popular choice for performance critical code. As a result, there are many diverse C and C++ APIs available for you to use in your own applications, some of which I listed earlier. I'll therefore concentrate on aspects of producing good APIs in C++ and include copious source code examples to illustrate these concepts better. This means that I'll deal with C++ specific topics such as templates, encapsulation, inheritance, namespaces, operators, const correctness, memory management, concurrency, the C++ Standard Library, and the pimpl idiom.

As I noted earlier, this edition of the book includes detailed coverage of all revisions of the C++ standard from C++98 through C++23. I've dedicated a chapter to covering the relevant features for each specific revision, broken out as separate sections. That way, if you're using a certain compiler version, say C++17, then you can easily review all of the API features for that version of the standard without getting distracted by features that aren't relevant to your development environment.

Conventions

Although it is more traditional to employ the term user to mean a person who uses a software application, such as a user of Microsoft Word or Mozilla Firefox, in the context of API design I will apply the term to mean a software developer who is creating an application and is using an API to achieve this. In other words, I will generally be talking about API users and not application users. The term client will be used synonymously in this regard. Note that the term client, in addition to referring to a human user of your API, can also refer impersonally to other pieces of software that must call methods in your API.

Whereas many file format extensions are used to identify C++ source files, such as `.cpp`, `.cc`, `.cxx`, `.h`, `.hh`, and `.hpp`, I will standardize on the use of `.cpp` and `.h` throughout this book. I will also use the terms module and component interchangeably to mean a single `.cpp` and `.h` file pair (except where module is meant to refer specifically to the new modules feature of C++20). These are notably not equivalent to a class, because a component or module may contain multiple classes. I will use the term library to refer to a physical collection, or package, of components (i.e., library > module/component > class).

The term method, although generally understood in the object-oriented programming community, is not strictly a C++ term. It originally evolved from the Smalltalk language. The equivalent C++ term is member function, although some engineers prefer the more specific definition of virtual member function. I am not particularly concerned with the subtleties of these terms in this book, so I will use method and member function interchangeably. Similarly, although the term data member is the more correct C++ expression, I will treat the term member variable as a synonym.

In terms of typographical conventions, I will use a fixed-width font to typeset all source code examples, as well as any filenames or language keywords that may appear in the text. Also, I will prefer upper camel case for all class and function names in the examples that I present (i.e., `CamelCase` instead of `camelCase` or `snake_case`). However, obviously I will preserve the case for any external code that I reference, such as `std::for_each()`. I follow the convention of using an m prefix in front of data members (e.g., `mMemberVar`) and s in front of static variables (e.g., `sStaticVar`).

The source examples within the book are often only code snippets and are not meant to show fully functional samples. I will also often strip comments from the example code in the book. This is done for reasons of brevity and clarity. In particular, I will often omit any preprocessor guard statements around a header file. I will assume that the reader is aware that every C/C++ header should enclose all of its content within include guard statements (or `#pragma once`) and that it's good practice to contain all of your API declarations within a consistent namespace. In other words, it should be assumed that each header file that I present is implicitly surrounded by code such as:

```
#ifndef MY_MODULE_H
#define MY_MODULE_H

// required #include files...

namespace apibook {

// API declarations ...

}

#endif // MY_MODULE_H
```

TIP: I will also highlight various API design tips and key concepts throughout the book. These callouts are provided to let you search quickly for a concept you wish to reread. Or if you are particularly pressed for time, you could simply scan the book for these tips and then read the surrounding text to gain greater insight for those topics that interest you the most.

Book website

This book also has a supporting website, https://APIBook.com/. On this site you can find general information about the book as well as supporting material such as the complete set of source code examples contained within the text. Feel free to download and play with these samples yourself. They were designed to be as simple as possible while still being useful and illustrative. I have used the cross-platform CMake build system to facilitate compiling and linking the examples, so they should work on Windows, Mac OS X, and UNIX operating systems.

I will publish any information about new revisions of this book and any errata on this website. You can also find my contact information on the website, if you have any corrections or suggestions you want to share with me. If I make a change to the book based on your feedback, I will attribute your contribution on the website and in the acknowledgments section of a subsequent printing of the book.

Acknowledgments

This book was significantly improved by the technical review and feedback of many of my friends and colleagues. I'm indebted to them for taking the time to read early versions of the manuscript and provide thoughtful insights and suggestions. Thank you, Paul Strauss, Eric Gregory, Rycharde Hawkes, Nick Long, James Chalfant, Brett Levin, Marcus Marr, Jim Humelsine, and Geoff Levner.

Many readers of earlier printings of this book have also sent me feedback over the years, most of which I've incorporated into this version here. I would like to thank each of them for their valued contribution: Javier Estrada, Ivan Komarov, Dan Smith, Alexandros Gezerlis, QiFo Lin, Çağdaş Çalık, Markus Geimer, Roger Orr, Chris Bond, Philip Koshy, Andrea Bigagli, Jiang Zhiqiang, William Sommerwerck, Mark Jones, Yilmaz Arslanoglu, Fernando Rozenblit, Sören Sprößig, and Charles Garraud.

My passion for good application programming interface design has been developed by working with many great software engineers and managers. Working at several different organizations, I've been exposed to a range of design perspectives, software development philosophies, and problem-solving approaches. Throughout these varied experiences, I've had the privilege to work alongside and learn from many uniquely talented individuals, including:

- **SRI International**: Bob Bolles, Adam Cheyer, Elizabeth Churchill, David Colleen, Brian Davis, Michael Eriksen, Jay Feuquay, Marty A. Fischler, Aaron Heller, Lee Iverson, Jason Jenkins, Luc Julia, Yvan G. Leclerc, Pat Lincoln, Chris Marrin, Ray C. Perrault, and Brian Tierney.
- **Pixar Animation Studios**: Brad Andalman, David Baraff, Ian Buono, Gordon Cameron, Ed Catmull, Chris Colby, Bena Currin, Gareth Davis, Tony DeRose, Mike Ferris, Kurt Fleischer, Sebastian Grassia, Eric Gregory, Tara Hernandez, Paul Isaacs, Oren Jacob, Michael Kass, Chris King, Brett Levin, Tim Milliron, Alex Mohr, Cory Omand, Eben Osbty, Allan Poore, Chris Shoeneman, Patrick Schork, Paul Strauss, Kiril Vidimče, Adam Woodbury, David Yu, Dirk van Gelder, and Brad West.
- **The Bakery Animation Studio**: Sam Assadian, Sebastien Guichou, Arnauld Lamorlette, Thierry Lauthelier, Benoit Lepage, Geoff Levner, Nick Long, Erwan Maigret, and Bariş Metin.
- **Linden Lab**: Nat Goodspeed, Andrew de Laix, Howard Look, Brad Kittenbrink, Brian McGroarty, Adam Moss, Mark Palange, Jim Purbrick, and Kent Quirk.
- **PullString**: Renée Adams, Olena Anderson, Jeff Aydelotte, Shawn Blakesley, Michael Chann, Jeremy Cytryn, Firat Enderoglu, Michael Fitzpatrick, John Fuetsch, Paul Gluck, Laurel Goodhart, Lucas Ives, Patrick Ku, Maciej Kus, Brian Langner, Robin McDonald, Will Miller, Benjamin Morse, Daniel Sinto, Aurelio Tinio, and Ali Ziaei.

- **Apple**: Joe DeSimpliciis, Greg James, Thomas Karatzas, Aldo Longhi, Steven Martin, Tim Meighen, Dan Niemeyer, Antoine Raux, and Stanley Thalhammer.

I would especially like to acknowledge the great impact that Dr. Yvan G. Leclerc has made on my life from my early years at SRI International. Yvan was my first manager and a true friend. He taught me how to be a good manager of people, how to be a rigorous scientist and engineer, and at the same time how to enjoy life to its fullest. It is a great tragedy that incredible individuals such as Yvan are taken from us too soon.

Many thanks must also go to Elsevier/Morgan Kaufmann Publishers for all their work reviewing, copy editing, typesetting, and publishing this book. This work would quite literally not exist without their backing and energy. Specifically, I want to acknowledge the contribution of Todd Green, Robyn Day, André Cuello, and Melissa Revell for the first edition, as well as Chris Katsaropoulos, Palak Gupta, Prasanna Kalyanaraman, and Gregory Harris for the second edition.

Most importantly, I'd like to thank my wife, Genevieve Vidanes, for supporting and tolerating me while I spent many hours intently focused on a computer screen. Writing a book can be a big disruption to personal and family life, but she encouraged me throughout the process while also knowing exactly when to make me pause and take a break. Thank you, Genevieve, for your constant love and support. I couldn't do any of this without you on my side.

Introduction

What are APIs?

An application programming interface (API) provides an abstraction for a problem and specifies how clients should interact with software components that implement a solution to that problem. The components themselves are typically distributed as a software library, allowing them to be used in multiple applications. In essence, APIs define reusable building blocks that allow modular pieces of functionality to be incorporated into end-user applications.

An API can be written for yourself, for other engineers in your organization, or for the development community at large. It can be as small as a single function or involve hundreds of classes, methods, free functions, data types, enumerations, and constants. Its implementation can be proprietary or open source. The important underlying concept is that an API is a well-defined interface that provides a specific service to other pieces of software.

A modern application is typically built on top of many APIs, where some of these can also depend upon further APIs. This is illustrated in Fig. 1.1, which shows an example application that directly depends on the API for three libraries (Libraries 1–3), where two of those APIs depend on the API for a further two libraries (Libraries 4–5). For instance, an image viewing application may use an API for loading GIF images, and that API may itself be built upon a lower-level API for compressing and decompressing data.

API development is ubiquitous in modern software development. Its purpose is to provide a logical interface to the functionality of a component while also hiding any implementation details. For example, our API for loading GIF images may simply provide a `LoadImage()` method that accepts a filename and returns a 2D array of pixels. All of the file format and data compression details are hidden behind this simple interface. This concept is also illustrated in Fig. 1.1, where client code only accesses an API via its public interface, shown as the dark section at the top of each box.

Contracts and contractors

As an analogy, consider the task of building your own home. If you were to build a house entirely on your own, you would need to possess a thorough understanding of architecture, plumbing, electrical wiring, carpentry, masonry, and many other trades. You would also need to perform every task yourself and keep track of the minutest of details for every aspect of the project, such as whether you have enough wood for your

API Design for C++. https://doi.org/10.1016/B978-0-443-22219-1.00023-4

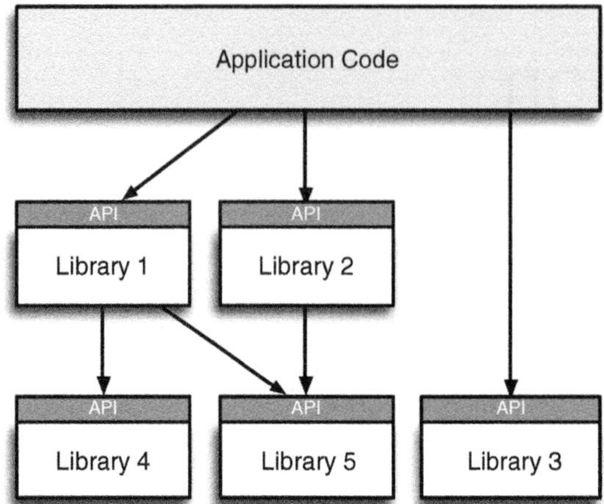

FIGURE 1.1 An application that calls routines from a hierarchy of application programming interfaces (APIs). Each box represents a software library where the *dark section* represents the public interface, or API, for that library while the *white section* represents the hidden implementation behind that API.

floorboards or whether you have the right fasteners to fit your screws. Finally, because you are the only person working on the project, you can perform only a single task at any point in time and hence the total time to complete the project could be very large.

An alternative strategy is to hire professional contractors to perform key tasks for you. You could hire an architect to design the plans for the house, a carpenter for all your woodwork needs, a plumber to install the water pipes and sewage system for your house, and an electrician to set up the power systems. Taking this approach, you negotiate a contract with each of your contractors—telling them what work you want done and agreeing upon a price. They then perform that work for you. If you're lucky, maybe you even have a good friend who is a contractor and they offer you their services for free. With this strategy, you are freed from the need to know everything about all aspects of building a house and instead you can take a higher-level supervisory role to select the best contractors for your purpose and ensure that the work of each individual contractor is combined to produce the vision of your ideal home.

The analogy to APIs is probably obvious: the house that you're building equates to a software program that you want to write, and the contractors provide APIs that abstract each of the tasks you need to perform and hide the implementation details of the work involved. Your task then resolves to selecting the appropriate APIs for your application and integrating them into your software. The analogy of having skilled friends who provide contracting services for free is meant to represent the use of freely available open source libraries in contrast to commercial libraries that require a licensing fee. The analogy could even be extended by having some of the contractors employing sub-contractors, which corresponds to certain APIs depending upon other APIs to perform their task.

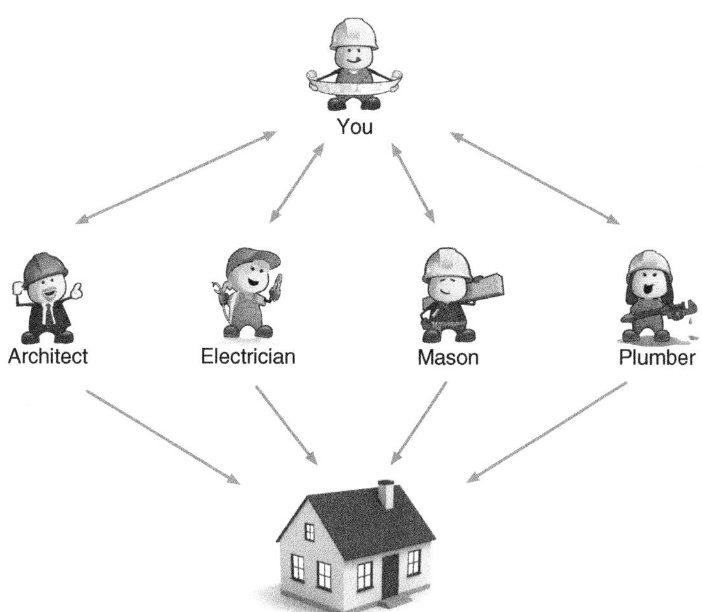

FIGURE 1.2 Using contractors to perform specialized tasks to build a house.

The contractor analogy is a common one in object-oriented programming. Early practitioners in the field talked about an object defining a binding contract for the services or behavior that it provides. An object then implements those services when asked to by a client program, potentially by subcontracting some of the work out to other objects behind the scenes (Snyder, 1986; Meyer, 1987) (Fig. 1.2).

APIs in C++

Strictly speaking, an API is simply a description of how to interact with a component. That is, it provides an abstraction and a functional specification for a component. In fact, many software engineers prefer to expand the acronym API as abstract programming interface instead of application programming interface.

In C++, this is traditionally embodied as one or more header (`.h`) files plus supporting documentation files. However as of C++20, we also have the concept of modules as an alternative to using headers. An implementation for a given API is often represented as a library file that can be linked into end-user applications. This can either be a static library such as a `.lib` file on Windows or `.a` on macOS and Linux, or a dynamic library such as a `.dll` file on Windows, `.dylib` on macOS, or `.so` on Linux.

A C++ API will therefore generally include the following elements:

1. **Headers/Modules**: A collection of `.h` header files or module interface units that define the interface and allow client code to be compiled against that interface.

Open source APIs also include the source code (`.cpp` files) for the API implementation.
2. **Libraries**: One or more static or dynamic library files that provide an implementation for the API (e.g., `.a`, `.lib`, `.dll`, `.dylib`, etc. files). Clients can link their code against these library files to add the functionality to their applications. (I'll go into more detail about the related terms library, framework, and software development kit [SDK] later in this chapter.)
3. **Documentation**: Overview information that describes how to use the API, often including automatically generated documentation for all the classes and functions in the API.

As an example of a well-known API, Microsoft's Windows API (often referred to as the Win32 API) is a collection of C functions, data types, and constants that enable programmers to write applications that run on the Windows platform. This includes functions for file handling, process and thread management, creating graphical user interfaces, talking to networks, and so on.

> *TIP: An API is a logical interface to a software component that hides the internal details required to implement it.*

The Win32 API is an example of plain C API rather than a C++ API. While you can use a C API directly from a C++ program, a good example of a specific C++ API is the C++ Standard Library. This contains a set of container classes, iterators for navigating over the elements in those containers, and various algorithms that act on those containers (Josuttis, 1999). For instance, the collection of algorithms includes high-level operations such as `std::search()`, `std::reverse()`, `std::sort()`, and `std::set_intersection()`. The C++ Standard Library therefore presents a logical interface to the task of manipulating collections of elements without exposing any of the internal details about how each algorithm is implemented.

What's different about API design?

Interfaces are the most important code that a developer writes. That's because problems in an interface are far more costly to fix than problems in the associated implementation code. As a result, the process of developing shared APIs demands more attention than standard application or graphical user interface (GUI) development. Of course, both should involve best design practices; however in the case of API development, these are critical to its success. Specifically, there are some key differentiating factors of API development:

- An API is an interface designed for developers, in much the same way that a GUI is an interface designed for end users. In fact, it's been said that an API is a user interface for programmers (Arnold, 2005). As such, your API could be used by thousands of developers around the world, and it will undoubtedly be used in ways that you never intended (Tulach, 2008). You must anticipate this in your design. A well-designed API can be your organization's biggest asset. Conversely a poor API can create a support nightmare and even turn your users toward your competitors, just as a buggy or difficult-to-use GUI might force an end user to switch to a different application.
- Multiple applications can share the same API. In Fig. 1.1, I showed that a single application can be composed of multiple APIs. However, any one of those APIs could also be reused in several other applications. This means that while problems in the code for any given application will affect only that one application, errors in an API can affect all applications that depend upon that functionality.
- You must strive for backward compatibility whenever you change an API. If you make an incompatible change to your interface, your clients' code may fail to compile, or worse, their code could compile but behave differently or crash intermittently. Imagine the confusion and chaos that would arise if the signature of the `printf()` function in the standard C library was different for different compilers or platforms. The simple "Hello World" program might not look so simple any more:

```
#include <stdio.h>
#ifdef _WIN32
#include <windows.h>
#endif
#ifdef __cplusplus
#include <iostream>
#endif

int main(int, char *argv[])
{
#if __STRICT_ANSI__
    printf("Hello World\n");
#elif defined(_WIN32)
    PrintWithFormat("Hello World\n");
#elif defined(__PRINTF_DEPRECATED__)
    fprintf(stdout, "Hello World\n");
#elif defined(__PRINTF_VECTOR__)
    const char *lines[2] = {"Hello World", NULL};
    printf(lines);
#elif defined(__cplusplus)
    std::cout << "Hello World" << std::endl;
#else
#error No terminal output API found
#endif
    return 0;
}
```

This may seem like a contrived example, but it's actually not that extreme. Take a look at the standard header files that come with your compiler and you will find declarations that are just as convoluted and inscrutable, or perhaps worse.

- Owing to the backward compatibility requirement, it is critical to have a change control process in place. During the normal development process, many developers may fix bugs or add new features to an API. Some of these may be junior engineers who do not fully understand all aspects of good API design. As a result, it's important to hold an API review before releasing a new version of the API. This involves one or more senior engineers checking that all changes to the interface are acceptable, have been made for a valid reason, and are implemented in the best way to maintain backward compatibility. Many open source APIs also enforce a change request process to gain approval for a change before it's added to the source code.
- APIs tend to live for a long time. There can be a large up-front cost to produce a good API because of the extra overhead of planning, design, versioning, and review that are necessary. However, if done well, the long-term cost can be substantially cheaper because you have the ability to make radical changes and improvements to your software without disrupting your clients. That is, your development velocity can be greater because of the increased flexibility that the API affords you.
- The need for good documentation is paramount when writing an API, particularly if you don't provide the source code for your implementation. Users can look at your header files to glean how to use it, but this doesn't define the behavior of the API, such as acceptable input values or error conditions. Well-written, consistent, and extensive documentation is therefore imperative for any good API.
- The need for automated testing is similarly very high. Of course, you should always test your code, but when you're writing an API you may have hundreds of other developers, and thousands of their users, depending upon the correctness of your code. If you are making major changes to the implementation of your API, you can be more confident that you will not break your clients' programs if you have a thorough suite of regression tests to verify that the API behavior has not changed.

Writing good APIs is difficult. While the necessary skills are founded upon general software design principles, they also require additional knowledge and development processes to address the points listed earlier. However, the principles and techniques of API design are rarely taught to engineers. Normally, these skills are gained only through experience—by making mistakes and learning empirically what does and does not work (Henning, 2009). This book is an attempt to redress this situation, to distill the strategies of industrial-strength, future-proof API design that have evolved through years of software engineering experience into a comprehensive, practical, and accessible format.

> *TIP: An API describes software used by other engineers to build their applications. As such, it must be well-designed, documented, regression tested, and stable between releases.*

Why should you use APIs?

The question of why you should care about APIs in your own software projects can be interpreted in two ways:

1. Why should you design and write your own APIs? Or,
2. Why should you use APIs from other providers in your applications?

I'll tackle both perspectives in the following sections as I present the various benefits of using APIs in your projects.

More robust code

If you're writing a module to be used by other developers, either for fellow engineers within your organization or for external customers of your library, then it would be a wise investment to create an API for them to access your functionality. Doing so will offer you the following benefits:

- **Hides implementation.** By hiding the implementation details of your module, you gain the flexibility to change the implementation at a future date without causing an upheaval for your users. Without doing so, you'll either restrict yourself in terms of the updates you can make to your code or you'll force your users to rewrite their code to adopt new versions of your library. If you make it too onerous for your clients to update to new versions of your software, then it's highly likely that they will either not upgrade at all or look elsewhere for an API that will not be as much work for them to maintain. Good API design can therefore significantly affect the success of your business or project.
- **Increases longevity.** Over time, systems that expose their implementation details tend to devolve into spaghetti code in which every part of the system depends upon the internal details of other parts of the system. As a result, the system becomes fragile, rigid, immobile, and viscous (Martin, 2000). This often results in organizations having to spend significant effort to evolve the code toward a better design, or simply rewrite it from scratch. By investing in good API design up-front and paying the incremental cost to maintain a coherent design, your software can survive for longer and cost less to maintain in the long run. I'll delve much more deeply into this point at the start of Chapter 4.
- **Promotes modularization.** An API is normally devised to address a specific task or use case. As such, APIs tend to define a modular grouping of functionality with a

coherent focus. Developing an application on top of a collection of APIs promotes loosely coupled and modular architectures in which the behavior of one module is not dependent upon the internal details of another module. If you work at a large company, this can have direct benefits for you if you're working on one such modular API. That's because it can be clearer if a bug is due to code that you own or for which another team is responsible. So you can spend less time triaging bugs that are unrelated to your code.

- **Reduces code duplication.** Code duplication is one of the cardinal sins of software engineering and should be stamped out whenever possible. By keeping all your code's logic behind a strict interface that all clients must use, you centralize the behavior in a single place. Doing so means that you need to update only one place to change the behavior of your API for all of the clients. This can help to remove duplication of implementation code throughout your code base. In fact, many APIs are created after discovering duplicated code and deciding to consolidate it behind a single interface. This is a good thing.
- **Removes hardcoded assumptions.** Many programs may contain hardcoded values that are copied throughout the code, for example, using the filename `myprogram.log` whenever data are written to a log file. Instead, APIs can be used to provide access to this information without replicating these constant values across the code base. For example, a `GetLogFilename()` API call could be used to replace the hardcoded "`myprogram.log`" string.
- **Easier to change the implementation.** If you have hidden all the implementation details of your module behind its public interface, then you can change those implementation details without affecting any code that depends upon the API. For example, you might decide to change a file parsing routine to use `std::string` containers instead of allocating, freeing, and reallocating your own `char *` buffers.
- **Easier to optimize.** Similarly, with your implementation details successfully hidden, you can optimize the performance of your API without requiring any changes to your clients' code. For example, you could add a caching solution to a method that performs some computationally intensive calculation. This is possible because all attempts to read and write your underlying data are performed via your API, so it becomes much easier to know when you must invalidate your cached result and recomputed the new value.

Code reuse

Code reuse is the use of existing software to build new software. It's one of the holy grails of modern software development. APIs provide a mechanism to enable code reuse.

In the early years of software development, it was common for a company to have to write all code for any application they produced. If the program needed to read GIF images or parse a text file, the company would have to write all that code in-house. Nowadays, with the proliferation of good commercial and open source libraries, it makes much more sense simply to reuse code that others have written. For example, there are various open source image reading APIs and file format parsing APIs that you can download and use in your applications today. These libraries have been refined and debugged by many developers around the world and have been battle-tested in many other programs.

In essence, software development has become a lot more modular with the use of distinct components that form the building blocks of an application and talk together via their published APIs. The benefit of this approach is that you don't need to understand every detail of every software component, in the same way that for the earlier house building analogy you can delegate many details to professional contractors. This can translate into faster development cycles, because you can either reuse existing code or decouple the schedule for various components. It also allows you to concentrate on your core business logic rather than having to spend time reinventing the wheel.

One of the difficulties in achieving code reuse, however, is that you must often come up with a more general interface than you originally intended. That's because other clients may have additional expectations or requirements. Effective code reuse therefore follows from a deep understanding of the clients of your software and designing a system that integrates their collective interests with your own.

SIDEBAR: C++ APIs and the Web

The trend toward applications that depend upon third-party APIs is particularly popular in the field of cloud computing. Here, web applications rely more and more on web services (APIs) to provide core functionality. In the case of web mash-ups, the application itself is sometimes simply a repackaging of multiple existing services to provide a new service, such as combining the Google Maps API with a local crimes statistics database to provide a map-based interface to the crime data.

In fact, it's worth taking a few moments to highlight the importance of C++ API design in web development. A superficial analysis might conclude that server-side web development is confined to scripting languages such as PHP, Perl, or Python (the "P" in the popular LAMP acronym), or .NET languages based upon Microsoft's Active Server Pages technology. This may be true for small-scale web development. However, it is noteworthy that many large-scale web services use a C++ backend to deliver optimal performance.

In fact, Facebook developed a product called HipHop to convert their PHP code into C++ to improve the performance of their social networking site. The Figma web application also used C++ compiled to WebAssembly to optimize their website. C++ API design therefore has a role to play in scalable web service development. Additionally, if you develop your core APIs in C++, not only can they form a high-performance web service, your code can also be reused to deliver your product in other forms, such as desktop or mobile phone versions.

As an aside, one potential explanation for this shift in software development strategy is the result of the forces of globalization (Friedman, 2007; Wolf, 2004). In effect, the convergence of the Internet, standard network protocols, and web technologies has created a leveling of the software playing field. This has enabled companies and individuals all over the world to create, contribute to, and compete with large complex software projects. This form of globalization promotes an environment where companies and developers anywhere in the world can forge a livelihood out of developing software subsystems. Other organizations in different parts of the world can then build end-user applications by assembling and augmenting these building blocks to solve specific problems. In terms of our focus here, APIs provide the mechanism to enable this globalization and componentization of modern software development.

Parallel development

Even if you're writing in-house software, your fellow engineers will very likely need to write code that uses your code. If you use good API design techniques, you can simplify their lives, and by extension your own (because you won't have to answer as many questions about how your code works or how to use to it). This becomes even more important if multiple developers are working in parallel on code that depends upon each other.

For example, let's say that you are working on a string encryption algorithm that another developer wants to use to write data out to a configuration file. One approach would be to have the other developer wait until you're finished your work. Then they can use it in their file writer module. However, a far more efficient use of time would be for the two of you to meet early on and agree upon an appropriate API. Then you can put that API in place with placeholder functionality that your colleague can start calling immediately, such as:

```
#include <string.h>

class StringEncryptor
{
public:
    /// set the key to use for the Encrypt() and Decrypt() calls
    void SetKey(const std::string &key);

    /// encrypt an input string based upon the current key
    std::string Encrypt(const std::string &str) const;

    /// decrypt a string using the current key - calling
    /// Decrypt() on a string returned by Encrypt() will
    /// return the original string for the same key.
    std::string Decrypt(const std::string &str) const;
};
```

You can then provide a simple implementation of these functions, so that at least the module will compile and link. For example, the associated `.cpp` file might look like:

```
void StringEncryptor::SetKey(const std::string &key)
{
}

std::string StringEncryptor::Encrypt(const std::string &str)
{
    return str;
}

std::string StringEncryptor::Decrypt(const std::string &str)
{
    return str;
}
```

In this way, your colleague can use this API and proceed with the work without being held up by your progress. For the time being, your API will not actually encrypt any strings, but that's just a minor implementation detail! The important point is that you have a stable interface—a contract—on which you both agree, and that it behaves appropriately, such as `Decrypt(Encypt("Hello")) == "Hello"`. When you finish your work and update the `.cpp` file with the correct implementation, your colleague's code will simply work without any further changes required on his or her part.

In reality, it's likely that there will be interface issues that you didn't anticipate before you started writing the code and you will probably have to iterate on the API a few times to get it just right. However, for the most part, the two of you can work in parallel with minimal holdups.

This approach also encourages test-driven, or test-first, development. By stubbing out the API early on, you can write unit tests to validate the desired functionality and run these continuously to make sure that you haven't broken your contract with your colleague.

Scaling this process up to an organizational level, your project could have separate teams that may be remote from each other, even working to different schedules. But by defining each team's dependencies upfront, and creating APIs to model these, each team can work independently and with minimal knowledge of how the other teams are implementing their work behind the API. This efficient use of resources, and the corresponding reduction in redundant communication, can correlate to significant overall cost savings for an organization.

When should you avoid APIs?

Designing and implementing an API usually requires more work than writing normal application code. That's because the purpose of an API is to provide a robust and stable interface for other developers to use. As such, the level of quality, planning, documentation, testing, support, and maintenance is far higher for an API than for software that's meant to be used only within a single application.

As a result, if you're writing an internal module that doesn't require other clients to communicate with it, then the extra overhead of creating and supporting a stable public interface for your module may not be worth the effort. However, this is not a reason to write sloppy code. Spending the extra time to adhere to the principles of API design will not be wasted effort in the long run.

On the flip side of the coin, consider that you're a software developer who wants to use a third-party API in your application. The previous section discussed several reasons why you might want to reuse external APIs in your software. However, there may be cases in which you wish to avoid using a particular API and pay the cost to implement the code yourself, or look for an alternate solution. For example:

- **License restrictions.** An API may provide everything you need for functionality, but the license restrictions may be prohibitive for your needs. For example, if you want to use an open source package that's distributed under the GNU General Public License (GPL), then you are required to release any derived works under the GPL as well. This means that using this package in your program would require you to release the entire source code for your application, a constraint that may not be acceptable for a commercial application. Other licenses, such as the GNU Lesser GPL are more permissive and tend to be more common for software libraries. Another licensing aspect is that the dollar cost for a commercial API may be too high for your project, or the licensing terms may be too restrictive, such as requiring a license fee per developer or even per user.
- **Functionality mismatch.** An API may appear to solve a problem that you have but may do it in a way that doesn't match the constraints or functional requirements of your application. For example, perhaps you're developing an image processing tool and you want to provide a Fourier transform capability. There are many implementations of the fast Fourier transform (FFT) available, but many of these are 1D algorithms, whereas you require a 2D FFT because you are dealing with 2D image data. Additionally, many 2D FFT algorithms work only on data sets with dimensions that are a power of 2 (e.g., 256 × 256 or 512 × 512 pixels). Furthermore, perhaps the API that you found doesn't work on the platforms that you must support, or perhaps it doesn't match the performance criteria that you've specified for your application.
- **Lack of source code.** Although there are many open source APIs, sometimes the best API for your case may be a closed source offering. That is, only the header files for the interface are made available to you, but the underlying C++ source files are not distributed with the library. This has several important implications. Among these is the fact that if you encounter a bug in the library, you're unable to inspect the source code to understand what might be going wrong. Reading the

source can be a valuable technique for tracking down a bug and potentially discovering a work-around for the issue.

Furthermore, without access to the source code for an API, you lose the ability to change the source to fix a bug. This means that the schedule for your software project could be adversely affected by unanticipated problems in a third-party API you're using and by the time spent waiting for the owners of that API to address your bug reports and distribute a fixed patch.

- **Lack of documentation.** An API may appear to fulfill a need that you have in your application, but if the API has poor or nonexistent documentation, then you may decide to look elsewhere for a solution. Perhaps it's not obvious how to use the API, perhaps you cannot be sure how the API will behave under certain situations, or perhaps you simply don't trust the work of an engineer who hasn't taken the time to explain how their code should be used.

While I've tried to play devil's advocate here and come up with cases where you don't need to worry about API design, over the course of my career as a software engineer, I've come to appreciate that everything is API design. For example, in a large organization with a lot of internal code, it's tempting to think of everything as just "the code," because it's easy to change anything you want. But not all code changes should be treated equally. If you don't have strong API boundaries and processes in place, then treating all of your software as one giant ball of code will quickly devolve into chaos. As Alex Martelli phrased it, the worst API is no API at all (Martelli, 2011). If you take one thing from this book, it should be to treat every software problem as an API design problem first and foremost.

API examples

APIs are everywhere. Even if you've been programming for only a short amount of time, the chances are that you've written code to use an API or two, and maybe you've also written one yourself.

Layers of APIs

An API can be any size, from a single function to a large collection of classes. It can also provide access to functionality at any architectural level, from low-level operating system (OS) calls all the way up to GUI tool kits. The following list presents various common APIs, many of which you've probably heard of already, to give you an appreciation for how prevalent API development is:

- **OS APIs.** Every OS must provide a set of standard APIs to allow programs to access OS-level services. For example, the POSIX API defines functions such as `fork()`, `getpid()`, and `kill()` for managing UNIX-style processes, whereas Microsoft's

Win32 API includes functions such as `CreateProcess()`, `GetCurrentProcess()`, and `TerminateProcess()` for managing Windows processes. These are stable low-level APIs that should never change; otherwise many programs could break!

- **Language APIs.** The C language is supported by the C Standard Library, implemented as the libc library and associated man pages. It includes familiar functions such as `printf()`, `scanf()`, and `fopen()`. The C++ language also offers the C++ Standard Library, which provides an API for various container classes (e.g., `std::string`, `std::vector`, `std::set`, and `std::map`), iterators (e.g., `std::vector<double>::iterator`), and generic algorithms (e.g., `std::sort`, `std::for_each`, and `std::set_union`). For example, the following code snippet uses the C++ Standard Library API to iterate through all elements in a vector and print them out (although in modern code, using `auto` within a range-based `for` loop would be simpler):

```
#include <iostream>
#include <vector>

void PrintVector(const std::vector<float> &vec)
{
    std::vector<float>::const_iterator it;
    for (it = vec.begin(); it != vec.end(); ++it) {
        std::cout << *it << std::endl;
    }
}
```

- **Image APIs.** Gone are the days when developers needed to write their own image reading and writing routines. There is now a wide range of open source packages out there for you to download and use in your own programs. For example, there's the popular libjpeg library that provides an implementation of a JPEG/JFIF decoder and encoder. There's the extensive libtiff library for reading and writing various flavors of TIFF files. And there's the libpng library for handling PNG format images. All of these libraries define APIs that let you write code to read and write the image formats without having to know anything about the underlying file formats. For example, the follow code snippet uses the libtiff API to find the dimensions of a TIFF image:

```
TIFF *tif = TIFFOpen("image.tiff", "r");
if (tif) {
    uint32 w, h;
    TIFFGetField(tif, TIFFTAG_IMAGEWIDTH, &w);
    TIFFGetField(tif, TIFFTAG_IMAGELENGTH, &h);
    printf("Image size = %d x %d pixels\n", w, h);
    TIFFClose(tif);
}
```

- **3D graphics APIs.** The two classic real-time 3D graphics APIs are OpenGL and DirectX. These let you define 3D objects in terms of small primitives, such as triangles or polygons; specify the surface properties of those primitives, such as color, normal, and texture; and define the environment conditions, such as lights, fog, and clipping panes. Thanks to standard APIs such as these, game developers can write 3D games that will work on graphics cards old and new, from many different manufacturers. That's because each graphics card manufacturer provides drivers that provide the implementation details behind the OpenGL or DirectX API. Before the widespread use of these APIs, a developer had to write a 3D application for a specific piece of graphics hardware, and this program would probably not work on another machine with different graphics hardware. These APIs also enable a host of higher-level scene graph APIs, such as OpenSceneGraph, OpenSG, and OGRE. The following code segment shows the classic example of rendering a triangle, with a different color for each vertex, using the OpenGL API:

```
glClear(GL_COLOR_BUFFER_BIT);
glBegin(GL_TRIANGLES);
    glColor3f(0.0, 0.0, 1.0);   /* blue */
    glVertex2i(0, 0);
    glColor3f(0.0, 1.0, 0.0);   /* green */
    glVertex2i(200, 200);
    glColor3f(1.0, 0.0, 0.0);   /* red */
    glVertex2i(20, 200);
glEnd();
glFlush();
```

- **GUI APIs.** Any application that wants to open its own window needs to use a GUI tool kit. This is an API that provides the ability to create windows, buttons, text fields, dialogues, icons, menus, and so on. The API will normally also provide an event model, to allow the capturing of mouse and keyboard events. Some popular C/C++ GUI APIs include the wxWidgets library, Qt Company's Qt API, GTK+, and X/Motif. It used to be the case that if a company wanted to release an application on more than one platform, such as Windows and Mac, it would have to rewrite the user interface code using a different GUI API for each platform, or it would have to develop its own in-house cross-platform GUI tool kit. However, these days most modern GUI tool kits are available for multiple platforms including Windows, Mac, and Linux, which makes it far easier to write cross-platform applications. As a sample of a modern cross-platform GUI API, the following complete program shows a bare minimum Qt program that pops up a window with a Hello World button:

```
#include <QApplication>
#include <QPushButton>

int main(int argc, char *argv[])
{
    QApplication app(argc, argv);

    QPushButton hello("Hello world!");
    hello.resize(100, 30);

    hello.show();
    return app.exec();
}
```

SIDEBAR: C++ Standard Library versus Standard Template Library

The C++ Standard Library is the official name of the collection of classes and functions written in the core language and included in the C++ ISO standard.

There is also the older term Standard Template Library (STL), which refers to the library written by Alexander Stepanov before C++ was standardized.

Parts of the STL were standardized in the C++ Standard Library, but it's important to note that the two APIs are not the same. I will refer to the C++ Standard Library throughout this book because it's part of the official ISO standard.

Of course, this list is just a brief cross-section of all possible APIs that are out there. You'll also find APIs to let you access data over networks, to parse and generate XML or YAML files, to help you write multithreaded programs, or to solve complex mathematical problems. The point of the list was simply to demonstrate the breadth and depth of APIs that have been developed to help you build your applications, and to give you a flavor for what code based upon these APIs looks like.

TIP: APIs are used everywhere in modern software development, from OS- and language-level APIs to image, audio, graphics, concurrency, network, XML, mathematics, web browsing, or GUI APIs.

A real-life example

The previous list of API examples was purposefully arranged by architectural level to show the range of APIs that you might use when building an application. You will often use APIs from several architectural levels when building a large software product. For example, Fig. 1.3 presents an example architecture diagram for the Second Life Viewer developed by Linden Lab. This is a large open source program that lets users interact

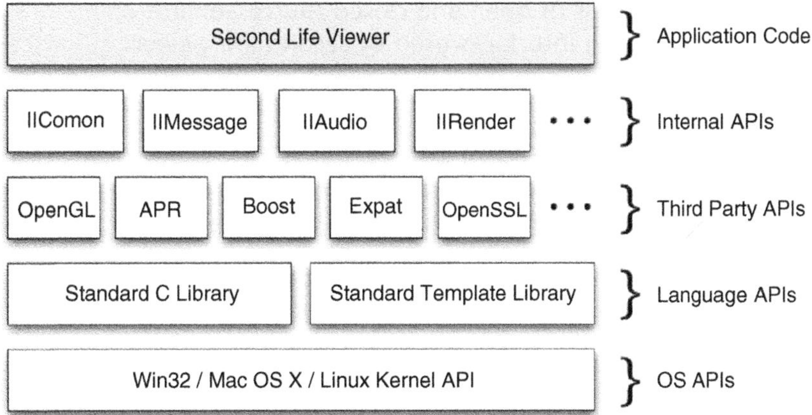

FIGURE 1.3 Architecture diagram for the second life viewer. *API*, application programming interface; *APR*, Apache Portable runtime; *OS*, operating system.

with each other in an online 3D virtual world, with the ability to perform voice chat and text messaging among users. The diagram demonstrates the use and layering of APIs in a large C++ software project.

Of note is the layer of internal APIs, by which I mean the set of modules that a company develops in-house for a particular product, or suite of products. Although Fig. 1.3 simply shows these as a single layer for the purpose of simplicity, the set of internal APIs will form an additional stack of layers, from foundation-level routines that provide an in-house string, dictionary, file IO, threading routines, and so on to APIs that provide the core business logic of the application, all the way up to custom GUI APIs for managing the application's user interface.

Obviously, Fig. 1.3 doesn't provide an exhaustive list of all APIs that are used in this application. It simply shows a few examples for each layer of the architecture. However, Table 1.1 present the set of third-party dependencies for the application, to give you an idea of how many open source and commercial closed source dependencies a contemporary software project is built upon. When you factor in system and OS libraries as well, this list grows even further.

Libraries, frameworks, and software development kits

I briefly introduced the term library earlier in this chapter. Here I'll go into more details about libraries and discuss some related terms such as framework and SDK.

A software library is a collection of data and compiled code for one or more architectures. It's what stores the implementation code for your API and is the artifact that your clients link into their programs. Common types of library include static libraries (e.g., `.a` or `.lib`), where the library code is linked directly into client code at link time, and dynamic libraries (e.g., `.so`, `.dll`, or `.dylib`), where the library code is loaded and linked

Table 1.1 List of open and closed source application programming interfaces used by second life viewer.

API	Description
APR	Apache Portable runtime
Boost	Set of portable C++ source libraries
c-ares	Asynchronous DNS resolution library
cURL	Client URL request library
Expat	Stream-oriented XML Parser
FMOD	Commercial audio engine and mp3 stream decoder
FreeGLUT	Open source version of the OpenGL Utility tool kit (GLUT)
FreeType	Font rasterization engine.
glh_linear	C++ OpenGL helper library
Jpeglib	JPEG decoder library
KDU	Commercial Kakadu (KDU) JPEG-2000 decoder library
Libpng	PNG image library
Llqtwebkit	Qt's WebKit embeddable web browser
Ogg Vorbis	Compressed audio format library
OpenGL	3D graphics rendering engine
Openjpeg	Open-source JPEG-2000 library, alternative to KDU
OpenSSL	Secure sockets layer (SSL) library
Quicktime	Library for playing video clips
Vivox	Commercial voice chat library
Xmlrpc-epi	XML-RPC protocol library
Zlib	Lossless data compression library

into client code at runtime. Most OSs have a dynamic library search path to influence how they find these files at runtime, and compilers can embed hints into the compiled code about where to find any dynamic library dependencies. I cover these details in Appendix A on Libraries.

You may choose to distribute your software only as binary library files and public headers, or you may choose to adopt an open source model where clients can directly build the library files themselves from your source code.

A library file is normally accompanied by a set of header files (`.h`) that describe the elements of your API (i.e., the classes, functions, templates, enums, namespaces, etc. that make up your interface). These header files are needed for clients to compile their code against your interface. It's also worth noting that C++20 introduced an alternative to the use of headers with the concept of modules, which I will cover in more detail in the chapter on C++ Revisions. Altogether, a library and its associated header files or modules are often referred to as an SDK. An SDK may also include other resources to help you use an API, such as documentation, example source code, and supporting tools.

As an example, Apple publishes various iOS APIs for developing applications that run on the iPhone, iPod Touch, and iPad devices. Examples include the UIKit user interface

API, the WebKit API to embed web browser functionality in your applications, and the Core Audio API for audio services. Apple also provides the iOS SDK, which is a downloadable installer that contains the headers and libraries that implement the various iOS APIs. These are the files that you compile and link against to give your programs access to the underlying functionality of the APIs. The iOS SDK includes API documentation, sample code, various templates for Apple's Integrated Development Environment called XCode, and the iPhone Simulator that lets you run iPhone apps on your desktop computer.

The term framework can have a few interpretations. On Apple platforms, there is a `.framework` asset type that is essentially a bundle containing different versions of a library and its associated headers. However, the term can also imply a larger collection of APIs. For example, the Qt Framework is composed of many separate modules that offer different APIs for different tasks. Furthermore, many engineers subscribe to the definition that the key difference between a library and a framework is that a framework uses inversion of control. This means that the framework has some built-in behavior and that you can register code with the framework, but it decides when to call your code. A good example of this is a UI framework in which the framework owns the main event loop, but you can register functions and classes to handle specific events such as button presses. In other words, a library is a collection of functionality that you can call from your code, whereas a framework is a collection of functionality that calls your code.

File formats and network protocols

There are several other forms of communication "contracts" that are commonly used in computer applications. One of the most familiar is the file format. This is a way to save in-memory data to a file on disk using a well-known layout of those data. For example, the JPEG File Interchange Format (JFIF) is an image file format for exchanging JPEG-encoded imagery, commonly given the `.jpg` or `.jpeg` file extension. The format of a JFIF file header is shown in Table 1.2.

Given the format for a data file, such as this JFIF/JPEG format, any program can read and write image files in the format. This allows the easy interchange of image data among different users and the proliferation of image viewers and tools that can operate on those images.

Similarly, client/server applications, peer-to-peer applications, and middleware services work by sending data back and forward using an established protocol, usually over a network socket. For example, the Subversion version control system uses a client/server architecture in which the master repository is stored on the server and individual clients synchronize their local clients with the server (Rooney, 2005). To make this work, the client and the server must agree upon the format of data that are transmitted across the network. This is known as the client/server protocol or line protocol. If the client sends a data stream that does not conform to this protocol, then the server will not be

Table 1.2 JPEG File Interchange Format header specification.

Field	Byte size	Description
APP0 marker	2	Always 0xFFE0
Length	2	Length of segment excluding APP0 marker
Identifier	5	Always 0x4A46494600 ("JFIF\0")
Version	2	0x0102
Density units	1	Units for pixel density fields, 0 = no units
X density	2	Integer horizontal pixel density
Y density	2	Integer vertical pixel density
Thumbnail width (w)	1	Horizontal size of embedded thumbnail
Thumbnail height (h)	1	Vertical size of embedded thumbnail
Thumbnail data	3 × w × h	Uncompressed 24-bit RGB raster data

able to understand the message. It is therefore critical that the specification of the client/server protocol is well-defined and that both the client and server conform to the specification.

Both of these cases are conceptually similar to an API in that they define a standard interface, or specification, for information to be exchanged. Also, any changes to the specification must consider the impact on existing clients. Despite this similarity, file formats and line protocols are not actually APIs because they are not programming interfaces for code that you link into your application. However, a good rule of thumb is that whenever you have a file format or a client/server protocol, you should also have an associated API to manage changes to that specification.

> *TIP: Whenever you create a file format or client/server protocol, you should also create an API for it. This allows the details of the specification, and any future changes to it, to be centralized and hidden.*

For example, if you specify a file format for your application's data, you should also write an API to allow reading and writing files in that format. For one thing, this is simply good practice, so that knowledge of the file format is not distributed throughout your application. More importantly, having an API allows you easily to change the file format in the future without having to rewrite any code outside the API implementation. Finally, if you do end up with multiple different versions of a file format, then your API can abstract that complexity away so that it can read and write data in any version of the format, or it can know if the format is written with a newer version of the API and take appropriate steps. In essence, the actual format of the data on disk becomes a hidden implementation detail with which your application does not need to be concerned. I'll discuss this more when talking about data-driven APIs in the Styles chapter.

This advice applies just as well to client/server applications, in which the definition of a common protocol, and a common API to manage that protocol, can allow the client and server teams to work relatively independently of each other. For instance, you may begin using user datagram protocol (UDP) as the transport layer for part of your system but later decide to switch to transmission control protocol (TCP) (as indeed happened with the Second Life code base). If all network access had already been abstracted behind an appropriate API, then such a major implementation change would have little to no impact on the rest of the system.

About this book

Now that I've covered the basics of what an API is, and the pros and cons of API development, I'll dive into details such as how to design good APIs, how to implement them efficiently in C++, and how to version them without breaking backward compatibility. The progression of chapters in this book roughly follows the standard evolution of an API, from initial design through implementation, versioning, documentation, and testing.

Chapter 2: *Qualities*

I begin the main text with a chapter that answers the question, what is a good API? This will cover a wide gamut of qualities that you should be aware of when designing your APIs, such as information hiding, minimal completeness, and loose coupling. As I do throughout the book, I illustrate these concepts with many C++ source code examples to show how they relate to your own projects.

Chapter 3: *Patterns*

The next couple of chapters tackle the question, how do you design a good API? Accordingly, Chapter 3 looks at some specific design patterns and idioms that are particularly helpful in API design. These include the pimpl idiom, Singleton, Factory Method, Proxy, Adapter, Façade, and Observer.

Chapter 4: *Design*

Continuing the topic of how to design a good API, Chapter 4 discusses functional requirement gathering and use case modeling to drive the design of a clean and usable interface, as well some techniques of object-oriented analysis and object-oriented design. This chapter includes a discussion on many of the problems that large software projects face. These observations are taken from real-world experiences and provide insight into the issues that arise when doing large-scale API development.

Chapter 5: *Styles*

The next few chapters focus on creating high-quality APIs with C++. This is a deep and complex topic and is, of course, the specific focus of this book. I therefore begin by describing various styles of C and C++ APIs that you could adopt in your projects, such as flat C APIs, object-oriented APIs, template-based APIs, and data-driven APIs.

Chapter 6: *C++ Usage*

Next, I discuss various core C++ language features that can affect good API design. This includes numerous important issues such as good constructor and operator style, namespaces, pointer versus reference parameters, the use of friends, and how to export symbols in a dynamic library. These are best practices that you can use for any version of the C++ standard.

Chapter 7: *C++ Revisions*

Since the initial C++98 ISO standard was published, there have been several revisions to the standard, including C++03, C++11, C++14, C++17, C++20, and C++23. Most of these revisions introduced new features that can affect API design. This chapter focuses on each of these revisions in turn and identifies the API features that you can take advantage of in your own projects.

Chapter 8: *Performance*

In this chapter, I analyze performance issues in APIs and show you how to build high-performing APIs in C++. This involves the use of const references, forward declares, data member clustering, and inlining. I also present various tools that can help you to assess the performance of your code.

Chapter 9: *Concurrency*

The initial C++98 and C++03 standards did not consider how code should behave in a multithreaded environment. With the advent of C++11, the language now has broad support for writing concurrent and parallel code. This chapter covers topics such as data races, race conditions, thread safety, and reentrancy in C++ code.

Chapter 10: *Versioning*

With the foundations of API design in hand, I start to expand into more complex aspects, beginning with API versioning and how to maintain backward compatibility. This is one of the most important—and difficult—aspects of robust API design. Here I will define the various terms backward, forward, functional, source, and binary/application binary interface compatibility, and describe how to evolve an API with minimal impact to your clients.

Chapter 11: *Documentation*

Next, I dedicate a chapter to the topic of API documentation. An API is ill-defined without proper supporting documentation, so I present good techniques for commenting and documenting your API, with specific examples using the excellent Doxygen tool.

Chapter 12: *Testing*

The use of extensive testing lets you evolve an API with the confidence that you're not breaking your clients' programs. Here I present various types of automated testing, including unit, integration, and performance tests, and present examples of good testing methodologies for you to use in your own projects. This covers topics such as test-driven development, stub and mock objects, testing private code, and contract programming.

Chapter 13: *Objective-C and Swift*

The Objective-C and Swift programming languages have become the de facto mechanisms to develop apps for Apple's ecosystem of products. But you may also want to integrate your C++ code into these apps. I will therefore cover how Objective-C++ and C++ code can coexist and how a C++ API can be accessed from Swift code.

Chapter 14: *Scripting*

In addition to integrating your C++ code into other compiled languages, you may wish to integrate it into different scripting languages, perhaps so that power users of your application can write scripts to perform custom actions. I therefore talk about how to create script bindings for a C++ API so that it can be called from languages such as Python and Ruby.

Chapter 15: *Extensibility*

Another advanced topic is that of user extensibility: creating an API that allows programmers to write custom C++ plug-ins that extend the basic functionality you ship with the API. This can be a critical mechanism to promote adoption of your API and to help it survive in the long term. Additionally, I cover how to create extensible interfaces using inheritance and templates.

Appendix A: *Libraries*

The book concludes with an appendix on how to create static and dynamic libraries. You must be able to create libraries for your code to be used by others. Also, there are interface design issues to consider when creating dynamic libraries, such as the set of symbols that you export publicly. So I discuss the differences between static and shared libraries and demonstrate how you can make your compiler produce these libraries to allow the reuse of your code in other applications.

Qualities

This chapter aims to answer the question, what are the basic qualities of a good application programming interface (API)? Most developers would agree that a good API should be elegantly designed but still highly usable. It should be a joy to use but also fade into the background (Henning, 2009). These are fine qualitative statements, but what are the specific design aspects that enable these? Obviously, every API is different. However, there are certain qualities that promote high-quality API design and should be adhered to whenever possible, as well as many that make for poor designs and should be avoided.

There are no absolutes in API design: you cannot apply a fixed set of rules to every situation. However, while there may be individual cases where you decide to deviate from certain advice in this chapter, you should do so only after reasoned and considered judgment. The guidance here should form the bedrock of your API design decisions.

In this chapter, I will concentrate on generic, language-neutral qualities of an API, such as information hiding, consistency, and loose coupling. I will provide a C++ context for each of these concepts, but overall, the advice in this chapter should be useful to you whether you are working on a C++, Java, C#, or Python project. In later chapters, I will deal with C++ specific issues, such as const correctness, namespaces, and constructor usage.

Many of the topics in this chapter also provide a jumping-off point into deeper treatments later in the book. For example, whereas I mention the use of the Pimpl idiom as a solution for hiding internal details in C++, I dedicate more space to this important topic in the following chapter on design patterns.

Model the problem domain

An API is written to solve a particular problem or perform a specific task. So, first and foremost, the API should provide a coherent solution for that problem, and it should be formulated in a way that models the actual domain of the problem. For example, it should provide a good abstraction of the problem area and it should model the key objects of that domain. Doing so can make the API easier for your users to employ and understand because it will more closely correlate with their preexisting knowledge and expectations.

Provide a good abstraction

An API should provide a logical abstraction for the problem that it solves. That is, it should be formulated in terms of high-level concepts that make sense in the chosen

API Design for C++. https://doi.org/10.1016/B978-0-443-22219-1.00014-3

problem domain, rather than exposing low-level implementation issues. You should be able to give your API documentation to a nonprogrammer, and that person should be able to understand the concepts of the interface and how it's meant to work.

Furthermore, it should be apparent to the nontechnical reader that the group of operations provided by the API makes sense and belongs together as a unit. Each class should have a central purpose, and that should be reflected in the name of the class and its methods. In fact, it's good practice to have another person review your API early on to make sure that it presents a logical interface to fresh eyes.

Coming up with a good abstraction is no simple task. I dedicate most of Chapter 4 to this complex topic. However, it should be noted that there is no single correct abstraction for any given problem. Most APIs can be modeled in several different ways, each of which may provide a good abstraction and a useful interface. The key point is that there is some consistent and logical underpinning to your API.

For example, let's consider an API for a simple address book program. Conceptually, an address book is a container for the details of multiple people. It seems logical, then, that our API should provide an `AddressBook` object that contains a collection of `Person` objects, where a `Person` object describes the name and address of a single contact. Furthermore, you want to be able to perform operations such as adding a person to the address book or removing them. These are operations that update the state of the address book, and so logically they should be part of the `AddressBook` object. This initial design can be represented visually using a notation called Unified Modeling Language (UML).

If you're not familiar with UML, Fig. 2.1 shows an `AddressBook` object that contains a one-to-many composition of `Person` objects as well as two operations, `AddPerson()` and `DeletePerson()`. The `Person` object contains a set of public attributes to describe a single person's name and address. I will refine this design in a moment, but for now it serves as an initial logical abstraction of the problem domain.

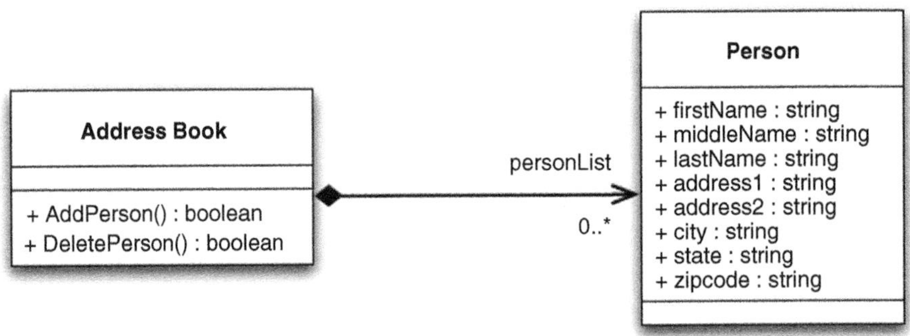

FIGURE 2.1 High-level Unified Modeling Language abstraction of an address book application programming interface.

SIDEBAR: Unified Modeling Language class diagrams

The UML specification defines a collection of visual notations to model object-oriented software systems (Booch et al., 2005). *In this book, I will often use the UML class diagram to depict class designs, as in* Fig. 2.1. *In these diagrams, a single class is represented with a box that is segmented into three parts:*

1. *The upper section contains the class name.*
2. *The middle section lists the attributes of the class.*
3. *The lower section enumerates the methods of the class.*

Each of the entries in the middle and lower sections of the class box can be prefixed with a symbol to indicate the access level, or visibility, for that attribute or method:

+ *indicates a public class member.*
– *indicates a private class member.*
indicates a protected class member.

Relationships between classes are illustrated with various styles of connecting lines and arrowheads. Some of the common relationships that are possible in a UML class diagram are:

Association*: A simple dependency between two classes in which neither owns the other, shown as a solid line. Association can be directional, indicated with an open arrowhead as >.*
Aggregation*: A has-a, or whole/part, relationship in which neither class owns the other, shown as a line with a hollow diamond.*
Composition*: A has-a relationship in which the lifetime of the part is managed by the whole. This is represented as a line with a filled diamond.*
Generalization*: A subclass relationship between classes, shown as a line with a hollow triangle arrowhead.*

Each side of a relationship can also be annotated to define its multiplicity. This lets you specify whether the relationship is one-to-one, one-to-many, or many-to-many. Some common multiplicities include:

0..1 = Zero or one instances
1 = Exactly one instance
0.. = Zero or more instances*
1.. = One or more instances*

Model the key objects

An API should also model the key objects for the problem domain. This process is often called object-oriented design or object modeling because it aims to describe the hierarchy of objects in the specific problem domain. The goal of object modeling is to identify the collection of major objects, the operations they provide, and how they relate to each other.

Once again, there is no single correct object model for a given problem domain. Instead, the task of creating an object model should be driven by the specific requirements for the API. Different demands on an API may require a different object model to represent those demands most effectively. For example, continuing our address book example, let's assume that we've received the following requirements for our API:

1. Each person may have multiple addresses.
2. Each person may have multiple telephone numbers.
3. Telephone numbers can be validated and formatted.
4. An address book may contain multiple people with the same name.
5. An existing address book entry can be modified.

These requirements will have a large impact on the object model for the API. Our original design in Fig. 2.1 supports only a single address per person. To support more than one address, you could add extra fields to the `Person` object (e.g., `HomeAddress1`, `WorkAddress1`), but this would be a brittle and inelegant solution. Instead, you could introduce an object to represent an address (e.g., `Address`) and allow a `Person` object to contain multiples of these.

The same is true of telephone numbers: you can factor these into their own object (e.g., `TelephoneNumber`) and allow the `Person` object to hold multiple of these. Another reason to create an independent `TelephoneNumber` object is that we need to support operations such as `IsValid()`, to validate a number, and `GetFormattedNumber()`, to return a nicely formatted version of the number. These are operations that naturally operate on a telephone number, not a person, which suggests that telephone numbers should be represented by their own first-class objects.

The requirement that multiple `Person` objects may hold the same name essentially means that a person's name cannot be used to identify a unique instance of the `Person` object. You therefore need some way to identify a unique `Person` instance (e.g., so that you can locate and update an existing entry in the address book). One way to satisfy this requirement would simply be to generate a universally unique identifier (UUID) for each person. Putting all this together, you might conclude that the key objects for our address book API are:

- **Address Book**: Contains zero of more `Person` objects, with operations such as `AddPerson()`, `DeletePerson()`, and `UpdatePerson()`.
- **Person**: Fully describes the details for a single person, including zero or more addresses and telephone numbers. Each person is differentiated by a UUID.
- **Address**: Describes a single address, including an address type field such as Home or Work.
- **TelephoneNumber**: Describes a single telephone number, including a phone type field such as Home or Cell. This object supports operations such as `IsValid()` and `GetFormattedNumber()`.

This updated object model can be represented again as a UML diagram, as shown in Fig. 2.2.

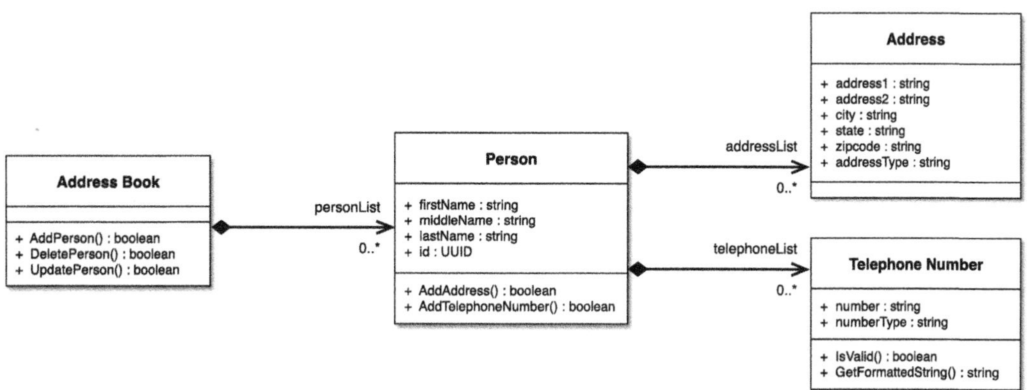

FIGURE 2.2 Unified Modeling Language diagram of the key objects in our address book application programming interface. *UUID*, universally unique identifier.

It's important to note that the object model for an API may need to change over time. As new requirements are received, or new functionality is added, the optimal breakdown of classes and methods to meet those needs may change. It's therefore always wise to reevaluate your object model in the light of new requirements, to assess whether your object model would benefit from a redesign, too. For example, you may anticipate the need for international addresses and decide to create a more general `Address` object to handle this capability. However, don't go over the top and try to create an object model that is more general than you need. Make sure you read the upcoming section on being minimally complete, too!

Solve the core problems

The main reason to develop an API is to solve some related set of problems. So, it must be clear to your users how they can apply your design to achieve that goal. We just saw how you can model the key objects for your problem domain, but the users of your API also need to know how to create those objects and how to call the methods on those objects to complete their tasks.

I introduced the concept of the UML class diagram earlier to show the relationships among different objects. Another UML diagram type is the sequence diagram. This shows the sequential interactions between objects that produce some desired output. In other words, it can show how your design can be used to achieve certain goals. For example, in the address book design described previously, I introduced objects for `AddressBook`, `Person`, and `TelephoneNumber`. But how would a client use these to change the type of a person's telephone number from a personal to a work number? Fig. 2.3 presents one possible sequence diagram to show how you could do this.

A sequence diagram represents the set of objects from left to right across the top of the diagram. The first message typically starts at the top left of the diagram, such as `AddressBook::GetInstance()` in our example. Each subsequent message appears below the

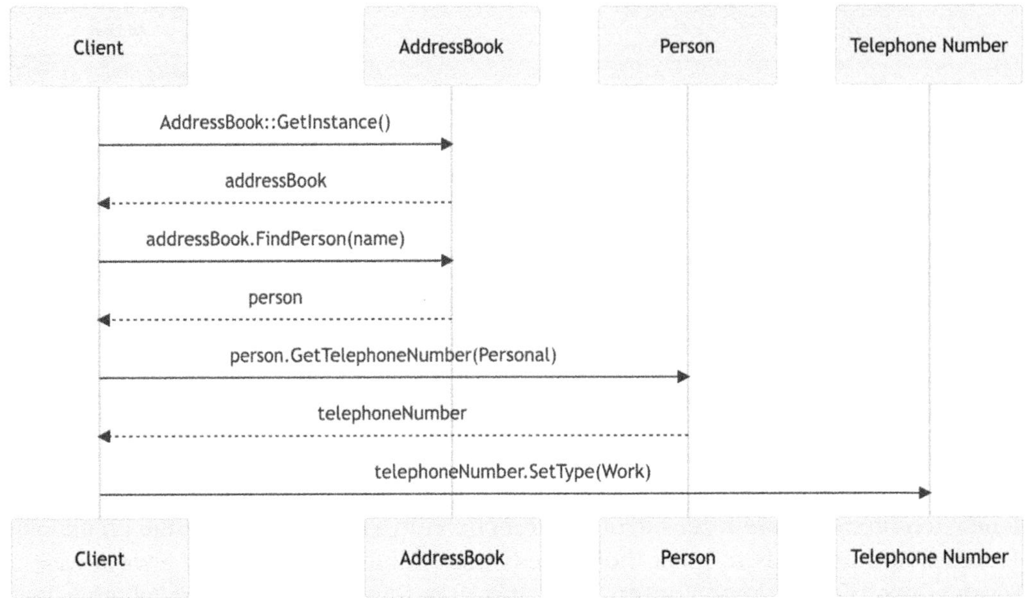

FIGURE 2.3 Unified Modeling Language sequence diagram to change a telephone number type.

previous one. An object that sends a message to another object is shown as a solid line and arrowhead (or an open arrowhead for an asynchronous message) and a return result is shown with a dotted line and open arrowhead.

So, Fig. 2.3 starts with a call to `AddressBook::GetInstance()` to get an instance of the address book. Then you would call an `AddressBook::FindPerson()` method to find a specific person. You can then call `GetTelephoneNumber()` on that `Person` object to get the personal telephone number (we'll assume there's only one, to keep the example simple). And then you can call a `SetType()` method on the telephone number to set it to a work phone number type. With this information, we can see that we need to add some new functions to our objects to support this use case. This lets us build out our API to ensure it solves the core tasks for the problem domain.

As an aside, there are several tools that let you create these UML class and sequence diagrams. For example, I used the Mermaid charting tool at https://mermaid.js.org/ to create the sequence diagram in Fig. 2.3 using the following Markdown description:

```
sequenceDiagram
    Client->>AddressBook: AddressBook::GetInstance()
    AddressBook-->>Client: addressBook
    Client->>AddressBook: addressBook.FindPerson(name)
    AddressBook-->>Client: person
    Client->>Person: person.GetTelephoneNumber(Personal)
    Person-->>Client: telephoneNumber
    Client->>Telephone Number: telephoneNumber.SetType(Work)
```

Hide implementation details

The primary reason for creating an API is to hide any implementation details so that these can be changed later without affecting existing clients. Therefore, the most important quality of an API is that it actually achieves this purpose. That is, any internal details—those parts that are most likely to change—must be kept secret from the client of the API. David L. Parnas referred to this concept as information hiding (Parnas, 1972).

There are two main categories of techniques relating to this goal: physical and logical hiding. Physical hiding means that the private source code is simply not made available to users. Logical hiding entails the use of language features to limit access to certain elements of the API.

Physical hiding: Declaration versus definition

In C and C++, the words declaration and definition are precise terms with very specific meanings. A declaration simply introduces a name and its type to the compiler without allocating any memory for it. In contrast, a definition provides details of a type's structure, or allocates memory in the case of variables. (The term function prototype, as used by C programmers, is equivalent to the term function declaration.) For example, the following are all declarations:

```
extern int i;
class MyClass;
void MyFunc(int value);
```

whereas the following are all definitions:

```
int i = 0;

class MyClass
{
public:
    float x, y, z;
};

void MyFunc(int value)
{
    printf("In MyFunc(%d).\n", value);
}
```

TIP: A declaration introduces the name and type of a symbol to the compiler. A definition provides the full details for that symbol, be it a function body or a region of memory.

In terms of classes and methods, the following code introduces a class definition with a single method declaration:

```
class MyClass
{
public:
    void MyMethod();
};
```

The implementation (body) of a method is provided in its definition:

```
void MyClass::MyMethod()
{
    printf("In MyMethod() of MyClass.\n");
}
```

Generally speaking, you provide declarations in your `.h` files and the associated definitions in your `.cpp` files. However, it's also possible to provide a definition for a method at the point where you declare it in a `.h` file, e.g.,

```
class MyClass
{
public:
    void MyMethod()
    {
        printf("In MyMethod() of MyClass.\n");
    }
};
```

This technique implicitly requests the compiler to inline the `MyMethod()` member function at all points where it's called. In terms of API design, this is therefore a bad practice because it exposes the code for how the method has been implemented and directly inlines the code into your clients' programs. You should therefore strive to limit your API headers to provide only declarations. In later chapters, I will discuss exceptions to this rule to support templates and conscious acts of inlining.

TIP: Physical hiding means storing internal details in a separate file (.cpp) from the public interface (.h).

Note that I will sometimes inline function implementations like this in the code examples of this book. However, this is done solely for the purposes of clarity and simplicity and should be avoided for any real-world APIs that you develop.

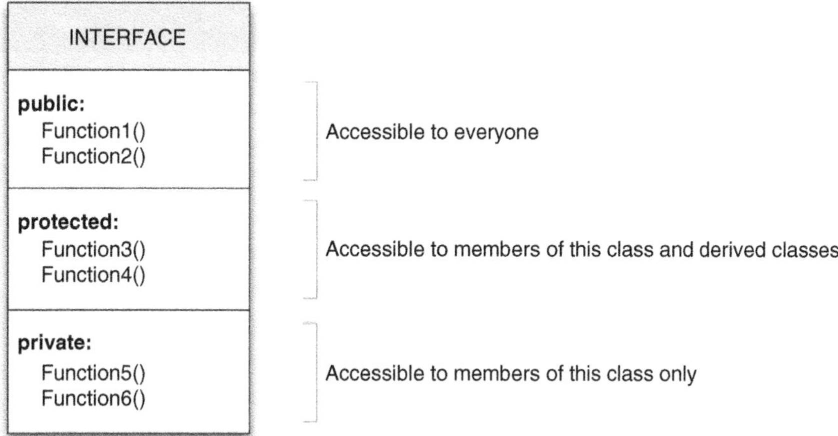

FIGURE 2.4 The three access levels for C++ classes.

In the later chapter on C++ Revisions, I will cover the concept of modules, which was introduced in C++20. This feature provides an alternative to the use of header files. However, you can still perform physical hiding with modules by putting your declarations in your module interface unit and definitions in your module implementation unit.

Logical hiding: Encapsulation

The object-oriented concept of encapsulation provides a mechanism for limiting the access to members of an object. In C++, this is implemented using the following access control keywords for classes and structs (classes and structs are functionally equivalent, differing only in their default access level). These access levels are illustrated in Fig. 2.4.

- **Public**: Members are accessible outside the class/struct. This is the default access level for structs.
- **Private**: Members are accessible only within the specific class in which they are defined. This is the default access level for classes.
- **Protected**: Members are accessible only within the specific class and any derived classes.

> *SIDEBAR: Encapsulation in other languages*
>
> *Whereas C++ provides the public, private, and protected access controls for your class members, other object-oriented languages provide different levels of granularity. For example, in the Smalltalk language all instance variables are private and all methods are public, whereas the Java language provides public, private, protected, and package-private levels of visibility.*

Continued

SIDEBAR: Encapsulation in other languages—cont'd

Package-private in Java means that a member can be accessed only by classes within the same package. This is the default visibility in Java. Package-private is a great way to allow other classes in a JAR (Java Archive) file to access internal members without exposing them globally to your clients. For example, it's particularly useful for unit tests that need to verify the behavior of private methods.

C++ does not have the concept of package-private visibility. Instead, it uses the more permissive notion of friendship to allow named classes and functions to access protected and private members of a class. Friendship can be used to enhance encapsulation, but if used carelessly it can also overexpose internal details to your clients.

Most likely, your users will not care about respecting your public API boundary. If you give them hooks into your internal workings that let them achieve what they want, they will use them to get their job done. Although this may appear to be good for them, because they were able to find a solution to their immediate problem, it may make it more difficult for you to change those implementation details in the future and so stifle your ability to improve and optimize your product.

TIP: Encapsulation is the process of separating the public interface of an API from its underlying implementation.

As an example of the lengths to which some users will go, the multiplayer first-person shooter game Counter-Strike has been a popular target for exploitation hacks since it appeared in around 2000. One of the most well-known of these is the wallhack. This is essentially a modified OpenGL driver that renders walls partially or fully transparent. This gives players the clear advantage that they can literally see through walls. Although you may not be creating a game or targeting gamers as your clients, the moral is that users will do whatever they can to get what they want. If some users will modify OpenGL graphics drivers to give them an advantage in a game, there are presumably others who will use your API's exposed internal details so that they can deliver the functionality that their boss has requested.

To illustrate this with a more directly applicable example, Roland Faber reports some of the difficulties that occurred at Siemens when one group decided to rely upon the internal details of an API from another group in the same company (Faber, 2010):

> *A team outside of Europe had to provide a remote control for a user interface implemented in Germany. Because the automation interface was incomplete, they decided to use internal interfaces instead without notifying the architects. The system integration suffered from unexpected problems due to uncoordinated interface changes, and costly refactoring was unavoidable.*

The following sections will therefore discuss how to use the access control features of your programming language to provide the maximum level of information hiding for your APIs. The later chapter on C++ Usage also points out some cases in which certain C++ language features can affect encapsulation, such as friends and external linkage.

> *TIP: Logical hiding means using the C++ language features of protected and private to restrict access to internal details.*

Hide member variables

The term encapsulation is also often used to describe the bundling of data with the methods that operate on those data. This is implemented in C++ by having classes that can contain both variables and methods. However, in terms of good API design, you should never make member variables public. If the data members form part of the logical interface of the API, then you should instead provide getter and/or setter methods that provide indirect access to the member variables. For example, you should avoid writing:

```
class Vector3
{
public:
    double x, y, z;
};
```

Instead, you should prefer:

```
class Vector3
{
public:
    double GetX() const;
    double GetY() const;
    double GetZ() const;
    void SetX(double val);
    void SetY(double val);
    void SetZ(double val);

private:
    double mX, mY, mZ;
};
```

The latter syntax is obviously more verbose and involves more typing on your part as the programmer, but the extra few minutes spent doing this could save you hours, or even days further down the line should you decide to change the interface. Some of the additional benefits of using getter/setter routines, rather than directly exposing member variables, include:

- **Validation.** You can perform validation on the values to ensure that the internal state of the class is always valid and consistent. For example, if you have a method that lets clients set a new RGB color, you could check that each of the supplied red, green, and blue values are within the valid range (e.g., 0 to 255 or 0.0 to 1.0).

- **Lazy evaluation.** Calculating the value of a variable may incur a significant cost, which you would prefer to avoid until necessary. By using a getter method to access the underlying data value, you can defer the costly calculation until the value is actually requested.
- **Caching.** A classic optimization technique is to store the value of a frequently requested calculation and then directly return that value for future requests. For example, a machine's total memory size can be found on Linux by parsing the `/proc/meminfo` file. Rather than performing a file read for every request to find the total memory size, it would be better to cache the result after the first read and then simply return that cached value for future requests.
- **Extra computation.** If necessary, you can perform additional operations whenever the client tries to access a variable. For example, perhaps you always want to write the current state of a `UserPreferences` object to a configuration file on disk whenever the user changes the value of a preference setting.
- **Notifications.** Other modules may wish to know when a value has changed in your class. For example, if you are implementing a data model for a progress bar, the user interface code will want to know when the progress value has been updated so that it can update the GUI. You might therefore wish to issue a change notification as part of a setter method.
- **Debugging.** You may want to add debugging or logging statements so that you can track when variables are accessed or changed by clients. Or you may wish to add assert statements to enforce assumptions.
- **Synchronization.** You may release the first version of your API and then later find that you need to make it thread safe. The standard way to do this is to add mutex locking whenever a value is accessed. This would be possible only if you have wrapped the access to the data values in getter and setter methods.
- **Finer access control.** If you make a member variable public, then clients can read and write that value as they wish. However, by using getter/setter methods, you can provide a finer level of read/write control. For example, you can make the value be read-only by not providing a setter method.
- **Maintaining invariant relationships.** Some internal data values may depend upon each other. For example, in a car animation system you may calculate the velocity and acceleration of the car based upon the time it takes to travel between key frames. You can calculate velocity based upon the change in position over time, and acceleration based upon the change in velocity over time. However, if a client can access your internal state for this calculation, they could change the acceleration value so that it does not correlate to the car's velocity, thus producing unexpected results.

On the other hand, if the member variables are not actually part of the logical interface—that is, they represent internal details that are not relevant to the public interface—then they should simply be hidden from the interface. For example, consider the following definition for a stack of integers:

```
class IntegerStack
{
public:
    static const int MAX_SIZE = 100;
    void Push(int val);
    int Pop();
    bool IsEmpty() const;
    int mStack[MAX_SIZE];
    int mCurSize;
};
```

Clearly this is a bad API because it exposes the way that the stack has been (poorly) implemented as a fixed array of integers, and it reveals the internal state of the stack via the `mCurSize` variable. If at some future date you decided to improve the implementation of this class, for example by using an `std::vector` or `std::list` rather than a fixed-size statically allocated array, then you may find this difficult to do. That's because you've exposed the existence of the `mStack` and `mCurSize` variables, and so client code could be relying on the ability to access these variables directly. By changing your implementation, you could break your clients' code.

Instead, these member variables should be hidden from the start so that client code cannot access them:

```
class IntegerStack
{
public:
    void Push(int val);
    int Pop();
    bool IsEmpty() const;

private:
    static const int MAX_SIZE = 100;
    int mStack[MAX_SIZE];
    int mCurSize;
};
```

I have stated that member variables should never be public, but can they be declared as protected? If you make a variable protected, then it can be accessed directly by any clients that subclass your class, and then exactly the same arguments apply as for the public case. As such, you should never make your member variables protected, either. As Alan Snyder states, inheritance severely compromises the benefits of encapsulation in object-oriented programming languages (Snyder, 1986).

TIP: Data members of a class should always be declared private, never public or protected.

The only semiplausible argument for exposing member variables is for performance reasons. Executing a C++ method call incurs the overhead of pushing the method's parameters and return address onto the call stack, as well as reserving space for any local variables in the routine. Then when the method completes, the call stack must be unwound again. The cost to perform these actions may be noticeable for performance-critical regions of code, such as within a tight loop performing operations on a large number of objects. Code that directly accesses a public member variable may be two to three times faster than code that has to go through getter/setter methods.

However, even in these cases, you should never expose member variables. First, the overhead of a method call will very likely be insignificant for practically all of your API calls. Even if you are writing performance-critical APIs, the careful use of inlining combined with a modern optimizing compiler will normally completely eradicate the method call overhead, giving you all of the performance benefits of directly exposing member variables. If you're still concerned, then try timing your API with inlined getter/setters and then with public member variables. The accompanying source code for this book includes a sample program to do just this. See https://APIBook.com/ to download this code and try it out yourself. I'll also discuss this issue further in the chapter on Performance.

Hide implementation methods

In addition to hiding all member variables, you should hide all methods that don't need to be public. This is the principle of information hiding: segregating the stable interface for a class from the internal design decisions used to implement it. Early studies of several large programs found that those using information hiding techniques were four times easier to modify than programs that didn't (Korson and Vaishnavi, 1986). Although your own mileage may vary, it should be clear that hiding the internal details of your API will make for more maintainable and evolvable software.

The key point to remember is that a class should define what to do, not how it's done. For example, let's consider a class that lets you download a file from a remote http server:

```
#include <stdio.h>
#include <string>
#include <sys/socket.h>
#include <unistd.h>

class URLDownloader
{
public:
    URLDownloader();
    bool DownloadToFile(const std::string &url,
                        const std::string &localFile);
    bool SocketConnect(const std::string &host, int port);
    void SocketDisconnect();
    bool IsSocketConnected() const;
    int GetSocket() const;
    bool SocketWrite(const char *buffer, size_t bytes);
    size_t SocketRead(char *buffer, size_t bytes);
    bool WriteBufferToFile(char *buffer,
                           const std::string &filename);

private:
    int mSocketID;
    struct sockaddr_in mServAddr;
    bool mIsConnected;
};
```

All of the member variables have been correctly declared as private, which is a good start. However, several implementation-specific methods have been exposed, such as routines to open and read data from a socket and a routine to write the resulting in-memory buffer to a file on disk. The client doesn't need to know any of this. All the client wants to do is specify a URL and then magically have a file created on disk with the contents of that remote location.

There's also a particularly egregious method in there: `GetSocket()`. This is a public method that returns access to a private member variable. By calling this method, a client can get the underlying socket handle and could directly manipulate that socket without the knowledge of the `URLDownloader` class. What's particularly disturbing is that `GetSocket()` is declared as a const method, meaning that it doesn't modify the state of the class. Whereas this is strictly true, the client could then use the returned integer socket handle to modify the state of the class. The same kind of leaking of internal state can happen if you return a non-const pointer or reference to one of your private member variables. Doing so lets your clients modify the state of your objects without going through your API.

> *TIP: Never return non-const pointers or references to private data members. This breaks encapsulation.*

Obviously a better design for our `URLDownloader` class would be to make every method private, except for the constructor and the `DownloadToFile()` method. Everything else is implementation detail. You are then free to change the implementation without affecting any clients of the class.

There is still something unsatisfying about this situation, though. I have hidden the implementation details from the compiler's point of view, but a person can still look at the header file and see all of the internal details for your class. In fact, this is very likely because you must distribute the header file to allow clients to compile their code against your API. Also, you must `#include` all header files needed for the private members of your class, even though they're not dependencies of the public interface. For example, the `URLDownloader` header needs to `#include` all of the platform-specific socket headers.

This is an unfortunate limitation of the C++ language: all public, protected, and private members of a class must appear in the declaration for that class. Ideally, the header for our class would look like:

```
#include <string>

class URLDownloader
{
public:
    URLDownloader();
    bool DownloadToFile(const std::string &url,
                        const std::string &localFile);
};
```

Then all the private members could be declared somewhere else, such as in the `.cpp` file. However, this is not possible with C++ (this is so that the size of all objects can be known at compile time). Nevertheless, there are still ways to hide private members from your public header files (Headington, 1995). One popular technique is called the Pimpl idiom. This involves isolating all of a class's private data members inside a separate implementation class or struct in the `.cpp` file. The `.h` file then needs only to contain an opaque pointer to this implementation class. I'll discuss this extremely valuable technique in more detail in the Patterns chapter, coming up next.

I strongly urge you to adopt the Pimpl idiom in your APIs so that all implementation details can be completely hidden from your public header files. However, if you decide against this direction, you should at least attempt to remove private methods from the header when they are not necessary by moving them to the `.cpp` file and converting them to static functions (Lakos, 1996). This can be done when the private method accesses only public members of the class, or if it accesses no members of the class at all (such as a routine that accepts a filename string and returns the extension for that filename). Many engineers feel that just because a class uses a private method, it must be included in the class declaration. However, this simply exposes more implementation details than necessary.

TIP: Prefer declaring private functionality as static functions within the .cpp file rather than exposing them in public headers as private methods. (Using the Pimpl idiom is even better, though).

Hide implementation classes

In addition to hiding the internal methods and variables for your classes, you should endeavor to hide any actual classes that are purely implementation detail. Most programmers are used to hiding methods and variables, although many forget also to consider that not all classes are public. Indeed, some classes are needed only for your implementation and should not be revealed as part of the public interface of your API.

For example, consider a simple `Fireworks` class: an interface that lets you specify the location of a fireworks animation on the screen and lets you control the color, speed, and number of fire particles. Clearly the API will need to keep track of each particle of the firework effect, so that it can update each particle's position per frame. This implies that a `FireParticle` class should be introduced to contain the state for a single fire particle. However, clients of the API never need to access this class; it's purely required for the API's implementation.

We could therefore use either physical or logical hiding to remove the class from the public API. In this case, physical hiding would mean putting the implementation of the class in a private header file (i.e., in a header file that's not included with your set of public API headers). Then we forward declare the class in the public header so that the name is known to the compiler, e.g.,

```
// fireworks.h (public header)
#include <vector>

class FireParticle;

class Fireworks
{
public:
    Fireworks();

    void SetOrigin(double x, double y);
    void SetColor(float r, float g, float b);
    void SetGravity(float g);
    void SetSpeed(float s);
    void SetNumberOfParticles(int num);

    void Start();
    void Stop();
    void NextFrame(float dt);

private:
    double mOriginX, mOriginY;
    float mRed, mGreen, mBlue;
    float mGravity;
    float mSpeed;
    bool mIsActive;
    std::vector<FireParticle *> mParticles;
};
```

```
// fireparticle.h (private header)
class FireParticle
{
public:
    double mX, mY;
    double mVelocityX, mVelocityY;
    double mAccelerationX, mAccelerationY;
    double mLifeTime;
};
```

Alternatively, you can also use logical hiding to declare the class as private, so clients can't access the class from their code. This can be done by nesting it within another class's private section, e.g.,

```
#include <vector>

class Fireworks
{
public:
    Fireworks();

    void SetOrigin(double x, double y);
    void SetColor(float r, float g, float b);
    void SetGravity(float g);
    void SetSpeed(float s);
    void SetNumberOfParticles(int num);

    void Start();
    void Stop();
    void NextFrame(float dt);

private:
    class FireParticle
    {
    public:
        double mX, mY;
        double mVelocityX, mVelocityY;
        double mAccelerationX, mAccelerationY;
        double mLifeTime;
    };

    double mOriginX, mOriginY;
    float mRed, mGreen, mBlue;
    float mGravity;
    float mSpeed;
    bool mIsActive;
    std::vector<FireParticle *> mParticles;
};
```

Notice that I didn't use getter/setter methods for the `FireParticle` class. You could certainly do so if you wanted to, but it's not strictly necessary because the class is not accessible from the public interface.

Again, you could also hide the entire `FireParticle` symbol from appearing in the header file so that it's hidden even from casual inspection of the header file. I will discuss how to do this in the next chapter.

Minimally complete

A good API should be minimally complete. That is, it should be as small as feasible, but no smaller.

It's perhaps obvious that an API should be complete: that it provides clients with all the functionality they need. However, it may be less obvious what that functionality is. To answer this question, you should perform requirements gathering and use case modeling early on so that you understand what the API is expected to do. You can then assert that it delivers on those expectations. I will talk more about requirements and use cases in the chapter on Design.

Less obvious is the apparent contradiction for an API to be minimal. However, this is one of the most important qualities for which you can plan, and one that has a massive impact on the long-term maintenance and evolution of your API. In a very real sense, the decisions you make today will constrain what you can do tomorrow. It also has a large impact on the ease of use of the API, because a compact interface is one that can easily fit inside the head of your users (Blanchette, 2008). I will therefore dedicate the following sections to discuss various techniques to keep your API minimal, and why this is a good thing.

TIP: Remember Occam's razor: Pluralitas non est ponenda sine *necessitate (plurality should not be posited without necessity).*

Don't overpromise

Every public element in your API is a promise—a promise that you'll support that functionality for the lifetime of the API. You can break that promise, but doing so may frustrate your clients and cause them to rewrite their code. Even worse, they may decide to abandon your API because they have grown weary of continually having to fix their code because you can't keep your API stable. Or they may simply be unable to use your API anymore because you've removed functionality that supported their unique use case.

The key point is that once you release an API and have clients using it, adding new functionality is easy but removing functionality is really difficult. The best advice then is: when in doubt, leave it out (Bloch, 2008; Tulach, 2008).

This advice can be counter to the best intentions of the API designer. As an engineer, you want to provide a flexible and general solution to the problem you're solving. There is a temptation to add extra levels of abstraction or generality to an API because you think it might be useful in the future. You should resist this temptation for the following reasons:

1. The day may never come when you need the extra generality.
2. If that day does come, you may have more knowledge about the use of your API and a different solution may present itself from the one you originally envisioned.
3. If you do need to add the extra functionality, it'll be easier to add it to a simple API than a complex one.

As a result, you should try to keep your APIs as simple as you can: minimize the number of classes you expose and the number of public members in those classes. As a bonus, this will also make your API easier to understand, easier for your users to keep a mental model of the API in their heads, and easier for you to evolve.

TIP: When in doubt, leave it out! Minimize the number of public classes and functions in your API.

Don't repeat yourself

The don't repeat yourself (DRY) principle is about reducing the repetition of information in your code. It's typically described as requiring a single, unambiguous, and authoritative representation for any given piece of information in a system (Hunt and Thomas, 1999).

This general concept is also expressed by several other similar design principles, such as the single source of truth principle, which aims to define data in only one place, and the single responsibility principle, which states that a class should have only one reason to change and shouldn't be responsible for multiple unrelated tasks.

Taken together, these principles are focused on reducing duplication by defining data and behavior in one place. As such, they offer ways to achieve minimal completeness.

However, it's also worth noting that these principles can be taken to extremes if they're applied blindly. Trying to remove every piece of code duplication in your project can lead to overly complex code or unnecessary coupling between classes. So as with most rules of thumb, they should be applied thoughtfully, not rigidly. However, they can be used to help you identify places where a better design might be possible.

Some common ways to address code duplication are:

1. **Function Refactoring:** The simplest solution is just to pull out the duplicated code into a single function that can be called from all of the places that need the functionality. If this is part of your implementation code, then this may not even affect your public API.
2. **Class Abstractions:** This involves creating more general levels of abstractions to represent the concepts in your system so that shared functionality can be pushed further down the inheritance hierarchy. This is particularly relevant to API design if it concerns the set of public classes and inheritance hierarchies that you expose to your clients.
3. **Dependency Injection:** This is where the functionality is defined in a single object and that object is passed into the method calls where the functionality is needed. This may affect your API if you need to update various public API method calls to pass in the object.
4. **Automation:** In cases where duplication is unavoidable, you can define a single source of truth for the functionality and write automation tools to generate the other code from that source of truth. This lets you keep all duplicated code sections in sync by running your automation scripts.

We can take the address book example from earlier in this chapter to illustrate this concept. We identified that each person in the address book may have multiple telephone numbers. One solution would be just to add work and home phone numbers directly to the person object. However, if we wanted to offer various functions to access or format each of these phone numbers, or if we wanted to associate phone numbers with other objects such as addresses, we can see that it may make more sense to represent phone numbers as their own objects in the system. That way, all functionality for manipulating phone numbers can be contained in one place.

A good way to think about this is to ask yourself: if you needed to make a change to how something works, would you have to update multiple places in your code to make that change? If so, then you may want to consider refactoring your design. Again, though, this shouldn't be taken as a hard and fast rule. You need to consider the impact on the simplicity and usability of your design as well. For example, perhaps you can instead apply an automation solution behind the scenes to keep information in sync without affecting your API design. (We'll talk about intentional redundancy later in this chapter.)

Martin Fowler suggested that you try to refactor code only when you repeat something a third time (Fowler et al., 1999). This is encapsulated in the acronym WET, or write everything twice, meaning that it's okay to duplicate something twice, but no more. There is also another acronym called AHA, avoid hasty abstractions, which represents the tension between refactoring duplicated code and adding more complex abstractions to your interface. There are so many acronyms for duplicated code!

As a final note on reducing duplication, it's possible to interpret the DRY and related principles to mean that there should be only one way to do something. However, that's not always a good goal. Consider the Standard Library container classes that let you search for an element using a `find()` method. This can be used to discover if an element exists in the container by comparing the result to the `end()` iterator, but the code to do this is verbose and has the potential for you to get the logic of the comparison wrong. As of C++20, there is now the `contains()` method, which is much simpler to use, although arguably it is not adding anything you couldn't already do. However, if this new function is just implemented in terms of existing API functions, there is still only one place to update if a change to the behavior is needed. We will delve deeper into this idea of convenience API calls in the next section.

Convenience APIs

Keeping an API as small as possible can be a difficult task. There's a natural tension between reducing the number of functions in your API and making the API easy to use for a range of clients. This is an issue that most API designers face: whether to keep the API pure and focused or to allow convenience wrappers. (By the term convenience wrappers, I mean utility routines that encapsulate multiple API calls to provide simpler higher-level operations).

On the one hand, there's the argument that an API should provide only one way to perform one task. This ensures that the API is minimal, singularly focused, consistent, and easy to understand. It also reduces the complexity of the implementation, with all the associated benefits of greater stability and easier debugging and maintenance. Grady Booch refers to this as primitiveness, or the quality to which a method needs access to the internal details of a class to be implemented efficiently, as opposed to nonprimitive methods that can be entirely built on top of primitive ones, without access to any internal state (Booch et al., 2007).

However, on the other hand, there is also the argument that an API should make simple things easy to do. Clients should not be required to write lots of code to perform basic tasks. Doing so can give rise to blocks of boilerplate code that get copied and pasted to other parts of their source code, and whenever blocks of code are copied there is the potential for code divergence and bugs. Also, you may wish to target your API to a range of clients, from those who want lots of control and flexibility to those who just want to perform a single task as simply as possible.

Both goals are useful and desirable. Fortunately, they do not need to be mutually exclusive. There are several ways you can provide higher-level convenience wrappers for your core API's functionality without diluting its primary purpose. The important point is that you don't mix your convenience API in the same classes as your core API. Instead, you can produce supplementary classes that wrap certain public functionality of your core API. These convenience classes should be fully isolated from your core API (e.g., in different source files or even completely separate libraries). This has the additional benefit of ensuring that your convenience API depends only on the public interface of your core API, and not on any internal methods or classes. Let's look at a real-world example of this.

The OpenGL API provides platform-independent routines for writing 2D and 3D applications. It operates on the level of simple primitives such as points, lines, and polygons that are transformed, lit, and rasterized into a frame buffer. The OpenGL API is extremely powerful, but it's also aimed at a very low level. For example, creating a sphere in OpenGL would involve explicitly modeling it as a surface of small flat polygons, as demonstrated in the following code snippet:

```
for (int i = 0; i <= stacks; ++i) {
    GLfloat stack0 = ((i - 1.0) / stacks - 0.5) * M_PI;
    GLfloat stack1 = ((GLfloat) i / stacks - 0.5) * M_PI;
    GLfloat z0 = sin(stack0);
    GLfloat z1 = sin(stack1);
    GLfloat r0 = cos(stack0);
    GLfloat r1 = cos(stack1);

    glBegin(GL_QUAD_STRIP);
    for (int j = 0; j <= slices; ++j) {
        GLfloat slice = (j - 1.0) * 2 * M_PI / slices;
        GLfloat x = cos(slice);
        GLfloat y = sin(slice);

        glNormal3f(x * r0, y * r0, z0);
        glVertex3f(x * r0, y * r0, z0);
        glNormal3f(x * r1, y * r1, z1);
        glVertex3f(x * r1, y * r1, z1);
    }
    glEnd();
}
```

However, most OpenGL implementations also include the OpenGL Utility Library (GLU). This is an API built on top of the OpenGL API that provides higher-level functions, such as mip-map generation, coordinate conversions, quadric surfaces, polygon tessellation, and simple camera positioning. These functions are defined in a library completely separate from the OpenGL library, and the functions all begin with the `glu` prefix to differentiate them from the core OpenGL API. For example, the following code snippet shows how easy it is to create a sphere using GLU:

```
GLUquadric *qobj = gluNewQuadric();
gluSphere(qobj, radius, slices, stacks);
gluDeleteQuadric(qobj);
```

This is a great example of how to maintain the minimal design and focus of a core API while providing additional convenience routines that make it easier to use that API. In fact, there are other APIs built on top of OpenGL that provide further utility classes, such as Mark Kilgard's OpenGL Utility Toolkit (GLUT). This API offers routines to create various solid and wireframe geometric primitives (including the famous Utah teapot) as well as simple window management functions and event processing. Fig. 2.5 shows the relationships among GL, GLU, and GLUT.

Ken Arnold refers to this concept as progressive disclosure, meaning that your API should present the basic functionality via an easy-to-use interface while reserving advanced functionality for a separate layer (Arnold, 2005). He notes that this concept is often seen in GUI designs in the form of an Advanced or Export button that discloses further complexity. In this way, you can still provide a powerful API while ensuring that the expert use cases don't obfuscate the basic workflows.

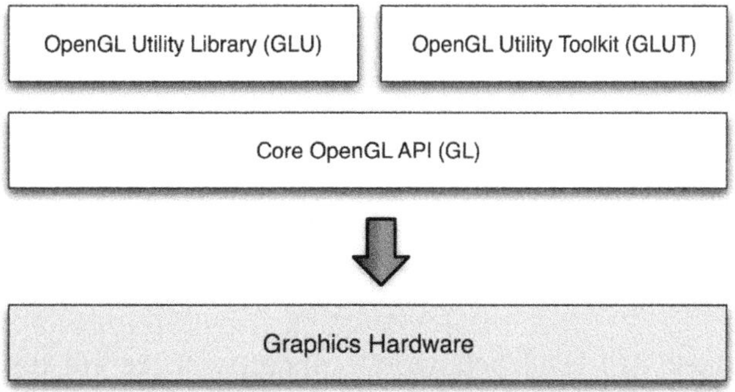

FIGURE 2.5 An example of a core application programming interface (API) (OpenGL) that is separated from the convenience APIs that are layered on top of it (OpenGL Utility Library and OpenGL Utility Toolkit).

> *TIP: Add convenience APIs as separate modules or libraries that sit on top of your minimal core API.*

Add virtual functions judiciously

One subtle way that you can expose more functionality than you intended is through inheritance: that is, by making certain member functions virtual. Doing so allows your clients to subclass your class and reimplement any virtual methods. Although this can be very powerful, you should be aware of the potential pitfalls:

- You can implement seemingly innocuous changes to your base classes that have a detrimental impact on your clients. This can happen because you evolve the base class in isolation, without knowing all of the ways that your clients have built upon your virtual API. This is often referred to as the fragile base class problem (Blanchette, 2008).
- Your clients may use your API in ways that you never intended or imagined. This can result in the call graph for your API executing code that you do not control and that may produce unexpected behavior. As an extreme example, there is nothing to stop a client calling `delete this` in an overridden method. This may even be a valid thing to want to do, but if you did not design your implementation to allow this, then your code will very likely crash.
- Clients may extend your API in incorrect or error-prone ways. For example, you may have a thread-safe API but, depending upon your design, a client could override a virtual function and provide an implementation without performing the appropriate mutex locking operations, opening the potential for difficult-to-debug race conditions. (There's a design pattern called Thread-Safe Interface that addresses this specific concern, which I describe in the Concurrency chapter).
- Overridden functions may break the internal integrity of your class. For example, the default implementation of a virtual method may call other methods in the same class to update its internal state. If an overridden method does not perform these same calls, then the object could be left in an inconsistent state and behave unexpectedly or crash.

In addition to these API-level behavioral concerns, there are the standard matters about which you should be aware when using virtual functions in C++:

- Virtual function calls must be resolved at runtime by performing a vtable lookup, whereas nonvirtual function calls can be resolved at compile time. This can make virtual function calls slower than nonvirtual calls. In reality, this overhead may be negligible, particularly if your function does nontrivial work or if it is not called frequently.

- The use of virtual functions increases the size of an object, typically by the size of a pointer to the vtable. This may be an issue if you wish to create a small object that requires a very large number of instances. Again, in practice this will likely be insignificant compared with the amount of memory consumed by your various member variables.
- Adding, reordering, or removing a virtual function will break binary compatibility. This is because a virtual function call is typically represented as an integer offset into the vtable for the class. Therefore changing its order, or causing the order of any other virtual functions to change, means that existing code will need to be recompiled to ensure that it still calls the right functions.
- Virtual functions cannot always be inlined. You may reasonably think that it does not make sense to declare a virtual as inline at all, because virtual functions are resolved at runtime whereas inlining is a compile-time optimization. However, there are certain constrained situations in which a compiler can inline a virtual function. All the same, these are far fewer than the cases in which a nonvirtual function can be inlined. (Remember that the `inline` keyword in C++ is merely a hint to the compiler).
- Overloading is tricky with virtual functions. A symbol declared in a derived class will hide all symbols with the same name in the base class. Therefore, a set of overloaded virtual functions in a base class will be hidden by a single overriding function in a subclass. There are ways to get around this (Dewhurst, 2002), but it's better simply to avoid overloading virtual functions.

Ultimately, you should allow overriding only if you explicitly intend for this to be possible. A class with no virtual functions tends to be more robust and requires less maintenance than one with virtual functions. As a rule of thumb, if your API does not internally call a particular method, then that method probably should not be virtual. You should also allow subclassing only in situations where it makes sense: where the potential subclasses form an is-a relationship with the base class.

If you still decide that you want to allow subclassing, make sure that you design your API to allow it safely. Remember the following rules:

1. Always declare your destructor to be virtual if there are any virtual functions in your class. This is so that subclasses can free any extra resources that they may have allocated.
2. Always document how methods of your class call each other. If clients want to provide an alternative implementation for a virtual function, they will need to know which other methods need to be called within their implementation to preserve the internal integrity of the object. Or use a design pattern such as Thread-Safe Interface that encodes the dependency.
3. Never call virtual functions from your constructor or destructor. These calls will never be directed to a subclass (Meyers, 2005). This does not affect the appearance of your API, but it's a good rule to know all the same.

> *TIP: Avoid declaring functions as overridable (virtual) until you have a valid and compelling need to do so.*

Easy to use

A well-designed API should make simple tasks easy and obvious. For example, it should be possible for a client to look at the method signatures of your API and be able to glean how to use it without any additional documentation. There have been various studies showing that APIs that are more usable can improve the productivity of developers using them (Clarke, 2004).

This API usability quality follows closely from the quality of minimalism: if your API is simple, it should be easy to understand. Similarly, it should follow the rule of least surprise, which posits that an interface should behave in a way that most users would expect. This is done by employing existing models and patterns so that the user can focus on the task at hand, rather than being distracted by the novelty or involution of your interface (Raymond, 2003).

Of course, this is not an excuse for you to ignore the need for good supporting documentation. In fact, it should make the task of writing documentation easier. As we all know, a good example can go a long way. Providing sample code can greatly aid the ease of use of your API. Good developers should be able to read example code written using your API and understand how to apply it to their own tasks.

The following sections will discuss various aspects and techniques to make your API easier to understand and ultimately simpler to use. Before I do so, though, it should be noted that an API may provide additional complex functionality for expert users that is not so easy to use. However, this should not be done at the expense of keeping the simple case easy.

Discoverable

A discoverable API is one in which users can work out how to use the API on their own without any accompanying explanation or documentation. To illustrate this with a counterexample from the field of user interface design, the Start button in Windows XP does not provide a very discoverable interface for locating the option to shut down the computer. Likewise, the Restart option is accessed rather unintuitively by clicking on the Turn Off Computer button.

Discoverability does not necessarily lead to ease of use. For example, it's possible for an API to be simple for a first-time user to learn but cumbersome for an expert user to use on a regular basis. However, in general, discoverability should help you produce a more usable interface.

There are several ways you can promote discoverability when you design your APIs. Devising an intuitive and logical object model is one important way, as is choosing good names for your classes and functions. Indeed, coming up with clear, descriptive, and appropriate names can be one of the most difficult tasks in API design. I will present specific recommendations for class and function names in Chapter 4 when I discuss API design techniques. Avoiding the use of abbreviations can also be a factor in discoverability (Blanchette, 2008), so that users don't have to remember if your API employs `GetCurrentValue()`, `GetCurrValue()`, `GetCurValue()`, or `GetCurVal()`.

Difficult to misuse

A good API, in addition to being easy to use, should also be difficult to misuse. Scott Meyers suggests that this is the most important general interface design guideline (Meyers, 2004). Some of the most common ways to misuse an API include passing the wrong arguments to a method and passing illegal values to a method. These can happen when you have multiple arguments of the same type and the user forgets the correct order of the arguments, or when you use an `int` to represent a small range of values instead of a more constrained `enum` type (Bloch, 2008). For example, consider the following method signature:

```
std::string FindString(const std::string &text,
                       bool search_forward,
                       bool case_sensitive);
```

It would be easy for users to forget whether the first `bool` argument is the search direction or the case sensitivity flag. Passing the flags in the wrong order would result in unexpected behavior and probably cause users to waste a few minutes debugging the problem, until they realized that they had transposed the `bool` arguments. However, you could design the method so that the compiler catches this kind of error for them, by introducing a new `enum` type for each option, or preferably an `enum class` as of C++11. For example:

```
enum class SearchDirection {
    Forward,
    Backward
};

enum class CaseSensitivity {
    Sensitive,
    Insensitive
};

std::string FindString(const std::string &text,
                       SearchDirection direction,
                       CaseSensitivity case_sensitivity);
```

Not only does this mean that users can't mix up the order of the two flags, because it would generate a compilation error, but also the code they would write is now more self-descriptive. Compare:

```
FindString(text, true, false);
```

with

```
FindString(text, SearchDirection::Forward, CaseSensitivity::Insensitive);
```

There are also many APIs that use integers to represent a set of options where an enum would be a better choice. For example, the OpenCV API has a C function called `cvFlip()` that accepts an integer `flip_mode` parameter. A value of 0 means flip around the X axis, a value of 1 means flip around the Y axis, and a value of −1 means flip around both axes. An easier to use solution might have been to introduce an enum with `XAxis`, `YAxis`, and `BothAxes` values.

TIP: Prefer the use of enums over Booleans and integers to improve code readability.

For more complex cases in which an `enum` is insufficient, you could even introduce new classes to ensure that each argument has a unique type. For example, Scott Meyers illustrates this approach with the use of a `Date` class that is constructed by specifying three integers (Meyers, 2004, 2005):

```
class Date
{
public:
    Date(int year, int month, int day);
    ...
};
```

Meyers notes that in this design clients could pass the year, month, and day values in the wrong order, and they could also specify illegal values, such as a month of 13. To get around these problems, he suggests the introduction of specific classes to represent a year, month, and day value, such as:

```
class Year
{
public:
    explicit Year(int y) :
        mYear(y)
    {}
    int GetYear() const { return mYear; }

private:
    int mYear;
};

class Month
{
public:
    explicit Month(int m) :
        mMonth(m)
    {}
    int GetMonth() const { return mMonth; }
    static Month Jan() { return Month(1); }
    static Month Feb() { return Month(2); }
    static Month Mar() { return Month(3); }
    static Month Apr() { return Month(4); }
    static Month May() { return Month(5); }
    static Month Jun() { return Month(6); }
    static Month Jul() { return Month(7); }
    static Month Aug() { return Month(8); }
    static Month Sep() { return Month(9); }
    static Month Oct() { return Month(10); }
    static Month Nov() { return Month(11); }
    static Month Dec() { return Month(12); }

private:
    int mMonth;
};

class Day
{
public:
    explicit Day(int d) :
        mDay(d)
    {}
    int GetDay() const { return mDay; }

private:
    int mDay;
};
```

Now, the constructor for the `Date` class can be expressed in terms of these new `Year`, `Month`, and `Day` classes:

```
class Date
{
public:
    Date(const Year &y, const Month &m, const Day &d);
    ...
};
```

Using this design, clients can create a new Date object with the following unambiguous and easy to understand syntax. Also, any attempts to specify the values in a different order will result in a compile-time error:

```
Date birthday(Year(1976), Month::Jul(), Day(7));
```

TIP: Avoid functions with multiple parameters of the same type.

C++11 introduced the concept of user-defined literals that can also be employed to help your clients not misuse your APIs. Using this feature, you can define custom literal suffixes for different units and provide code to convert between those different units. For example, you can allow your clients to specify a temperature literal value as 100.0_degC or 100.0_degF and define how to convert these literals into the values your API expects. For more details on this feature, refer to the C++11 section on User Defined Literals in the C++ Revisions chapter.

Consistent

A good API should apply a consistent design approach so that its conventions are easy to remember, and therefore easy to adopt (Blanchette, 2008). This applies to all aspects of API design such as naming conventions, parameter order, the use of standard patterns, memory model semantics, the use of exceptions, and error handling.

In terms of the first of these, consistent naming conventions imply the reuse of the same words for the same concepts across the API. For example, if you have decided to use the verb pairs Begin and End, you should not mingle the terms Start and Finish. As another example, the Qt3 API mixes the use of abbreviations in several of its method names, such as prevValue() and previousSibling(). This is another example of why the use of abbreviations should be avoided wherever possible.

The use of consistent method signatures is an equally critical design quality. If you have several methods that accept similar argument lists, you should endeavor to keep a consistent number and order for those arguments. To give a counterexample, I refer you to the following C library functions:

```
void bcopy(const void *s1, void *s2, size_t n);
char *strncpy(char *restrict s1, const char *restrict s2, size_t n);
```

Both functions involve copying n bytes of data from one area of memory to another. However, the bcopy() function copies data from s1 into s2, whereas strncpy() copies from s2 into s1. This can give rise to subtle memory errors if a developer were to decide to switch usage between the two functions without a close reading of the respective man pages. To be fair, there is a clue to the conflicting specifications in the function signatures: note the use of the const pointer in each case. However, this could easily be missed and won't be caught by a compiler if the source pointer is not declared to be const.

Also note the inconsistent use of the words copy and cpy.

Let's take another example from the C Standard Library. The familiar malloc() function is used to allocate a contiguous block of memory, and the calloc() function performs the same operation with the addition that it initializes the reserved memory with 0 bytes. However, despite their similar purpose, they have different function signatures:

```
void *calloc(size_t count, size_t size);
void *malloc(size_t size);
```

The malloc() function accepts a size in terms of bytes, whereas calloc() allocates (count * size) bytes. In addition to being inconsistent, this violates the principle of least surprise. As a further example, the read() and write() standard C functions accept a file descriptor as their first parameter, whereas the fgets() and fputs() functions require the file descriptor to be specified last (Henning, 2009). And as David Hanson so eloquently put it, the C library function realloc() is a marvel of confusion (Hanson, 1996).

TIP: Use consistent function naming and parameter ordering.

These examples have been focused on a function or method level, but of course consistency is important at a class level, too. Classes that have similar roles should offer a similar interface. The C++ Standard Library is a great example of this. The std::vector, std::set, std::map, and even std::string classes all offer a size() method to return the number of elements in the container. They also all support the use of iterators, so that once you know how to iterate through an std::set you can apply the same knowledge to an std::map. This makes it easier to memorize the programming patterns of the API.

You get this kind of consistency for free through polymorphism: by placing the shared functionality into a common base class. However, often it doesn't make sense for all of your classes to inherit from a common base class, and you shouldn't introduce a base class purely for this purpose because it increases the complexity and class count for your interface. Indeed, it's noteworthy that the C++ Standard Library container classes do not inherit from a common base class. Instead, you should explicitly design for this by manually identifying the common concepts across your classes and using the same conventions to represent these concepts in each class. This is often referred to as static polymorphism.

You can also make use of C++ templates to help you define and apply this kind of consistency. For example, you could create a template for a 2D coordinate class and specialize it for integers, floats, and doubles. In this way you are assured that each type of coordinate offers the same interface. The following code sample provides a simple example of this:

```
template <class T>
class Coord2D
{
public:
    Coord2D(T x, T y) :
        mX(x),
        mY(y)
    {}

    T GetX() const { return mX; }
    T GetY() const { return mY; }

    void SetX(T x) { mX = x; }
    void SetY(T y) { mY = y; }

    void Add(T dx, T dy) { mX += dx; mY += dy; }
    void Multiply(T dx, T dy) { mX *= dx; mY *= dy; }

private:
    T mX;
    T mY;
};
```

With this template definition, you can create variables of type `Coord2D<int>`, `Coord2D<float>`, and `Coord2D<double>` and all of these will have the same interface.

A further aspect of consistency is the use of familiar patterns and standard platform idioms. When you buy a new car, you don't have to relearn how to drive. The concept of using brake and accelerator pedals, a steering wheel, and a gear stick (be it manual or automatic) is universal the world over. If you can drive one car, it's very likely that you can drive a similar one.

Likewise, the easiest APIs to use are those that require minimal new learning for your users. For example, most C++ developers are familiar with the Standard Library and its use of container classes and iterators. Therefore, if you decide to write an API with a similar purpose, it would be advantageous to mimic the patterns of the Standard Library because other developers will then find the use of your API to be familiar and easy to adopt.

Finally, an important aspect of consistency is to do something in one way, not two ways. After all, API design is about making decisions. However, sometimes there's no clear good way forward or you may feel compelled to incorporate conflicting feedback from multiple sources. But if you end up providing multiple ways to do the same thing, then the end product will be less usable and you'll have to support those multiple designs going forward (Martelli, 2011).

Orthogonal

In mathematics, two vectors are said to be orthogonal if they are perpendicular (at 90 degrees) to each other (i.e., their inner product is zero). This makes them linearly independent, meaning that there is no set of scalars that can be applied to the first vector to produce the second vector. To take a geographic analogy, the perpendicular directions east and north are independent of each other: no amount of walking east will take you north. Said in slightly more computer lingo, changes to your east coordinate have no effect on your north coordinate.

In terms of API design, orthogonality means that methods do not have side effects. Calling a method that sets a particular property should change only that property and not additionally change other publicly accessible properties. As a result, making a change to the implementation of one part of the API should have no effect on other parts of the API (Raymond, 2003). Striving for orthogonality produces APIs with behaviors that are more predictable and comprehensible. Furthermore, code that does not create side effects, or relies upon the side effects of other code, is much easier to develop, test, debug, and change because its effects are more localized and bounded (Hunt and Thomas, 1999).

TIP: An orthogonal API means that functions do not have side effects.

Let's look at a specific example. Perhaps you've stayed in a motel where the controls of the shower are very unintuitive. You want to be able to set the power and the temperature of the water, but instead you have a single control that seems to affect both properties in a complex and nonobvious way. This could be modeled using the following API:

```
class CheapMotelShower
{
public:
    float GetTemperature() const;  // units = Fahrenheit
    float GetPower() const;        // units = percent, 0..100
    void SetPower(float p);

private:
    float mTemperature;
    float mPower;
};
```

Just to illustrate this further, let's consider the following implementation for the public methods of this class:

```
float CheapMotelShower::GetTemperature() const
{
    return mTemperature;
}

float CheapMotelShower::GetPower() const
{
    return mPower;
}

void CheapMotelShower::SetPower(float p)
{
    if (p < 0) {
        p = 0;
    }
    if (p > 100) {
        p = 100;
    }
    mPower = p;
    mTemperature = 42.0f + sin(p / 38.0f) * 45.0f;
}
```

In this case you can see that setting the power of the water flow also affects the temperature of the water via a nonlinear relationship. As a result, it's not possible to achieve certain combinations of temperature and power, and naturally the preferred combination of hot water and full power is unattainable. Also, if you were to change the implementation of the `SetPower()` method, it would have the side effect of affecting the result of the `GetTemperature()` method. In a more complex system, this interdependence may be something that we as programmers forget about, or are simply unaware of, and so a simple change to one area of code may have a profound impact on the behavior of other parts of the system.

Instead, let's consider an ideal, orthogonal interface for a shower, in which the controls for temperature and power are independent:

```
class IdealShower
{
public:
    float GetTemperature() const;  // units = Fahrenheit
    float GetPower() const;        // units = percent, 0..100
    void SetTemperature(float t);
    void SetPower(float p);

private:
    float mTemperature;
    float mPower;
};

float IdealShower::GetTemperature() const
{
    return mTemperature;
}

float IdealShower::GetPower() const
{
    return mPower;
}

void IdealShower::SetTemperature(float t)
{
    if (t < 42) {
        t = 42;
    }
    if (t > 85) {
        t = 85;
    }
    mTemperature = t;
}

void IdealShower::SetPower(float p)
{
    if (p < 0) {
        p = 0;
    }
    if (p > 100) {
        p = 100;
    }
    mPower = p;
}
```

Two important factors to remember for designing orthogonal APIs are:

1. **Reduce redundancy.** Ensure that the same information is not represented in more than one way. There should be a single authoritative source for each piece of knowledge.
2. **Increase independence.** Ensure that there is no overlap of meaning in the concepts that are exposed. Any overlapping concepts should be decomposed into their basal components.

Another popular interpretation of orthogonal design is that different operations can all be applied to each available data type. This is a definition that is commonly used in the fields of programming language and CPU design. In the latter case, an orthogonal instruction set is one in which instructions can use any CPU register in any addressing mode, as opposed to a nonorthogonal design in which certain instructions can use only certain registers. Again, in terms of API design, the Standard Library provides an excellent example of this. It offers a collection of generic algorithms and iterators that can be used on any container. For example, the Standard Library `std::count` algorithm can be applied to any of the `std::vector`, `std::set`, or `std::map` containers. Hence the choice of algorithm is not dependent upon the container class being used.

Robust resource allocation

One of the trickiest aspects of programming in C++ is memory management. This is particularly so for developers who are used to memory managed languages such as Java or C#, in which objects are freed automatically by a garbage collector. By contrast, most C++ bugs arise from some kind of misuse of pointers or references, such as:

- **Null dereferencing**: trying to use `->` or `*` operators on a `nullptr`.
- **Double freeing**: calling `delete` or `free()` on a block of memory twice.
- **Accessing invalid memory**: trying to use `->` or `*` operators on a pointer that has not yet been allocated or that has already been freed.
- **Mixing allocators**: using `delete` to free memory that was allocated with `malloc()`, or using `free()` to return memory allocated with `new`.
- **Incorrect array deallocation**: using the `delete` operator, instead of `delete []`, to free an array.
- **Memory leaks**: not freeing a block of memory when you are finished with it.

These problems arise because it's not possible to tell whether a plain C++ pointer is referencing valid memory or if it's pointing to unallocated or freed memory. It's therefore reliant upon the programmer to track this state and ensure that the pointer is never dereferenced incorrectly. But as we know, programmers are fallible. However, many of these kinds of problems can be avoided through the use of managed (or smart) pointers, such as:

1. **Shared pointers.** These are reference counted pointers where the reference count can be incremented by one when a piece of code wants to hold onto the pointer, and decremented by one when it is finished using the pointer. When the reference count reaches zero, the object pointed to by the pointer is automatically freed. This kind of pointer can help to avoid the problems of accessing freed memory by ensuring that the pointer remains valid for the period that you want to use it.
2. **Weak pointers.** A weak pointer contains a pointer to an object, normally a shared pointer, but does not contribute to the reference count for that object. If you have a shared pointer and a weak pointer referencing the same object, and the shared pointer is destroyed, the weak pointer immediately becomes null. In this way, weak pointers can detect whether the object being pointed to has expired: if the reference count for the object to which it's pointing is zero. This helps to avoid the

dangling pointer problem where you can have a pointer that's referencing freed memory.

3. **Unique pointers.** These pointers support ownership of single objects and automatically deallocate their objects when the pointer goes out of scope. They are sometimes also called scoped pointers or auto pointers. Unique pointers are defined as owning a single object and as such they cannot be copied or shared but they can be moved to a new owner.

These smart pointers are not part of the original C++98 specification. However, support for them was added in C++11 with the `std::shared_ptr`, `std::weak_ptr`, and `std::unique_ptr` templates.

Employing these smart pointers can make your API much easier to use and less prone to the kind of memory errors listed earlier. For example, the use of `std::shared_ptr` can alleviate the need for users to free a dynamically created object explicitly. Instead, the object will automatically be deleted when it's no longer referenced. For example, consider an API that allows you to create instances of an object via a factory method called `CreateInstance()`:

```
#include <memory>

using MyObjectPtr = std::shared_ptr<MyObject>;

class MyObject
{
public:
    static MyObjectPtr CreateInstance();
    ~MyObject();
};
```

where the implementation of the factory method looks like:

```
MyObjectPtr MyObject::CreateInstance()
{
    return std::make_shared<MyObject>();
}
```

Given this API, a client could create instances of `MyObject` as:

```
int main(int argc, char *argv[])
{
    MyObjectPtr ptr = MyObject::CreateInstance();
    ptr = MyObject::CreateInstance();
    return 0;
}
```

In this example, two instances of `MyObject` are created, and both instances are destroyed when the `ptr` variable goes out of scope (at the end of the program in this case). If instead the `CreateInstance()` method simply returned a `MyObject *` raw pointer type, then the destructor would never get called in the previous example. The use of smart pointers can therefore make memory management simpler and hence make your API easier for clients to use.

In general, if you have a function that returns a pointer that your clients should delete, or if you expect the client to need the pointer for longer than the life of your object, then you should return it using a smart pointer, such as an `std::shared_ptr`. On the other hand, if ownership of the pointer will be retained by your object, then you can return a weak pointer. For example:

```
// ownership of MyObject is shared with the caller
std::shared_ptr<MyObject> GetObject() const;

// ownership of MyObject is retained by the API
std::weak_ptr<MyObject> GetObject() const;
```

TIP: Return a dynamically allocated object using a smart pointer wherever possible.

It's worth noting that these kinds of memory management issues are simply a specific case of the more general category of resource management. For instance, similar types of problems can be encountered with the manipulation of mutex locks or file handles. We can generalize the concepts of smart pointers to the task of resource management with the following observation: resource allocation is object construction, and resource deallocation is object destruction. This is often referred to with the acronym RAII (resource acquisition is initialization).

As an example, examine this code, which illustrates a classic synchronization bug:

```
void SetName(const std::string &name)
{
    mMutex.lock();

    if (name.empty()) {
        return;
    }
    mName = name;

    mMutex.unlock();
}
```

Obviously, this code will fail to unlock the mutex if an empty string is passed into the method. As a result, the program will deadlock on the next attempt to call the method because the mutex will still be locked. However, you could create a class called `ScopedMutex`, whose constructor locks the mutex and whose destructor unlocks it again. With such a class, you could rewrite this method as:

```
void SetName(const std::string &name)
{
    ScopedMutex locker(mMutex);

    if (name.empty()) {
        return;
    }

    mName = name;
}
```

Now you can be assured that the lock will be released whenever the method returns, because the mutex will be unlocked whenever the `ScopedMutex` variable goes out of scope. As a result, you do not need to check every return statement meticulously to ensure that it has explicitly freed the lock. As a bonus, the code is also much more readable. Support for scoped locking was adding in C++11 with the `std::lock_guard` and `std::scoped_lock` class templates.

The take-home point in terms of API design is that if your API provides access to the allocation and deallocation of some resource, then you should consider providing a class to manage this, where resource allocation happens in the constructor and deallocation happens in the destructor (and perhaps additionally through a public `Release()` method so that clients have more control over when the resource is freed).

> *TIP: Think of resource allocation and deallocation as object construction and destruction.*

One caveat to be aware of with RAII is that C++ constructors and destructors don't have return values, and you should also not trigger an exception in a destructor. So there may be some cases for which RAII is not the best fit (e.g., when you want to detect an error and retry some logic).

Platform independent

A well-designed C++ API should always avoid platform-specific `#if`/`#ifdef` lines in its public headers that produce different APIs on different platforms or that produce different APIs in debug versus release builds. If your API presents a high-level and logical model for your problem domain, as it should, there are very few cases when that API should be different for different platforms.

For example, let's take the case of an API that encapsulates the functionality offered by a mobile phone. Some mobile phones offer built-in GPS devices that can deliver the geographic location of the phone, but not all devices offer this capability. However, you should never expose this situation directly through your API, such as in this example:

```
class MobilePhone
{
public:
    bool StartCall(const std::string &number);
    bool EndCall();

#if defined TARGET_OS_IPHONE
    bool GetGPSLocation(double &lat, double &lon);
#endif
};
```

This poor design creates a different API on different platforms. Doing so forces the clients of your API to introduce the same platform specificity into their own applications. For example, in the previous case, your clients would have to guard any calls to `GetGPSLocation()` with precisely the same `#if` conditional statement; otherwise their code may fail to compile with an undefined symbol error on other platforms.

Furthermore, if in a later version of the API you also add support for another device class, say Windows Mobile, then you would have to update the `#if` line in your public header to include `_WIN32_WCE`. Then, your clients would have to find all instances in their code where they have embedded the `TARGET_OS_IPHONE` define and extend it to include `_WIN32_WCE`. This is because you have unwittingly exposed the implementation details of your API.

Instead, you should hide the fact that the function works only on certain platforms and provide a method to determine whether the implementation offers the desired capabilities on the current platform. For example:

```
class MobilePhone
{
public:
    bool StartCall(const std::string &number);
    bool EndCall();

    bool HasGPS() const;
    bool GetGPSLocation(double &lat, double &lon);
};
```

Now your API is consistent over all platforms and doesn't expose the details of which platforms support GPS coordinates. The clients can now write code to check whether the current device supports a GPS device, by calling `HasGPS()`, and if so they can call the `GetGPSLocation()` method to return the actual coordinate. The implementation of the `HasGPS()` method might look something like:

```
bool MobilePhone::HasGPS() const
{
#if defined TARGET_OS_IPHONE
    return true;
#else
    return false;
#endif
}
```

This is far superior to the original design because the platform-specific `#if` statement is now hidden in the `.cpp` file rather than being exposed to your clients in the header file.

TIP: Never put platform-specific `#if` or `#ifdef` statements into your public APIs. It exposes implementation details and makes your API behave differently on differently platforms.

One case where maintaining platform independence can be tricky is when you're writing an API to interface with a platform-specific resource, such as a routine that draws in a window and requires the appropriate window handle for the native operating system. One way to deal with this is to return the resource as a `void *` pointer (i.e., a pointer with no type). Although returning an untyped pointer is generally frowned upon, it's better than making your API be platform specific. As an example of this, the OpenCV API returns operating system–dependent handles as `void *`, requiring clients to cast this to the appropriate type in their application code (e.g., an `HWND` on Win32, `WindowRef` on macOS, or `Widget *` on X Windows). Here's an example of what that API looks like:

```
void *cvGetWindowHandle(const char *name);
const char *cvGetWindowName(void *window_handle);
```

Loosely coupled

In 1974, Wayne Stevens, Glenford Myers, and Larry Constantine published their seminal paper on structured software design. This paper introduced the two interrelated concepts of coupling and cohesion (Stevens et al., 1974), which I will define as:

- **Coupling.** A measure of the strength of interconnection among software components: that is, the degree to which each component depends upon other components in the system.
- **Cohesion.** A measure of how coherent or strongly related the various functions of a single software component are.

Good software designs tend to correlate with low (or loose) coupling and high cohesion: that is, a design that minimizes the functional relatedness and connectedness between different components. Achieving this goal allows components to be used, understood, and maintained independently of each other.

TIP: Good APIs exhibit loose coupling and high cohesion.

Steve McConnell presents a particularly effective analogy for loose coupling. Model railroad cars use simple hook or knuckle couplers to connect cars together. These allow easy linking of train cars—normally just by pushing two cars together—with a single point of connection. This is therefore an example of a loosely coupled system. Imagine how much more difficult it would be to connect the cars if you had to employ several types of connections, perhaps involving screws and wires, or if only particular cars could interface with certain other kinds of cars (McConnell, 2004).

One way to think of coupling is that given two components, A and B, how much code in B must change if A changes. There are various measures that can be used to evaluate the degree of coupling between components:

- **Size.** Relates to the number of connections between components, including the number of classes, methods, and arguments per method. For example, a method with fewer parameters is more loosely coupled to components that call it.
- **Visibility.** Refers to the prominence of the connection between components. For example, changing a global variable to affect the state of another component indirectly is a poor level of visibility.
- **Intimacy.** Refers to the directness of the connection between components. If A is coupled to B, and B is coupled to C, then A is indirectly coupled to C. Another example is that inheriting from a class is a tighter coupling than including that class as a member variable (composition), because inheritance also provides access to all protected members of the class.
- **Flexibility.** Relates to the ease of changing the connections between components. For example, if the signature of a method in A needs to change so that B can call it, flexibility describes how easy it is to change that method and all dependent code.

One particularly abhorrent form of tight coupling that you should always avoid is having two components that directly or indirectly depend upon each other (i.e., a dependency cycle or circular dependency). This makes it difficult or impossible to reuse a component without including all of its circularly dependent components. I discuss this form of tight coupling further in Chapter 4.

In the following sections, I will present various techniques to reduce coupling between the classes and methods within your API (inter-API coupling).

However, there is also the interesting question of how your API design decisions affect the cohesion and coupling of your clients' applications (intra-API coupling). Your API is designed to solve a specific problem, and that unity of purpose should translate well into your API being a component with high cohesion in your clients' programs. However, in terms of coupling, the larger your API is—the more classes, methods, and arguments you expose—the more ways in which your API can be accessed and coupled to your clients' applications. As such, the quality of being minimally complete can also contribute toward loose coupling. There is also the issue of how much coupling with other components you have designed into your API. For example, the libpng library depends upon the libz library. This coupling is exposed at compile time, via a reference to the `zlib.h` header in `png.h`, and also at link time. This requires clients of the libpng to be aware of the libz dependency and ensure that they also build and link against this additional library.

Coupling by name only

If class A only needs to know the name of class B (i.e., it does not need to know the size of class B or call any methods in the class), then class A does not need to depend upon the full declaration of B. In these cases, you can use a forward declaration for class B rather than including the entire interface, and so reduce the coupling between the two classes (Lakos, 1996). For example:

```
class MyObject;  // only need to know the name of MyObject

class MyObjectHolder
{
public:
    MyObjectHolder();

    void SetObject(const MyObject &obj);
};
```

In this case, `MyObjectHolder` only uses a reference to `MyObject`, and as such it doesn't need to `#include "MyObject.h"` to get the full declaration of `MyObject`. The same would be true if a pointer to `MyObject` was used instead. By using a forward declaration, the `MyObjectHolder` class can be decoupled from the physical implementation of `MyObject`.

TIP: Use a forward declaration for a class unless you actually need to #include its full definition.

Reducing class coupling

Scott Meyers recommends that whenever you have a choice, you should prefer declaring a function as a nonmember nonfriend function rather than as a member function (Meyers, 2000). Doing so improves encapsulation and reduces the degree of coupling of those functions to the class. For example, consider the following class snippet that provides a `PrintName()` member function to output the value of a member variable to stdout. The function uses a public getter method, `GetName()`, to retrieve the current value of the member variable:

```
// myobject.h
class MyObject
{
public:
    void PrintName() const;
    std::string GetName() const;
    ...

protected:
    ...

private:
    std::string mName;
    ...
};
```

Following Meyers' advice, you should prefer this representation:

```
// myobject.h
class MyObject
{
public:
    std::string GetName() const;
    ...

protected:
    ...

private:
    std::string mName;
    ...
};

void PrintName(const MyObject &obj);
```

This latter form reduces coupling because the free function `PrintName()` can access only the public methods of `MyObject` (and only the const public methods in this particular case), whereas in the form where `PrintName()` is a member function, it can also access all of the private and protected member functions and data members of `MyObject` as well as any protected members of any bases classes, if it had any. Preferring the nonmember nonfriend form therefore means that the function is not coupled to the internal details of the class. It is therefore far less likely to break when the internal details of `MyObject` are changed (Tulach, 2008).

This technique also contributes to more minimally complete interfaces, in which a class contains only the minimal functionality required to implement it, whereas functionality that's built on top of its public interface is declared outside the class (as in the example of convenience APIs I discussed earlier). It's worth noting that this happens a lot in the C++ Standard Library, where algorithms such as `std::for_each()` and `std::unique()` are declared outside each container class.

To convey the conceptual relatedness of `MyObject` and `PrintName()` better, you could declare both of these within a single namespace. Alternatively, you could declare `PrintName()` within its own namespace, such as `MyObjectHelper`, or even as a static function within a new helper class called `MyObjectHelper`. As I've already covered in the section on convenience APIs, this helper namespace can and should be contained in a separate module. For example:

```
// myobjecthelper.h
namespace MyObjectHelper {

void PrintName(const MyObject &obj);

}
```

TIP: Prefer using nonmember nonfriend functions instead of member functions to reduce coupling.

Intentional redundancy

Normally, good software engineering practice aims to remove redundancy: to ensure that each significant piece of knowledge or behavior is implemented only once (Pierce, 2002). However, reuse of code implies coupling, and sometimes it's worth the cost to add a small degree of duplication to sever an egregious coupling relationship (Parnas, 1979). This intentional duplication may take the form of code or data redundancy.

As an example of code redundancy, consider two large components that are dependent upon each other. However, when you investigate the dependency more deeply, it turns out that it resolves to one component relying on a trivial piece of functionality in the other, such as a function to calculate minimum or maximum. Standard practice would be to find a lower-level home for this functionality, so that both components may depend upon that smaller piece of functionality instead of upon each other. However, sometimes this kind of refactoring doesn't make sense (e.g., if the functionality is simply not generic enough to be demoted to a lower level in the system). Therefore, in certain cases it may make sense to duplicate the code to avoid the coupling (Lakos, 1996).

In terms of adding data redundancy, consider the following API for a text chat system that logs each message sent by a user:

```
#include "ChatUser.h"
#include <string>
#include <vector>

class TextChatLog
{
public:
    bool AddMessage(const ChatUser &user, const std::string &msg);

    int GetCount() const;
    std::string GetMessage(int index);

private:
    struct ChatEvent
    {
        ChatUser mUser;
        std::string mMessage;
        size_t mTimestamp;
    };

    std::vector<ChatEvent> mChatEvents;
};
```

This design accepts individual text chat events, taking an object that describes users and the message that they typed. This information is augmented with the current timestamp and added to an internal list. The `GetCount()` method can then be used to find how many text chat events have occurred and the `GetMessage()` method will return a formatted version of a given chat event, such as:

```
Joe Blow [09:46]: What's up?
```

However, the `TextChatLog` class is clearly coupled with the `ChatUser` class, which may be a very heavy class that pulls in many other dependencies. You may therefore decide to investigate this situation and find that `TextChatLog` only ever uses the name of the user.

That is, it keeps the `ChatUser` object around but only calls the `ChatUser::GetName()` method. One solution to remove the coupling between the two classes would therefore be simply to pass the name of the user into the `TextChatLog` class, as in this refactored version:

```
#include <string>
#include <vector>

class TextChatLog
{
public:
    bool AddMessage(const std::string &user, const std::string &msg);

    int GetCount() const;
    std::string GetMessage(int index);

private:
    struct ChatEvent
    {
        std::string mUserName;
        std::string mMessage;
        size_t mTimestamp;
    };

    std::vector<ChatEvent> mChatEvents;
};
```

This creates a redundant version of the user's name (it is now stored in the `TextChatLog` as well as the `ChatUser` classes), but it breaks the dependency between the two classes. It may also have the additional benefit of reducing the memory overhead of `TextChatLog`, if `sizeof(std::string) < sizeof(ChatUser)`.

Nonetheless, even though this is an intentional duplication, it's still duplication and should be undertaken with great care, consideration, and excellent comments. For example, if the chat system were updated to allow users to change their names, and it was decided that when this happens all previous messages from that user should update to show the new name, then you may have to revert to the original tightly coupled version (or risk more coupling by requiring `ChatUser` to inform `TextChatLog` whenever a name change occurs).

TIP: Data redundancy can sometimes be justified to reduce coupling between classes.

Manager classes

A manager class is one that owns and coordinates several lower-level classes. This can be used to break the dependency of one or more classes upon a collection of low-level classes. This is often implemented using a Façade design pattern, which I'll cover in the chapter on Patterns, or a Mediator design pattern. The difference between the two is that a Façade exposes only existing functionality in a different way whereas a Mediator adds new functionality.

For example, consider a structured drawing program that lets you create 2D objects, select objects, and move them around a canvas. The program supports several kinds of input devices to let users select and move objects, such as a mouse, tablet, and joystick. A naive design would require both the select and move operations to know about each kind of input device, as shown in this UML diagram (Fig. 2.6).

Alternatively, you could introduce a manager class to coordinate access to each of the specific input device classes. In this way, the `SelectObject` and `MoveObject` classes need only to depend upon this single manager class, and then the manager class needs to depend upon only the individual input device classes. This may also require creating some form of abstraction for the underlying classes. For example, notice that `MouseInput`, `TabletInput`, and `JoystickInput` each have a slightly different interface. Our manager class could therefore put in place a generic input device interface that abstracts away the specifics of a particular device. The improved, more loosely coupled design is shown in Fig. 2.7.

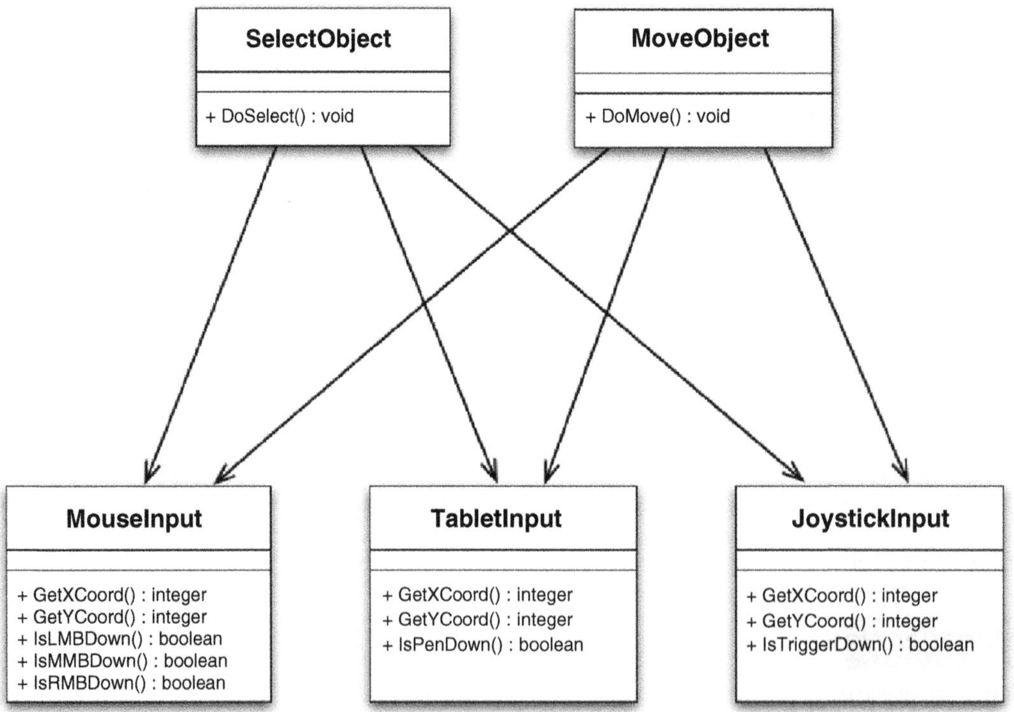

FIGURE 2.6 Multiple high-level classes, each of which is coupled to several low-level classes.

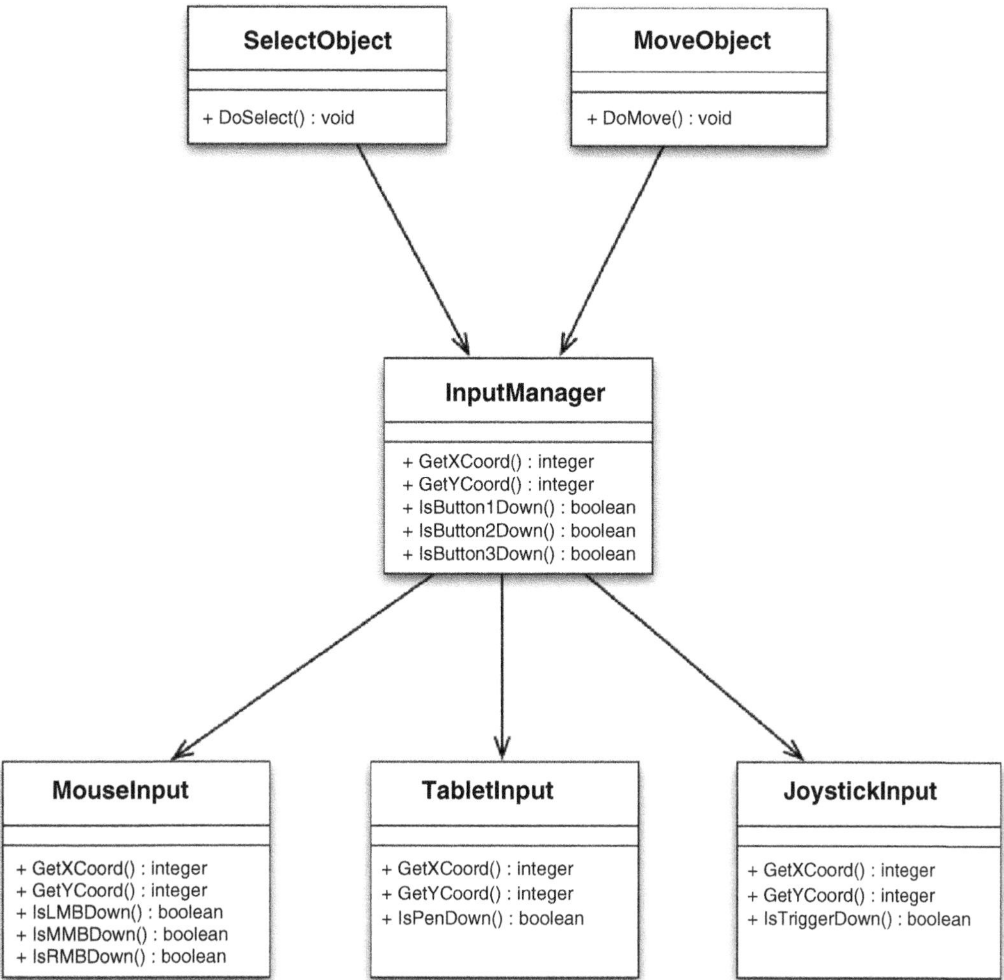

FIGURE 2.7 Using a manager class to reduce coupling to lower-level classes.

Note that this design also scales well, too. This is because more input devices can be added to the system without introducing further dependencies for `SelectObject` or `MoveObject`. Also, if you decided to add additional manipulation objects, such as `RotateObject` and `ScaleObject`, they need only a single dependency on `InputManager`, rather than each introducing further coupling to the underlying device classes.

> *TIP: Manager classes can reduce coupling by encapsulating several lower-level classes. They can be implemented using the Façade or Mediator design patterns.*

Callbacks, observers, and notifications

The final technique that I'll present to reduce coupling within an API relates to the problem of notifying other classes when some event occurs.

Imagine an online 3D multiplayer game that allows multiple users to play against each other. Internally, each player may be represented with a UUID such as e5b43bba-fbf2-4f91-ac71-4f2a12d04847. However, users want to see the names of other players, not inscrutable UUID strings. The system therefore implements a player name cache, `NameCache`, to store the correspondence between the UUID and the human-readable name.

Now, let's say the class that manages the pregame lobby, `PreGameLobby`, wants to display the name of each player. The set of operations may proceed as:

1. The `PreGameLobby` class calls `NameCache::RequestName()`,
2. The `NameCache` sends a request to the game server for the name of the player with this UUID,
3. The `NameCache` receives the player name information from the server, and
4. The `NameCache` calls `PreGameLobby::SetPlayerName()`.

However, in this case `PreGameLobby` depends upon `NameCache` to call the `RequestName()` method, and `NameCache` depends upon `PreGameLobby` to call the `SetPlayerName()` method. This is an overly brittle and tightly coupled design. Consider what would happen if the in-game system also needed to know a player's name to display it above the user. Would you then extend `NameCache` also to call the `InGame::SetPlayerName()` method, further tightening the coupling?

A better solution would be to have the `PreGameLobby` and `InGame` classes register interest in updates from the `NameCache`. Then the `NameCache` can notify any interested parties without having a direct dependency upon those modules. There are several ways that this can be done, such as callbacks, observers, and notifications. I will go into the details of each of these in a moment, but first, here are a number of general issues to understand when using any of these schemes:

- **Reentrancy.** When writing an API that calls out to unknown user code, you have to consider that this code may call back into your API. In fact, the client may not even realize that this is happening. For example, if you are processing a queue of objects and you issue a callback as you process each individual object, it is possible that the callback will attempt to modify the state of the queue by adding or removing objects. At a minimum, your API should guard against this behavior with a coding error. However, a more elegant solution would be to allow this reentrant behavior and implement your code such that it maintains a consistent state.
- **Lifetime management.** Client code should have a clean way to disconnect from your API (i.e., to declare that they're no longer interested in receiving updates). This is particularly important when the client object is deleted, because further attempts to send messages to it could cause a crash. Similarly, your API may wish to guard against duplicate registrations, to avoid calling the same client code multiple times for the same event.

- **Event ordering.** The sequence of callbacks or notifications should be clear to the user of your API. For example, the Cocoa API does a good job of making it clear whether a notification is sent before or after an event by using names such as `willChange` and `didChange`. On the other hand, the Qt tool kit is less specific about this: a changed signal is sometimes sent before the object in question is actually updated.

These points highlight the general issue that you should always make it clear to your clients what they can and cannot do—what assumptions they can and cannot make about your API—within their callback code. This can be done through your API documentation, or it can be enforced more explicitly by giving the callback a limited interface that exposes only a safe subset of all potential operations.

Callback functions

In plain C, a callback is simply a pointer to a function within module A that's passed to module B so that B can invoke the function in A at an appropriate time. Module B knows nothing about module A and has no include or link dependencies upon A. This makes callbacks particularly useful to allow low-level code to execute high-level code that it cannot have a link dependency on. As such, callbacks are a popular technique to break cyclic dependencies in large projects.

It's also sometimes useful to supply a closure with the callback function. This is a piece of data that module A passes to B, and which module B includes in the function callback to A. This is a way for module A to pass through some state that will be important to receive in the callback function.

The following header file shows how you can define a simple callback API in C++11 syntax:

```
#include <string>

class ModuleB
{
public:
    using CallbackType = void (*)(const std::string &, void *);

    ModuleB() :
        mCallback(nullptr),
        mClosure(nullptr)
    {}

    void SetCallback(CallbackType cb, void *closure)
    {
        mCallback = cb;
        mClosure = closure;
    }

    void InvokeCallback(const std::string &name)
    {
        if (mCallback) {
            (*mCallback)(name, mClosure);
        }
    }

private:
    CallbackType mCallback;
    void *mClosure;
};
```

(A more sophisticated example would support adding multiple callbacks to ModuleB, perhaps storing them in an std::vector, and then invoking each registered callback in turn).

There are a few problems with this approach. One is that for object-oriented C++ programs, it's nontrivial to use a nonstatic (instance) method as a callback. This is because the this-pointer of the object also needs to be passed along. C++11 offers a couple of solutions for this. You can use std::bind to wrap up the method call with the object instance. Or you can use std::function to define the callback type and then use a lambda that captures the object. Here's an example of the latter:

```
class ModuleB
{
public:
    using CallbackType = std::function<void (const std::string &,
void *)>;

    ModuleB() :
      mCallback(nullptr),
      mClosure(nullptr)
    {}

    void SetCallback(CallbackType cb, void *closure)
    {
        mCallback = cb;
        mClosure = closure;
    }

    void InvokeCallback(const std::string &name)
    {
        if (mCallback) {
            mCallback(name, mClosure);
        }
    }

private:
    CallbackType mCallback;
    void *mClosure;
};

// use a lambda to pass an object instance method as a callback
ModuleA a;
ModuleB b;
b.SetCallback([&a](const std::string &name, void *data){
    a.CallbackMethod(name, data);
}, nullptr);
b.InvokeCallback("hello world");
```

Another problem with this approach is the use of the untyped void * parameter to pass data into the callback function. If you need to work with objects, or pass state into a callback, then it would be better to use an object-based solution such as the observer pattern.

Observers

Callback functions present a solution that work well in plain C programs, but a more object-oriented solution is to use the concept of observers. This is a software design pattern in which an object maintains a list of its dependent objects (observers) and notifies them by calling one of their methods. This is a very important pattern for minimizing coupling in API design. In fact, I have dedicated an entire section to it in the Patterns chapter. I will defer detailed treatment of observers until then, but I will briefly cover the topic here to offer a comparison with the earlier callback function approach.

The general approach to implement the observer pattern is to define a base class, often an abstract base class, which declares the callback method (or methods) that can be called. You can then provide an implementation of this interface with your specific callback functionality. You can also store any state that's specific to your needs in the data members of your observer class. For example:

```
class Observable
{
public:
    virtual ~Observable() {}
    virtual void CallbackMethod(const std::string &name) = 0;
};

class MyObserver : public Observable
{
public:
    MyObserver(int data) :
        mData(data)
    {}

    void CallbackMethod(const std::string &name)
    {
        std::cout << name << ": " << mData << "\n";
    }

private:
    int mData;
};
```

Then the module that registers and invokes your observer might look as follows. Note that I store the observer object using a shared pointer so that the code is robust to the life cycle of your observer:

```
class ModuleB
{
public:
    void SetObserver(std::shared_ptr<Observable> cb)
    {
        mObserver = cb;
    }

    void InvokeObserver(const std::string &name)
    {
        if (mObserver) {
            mObserver->CallbackMethod(name);
        }
    }

private:
    std::shared_ptr<Observable> mObserver;
};

// register an instance of Observable with ModuleB
ModuleB b;
b.SetObserver(std::make_shared<MyObserver>(2));
b.InvokeObserver("hello world");
```

You can also use a function object, or functor, to define your callback function. This is simply a type that implements `operator()`. This has the benefit of making the object behave more like a callback function, if that is desired. For example:

```
class FunctorType
{
public:
    virtual ~FunctorType() {}
    virtual void operator()(const std::string &name) = 0;
};

class MyFunctor : public FunctorType
{
public:
    MyFunctor() :
        mData(0)
    {}

    void SetData(int data)
    {
        mData = data;
    }

    void operator()(const std::string &name)
    {
        std::cout << name << ": " << mData << "\n";
    }

private:
    int mData;
};
```

Notifications

Callbacks and observers tend to be created for a particular task, and the mechanism to use them is normally defined within the objects that need to perform the actual callback. An alternative solution is to build a centralized mechanism to send notifications, or events, between unconnected parts of the system. The sender does not need to know about the receiver beforehand, and so this can be used to reduce coupling between the sender and receiver. There are several kinds of notification schemes, but one particularly popular one is signals and slots.

The signals and slots concept was introduced by the Qt library as a generic way to allow any event, such as a button click or a timer event, to be sent to any interested method to act upon it. However, there are now several alternative implementations of signals and slots available for use in plain C++ code, including Boost's `boost::signals` and `boost::signals2` library.

Signals can be thought of simply as callbacks with multiple targets (slots). All slots for a signal are called when that signal is invoked, or emitted. To give a concrete example, the following code snippet uses `boost::signal` to create a signal that takes no arguments. You then connect a simple object to the signal. Finally, you emit the signal, which causes `MySlot::operator()` to be called, resulting in a message being printed to stdout:

```
class MySlot
{
public:
    void operator()() const
    {
        std::cout << "MySlot called!" << std::endl;
    }
};

// Create an instance of our MySlot class
MySlot slot;

// Create a signal with no arguments and a void return value
boost::signal<void()> signal;

// Connect our slot to this signal
signal.connect(slot);

// Emit the signal and thereby call all of the slots
signal();
```

In practical use, a low-level class could therefore create and own a signal. It then allows any unconnected classes to add themselves as slots to this signal. Then the low-level class can emit its signal at any appropriate time and all connected slots will be called.

Stable, documented, and tested

A well-designed API should be stable and future-proof. In this context, stable does not necessarily mean that the API never changes, but rather that the interface should be versioned and should not change incompatibly from one version to the next. Related to this, the term future-proof means that an API should be designed to be extensible so that it can be evolved elegantly rather than changed chaotically.

A good API should also be well-documented so that users have clear information about the capabilities, behavior, best practices, and error conditions for the API. Finally, there should be extensive automated tests written for the implementation of the API, so that new changes can be made with the confidence that existing use cases have not been broken.

These topics have been condensed into a single section at the end of the chapter not because they are of minor concern or importance; quite the opposite, in fact. These issues are so fundamentally important to the development of high-quality, robust, and easy-to-use APIs that I have dedicated entire chapters to each topic.

I've included this placeholder section here because I feel that a chapter on API qualities should at least reference these important topics. However, for specific details, I refer you to the separate chapters on Versioning, Documentation, and Testing for complete coverage of each respective topic.

3 Patterns

In the previous chapter, I discussed the qualities that differentiate a good application programming interface (API) from a bad one. In the next couple of chapters, I'll focus on the techniques and principles of building high-quality APIs. In this chapter, I will cover several useful design patterns and idioms that relate to C++ API design.

A design pattern is a general solution to a common software design problem. The term was made popular by the book *Design Patterns: Elements of Reusable Object-Oriented Software*, also known as the *Gang of Four* book (Gamma et al., 1994). That book introduced a collection of 23 generic design patterns organized into three main categories, which are enumerated in Table 3.1.

Since the initial publication of the design pattern book in 1994, several more design patterns have been added to this list, including an entire new categorization of Concurrency design patterns. The original authors have also suggested an improved categorization of Core, Creational, Periphery, and Other (Gamma et al., 2009).

However, it's not the intent of this API book to cover all of these design patterns. There are plenty of other books on the market that focus solely on that topic. Instead, I will concentrate on those design patterns that are of particular importance to the design of high-quality APIs, and discuss their practical implementation in C++. I will also cover C++ idioms that may not be considered true generic design patterns, but which are nevertheless important techniques for C++ API design. Specifically, I will go into details for these techniques:

- **Pimpl idiom.** This technique lets you completely hide internal details from your public header files. Essentially, it lets you move private member data and functions to the `.cpp` file. It is therefore an indispensable tool for creating well-insulated APIs.
- **Singleton and Factory Methods.** These are two very common creational design patterns that are good to understand deeply. Singleton is useful when you want to enforce that only one instance of an object is ever created. It has some rather tricky implementation aspects in C++ that I'll explore. The Factory Method pattern provides a generalized way to create instances of an object and can be a great way to hide implementation details for a derived class.
- **Proxy, Adapter, and Façade.** These structural patterns describe various solutions for wrapping an API on top of an existing incompatible or legacy interface. This is often the entire goal of writing an API: to improve the interface of some poorly designed ball of code. The Proxy and Adapter patterns provide a one-to-one

API Design for C++. https://doi.org/10.1016/B978-0-443-22219-1.00019-2

Table 3.1 The 23 design patterns introduced by Gamma et al. (1994).

Creational patterns	
Abstract Factory	Encapsulates a group of related factories.
Builder	Separates a complex object's construction from its representation.
Factory Method	Lets a class defer instantiation to subclasses.
Prototype	Specifies a prototypical instance of a class that can be cloned to produce new objects.
Singleton	Ensures a class has only one instance.
Structural patterns	
Adapter	Converts the interface of one class into another interface.
Bridge	Decouples an abstraction from its implementation so that both can be changed independently.
Composite	Composes objects into tree structures to represent part–whole hierarchies.
Decorator	Adds additional behavior to an existing object in a dynamic fashion.
Façade	Provides a unified higher-level interface to a set of interfaces in a subsystem.
Flyweight	Uses sharing to support large numbers of fine-grained objects efficiently.
Proxy	Provides a surrogate or placeholder for another object to control access to it.
Behavioral patterns	
Chain of Responsibility	Gives more than one receiver object a chance to handle a request from a sender object.
Command	Encapsulates a request or operation as an object, with support for undoable operations.
Interpreter	Specifies how to represent and evaluate sentences in a language.
Iterator	Provides a way to access the elements of an aggregate object sequentially.
Mediator	Defines an object that encapsulates how a set of objects interact.
Memento	Captures an object's internal state so that it can be restored to the same state later.
Observer	Allows a one-to-many notification of state changes between objects.
State	Allows an object to appear to change its type when its internal state changes.
Strategy	Defines a family of algorithms, encapsulates each one, and makes them interchangeable at run time.
Template Method	Defines the skeleton of an algorithm in an operation, deferring some steps to subclasses.
Visitor	Represents an operation to be performed on the elements of an object structure.

mapping of new classes to preexisting classes, whereas the Façade provides a simplified interface to a larger collection of classes.

- **Observer.** This behavioral pattern can be used to reduce direct dependencies between classes. It allows conceptually unrelated classes to communicate by allowing one class (the observer) to register for notifications from another class (the subject). As such, this pattern is an important aspect of designing loosely coupled APIs.

In addition to covering these patterns and idioms in this chapter, I will discuss the Thread-Safe Interface pattern in Chapter 9 (Concurrency) and the Visitor behavioral pattern in Chapter 15 (Extensibility).

Pimpl idiom

The term pimpl was first introduced by Jeff Sumner as shorthand for pointer to implementation (Sutter, 1999). This technique can be used to avoid exposing private details in your header files (Fig. 3.1). It's therefore an important mechanism to help you maintain a strong separation between your API's interface and implementation (Sutter and Alexandrescu, 2004). Although pimpl is not strictly a design pattern (it's a work-around to C++ specific limitations), it is an idiom that can be considered a special case of the Bridge design pattern.

If you change one programming habit after reading this book, I hope you'll choose to pimpl more API code.

> *TIP: Use the pimpl idiom to keep implementation details out of your public header files.*

Using pimpl

Pimpl relies on the fact that it's possible to define a data member of a C++ class that is a pointer to a forward declared type: that is, where the type has been introduced only by name and has not yet been fully defined, thus allowing us to hide the definition of that type within the `.cpp` file. This is often called an opaque pointer because the user cannot see the details for the object being pointed to. In essence, pimpl is a way to employ both logical and physical hiding of your private data members and functions.

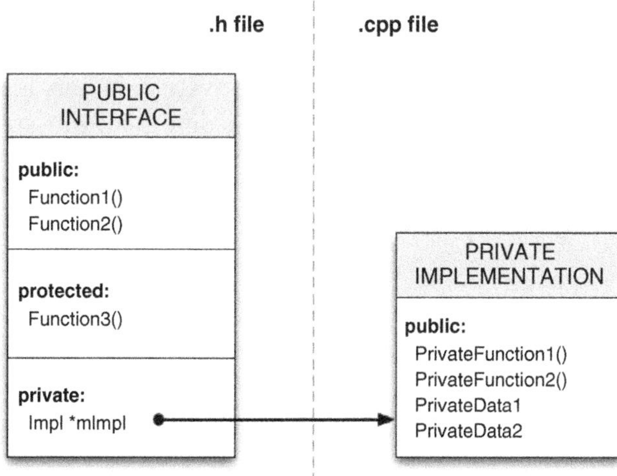

FIGURE 3.1 The pimpl idiom, in which a public class has a private pointer to a hidden implementation class.

Let's look at an example to illustrate this. Consider the following API for an autotimer, a named object that prints out how long it was alive when it's destroyed:

```
// autotimer.h
#ifdef _WIN32
#include <windows.h>
#else
#include <sys/time.h>
#endif

#include <string>

class AutoTimer
{
public:
    /// Create a new timer object with a human-readable name
    explicit AutoTimer(const std::string &name);
    /// On destruction, the timer reports how long it was alive
    ~AutoTimer();

private:
    // Return how long the object has been alive
    double GetElapsed() const;

    std::string mName;
#ifdef _WIN32
    DWORD mStartTime;
#else
    struct timeval mStartTime;
#endif
};
```

This API violates a number of the important qualities I presented in the previous chapter. For example, it includes platform-specific defines and it makes the underlying implementation details of how the timer is stored on different platforms visible to anyone looking at the header file. To be fair, the API does a good job of exposing only the necessary methods as public (i.e., the constructor and destructor) and marking the remaining methods and data members as private. However, C++ requires you to declare these private members in the public header file, and this is why you have to include the platform-specific `#if` directives.

What you really want to do is to hide all of the private members in the `.cpp` file. Then you wouldn't need to include any of those bothersome platform specifics. The pimpl idiom lets you do this by placing all of the private members into a class (or struct) that is forward declared in the header but defined in the `.cpp` file. For example, you could recast the previous header using pimpl as:

```
// autotimer.h
#include <string>

class AutoTimer
{
public:
    explicit AutoTimer(const std::string &name);
    ~AutoTimer();

private:
    class Impl;
    Impl *mImpl;
};
```

Now the API is much cleaner! There are no platform-specific preprocessor directives, and the reader cannot see any of the class's private members by looking at the header file.

The implication, however, is that our `AutoTimer` constructor must now allocate an object of type `AutoTimer::Impl` and then destroy it in its destructor. Also, all private members must be accessed via the `mImpl` pointer. However, for most practical cases, the benefit of presenting a clean implementation-free API far outweighs these costs.

To be complete, let's see what the underlying implementation looks like to work with this pimpled class. The resulting `.cpp` file looks a little bit messy owing to the platform-specific `#ifdef` lines, but the important thing is that this messiness is completely contained in the `.cpp` file now:

```
// autotimer.cpp
#include "autotimer.h"
#include <iostream>
#if _WIN32
#include <windows.h>
#else
#include <sys/time.h>
#endif

class AutoTimer::Impl
{
public:
    double GetElapsed() const
    {
#ifdef _WIN32
        return (GetTickCount() - mStartTime) / 1e3;
#else
        struct timeval end_time;
        gettimeofday(&end_time, nullptr);
        double t1 = mStartTime.tv_usec / 1e6 + mStartTime.tv_sec;
        double t2 = end_time.tv_usec / 1e6 + end_time.tv_sec;
        return t2 - t1;
#endif
    }

    std::string mName;
#ifdef _WIN32
    DWORD mStartTime;
#else
    struct timeval mStartTime;
#endif
};

AutoTimer::AutoTimer(const std::string &name) :
    mImpl(new AutoTimer::Impl())
{
    mImpl->mName = name;
#ifdef _WIN32
    mImpl->mStartTime = GetTickCount();
#else
    gettimeofday(&mImpl->mStartTime, nullptr);
#endif
}

AutoTimer::~AutoTimer()
{
    std::cout << mImpl->mName << ": took " << mImpl->GetElapsed()
              << " secs" << std::endl;
    delete mImpl;
}
```

Here you see the definition of the AutoTimer::Impl class, containing all of the private methods and variables that were originally exposed in the header. Note also that the AutoTimer constructor allocates a new AutoTimer::Impl object and initializes its members, and the destructor deallocates this object.

In the previous design, I declared the Impl class as a private nested class within the AutoTimer class. Declaring it as a nested class avoids polluting the global namespace with this implementation-specific symbol, and declaring it as private means that it does not pollute the public API of your class. However, declaring it to be private imposes the limitation that only the methods of AutoTimer can access the members of the Impl. Other classes or free functions in the .cpp file will not be able to access Impl. As an alternative, if this poses too much of a limitation, you could instead declare the Impl class to be a public nested class, as in this example:

```
// autotimer.h
class AutoTimer
{
public:
    explicit AutoTimer(const std::string &name);
    ~AutoTimer();

    // allow access from other classes/functions in autotimer.cpp
    class Impl;

private:
    Impl *mImpl;
};
```

TIP: When using the pimpl idiom, prefer to use a private nested implementation class. Only use a public nested Impl class (or a public nonnested class) if other classes or free functions in the .cpp must access Impl members.

Another design question worth considering is how much logic to locate in the Impl class. Some options include:

1. Only private member variables,
2. Private member variables and methods, or
3. All methods of the public class, such that the public methods are simply thin wrappers on top of equivalent methods in the Impl class.

Each of these options may be appropriate under different circumstances. However, in general, I recommend option 2: putting all private member variables and private methods in the `Impl` class. This lets you maintain the encapsulation of data and methods that act upon those data and lets you avoid declaring private methods in the public header file. Note that I adopted this design approach in our previous example by putting the `GetElapsed()` method inside the `Impl` class. Herb Sutter notes a couple of caveats with this approach (Sutter, 1999):

1. You can't hide private virtual methods in the implementation class. These must appear in the public class so that any derived classes are able to override them.
2. You may need to add a pointer in the implementation class back to the public class so that the `Impl` class can call public methods.

Copy semantics

A C++ compiler will create a copy constructor and assignment operator for your class if you don't explicitly define them. However, these default constructors will perform only a shallow copy of your object. This is bad for pimpled classes because it means that if a client copies your object, then both objects will point to the same implementation object, `Impl`. However, both objects will attempt to delete this same `Impl` object in their destructors, which will most likely lead to a crash. There are two main options for dealing with this:

1. **Make your class uncopyable.** If you don't intend for your users to create copies of an object, then you can declare the object to be noncopyable. As of C++11, you can easily do this by deleting the default compiler-generated copy constructor and assignment operator functions using the = `delete` syntax. Alternatively, if you need a C++98 solution, you can declare a copy constructor and assignment operator with no implementation to prevent the compiler from generating its own versions. Additionally, you can declare these functions as private, so that attempts to copy the object will generate a compile error rather than a link error.
2. **Explicitly define the copy semantics.** If you do want your users to be able to copy your pimpled objects, then you can define your own copy constructor and assignment operator. These can then perform a deep copy of your object (i.e., create a copy of the `Impl` object instead of just copying the pointer). I cover how to write your own constructors and operators in the C++ Usage chapter later in this book. However, an even better solution is simply to use a smart pointer for your `Impl` pointer because these already have well-defined copy semantics.

The next code example provides an updated version of our `AutoTimer` API, where I've made the object noncopyable by deleting the copy constructor and assignment operator. The associated `.cpp` file doesn't need to change:

```
#include <string>

class AutoTimer
{
public:
    explicit AutoTimer(const std::string &name);
    ~AutoTimer();

    // Make this object be non-copyable
    AutoTimer(const AutoTimer &) = delete;
    AutoTimer &operator=(const AutoTimer &) = delete;

private:
    class Impl;
    Impl *mImpl;
};
```

In the next section, I'll show how you can use a smart pointer to define the copy semantics of your pimpled object explicitly and also make the memory management of the `Impl` object more robust.

Pimpl and smart pointers

One of the inconvenient and error-prone aspects of pimpl is the need to allocate and deallocate the implementation object. Every now and then you may forget to the delete the object in your destructor, or you may introduce bugs by accessing the `Impl` object before you've allocated it or after you've destroyed it. As a convention, you should therefore ensure that the very first thing your constructor does is to allocate the `Impl` object (preferably via its initialization list), and the very last thing your destructor does is to delete it.

Alternatively, you could rely upon smart pointers to make this more robust. That is, you could use a shared pointer or a unique pointer to hold the implementation object pointer. An `std::shared_ptr` would allow the object to be copied without incurring the double delete issues I identified earlier. Using a shared pointer would, of course, mean that any copy would point to the same `Impl` object in memory. If you need the copied object to have a copy of the `Impl` object, then you will still need to write your own copy constructor and assignment operators (or use a copy-on-write pointer, as described in the Performance chapter). Here's an example of using the pimpl idiom with a shared pointer:

```
#include <memory>
#include <string>

class AutoTimer
{
public:
    explicit AutoTimer(const std::string &name);
    ~AutoTimer();

private:
    class Impl;
    std::shared_ptr<Impl> mImpl;
};
```

Alternatively, you could use an `std::unique_ptr` instead of an `std::shared_ptr` as another way to make the object noncopyable, because a unique pointer is noncopyable by definition.

> *TIP: Think about the copy semantics of your pimpl classes, and favor the use of a shared or unique pointer to manage initialization and destruction of the implementation pointer.*

Using either a shared or unique pointer means that the `Impl` object will be automatically freed when the `AutoTimer` object is destroyed: you no longer need to delete it explicitly in the destructor. So the destructor of our `autotimer.cpp` file can now be reduced simply to:

```
AutoTimer::~AutoTimer()
{
    std::cout << mImpl->mName << ": took " << mImpl->GetElapsed()
              << " secs" << std::endl;
}
```

Advantages of pimpl

There are many advantages to employing the pimpl idiom in your classes. These include:

- **Information hiding.** Private members are now completely hidden from your public interface. This allows you to keep your implementation details hidden (and proprietary in the case of closed-source APIs). It also means that your public header files are cleaner and more clearly express the true public interface. As a result, they can be more easily read and digested by your users. One further benefit of information hiding is that your users cannot as easily use dirty tactics to gain access to your private members, such as doing the following, which is actually legal in C++ (Lakos, 1996):

```
#define private public   // make private members be public!
#include "yourapi.h"     // can now access your private members
#undef private           // revert to default private semantics
```

- **Reduced coupling.** As I showed in the `AutoTimer` example earlier, without pimpl, your public header files must include header files for all your private member variables. In our example, this meant having to include `windows.h` or `sys/time.h`. This increases the compile-time coupling of your API on other parts of the system.

Using pimpl, you can move those dependencies into the .cpp file and remove those elements of coupling.
- **Faster compiles.** Another implication of moving implementation-specific includes to the .cpp file is that the include hierarchy of your API is reduced. This can have a very direct effect on compile times (Lakos, 1996). I will detail the benefits of minimizing include dependencies in the Performance chapter.
- **Greater binary compatibility.** The size of a pimpled object never changes because your object is always the size of a single pointer. Any changes you make to private member variables (recall that member variables should always be private) will affect only the size of implementation class that is hidden inside of the .cpp file. This makes it possible to make major implementation changes without changing the binary representation of your object.

Disadvantages of pimpl

The primary disadvantage of the pimpl idiom is that you must now allocate and free an additional implementation object for every object that's created. This increases the size of your object by the size of a pointer and may introduce a performance hit for the extra level of pointer indirection required to access all member variables, as well as the cost for additional calls to new and delete. If you're concerned with the memory allocator performance, then you may consider using the Fast Pimpl idiom (Sutter, 1999) in which you overload the new and delete operators for your Impl class to use a more efficient small-memory fixed-size allocator.

There is also the extra developer inconvenience of prefixing all private member accesses with something like mImpl->. This can make the implementation code harder to read and debug owing to the additional layer of abstraction. This becomes even more complicated when the Impl class has a pointer back to the public class. However, these inconveniences are not exposed to users of your API and are therefore not a concern from the point of view of your API's design. They are a burden that you, the developer, must shoulder for all of your users to receive a cleaner and more efficient API. To quote a certain science officer and his captain: "The needs of the many outweigh the needs of the few." "Or the one."

One final issue to be aware of is that the compiler will no longer catch changes to member variables within const methods. This is because member variables now live in a separate object. Your compiler will check only that you don't change the value of the mImpl pointer in a const method, but not whether you change any members pointed to by mImpl. In effect, every member function of a pimpled class could be defined as const (except, of course, the constructor or destructor). This is demonstrated by the following const method that legally changes a variable in the Impl object:

```
void PimpledObject::ConstMethod() const
{
    mImpl->mName = "string changed by a const method";
}
```

This can be addressed by introducing two private functions to access the `Impl` pointer: one const and one nonconst. That way, the compiler will enforce that you use the const version in a const method. If the const function also returns a `const Impl` pointer, then you also get compiler const checking for your `Impl` methods. For example:

```
// autotimer.h
class AutoTimer
{
public:
    explicit AutoTimer(const std::string &name);
    ~AutoTimer();

    void ConstMember() const;
    void NonConstMember();

private:
    class Impl;
    std::shared_ptr<Impl> GetImpl() { return mImpl; }
    std::shared_ptr<const Impl> GetConstImpl() const { return mImpl; }
    std::shared_ptr<Impl> mImpl;
};

// autotimer.cpp
class AutoTimer::Impl
{
public:
    double GetElapsed() const;
    std::string mName;
    void ImplConstMember() const;
    void ImplNonConstMember();
};

Void AutoTimer::ConstMember() const
{
    GetConstImpl()->ImplConstMember();
}

Void AutoTimer::NonConstMember()
{
    GetImpl()->ImplNonConstMember();
}
```

Opaque pointers in C

Although I have focused on C++ so far, you can create opaque pointers in plain C, too. The concept is the same: you create a pointer to a struct that is only defined in a `.c` file. The next header files demonstrate what this might look like in C:

```
/* autotimer.h */
/* declare an opaque pointer to an AutoTimer structure */
typedef struct AutoTimer *AutoTimerPtr;

/* functions to create and destroy the AutoTimer structure */
AutoTimerPtr AutoTimerCreate(const char *name);
void AutoTimerDestroy(AutoTimerPtr ptr);
```

The associated .c file might then look like:

```
#include "autotimer.h"
#include <stdio.h>
#include <stdlib.h>
#include <string.h>

#if _WIN32
#include <windows.h>
#else
#include <sys/time.h>
#endif

struct AutoTimer
{
    char *mName;
#if _WIN32
    DWORD mStartTime;
#else
    struct timeval mStartTime;
#endif
} AutoTimer;

AutoTimerPtr AutoTimerCreate(const char *name)
{
    AutoTimerPtr ptr = malloc(sizeof(AutoTimer));
    if (ptr) {
        ptr->mName = strdup(name);
#if _WIN32
        ptr->mStartTime = GetTickCount();
#else
        gettimeofday(&ptr->mStartTime, NULL);
#endif
    }
    return ptr;
}

static double GetElapsed(AutoTimerPtr ptr)
{
#if _WIN32
    return (GetTickCount() - ptr->mStartTime) / 1e3;
#else
    struct timeval end_time;
    gettimeofday(&end_time, nullptr);
    double t1 = ptr->mStartTime.tv_usec / 1e6 +
                ptr->mStartTime.tv_sec;
    double t2 = end_time.tv_usec / 1e6 + end_time.tv_sec;
    return t2 - t1;
#endif
}

void AutoTimerDestroy(AutoTimerPtr ptr)
{
    if (ptr) {
        printf("%s: took %f secs\n", ptr->mName, GetElapsed(ptr));
        free(ptr->mName);
        free(ptr);
    }
}
```

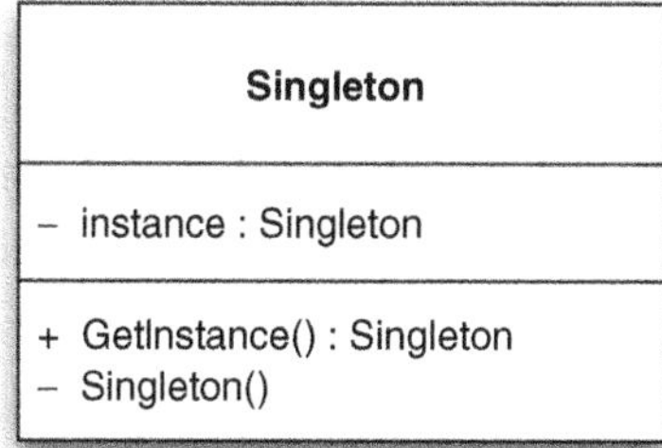

FIGURE 3.2 Unified Modeling Language (UML) diagram of the Singleton design pattern.

Singleton

The Singleton design pattern (Gamma et al., 1994) is used to ensure that a class only ever has one instance. The pattern also provides a global point of access to that single instance (Fig. 3.2). You can think of a singleton as a more elegant global variable. However, it offers several advantages over the use of global variables because it:

1. Enforces that only one instance of the class can be created.
2. Provides control over the allocation and destruction of the object.
3. Allows support for thread-safe access to the object's global state.
4. Avoids polluting the global namespace.
5. Avoids the nondeterministic initialization problem of nonlocal static constructors.

The Singleton pattern is useful for modeling resources that are inherently singular in nature, such as a class to access the system clock, the global clipboard, or the keyboard. It's also useful for creating manager classes that provide a single point of access to multiple resources, such as a thread manager or an event manager. On the other hand, the singleton is still essentially a way to add global variables to your system, albeit in a more manageable fashion. It can therefore introduce global state and dependencies into your API that are difficult to refactor later, as well as making it difficult to write unit tests that exercise isolated parts of your code.

I have decided to cover the concept of singletons here partly because they offer a useful and common API design technique. However, another reason is that they are intricate to implement robustly in C++, and so it's worth discussing some of the implementation details. Also, many programmers tend to overuse the Singleton pattern, so I wanted to highlight some of the disadvantages of singletons as well as provide alternative techniques.

> *TIP: A Singleton is a more elegant way to maintain global state, but you should always question whether you need global state.*

Implementing singletons in C++

The Singleton pattern involves creating a class with a static method that returns the same instance of the class every time it's called. This static method is often called `GetInstance()`, or similar. There are several C++ language features to consider when designing a singleton class:

- You don't want clients to be able to create new instances. This can be done by declaring the default constructor to be private, thus preventing the compiler from automatically creating it as public.
- You want the singleton to be noncopyable, to enforce that a second instance cannot be created. As you've seen earlier, this can be done by declaring a private copy constructor and a private assignment operator, or by using the `delete` specifier as of C++11.
- You want to prevent clients from being able to delete the singleton instance. This can be done by declaring the destructor to be private. (Note, however, that some compilers, such as Borland 5.5 and Visual Studio 6, incorrectly produce an error if you try to declare a destructor as private).
- The `GetInstance()` method could return either a pointer or a reference to the singleton class. However, if you return a pointer, clients could potentially delete the object. You should therefore prefer returning a reference.

The general form of a singleton as of C++11 can therefore be given as:

```
class Singleton
{
public:
    static Singleton &GetInstance();

    Singleton(const Singleton &) = delete;
    Singleton &operator=(const Singleton &) = delete;

private:
    Singleton();
    ~Singleton();
};
```

Then user code can request a reference to the singleton instance as:

```
Singleton &obj = Singleton::GetInstance();
```

Note that making the constructor and destructor private also means that clients cannot create subclasses of the singleton. However, if you wish to allow this, you can, of course, simply declare them to be protected instead.

> *TIP: Declare the constructor, destructor, copy constructor, and assignment operator to be private (or protected) to enforce the Singleton property. Or delete the copy constructor and assignment operator with the* `delete` *specifier.*

In terms of implementation, one of the areas to be very careful about is how the singleton instance is allocated. The important C++ initialization issue to recognize is explained by Scott Meyers as:

> *The relative order of initialization of non-local static objects in different translation units is undefined.*
>
> *(Meyers, 2005).*

This means that it would be dangerous to initialize our singleton using a nonlocal static variable. A nonlocal object is one that is declared outside a function or method. Static objects include global objects, objects declared as static inside a class or function, and objects defined at file or namespace scope. As a result, one way to initialize our singleton would be to create a static variable inside a method of our class, as:

```
Singleton &Singleton::GetInstance()
{
    static Singleton instance;
    return instance;
}
```

One nice property of this approach is that the instance will be allocated only when the `GetInstance()` method is first called. This means that if the singleton is never requested, the object is never allocated. However, on the downside, Andrei Alexandrescu notes that this technique relies on the standard last in first out deallocation behavior of static variables, which can result in singletons being deallocated before they should in situations where singletons call other singletons in their destructors. As an example of this problem, consider two singletons: `Clipboard` and `LogFile`. When `Clipboard` is instantiated, it also instantiates `LogFile` to output some diagnostic information. At program exit, `LogFile` is destroyed first because it was created last, then `Clipboard` is destroyed. However, the `Clipboard` destructor tries to call `LogFile` to log the fact that it is being destroyed, but `LogFile` has already been freed. This will most likely result in a crash on program exit.

In his *Modern C++ Design* book, Alexandrescu presents several solutions to this destruction order problem, including resurrecting the singleton if it is needed after it has been destroyed, increasing the longevity of a singleton so that it can outlive other singletons, and simply not deallocating the singleton (i.e., relying on the operating system to free all allocated memory and close any file handles). If you find yourself needing to implement one of these solutions, I refer you to this book for the details (Alexandrescu, 2001).

Another issue to be aware of when working with singletons is that processes have their own address space, so each process that loads a given library will have its own instances of any static variables. Also, static variables defined within plugins that are opened at run time with `dlopen` may not share the same memory address as other plugins or even the main program, so the use of singletons in plugins can be unsafe.

Singletons and thread safety

The C++98 and C++03 standards did not specify how code should behave in a multi-threaded environment. In particular, the initialization of static variables was not thread-safe. Under these early versions of the standard, if two threads happened to call our `GetInstance()` method at the same time, then the instance could be constructed twice, or it could be used by one thread before it had been fully initialized by the other thread.

This issue was addressed in the C++11 standard, which states that only one thread can enter the initialization of a static and that the compiler must not introduce any deadlocks around its initialization. So the previous implementation of `GetInstance()` is thread-safe as of C++11.

In the first edition of this book, I spent several pages describing thread-safe solutions for Singletons in C++03 by introducing mutex locking and considering optimizations such as the Double Check Locking Pattern. However, at this point, C++11 is well-established, and if you want to write thread-safe code in general, you should be using at least C++11. In this edition of the book, I will therefore simply state that if you want to have thread-safe Singletons, you should use a C++11 (or later) compiler.

TIP: If you use Singletons, you should use a C++11 (or later) compiler for thread safety.

Singleton versus dependency injection

Dependency injection is a technique in which an object is passed into a class (injected), rather than having the class create and store the object itself. Martin Fowler coined the term in 2004 as a more specific form of the Inversion of Control concept. As a simple example, consider this class that depends upon a database object:

```
class MyClass
{
public:
    MyClass() :
        mDatabase(new Database("mydb", "localhost", "user", "pass"))
    {}

private:
    Database *mDatabase;
};
```

The problem with this approach is that if the Database constructor is changed, or if someone changes the password for the account called user in the live database, then you will have to change MyClass to fix the problem. Also, it's difficult to unit test this object because the database implementation is embedded within it. And from an efficiency point of view, every instance of MyClass will create a new Database instance. As an alternative, you can use dependency injection to pass a preconfigured Database object into MyClass, as:

```
class MyClass
{
public:
    MyClass(std::shared_ptr<Database> db) :
        mDatabase(db)
    {}

private:
    std::shared_ptr<Database> mDatabase;
};
```

In this way, MyClass no longer needs to know how to create a Database instance. Instead, it gets passed an already constructed and configured Database object for it to use. I've changed the raw pointer to a shared pointer so that the dependency can be based on an abstract interface instead of a concrete object, and because MyClass no longer controls the lifetime of the database object. For example, if we have a MyDatabase concrete subclass of the abstract Database type, then the MyClass object can be created as:

```
MyClass(std::make_shared<MyDatabase>());
```

Or, even better, you can use a Factory Method (which I'll discuss later in this chapter) to create the instance, such as:

```
MyClass(DatabaseFactory::Create());
```

This example demonstrates constructor injection (i.e., passing the dependent object via the constructor), but you could just as easily pass in dependencies via a setter member function.

Dependency injection can be viewed as a way to avoid the proliferation of singletons by encouraging interfaces that accept the single instance as an input rather than requesting it internally via a GetInstance() method. This also enables better unit testing because the dependencies of an object can be substituted with stub or mock versions for the purposes of testing (e.g., by passing in a TestDatabase instance within your test code). I'll discuss this further in the Testing chapter.

> *TIP: Dependency injection reduces coupling between objects and makes it easier to support unit testing.*

Singleton versus Monostate

Most of the problems that are associated with the Singleton pattern derive from the fact that it's designed to hold and control access to global state. However, if you don't need to control when the state is initialized, or if you don't need to store state in the singleton object itself, then there are other techniques you can use, such as the Monostate design pattern.

The Monostate pattern allows multiple instances of a class to be created in which all of those instances use the same static data. For instance, here's a simple case of the Monostate pattern:

```
// monostate.h
class Monostate
{
public:
    int GetTheAnswer() const { return sAnswer; }

private:
    static int sAnswer;
};

// monostate.cpp
int Monostate::sAnswer = 42;
```

In this example, you can create multiple instances of the `Monostate` class, but all calls to the `GetTheAnswer()` method will return the same result, because all instances share the same static variable `sAnswer`. You could also hide the declaration of the static variable from the header completely just by declaring it as a file-scope static variable in `monostate.cpp` instead of a private class static variable. Static members do not contribute to the per instance size of a class, so doing this will have no physical impact on the API other than to hide implementation details from the header.

Some benefits of the Monostate pattern are that it:

- Allows multiple instances to be created.
- Offers transparent usage, because no special `GetInstance()` method is needed.
- Exhibits well-defined creation and destruction semantics, by using static variables.

As Robert C. Martin notes, Singleton enforces the structure of singularity by allowing only one instance to be created, whereas Monostate enforces the behavior of singularity by sharing the same data for all instances (Martin, 2002).

> *TIP: Consider using Monostate instead of Singleton if you don't need lazy initialization of global data or if you want the singular nature of the class to be transparent.*

As a further real-world example, the Second Life source code uses the Monostate pattern for its `LLWeb` class. This example uses a version of Monostate in which all member functions are declared static:

```
class LLWeb
{
public:
    static void InitClass();

    /// Load the given url in the user's preferred web browser
    static void LoadURL(const std::string &url);

    /// Load the given url in the Second Life internal web browser
    static void LoadURLInternal(const std::string &url);

    /// Load the given url in the operating system's web browser
    static void LoadURLExternal(const std::string &url);

    /// Returns escaped url (eg, " " to "%20")
    static std::string EscapeURL(const std::string &url);
};
```

In this case, `LLWeb` is simply a manager class that provides a single access point to the functionality for opening Web pages. The actual Web browser functionality itself is implemented in other classes. The `LLWeb` class does not hold any state itself, although, of course, internally any of the static methods could access static variables. Note that another way to implement this would be to create an `LLWeb` namespace with free functions within that namespace.

One of the drawbacks with this static method version of Monostate is that you cannot subclass any of the static methods, because static member functions cannot be virtual. Also, because you no longer instantiate the class, you cannot write a constructor or destructor to perform any initialization or cleanup. This is necessary in this case because `LLWeb` accesses dynamically allocated global state instead of relying on static variables that are initialized by the compiler. The creator of `LLWeb` got around this limitation by introducing an `initClass()` static method that requires a client program to initialize the class explicitly. A better design may have been to hide this call within the `.cpp` file and invoke it lazily from each of the public static methods. However, in that case, the same thread safety concerns I raised earlier would be applicable.

Singleton versus session state

In a recent retrospective interview, the authors of the original design patterns book stated that the only pattern they would consider removing from the original list is Singleton. This is because it's essentially a way to store global data and tends to be an indicator of poor design (Gamma et al., 2009).

Therefore, as a final note on the topic of singletons, I urge you to really think about whether a singleton is the correct pattern for your needs. It's often easy to think that you will only ever need a single instance of a given class. However, requirements change and code evolves, and in the future you may find that you need to support multiple instances of the class.

For example, consider that you're writing a simple text editor. You use a singleton to hold the current text style (e.g., bold, italics, underlined), because the user can only ever have one style active at one time. However, this restriction is valid only because of the initial assumption that the program can edit only one document at a time. In a later version of the program, you are asked to add support for multiple documents, each with their own current text style. Now you have to refactor your code to remove the singleton. Ultimately, singletons should be used only to model objects that are truly singular in their nature. For example, there is only one system clipboard, so it may still be reasonable to model the clipboard for the text editor as a singleton.

Often, it's useful to think about introducing a session or execution context object into your system early on. This is a single instance that holds all of the state for your code, rather than representing that state with multiple singletons. For example, in the text editor example, you might introduce a `Document` object. This will have accessors for things such as the current text style, but those objects do not have to be enforced as singletons. They are just plain classes that can be accessed from the `Document` class as `document->GetTextStyle()`. You can start off with a single `Document` instance, accessible by a call such as `Document::GetCurrent()`, for instance. You might even implement `Document` as a singleton to begin with. However, if you later need to add support for multiple contexts (i.e., multiple documents), then your code is in a much healthier state to support this because you have only one singleton to refactor instead of dozens. J.B. Rainsberger refers to this as a Toolbox Singleton, in which the application becomes the singleton, not the individual classes (Rainsberger, 2001).

> *TIP: There are several alternatives to the Singleton pattern, including dependency injection, the Monostate pattern, and the use of a session context.*

Factory Methods

A Factory Method is a creational design pattern that allows you to create objects without having to specify the specific C++ type of the object to create. In essence, a Factory Method is a generalization of a constructor. Constructors in C++ have several limitations, such as:

1. **No return result.** You cannot return a result from a constructor. This means that you can't signal an error during the initialization of an object by returning a `nullptr`, for instance (although you can signal an error by throwing an exception within a constructor).
2. **Constrained naming.** A constructor is easily distinguished because it is constrained to have the same name as the class in which it lives. However, this also limits its flexibility. For example, you cannot have two constructors that both take a single integer argument.
3. **Statically bound creation.** When constructing an object, you must specify the name of a concrete class that's known at compile time (e.g., you might write: `Foo *f = new Foo;` where `Foo` is a specific type that must be known by the compiler). There's no concept of dynamic binding at run time for constructors in C++.
4. **No virtual constructors.** You can't declare a virtual constructor in C++. As I have just noted, you must specify the exact type of the object to be constructed at compile time. The compiler therefore allocates the memory for that specific type and then calls the default constructor for any base classes (unless you explicitly specify a nondefault constructor in the initialization list). It then calls the constructor for the specific type itself. This is also why you can't call virtual methods from the constructor and expect them to call the derived override (because the derived class hasn't been initialized yet).

In contrast, Factory Methods circumvent all these limitations. At a basic level, a Factory Method is simply a normal method call that can return an instance of a class. However, they are often used in combination with inheritance, in which a derived class can override the Factory Method and return an instance of that derived class. It's also very common to implement factories using abstract base classes (ABCs) (DeLoura, 2001). So before I dive deeper into using Factory Methods, let's recap what ABCs are, as well as the related concept of interfaces.

Abstract base classes and interfaces

An ABC is a class that contains one or more pure virtual member functions. Such a class is not concrete and therefore cannot be instantiated using the `new` operator. Instead, it's used as a base class, in which derived classes can provide an implementation for each of the pure virtual methods.

There is also a related term called interface. Languages such as Java have an actual keyword called interface, but the term is less well-defined in the context of C++. Some C++ developers consider it to be a synonym for ABC. However, I believe it's generally accepted that an interface is a more specific version of an abstract class that only contains pure virtual functions (i.e., it has no implementation code or state), whereas the term ABC refers more generally to a class that can have one or more pure virtual functions but also nonpure virtual functions, implementation code, or data members. For example, consider the code:

```
// renderer.h
#include <string>

class IRenderer
{
public:
    virtual ~IRenderer() = default;
    virtual bool LoadScene(const std::string &filename) = 0;
    virtual void SetViewportSize(int w, int h) = 0;
    virtual void SetCameraPos(double x, double y, double z) = 0;
    virtual void SetLookAt(double x, double y, double z) = 0;
    virtual void Render() = 0;
};
```

This defines an interface to describe an extremely simple 3D graphics renderer. I've named the class with an I prefix to indicate that it's an interface. An alternative naming convention might be to use a suffix such as -able, -ible, or -ing (e.g., `Renderable`). The `= 0` suffix on the methods declares them to be pure virtual methods, meaning that they must be overridden in a derived class for that class to be concrete. The `IRenderer` class is therefore an interface because it contains only pure virtual functions, other than the defaulted virtual destructor.

Note that it's not strictly true to say that pure virtual methods provide no implementation. You can provide a default implementation for pure virtual methods in your `.cpp` file. For example, you could provide an implementation for `SetViewportSize()` in `renderer.cpp`, and then a derived class would be able to call `IRenderer::SetViewportSize()`, although it would still have to override the method explicitly as well. By our previous definition, we would then prefer to call this an ABC class rather than an interface because it provides an implementation for one or more of the virtual functions.

An ABC or interface is useful to describe abstract units of behaviors that can be shared by multiple classes. They specify a contract to which all concrete derived classes must conform. You might use an interface if you wanted to define only a set of rules for an object to follow without defining any specific behavior. Alternatively, you would use an ABC if you also wanted to provide a default implementation for those rules.

As with any class that has one or more virtual methods, you should always declare the destructor of an ABC or interface to be virtual. This code illustrates why this is important:

```
class IRenderer
{
public:
    // no virtual destructor declared
    virtual void Render() = 0;
};

class RayTracer : public IRenderer
{
public:
    RayTracer();
    ~RayTracer();
    void Render();   // provide implementation for ABC method
};

int main(int, char **)
{
    IRenderer *r = new RayTracer();
    // delete calls IRenderer::~IRenderer, not RayTraver::~RayTracer
    delete r;
}
```

Simple factory example

Now that I've reviewed the concepts of ABCs and interfaces, let's use them to provide a simple Factory Method. I'll continue with the previous `renderer.h` example and start by declaring the factory for objects of type `IRenderer`:

```
// rendererfactory.h
#include "renderer.h"
#include <string>

class RendererFactory
{
public:
    IRenderer *CreateRenderer(const std::string &type);
};
```

That's all there is to declaring a Factory Method: it's just a normal method that can return an instance of an object. Note that this method cannot return an instance of the specific type `IRenderer` because that's an abstract class and cannot be instantiated. However, it can return instances of derived classes. Also, you can use the string argument to `CreateRenderer()` to specify which derived type you want to create.

Let's assume that you've implemented three concrete derived classes of `IRenderer`: `OpenGLRenderer`, `DirectXRenderer`, and `MesaRenderer`. Let's further specify that you don't want users of your API to have any knowledge of the existence of these types: they must be completely hidden behind the API. Based on these conditions, you can provide an implementation of our Factory Method as:

```
// rendererfactory.cpp
#include "rendererfactory.h"
#include "directxrenderer.h"
#include "mesarenderer.h"
#include "openglrenderer.h"

IRenderer *RendererFactory::CreateRenderer(const std::string &type)
{
    if (type == "opengl") {
        return new OpenGLRenderer;
    }

    if (type == "directx") {
        return new DirectXRenderer;
    }

    if (type == "mesa") {
        return new MesaRenderer;
    }

    return nullptr;
}
```

This Factory Method can therefore return any of the three derived classes of `IRenderer`, depending upon the type string that the client passes in. This lets users decide which derived class to create at run time, not compile time, as a normal constructor requires you to do. This is an enormous advantage because it means that you can create different classes based upon user input, or upon the contents of a configuration file that's read at run time.

Also, note that the header files for the various concrete derived classes are included only in the factory's `.cpp` file. They do not appear in the `rendererfactory.h` public header. In effect, these are private header files and do not need to be distributed with your API. As such, users can never see the private details of your different renderers, and they can't ever see the specific types used to implement these different renderers. Users only ever specify a renderer via a string value, or an enum if you expect the set of renderers to be statically defined at compile time.

TIP: Use Factory Methods to provide more powerful class construction semantics and to hide subclass details.

This example demonstrates a perfectly acceptable Factory Method. However, one potential drawback is that it contains hardcoded knowledge of the available derived classes. If you add a new renderer to the system, you must edit `rendererfactory.cpp`. This is not terribly burdensome, and most important, it will not affect the public API. However, it does mean that you can't add support for new derived classes at run time. More specifically, it means that your users can't add new renderers to the system. I will address these concerns by presenting an extensible object factory.

Extensible factory example

To decouple the concrete derived classes from the Factory Method and allow new derived classes to be added at run time, you can update the factory class to maintain a map that associates type names to object creation callbacks (Alexandrescu, 2001). You can then allow new derived classes to be registered and unregistered using a couple of new method calls. The ability to register new classes at run time allows this form of the Factory Method pattern to be used to create extensible plugin interfaces for your API, as I will detail in Chapter 12.

One further issue to note is that the factory object must now hold state. As such, it would be best to enforce that only one factory object is ever created. This is why most factory objects are also singletons. In the interests of simplicity, however, I will use static methods and variables in our example here. Putting all of these points together, here's what our new object factory might look like:

```
// rendererfactory.h
#include "renderer.h"
#include <functional>
#include <map>
#include <string>

class RendererFactory
{
public:
    using CreateCallback = std::function<IRenderer *()>;

    static void RegisterRenderer(const std::string &type,
                                 CreateCallback cb);
    static void UnregisterRenderer(const std::string &type);

    static IRenderer *CreateRenderer(const std::string &type);

private:
    using CallbackMap = std::map<std::string, CreateCallback>;
    static CallbackMap mRenderers;
};
```

For completeness, the associated `.cpp` file might look like:

```
#include "rendererfactory.h"
#include <iostream>

// instantiate the static variable in RendererFactory
RendererFactory::CallbackMap RendererFactory::mRenderers;

void RendererFactory::RegisterRenderer(const std::string &type,
                                       CreateCallback cb)
{
    mRenderers[type] = cb;
}

void RendererFactory::UnregisterRenderer(const std::string &type)
{
    mRenderers.erase(type);
}

IRenderer *RendererFactory::CreateRenderer(const std::string &type)
{
    CallbackMap::iterator it = mRenderers.find(type);
    if (it != mRenderers.end()) {
        // call the creation callback to construct this derived type
        return (it->second)();
    }

    return nullptr;
}
```

A user of your API can now register (and unregister) new renderers in your system. The compiler will ensure that the user's new renderer conforms to your `IRenderer` interface (i.e., that it provides an implementation for all of the pure virtual methods in `IRenderer`). To illustrate this, the subsequent code shows how users could define their own renderer, register it with the object factory, then ask the factory to create an instance of it:

```
class UserRenderer : public IRenderer
{
public:
    ~UserRenderer() = default;
    bool LoadScene(const std::string &filename) { return true; }
    void SetViewportSize(int w, int h) {}
    void SetCameraPos(double x, double y, double z) {}
    void SetLookAt(double x, double y, double z) {}
    void Render() { std::cout << "User Render" << std::endl; }
    static IRenderer *Create() { return new UserRenderer; }
};

int main(int, char **)
{
    // register a new renderer
    RendererFactory::RegisterRenderer("user", UserRenderer::Create);

    // create an instance of our new renderer
    IRenderer *r = RendererFactory::CreateRenderer("user");
    r->Render();
    delete r;

    return 0;
}
```

One point worth noting here is that I added a `Create()` function to the `UserRenderer` class. This is because the register method of the factory needs to take a callback that returns an object. This callback doesn't have to be part of the `IRenderer` class (it could be a free function, for example). However, adding it to the `IRenderer` class is a good idea to keep all of the related functionality in the same place.

Note that in the previous extensible factory example, a renderer callback has to be visible to the `RegisterRenderer()` function at run time. However, this doesn't mean that you have to expose the built-in renderers of your API. These can still be hidden either by registering them within your API initialization routine or by using a hybrid of the simple factory and the extensible factory, in which the Factory Method first checks the type string against a few built-in names, and if none of those match it, then it checks for any names that have been registered by the user. This hybrid approach has the potentially desirable behavior that users cannot override your built-in classes.

I'll conclude the discussion of the Factory Method pattern by looking at a real-world example in a large C++ API. The Visualization ToolKit (VTK) is an open source package for manipulating and displaying scientific data (see https://vtk.org/). The base `vtkObject` has a protected constructor, so it cannot be created `new` or on the stack. The library then provides a `vtkObjectFactory` to create VTK objects. This contains a static method called `CreateInstance()` that creates an object from a list of registered `vtkObjectFactory` subclasses. Those subclasses override a `CreateObject()` protected method to implement the logic to create the object instance. The first time `CreateInstance()` is called, all shared libraries are loaded into the current process. The key functions of the `vtkObjectFactory` class are:

```
class vtkObjectFactory : public vtkObject
{
public:
    static vtkObject *CreateInstance(const char *vtkclassname, bool
isAbstract = false);
    static void CreateAllInstance(const char *vtkclassname,
vtkCollection *retList);
    static void RegisterFactory(vtkObjectFactory *);
    static void UnRegisterFactory(vtkObjectFactory *);
    static void UnRegisterAllFactories();
    static vtkObjectFactoryCollection *GetRegisteredFactories();
    ...

protected:
    virtual vtkObject *CreateObject(const char *vtkclassname);
};
```

API wrapping patterns

Writing a wrapper interface that sits on top of another set of classes is a relatively common API design task. For example, perhaps you're working with a large legacy code base, and rather than rearchitecting all of that code you decide to design a new cleaner API that hides the underlying legacy code (Feathers, 2004). Or perhaps you have written a C++ API and need to expose a plain C interface for certain clients. Or perhaps you have a third-party library dependency that you want your clients to be able to access, but you don't want to expose that library directly to them.

The downside of creating a wrapper API is the potential performance hit you may experience owing to the extra level of indirection, and the overhead of any extra state that needs to be stored at the wrapper level. However, this is often worth the cost to expose a higher-quality or more focused API, such as in the cases mentioned earlier.

There are several structural design patterns that deal with the task of wrapping one interface on top of another. I will describe three of these patterns in the following sections. These are, in increasing deviation between the wrapper layer and the original interface: Proxy, Adapter, and Façade.

The Proxy pattern

The Proxy design pattern provides a one-to-one forwarding interface to another class. Calling `FunctionA()` in the proxy class will cause it to call `FunctionA()` in the original class (Fig. 3.3). That is, the proxy class and the original class have the same interface. This can be thought of as a single-component wrapper, to use the terminology of Lakos (1996): that is, a single class in the proxy API maps to a single class in the original API.

This pattern is often implemented by making the proxy class store a copy of, or more likely a pointer to, the original class. Then the methods of the proxy class simply redirect to the method with the same name in the original object. A downside of this technique is the need to reexpose functions in the original object, a process that essentially equates to code duplication. This approach therefore requires diligence to maintain the integrity of the proxy interface when making changes to the original object. The following code provides a simple example of this technique:

```
class Proxy
{
public:
    Proxy() :
        mOrig(new Original)
    {}

    ~Proxy()
    {
        delete mOrig;
    }

    bool DoSomething(int value)
    {
        return mOrig->DoSomething(value);
    }

private:
    Original *mOrig;
};
```

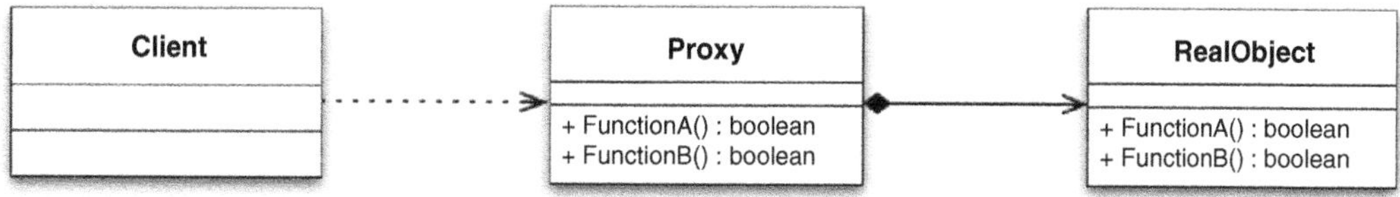

FIGURE 3.3 UML diagram of the Proxy design pattern.

An alternative solution is to augment this approach by using an abstract interface that's shared by both the proxy and original APIs. This allows the proxy object to be interchangeable with the original one, which can often be desirable. However, this approach requires you to be able to modify the original API if it doesn't already derive from an abstract interface. The code shown here demonstrates this approach:

```
class IOriginal
{
public:
    virtual bool DoSomething(int value) = 0;
};

class Original : public IOriginal
{
public:
    bool DoSomething(int value);
};

class Proxy : public IOriginal
{
public:
    Proxy() :
        mOrig(new Original)
    {}

    ~Proxy()
    {
        delete mOrig;
    }

    bool DoSomething(int value)
    {
        return mOrig->DoSomething(value);
    }

private:
    Original *mOrig;
};
```

> *TIP: A Proxy provides an interface that forwards function calls to another interface of the same form.*

A Proxy pattern is useful to modify the behavior of the `Original` class while still preserving its interface. This is particularly useful if the `Original` class is in a third-party library and hence not easily modifiable directly. Some use cases for the Proxy pattern include:

1. **Implement lazy instantiation of the Original object.** In this case, the `Original` object is not actually instantiated until a method call is invoked. This can be useful if instantiating the `Original` object is a heavyweight operation that you wish to defer until absolutely necessary.
2. **Implement access control to the Original object.** For example, you may wish to insert a permissions layer between the `Proxy` and the `Original` objects to ensure that users can call only certain methods on the `Original` object if they have the appropriate permission.
3. **Support debug or dry run modes.** This lets you insert debugging statements into the `Proxy` methods to log all calls to the `Original` object. Or you can stop the forwarding to certain `Original` methods with a flag to let you call the `Proxy` in a dry run mode, such as to turn off writing the object's state to disk.
4. **Make the Original class be thread-safe.** This can be done by adding mutex locking to the relevant methods that are not thread-safe. Although this may not be the most efficient way to make the underlying class thread-safe, it's a useful stopgap if you cannot modify `Original`.
5. **Support resource sharing.** You could have multiple `Proxy` objects share the same underlying `Original` class. For example, this could be used to implement reference counting or copy-on-write semantics. This case is actually another design pattern, called the Flyweight pattern, in which multiple objects share the same underlying data to minimize memory footprint.
6. **Protect against future changes in the Original class.** In this case, you anticipate that a dependent library will change in the future, so you create a Proxy wrapper around that API that directly mimics the current behavior. When the library changes in the future, you can preserve the old interface via your Proxy object and simply change its underlying implementation to use the new library methods. At that point, you will no longer have a Proxy object, but an Adapter, which is a nice segue to our next pattern.

The Adapter pattern

The Adapter design pattern translates the interface for one class into a compatible but different interface (Fig. 3.4). This is therefore similar to the Proxy pattern in that it's a single-component wrapper. However, the interface for the Adapter class and the Original class may be different.

This pattern is useful to be able to expose a different interface for an existing API to allow it to work better with other code. As in the case for the Proxy pattern, the two interfaces in question could come from different libraries. For example, consider a

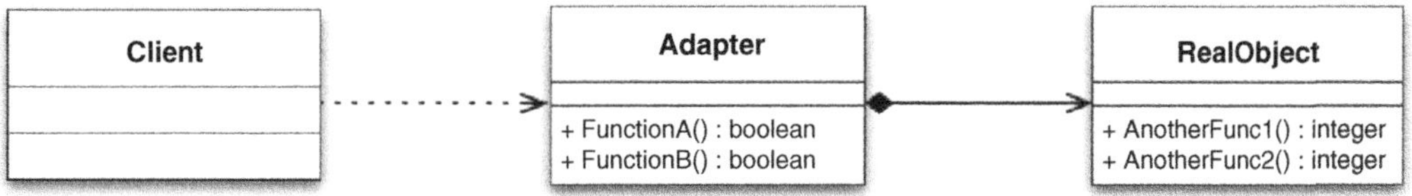

FIGURE 3.4 UML diagram of the Adapter design pattern.

geometry package that lets you define a series of primitive shapes. The parameters for certain methods may be in a different order from those that you use in your API, or they may be specified in a different coordinate system, or using a different convention such as (center, size) versus (bottom-left, top-right), or the method names may not follow your API's naming convention. You could therefore use an Adapter class to convert this interface into a compatible form for your API. For example,

```
class RectangleAdapter
{
public:
    RectangleAdapter() :
        mRect(new Rectangle)
    {}

    ~RectangleAdapter()
    {
        delete mRect;
    }

    void Set(float x1, float y1, float x2, float y2)
    {
        float w = x2 - x1;
        float h = y2 - y1;
        float cx = w / 2.0f + x1;
        float cy = h / 2.0f + y1;
        mRect->setDimensions(cx, cy, w, h);
    }

private:
    Rectange *mRect;
};
```

In this example, the `RectangleAdapter` uses a different method name and calling conventions to set the dimensions of the rectangle compared with the underlying `Rectangle` class. The functionality is the same in both cases. You're just exposing a different interface to allow you to work with the class more easily.

TIP: An Adapter translates one interface into a compatible but different interface.

It should be noted that Adapters can be implemented using composition (as in the previous example) or inheritance. These two flavors are often referred to as object adapters or class adapters, respectively. In the inheritance case, `RectangleAdapter` would derive from the `Rectangle` base class. This could be done using public inheritance if you wanted to also expose the interface of `Rectangle` in your adapter API, although more likely you would use private inheritance so that only your new interface is made public.

Some benefits of the Adapter pattern for API design include:

1. **Enforce consistency across your API.** As I discussed in the previous chapter, consistency is an important quality of good APIs. Using an Adapter pattern, you can collate multiple disparate classes, all of which have different interface styles, and provide a consistent interface to all of these. The result is that your API is more uniform and therefore easier to use.
2. **Wrap a dependent library of your API.** For example, your API may provide the ability to load a PNG image. You want to use the libpng library to implement this functionality, but you don't want to expose the libpng calls directly to the users of your API. This could be because you want to present a consistent and uniform API or because you want to protect against potential future API changes in libpng.
3. **Transform data types.** For example, consider that you have an API, `MapPlot`, that lets you plot geographic coordinates on a 2D map. `MapPlot` accepts only latitude and longitude pairs (using the WGS84 datum), specified as two double parameters. However, your API has a `GeoCoordinate` type that can represent coordinates in several coordinate systems, such as Universal Transverse Mercator or Lambert Conformal Conic. You could write an adapter that accepts your `GeoCoordinate` object as a parameter, converts this to geodetic coordinates (latitude, longitude) if necessary, and passes the two doubles to the `MapPlot` API.
4. **Provide interoperability between different languages.** For example, perhaps you've written an Objective-C API and you want to map the Objective-C classes and methods to C++ versions. Or perhaps you've written a plain C API and you want to create Adapter classes that wrap the C calls into C++ classes. It's open to debate whether this latter case can be strictly considered an Adapter pattern because design patterns are primarily concerned with object-oriented systems, but if you allow some flexibility in your interpretation of the term then you'll see that the concept is the same. The following code gives an example of a C++ adapter for a plain C API. (I'll discuss the differences between C and C++ APIs in more detail in the next chapter on Styles.)

```
class CppAdapter
{
public:
    CppAdapter()
    {
        mHandle = create_object();
    }

    ~CppAdapter()
    {
        destroy_object(mHandle);
    }

    void DoSomething(int value)
    {
        object_do_something(mHandle, value);
    }

private:
    CHandle *mHandle;
};
```

The Façade pattern

The Façade design pattern presents a simplified interface for a larger collection of classes. In effect, it defines a higher-level interface that makes the underlying subsystem easier to use (Fig. 3.5). To use Lakos' categorization, the Façade pattern is an example of a multicomponent wrapper (Lakos, 1996). Façade is therefore different from Adapter because Façade simplifies a class structure whereas Adapter maintains the same class structure.

As your API grows, so can the complexity of using that interface. The Façade pattern is a way to structure your API into subsystems to reduce this complexity and in turn make the API easier to use for most of your clients. A Façade might provide an improved interface while allowing access to the underlying subsystems. This is the same as the concept of convenience APIs that I described in the previous chapter, in which additional classes are added to provide aggregated functionality that make simple tasks easy. Alternatively, a Façade might completely decouple the underlying subsystems from the public interface so that these are no longer accessible. This is often called an encapsulating façade.

TIP: A Façade provides a simplified interface to a collection of other classes. In an encapsulating façade, the underlying classes are not accessible.

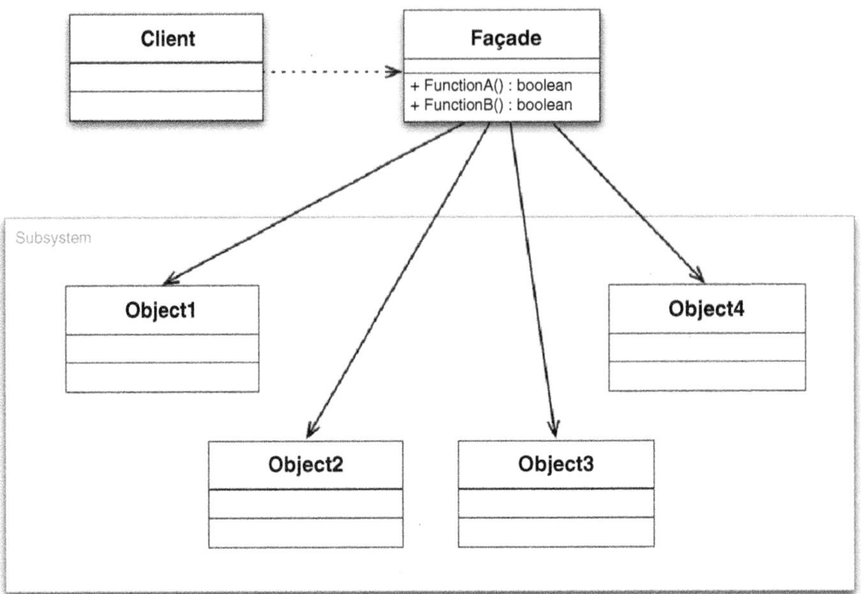

FIGURE 3.5 UML diagram of the Façade design pattern.

Let's look at an example to illustrate this pattern. Let's assume that you're on holiday and have checked into a hotel. You decide that you want to have dinner and then go to watch a show. To do so, you'll have to call a restaurant to make a dinner reservation, call the theater to book seats, and perhaps also arrange a taxi to pick you up from your hotel. You could express this in C++ as three separate objects with which you have to interact:

```
class Taxi
{
public:
    bool BookTaxi(int npeople, time_t pickup_time);
};

class Restaurant
{
public:
    bool ReserveTable(int npeople, time_t arrival_time);
};

class Theater
{
public:
    time_t GetShowTime();
    bool ReserveSeats(int npeople, int tier);
};
```

However, let's assume that you're staying in a high-end hotel and they have a helpful concierge who can assist you with all of this. In fact, the concierge will be able to find out the time of the show and then, using his local knowledge of the city, work out an appropriate time for your dinner and the best time to order your taxi. Translating this into terms of our C++ design, you now need only to interact with a single object with a far simpler interface:

```
class ConciergeFacade
{
public:
    time_t BookShow(int npeople, bool andRestaurant, bool andTaxi);
};
```

There are various useful applications of the Façade pattern in terms of API design:

1. **Hide legacy code.** Often you have to deal with old, decayed, legacy systems that are brittle to work with and no longer offer a coherent object model. In these cases, it can be easier to create a new set of well-designed APIs that sit on top of the old code. Then all new code can use these new APIs. Once all existing clients have been updated to the new APIs, the legacy code can be completely hidden behind the new façade (making it an encapsulating façade).
2. **Create convenience APIs.** As I discussed in the previous chapter, there is often a tension between providing general, flexible routines that provide more power versus simple easy-to-use routines that make the common use cases easy. A Façade is a way to address this tension by allowing both to coexist. In essence, a convenience API is a Façade. I used the example earlier of the OpenGL library, which provides low-level base routines, and the GLU library, which provides higher-level and easier-to-use routines that are built on top of the GL library.
3. **Support reduced- or alternate-functionality APIs.** By abstracting away the access to the underlying subsystems, it becomes possible to replace certain subsystems without affecting your client's code. This could be used to swap in stub subsystems to support demonstration or test versions of your API. It could also allow swapping in different functionality, such as using a different 3D rendering engine for a game or using a different image reading library. As a real-world example, the Second Life viewer can be built against the proprietary KDU JPEG-2000 decoder library. However, the open source version of the viewer is built against the slower OpenJPEG library.

Observer pattern

It's very common for objects to call methods in other objects. After all, achieving any nontrivial task normally requires several objects collaborating. However, to do this, an object A must know about the existence and interface of an object B to call methods on it. The simplest approach to doing this is for `A.cpp` to include `B.h` and then to call methods on that class directly. However, this introduces a compile-time dependency between A and B, forcing the classes to become tightly coupled. As a result, the generality of class A is reduced because it cannot be reused by another system without also pulling in class B. Furthermore, if class A also calls classes C and D, then changes to class A could affect all three of these tightly coupled classes. Additionally, this compile-time coupling means that users can't dynamically add new dependencies to the system at run time.

> *TIP: An Observer lets you decouple components and avoid cyclic dependencies.*

I will illustrate these problems, and how the Observer pattern helps, with reference to the popular Model–View–Controller (MVC) architecture.

Model–View–Controller

The MVC architectural pattern requires the isolation of business logic (the Model) from the user interface (the View), with the Controller receiving user input and coordinating the other two. MVC separation supports the modularization of an application's function and offers several benefits:

1. Segregation of Model and View components makes it possible to implement several user interfaces that reuse the common business logic core.
2. Duplication of low-level Model code is eliminated across multiple UI implementations.
3. Decoupling of Model and View code results in an improved ability to write unit tests for the core business logic code.
4. Modularity of components allows core logic developers and GUI developers to work simultaneously without affecting the other.

The MVC model was first described in 1987 by Steve Burbeck and Trygve Reenskaug at Xerox PARC and it remains a popular architectural pattern in applications and tool kits today. For example, modern UI tool kits such as Qt from The Qt Company, Apple's Cocoa, Java Swing, and Microsoft's Foundation Class library were all inspired by MVC. Taking the example of a single checkbox button, the current on/off state of the button is stored in the Model, the View draws the current state of the button on the screen, and the Controller updates the Model state and View display when the user clicks on the button.

> *TIP: The MVC architectural pattern promotes the separation of core business logic, or the Model, from the user interface, or View. It also isolates the Controller logic that effects changes in the Model and updates the View.*

The implication of MVC separation on code dependency means that View code can call Model code (to discover the latest state and update the UI), but the opposite is not true: Model code should have no compile-time knowledge of View code (because it ties the Model to a single View). Fig. 3.6 illustrates this dependency graph.

In a simple application, the Controller can effect changes to the Model based upon user input and communicate those changes to the View so that the UI can be updated. However, in real-world applications the View will normally also need to update to reflect additional changes to the underlying Model. This is necessary because changing one aspect of the Model may cause it to update other dependent Model state. This requires Model code to inform the View layer when state changes happen. However, as I've already stated, the Model code cannot statically bind and call the View code. This is where observers come in.

The Observer pattern is a specific instance of the Publish/Subscribe, or pub/sub, paradigm. These techniques define a one-to-many dependency between objects such that a publisher object can notify all subscribed objects of any state changes without directly depending upon them. The Observer pattern is therefore an important technique in terms of API design because it can help you to reduce coupling and increase code reuse.

Implementing the Observer pattern

The typical way to implement the Observer pattern is to introduce two concepts: the subject and the observer (also referred to as the publisher and subscriber). One or more observers register interest in the subject, and then the subject notifies all registered observers of any state changes. This is illustrated in Fig. 3.7.

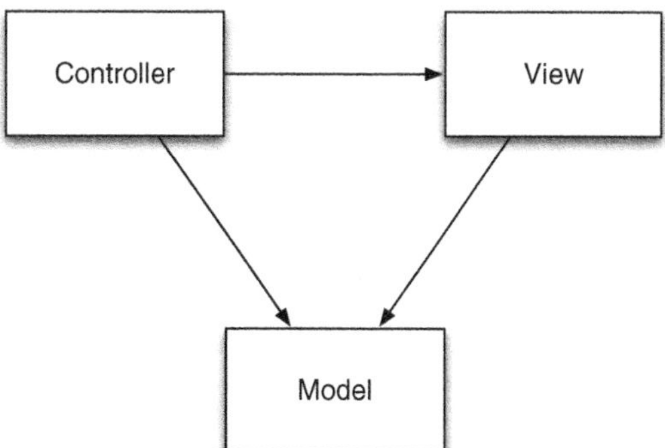

FIGURE 3.6 An overview of dependencies in the Model–View–Controller model. Both the Controller and the View depend upon the Model, but Model code has no dependency on Controller code or View code.

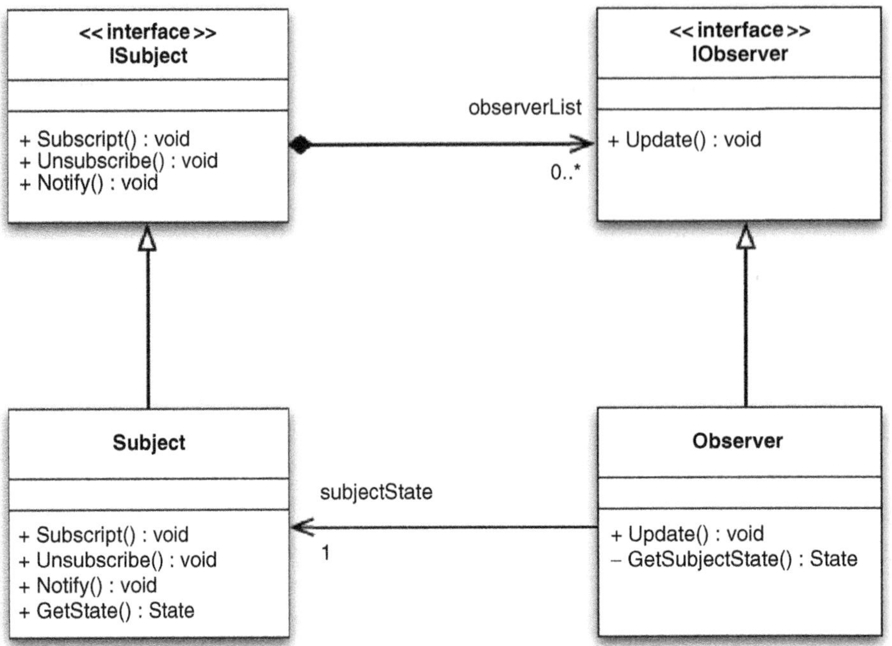

FIGURE 3.7 A UML representation of the Observer pattern.

This can be implemented using base classes to specify the abstract interface for both cases:

```
#include <map>
#include <vector>

enum class MessageType {
    Add,
    Remove
};

class IObserver
{
public:
    virtual ~IObserver() = default;
    virtual void Update(MessageType message) = 0;
};

class ISubject
{
public:
    ISubject();
    virtual ~ISubject();
    virtual void Subscribe(MessageType message, IObserver *observer);
    virtual void Unsubscribe(MessageType message, IObserver *observer);
    virtual void Notify(MessageType message);

private:
    using ObserverList = std::vector<IObserver *>;
    using ObserverMap = std::map<MessageType, ObserverList>;
    ObserverMap mObservers;
};
```

In this design, I've added support for the subject to be able to register and emit notifications for multiple different message types. This allows observers to subscribe only to the specific messages in which they're interested. For example, a subject that represents a stack of elements might wish to send out separate notifications when elements are added to or removed from that stack. Using these interfaces, you can define a minimal concrete subject class as:

```
#include "observer.h"

class MySubject : public ISubject
{
};
```

Finally, you can create observer objects simply by inheriting from the `IObserver` interface and implementing the `Update()` method. The following code demonstrates putting all these concepts together:

```
#include "subject.h"
#include <iostream>

class MyObserver : public IObserver
{
public:
    MyObserver(const std::string &str) :
        mName(str)
    {}

    void Update(MessageType message)
    {
        std::cout << mName << " received message\n";
    }

private:
    std::string mName;
};

int main(int, char **)
{
    MyObserver observer1("observer1");
    MyObserver observer2("observer2");
    MyObserver observer3("observer3");
    MySubject subject;

    subject.Subscribe(MessageType::Add, &observer1);
    subject.Subscribe(MessageType::Add, &observer2);
    subject.Subscribe(MessageType::Remove, &observer2);
    subject.Subscribe(MessageType::Remove, &observer3);

    subject.Notify(MessageType::Add);
    subject.Notify(MessageType::Remove);

    return 0;
}
```

This example demonstrates creating three separate observer classes and subscribes them for different combinations of the two messages defined by the earlier `MySubject` class. Finally, the calls to `subject.Notify()` cause the subject to traverse its list of observers that have been subscribed for the given message and calls the `Update()` method for each of them. The important point to note is that the `MySubject` class has no compile-time dependency on the `MyObserver` class. The relationship between the two classes is dynamically created at run time.

Of course, there may be a small performance cost for this flexibility: the cost of iterating through a list of observers before making the (virtual) function call. However, this cost is generally insignificant compared with benefits of reduced coupling and increased code reuse.

Push versus pull observers

There are many ways to implement the Observer pattern; the previous example is only one such method. However, I will note two major categories of observers: push-based and pull-based. This categorization determines whether all of the information is pushed to an observer via arguments to the `Update()` method, or whether the `Update()` method is simply used to send a notification about the occurrence of an event, and if the observer wishes to discover more details, then the subject object must be queried directly. As an example, a notification that the user has pressed the Return key in a text entry widget may pass the actual text that the user typed as a parameter of the `Update()` method (push), or it may rely on the observer calling a `GetText()` method on the subject to discover this information if it needs it (pull).

Fig. 3.7 illustrates a pull observer pattern because the `Update()` method has no arguments and the observer can query the subject for its current state. This approach allows you to use the same simple `IObserver` for all observers in the system. By comparison, a push-based solution would require you to define different abstract interfaces for each `Update()` method that requires a unique signature. A push-based solution is useful for sending small commonly used pieces of data along with a notification, such as the checkbox on/off state for a checkbox state change. However, it may be inefficient for larger pieces of data, such as sending the entire text every time a user presses a key in a text box widget.

I'll finish up the coverage of observers by providing a real-world example, referring again to the open source VTK. This library uses an observer pattern to allow client code to be called when various events happen in the system. The API for managing VTK observers looks like:

```
unsigned long AddObserver(unsigned long, vtkCommand *, float priority = 0.0f);
vtkTypeBool HasObserver(unsigned long);
vtkTypeBool HasObserver(unsigned long, vtkCommand *);
void RemoveObserver(vtkCommand *);
void RemoveObservers(unsigned long);
void RemoveObservers(unsigned long, vtkCommand *);
void RemoveAllObservers();
```

You can see from this set of functions that VTK observers are keyed on an `unsigned long`. However, the API includes overloaded versions of these functions that accept a `const char *` as a convenience. There's a `GetEventIdFromString()` function that's used internally to perform the mapping from a string to an `unsigned long` to support this.

The events themselves are defined in a large enum with values such as `vtkCommand::DeleteEvent`, `vtkCommand::RenderEvent`, and `vtkCommand::EnterEvent`. The observer object is modeled by a `vtkCommand` object, which looks like:

```
class vtkCommand : public vtkObjectBase
{
public:
    void UnRegister();
    virtual void Execute(vtkObject *caller, unsigned long eventid,
                         void *callData) = 0;

    static const char *GetStringFromEventId(unsigned long event);
    static unsigned long GetEventIdFromString(const char *event);
    ...
};
```

Observers are therefore expected to override the `Execute()` function to perform their custom work, and this function receives the object invoking the event, the event identifier, and any optional call data provided by the invoking object.

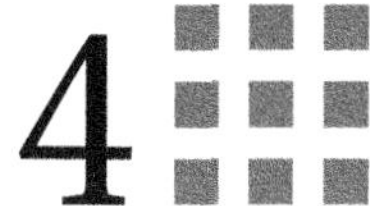

Design

In the preceding chapters, I laid the groundwork and developed the background to let you start designing your own application programming interfaces (APIs). I presented the various qualities that contribute to good API design and discussed standard patterns that apply to the design of maintainable APIs.

In this chapter, I put all of this information together and cover the specifics of high-quality API design, from overall architecture planning down to class design and individual function calls. However, good design is worth little if the API doesn't give your users the features they need. I will therefore also talk about defining functional requirements to specify what an API should do. I'll also cover the creation of use cases and user stories as a way to describe the behavior of the API from the user's point of view. These different analysis techniques can be used individually or together, but they should always precede any attempt to design the API: you can't design what you don't understand.

Fig. 4.1 shows the basic workflow for designing an API. This starts with an analysis of the problem, from which you can develop an initial design and then implement a solution for that design. This is a continual and iterative process: new requirements should trigger a reassessment of the design, as should changes from other sources such as major bug fixes. This chapter will focus on the first two stages of this process: analysis and design. The following chapters will deal with the remaining implementation issues such as C++ usage, documentation, and testing.

Before I jump into these design topics, however, I will spend some time looking at why good design is so important. This opening section is drawn from experience working on large code bases that have persisted for many years and have had dozens or hundreds of engineers working on them. The lessons learned from witnessing code bases evolve, or devolve, over many years offer compelling motivation to design them well from the start, and—just as important—to maintain high standards from then on. The consequences of not doing so can be very costly. To mix metaphors: good API design is a journey, not a first step.

A case for good design

This chapter focuses on the techniques that result in elegant API design. However, it's likely that you've worked on projects with code that doesn't live up to these grand ideals. You've probably worked with legacy systems that have weak cohesion, expose internal details, have no tests, are poorly documented, and exhibit nonorthogonal behavior. Despite this, some of these systems were probably well-designed when they were

API Design for C++. https://doi.org/10.1016/B978-0-443-22219-1.00008-8

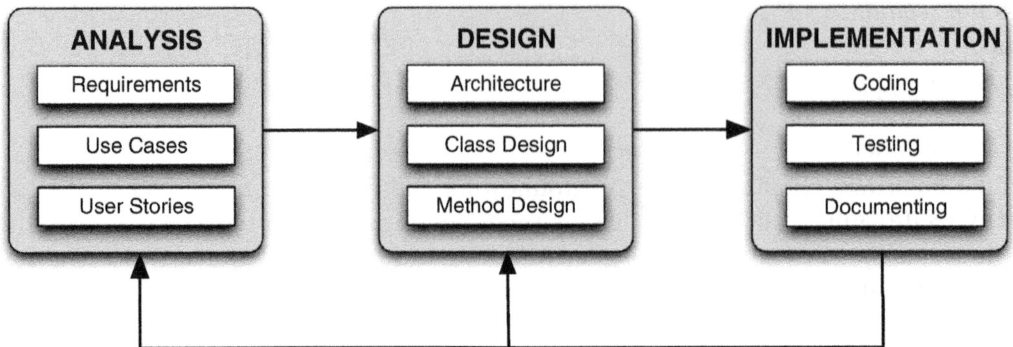

FIGURE 4.1 The stages of application programming interface development, from analysis to design and implementation.

originally conceived. However, over time the software decayed, becoming difficult to extend and requiring constant maintenance.

Accruing technical debt

All large successful companies grew from meager beginnings. A classic example of this is Hewlett Packard, which began with two electrical engineers in a Palo Alto garage in 1939 and eventually grew to become the first technology company in the world to post revenues exceeding $100 billion. The qualities that make a successful startup company are very different from those needed to maintain a multibillion (or trillion) dollar publicly traded corporation. Companies often go through several large organizational changes as they grow, and the same is true of their software practices.

A small software startup needs to get its product out as soon as possible to avoid being beaten to market by a competitor or running out of capital. The pressure on a software engineer in this environment is to produce a lot of software, and quickly. Under these conditions, the extra effort required to design and implement long-term APIs can be seen as an unaffordable luxury. This is a fair decision when the options are between getting to market quickly or your company perishing. I once spoke with a software architect for a small startup who forbade the writing of any comments, documentation, or tests because he felt that it would slow the development too much.

However, if a piece of software becomes successful, the pressure turns toward providing a stable, easy to use, and well-documented product. New requirements start appearing that necessitate the software to be extended in ways it was never meant to support. And all of this must be built upon the core of a product that was not designed to last for the long term. The result, in the words of Ward Cunningham, is the accrual of technical debt (Cunningham, 1992):

> *Shipping first time code is like going into debt. A little debt speeds development so long as it is paid back promptly with a rewrite. [...] The danger occurs when the debt*

is not repaid. Every minute spent on not-quite-right code counts as interest on that debt. Entire engineering organizations can be brought to a stand-still under the debt load of an unconsolidated implementation.

Steve McConnell expanded on this definition to note that there are two types of debt: unintentional and intentional. The former is when software designed with the best of intentions turns out to be error-prone, when a junior engineer writes low-quality code, or when your company buys another company whose software turns out to be a mess. The latter is when a conscious strategic decision is made to cut corners owing to time, cost, or resource constraints, with the intention that the right solution will be put in place after the deadline.

The problem, of course, is that there is always another important deadline, so it's perceived that there's never enough time to go back and perform the right fix. As a result, the technical debt gradually accrues: short-term glue code between systems lives on and becomes more deeply embedded, temporary hacks remain in the code and turn into features upon which clients depend, code that was meant only for last-minute demos is shipped as production code, coding conventions and documentation are ignored, and ultimately the original clean design degrades and becomes obfuscated. Robert C. Martin defined four warning signs that a code base is reaching this point (Martin, 2000). Here is a slightly modified version of those indicators:

- **Fragility.** Software becomes fragile when it has unexpected side effects, or when implementation details are exposed to the point that apparently unconnected parts of the system depend upon the internals of other parts of the system. The result is that changes to one part of the system can cause unexpected failures in seemingly unrelated parts of the code. Engineers are therefore afraid to touch the code and it becomes a burden to maintain.
- **Rigidity.** A rigid piece of software is one that is resistant to change. In effect, the design becomes brittle to the point that even simple changes cannot be implemented without great effort, normally requiring extensive, time-consuming, and risky refactoring. The result is a viscous code base for which efforts to make new changes are slowed significantly.
- **Immobility.** A good engineer will spot cases in which code can be reused to improve the maintainability and stability of the software. Immobile code is software that is immune to these efforts, making it difficult to be reused elsewhere. For example, the implementation may be too entangled with its surrounding code or it may be hardcoded with domain-specific knowledge.
- **Nontransferability.** If only a single engineer in your organization can work on certain parts of the code, then it can be described as nontransferable. Often the owner will be the developer who originally wrote the code, or the last unfortunate person who attempted to clean it up. For many large code bases, it's not possible for every engineer to understand every part of the code deeply, so having areas

that engineers cannot easily dive into and work with effectively is a bad situation for your project.

The result of these problems is that dependencies between components grow, causing conceptually unrelated parts of the code to rely upon each other's internal implementation details. Over time, this culminates in most program state and logic becoming global or duplicated (Fig. 4.2). This is often called spaghetti code or the big ball of mud (Foote and Yoder, 1997).

Paying back the debt

Ultimately, a company will reach the point where it has accrued so much technical debt that it spends more time maintaining and containing the legacy code base than adding new features for its customers. This often results in a next-generation project, to fix the problems with the old system. For example, when I met the software architect mentioned a few paragraphs back, the company had since grown and become successful, and his team was busily rewriting all of the code that he'd originally designed.

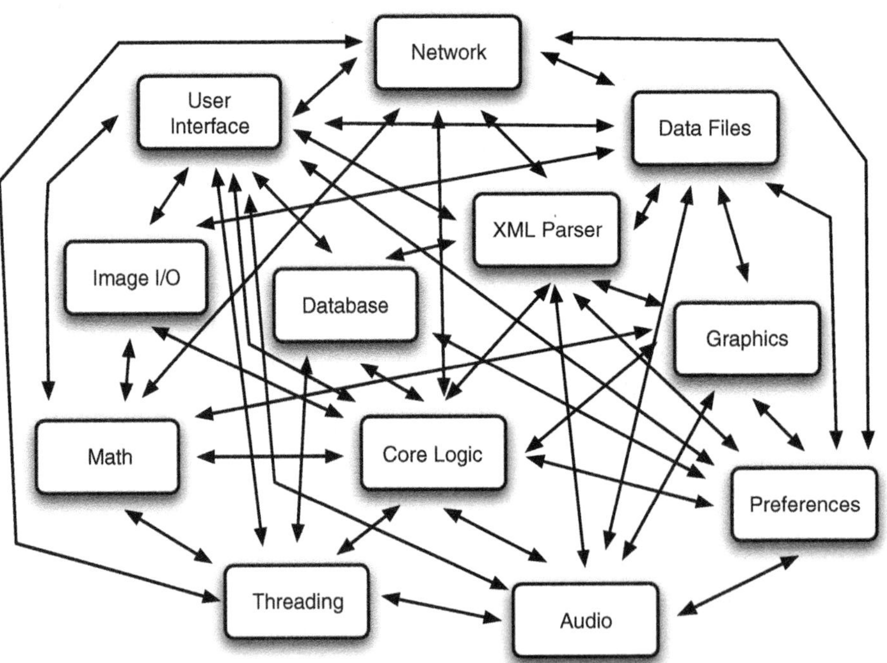

FIGURE 4.2 A decayed tightly coupled system that has devolved into a big ball of mud.

In terms of strategy, there are two extremes for such a next-generation project to consider:

1. **Evolution**: Design a system that incorporates all new requirements and then iteratively refactor the existing system until it reaches that point.
2. **Revolution**: Throw away the old code and then design and implement a completely new system from scratch.

Both options have their pros and cons, and both are difficult. I've heard the problem described as having to change the engine in a car, but the car is traveling at 100 mph and you can't stop the car.

In the evolution case, you always have a functional system that can be released. However, you still have to work within the framework of the legacy design, which may no longer be the best way to express the problem. New requirements may have fundamentally changed the optimal workflow for key use cases. A good way to go about the evolution approach is to hide the old ugly code behind new well-designed APIs (e.g., by using the wrapper patterns I presented in Chapter 3, such as Façade), incrementally update all clients to go through these cleaner APIs, and put the code under automated testing (Fowler et al., 1999; Feathers, 2004).

In the revolution case, you're freed from the shackles of old technology and can design a new tool with all of the knowledge you've learned from the old one (although as a pragmatic step, you may still harvest a few key classes from the old system to preserve critical behavior). You can also put new processes in place, such as requiring extensive unit test coverage for all new code, and use the opportunity to switch tools, such as adopting a new bug tracking system or source control management system. However, this option requires a lot more time and effort (i.e., money) to get to the point of a usable system, and in the meantime you either stop all development on the old tool or continue delivering new features in the old system, which keeps raising the bar for the new system to be successful. You must also be mindful of the second-system syndrome, in which the new system fails because it's overengineered and overly ambitious in its goal (Brooks, 1995).

In both cases, the need for a next-generation project introduces team dynamic issues and planning complexities. For example, do you keep a single team focused on both the new and old systems? This is desirable from a personnel point of view. However, short-term tactical needs tend to trump long-term strategic development, so it may be hard to sustain the next-generation project in the face of critical bug fixes and maintenance for the old one. Alternatively, if you split the development team in two, this can create a morale problem in which the developers working on the old system feel that they've been classed as second-rate developers and left behind to support a code base with no future.

Furthermore, the need for a technical restart can often instigate a business and company reorganization as well. This causes team structures and relationships to be reassessed and reshaped. It can also materially affect people's livelihoods, particularly when companies decide to downsize as part of refocusing the business. And all of this happens because of poor API design? Well, perhaps that's being a little dramatic. Reorganizations are a natural process in the growth of a company and can happen for many reasons: a structure that works for a startup with 10 people doesn't work for a successful business of 10,000. But the failure of software to react to the needs of the business is certainly one way that reorganizations can be triggered. For instance, in Jun. 2010, I was working at Linden Lab when it laid off 30% of its workforce and underwent a company-wide reorganization, primarily because the software couldn't be evolved fast enough to meet the company's revenue targets.

Design for the long term

Investing in a large next-generation effort to replace a decayed code base can cost a company millions of dollars. For example, just to pay the salary for a team of 20 developers, testers, writers, and managers for 1 year at an average salary of $100,000 would cost $2,000,000. However, the adoption of good design principles can help to avoid this drastic course of action. Let's start by enumerating some of the reasons why this scenario can arise:

1. A company simply doesn't create a good software design in the first place because of a belief that it will cost valuable time and money early on.
2. The engineers on the project are ignorant of good design techniques or believe that they don't apply to their project.
3. The code was never intended to last very long. For example, it was hastily written for a demo or it was meant to be throw-away prototype code.
4. The development process for the software project doesn't make technical debt visible, so knowledge of all of the parts of the system that need to be fixed gets lost or forgotten over time. (Agile processes such as Scrum attempt to keep debt visible by using a product backlog).
5. The system was well-designed at first, but its design gradually degraded over time owing to unregulated growth: for example, letting poor changes be added to the code even if they compromised the design of the system. In the words of Fred Brooks, the system loses its conceptual integrity (Brooks, 1995).
6. Changing requirements often necessitate the design to evolve, too, but the company continually postpones this refactoring work, either intentionally or unintentionally, in preference to short-term fixes, hacks, and glue code.
7. Bugs are allowed to exist for long periods of time. This is often caused by a drive to add new functionality continually without a focus on the overall quality of the product.
8. The code has no tests, so regressions creep into the system as engineers modify functionality, and parts of the code base ultimately turn into scary wastelands where engineers fear to make changes.

Let's tackle a few of these problems. First is the perception that good design slows you down too much. Truthfully, it may be the cheapest overall decision to write haphazardly structured software that gets you to market quicker and then to rewrite the code completely once you have a hold on the market. Also, certain aspects of writing good software can indeed appear to be more time-consuming, such as writing the extra code to pimpl your classes or writing automated tests to verify the behavior of your APIs. However, good design doesn't take as long as you might think, and it always pays off in the long run. Keeping a strong separation between interface and implementation pays dividends in the maintainability of your code, even in the short term, and writing automated tests gives you the confidence to change functionality rapidly without breaking existing behavior. It's noteworthy that Michael Feathers defines legacy code as code without tests, making the point that legacy doesn't have to mean old; you could be writing legacy code today (Feathers, 2004).

The beauty of APIs is that the underlying implementation can be as quick and dirty or as complete and elegant as you need. Good API design is about putting in place a stable logical interface to solve a problem. However, the code behind that API can be simple and inefficient at first. Then you can add more implementation complexity later without breaking that logical design. Related to this, APIs let you isolate problems to specific components. By managing the dependencies between components, you can limit the extent of problems. Conversely, in spaghetti code, in which each component depends upon the internals of other components, behavior becomes nonorthogonal, and bugs in one component can affect other components in nonobvious ways. The important message is therefore to take the time to put a good high-level design in place first, to focus on the dependencies and relationships between components. That is the primary focus of this chapter.

Another aspect of the problem is that if you don't continue to keep a high bar for your code quality, then as the code evolves the original design gradually decays. Cutting corners to meet a deadline is okay if you go back and do it right afterward. Remember to keep paying back your technical debt. Code tends to live for a lot longer than you think it will. It's good to remember this fact when you weaken an API, because you may have to support the consequences for a long time to come. It's therefore important to realize the impact of new requirements on the design of the API and to refactor your code to maintain a consistent and up-to-date design. It's equally important to enforce change control on your API so that it doesn't evolve in an unsupervised or chaotic fashion. I'll discuss ways to achieve these goals in Chapter 10, when I talk about API versioning.

Gathering functional requirements

The first step in producing a good design for a piece of software is to understand what it needs to do. It's amazing how much development time is wasted by engineers building the wrong thing. It's also quite eye-opening to see how often two engineers can hear the same informal description of a piece of work and come away with two completely different ideas about what it involves. This is not necessarily a bad thing: it's good to have minds that work differently to provide alternative perspectives and solutions. The problem is that the work was not specified in enough detail such that everyone involved could form a shared understanding and work toward the same goal. This is where requirements come in. There are several different types of requirements in the software industry, including:

- **Business requirements**: describe the value of the software in business terms (i.e., how it advances the needs of the organization)
- **Functional requirements**: describe the behavior of the software (i.e., what the software is supposed to accomplish)
- **Nonfunctional requirements**: describe the quality standards that the software must achieve (i.e., how well the software works for users)

I will concentrate primarily on functional and nonfunctional requirements in the following sections. However, it's still extremely important to ensure that the functionality of your software aligns with the strategic goals of your business. Otherwise, you run the risk of damaging the long-term success of your API.

What are functional requirements?

Functional requirements are simply a way to understand what to build so that you don't waste time and money building the wrong thing. In our diagram of the phases of software development (Fig. 4.1), functional requirements sit squarely in the analysis phase.

In terms of API development, functional requirements define the intended functionality for the API. These should be developed in collaboration with the clients of the API so that they represent the voice and needs of the user (Wiegers, 2003). Explicitly capturing requirements also lets you agree upon the scope of functionality with the intended users. Of course, the users of an API are also developers, but that doesn't mean that you should assume you know what they want just because you're a developer, too. At times, it may be necessary to second-guess or research requirements yourself. Nevertheless, you should still identify target users of your API—experts in the domain of your API—and drive the functional requirements from their input. For example, you can hold interviews, meetings, or use questionnaires to ask users:

- What tasks would they expect to achieve with the API?
- What would an optimal workflow be from their perspective?
- What are all the potential inputs, including their types and valid ranges?
- What are all the expected outputs, including type, format, and ranges?
- What file formats or protocols must be supported?
- What (if any) mental models do they have for the problem domain?
- What domain terminology do they use?

If you're revising or refactoring an existing API, you can also ask developers to comment on the code that they currently must write to use the API. This can help to identify cumbersome workflows and unused parts of an API. You can also ask them how they would prefer to use the API in an ideal world (Stylos et al., 2008).

TIP: Functional requirements are used to specify how your API should behave.

Functional requirements can also be supported by nonfunctional requirements. These are requirements that define the operational constraints of an API rather than how it behaves. These qualities can be just as critical to a user as the functionality provided by the API. Examples of nonfunctional requirements include aspects such as:

- **Performance.** Are there constraints on the speed of certain operations?
- **Platform compatibility.** On which platforms must the code run?
- **Security.** Are there data security, access, or privacy concerns?
- **Scalability.** Can the system handle real-world data inputs?
- **Flexibility.** Will the system need to be extended after release?
- **Usability.** Can the user easily understand, learn, and use the API?
- **Concurrency.** Does the system need to use multiple processors?
- **Cost.** How much will the software cost?

Example functional requirements

Functional requirements are normally managed in a requirements document in which each requirement is given a unique identifier and a description. A rationale for the requirement may also be provided to explain why it's necessary. It's typical to present the requirements as a concise list of bullet points and to organize the document into different themed sections so that requirements relating to the same part of the system can be colocated.

Good functional requirements should be simple, easy to read, unambiguous, verifiable, and free of jargon. It's important that they don't overly specify the technical implementation: functional requirements should document what an API should do and not how it does it. Very often, a client will tell you how a feature should be implemented. It's important instead to get them to tell you what the underlying problem is that they want to solve. There may be better ways to solve the problem than the one solution they're proposing.

To illustrate these points, here's an example list of functional requirements for a user interacting with an automated teller machine (ATM):

REQ 1.1. The system shall prevent further interaction if it's out of cash or is unable to communicate with the financial institution.

REQ 1.2. The system shall validate that the inserted card is valid for financial transactions on this ATM.
REQ 1.3. The system shall validate that the PIN number entered by the user is correct.
REQ 1.4. The system shall dispense the requested amount of money, if it is available, and debit the user's account by the same amount.
REQ 1.5. The system shall notify the user if the transaction cannot be completed. In that case, no money shall be taken from the user's account.

Maintaining the requirements

There's no such thing as stable requirements; you should always expect them to change over time. This happens for a variety of reasons, the most common of which is that users (and you) will have a clearer idea of how the system should function as you start building it. You should therefore make sure that you version and date your requirements, so that you can refer to a specific version of the document and know how old it is.

On average, 25% of a project's functional requirements will change during development, accounting for 70%–85% of the code that needs to be reworked (McConnell, 2004). While it's good to stay in sync with the evolving needs of your clients, you should also make sure that everyone understands the cost of changing requirements. Adding new requirements will cause the project to take longer to deliver. It may also require significant changes to the design, causing a lot of code to be rewritten.

In particular, you should be wary of falling into the trap of requirements creep. Any major changes to the functional requirements should trigger a revision of the schedule and costing for the project. In general, any new additions to the requirements should be evaluated against the incremental business value that they deliver. Assessing a new requirement from this pragmatic viewpoint should help to weigh the benefit of the change against the cost of implementing it. Another useful technique when being asked to add a large new requirement without a change to the delivery schedule is to ask which of the existing requirements should be removed so that the new one can be added. This can often help to focus the discussion on the relative priority of the new requirement versus the existing ones.

Creating use cases

A use case describes the behavior of an API based upon the interactions of a user or another piece of software (Jacobson, 1992). Use cases are essentially a form of functional requirement that specifically captures who does what with an API and for what purpose, rather than simply providing a list of features, behaviors, or implementation notes. Focusing on use cases helps you to design an API from the perspective of the client.

It is common to produce both a functional requirement document as well as a set of use cases. For example, use cases can be used to describe an API from the user's point of view, whereas functional requirements can be used to describe a list of features or the details of an algorithm. However, concentrating on just one of these techniques can often be sufficient, too. In that case, I recommend creating use cases because these resonate most closely with the way a user wants to interact with a system. When using both methods, you can either derive functional requirements from the use cases or vice versa. However, it's more typical to work with your users to produce use cases first and then to derive a list of functional requirements from these use cases.

TIP: Use cases describe the requirements for your API from the perspective of the user.

Ken Arnold uses the analogy of driving a car to illustrate the importance of designing an interface based upon its use rather than its implementation details. He notes that you are more likely to come up with a good experience for drivers by asking the question, "How does the user control the car?" instead of "How can the user adjust the rate of fuel pumped into each of the pistons?" (Arnold, 2005).

Developing use cases

Every use case describes a goal that an actor is trying to achieve. An actor is an entity external to the system that initiates interactions, such as a human user, a device, or another piece of software. Each actor may have different roles when interacting with the system. For example, a single actor for a database may take on the role of administrator, developer, or database user. A good way to approach the process of creating use cases is therefore to:

1. Identify all actors of the system and the roles that each have
2. Identify all goals that each role needs to accomplish
3. Create use cases for each goal.

Each use case should be written in plain language using the vocabulary of the problem domain. It should be named to describe the outcome of value to the actor. Each step of the use case should start with the role followed by an active verb. For example, continuing our ATM example, the following steps describe how to validate a user's PIN number:

Step 1. User inserts ATM card.
Step 2. System validates ATM card is valid for use with the ATM machine.
Step 3. System prompts the user to enter PIN number.
Step 4. User enters PIN number.
Step 5. System checks that the PIN number is correct.

Use case templates

A good use case represents a goal-oriented narrative description of a single unit of behavior. It includes a distinct sequence of steps that describes the workflow to achieve the goal of the use case. It can also provide clear preconditions and postconditions to specify the state of the system before and after the use case (i.e., to state the dependencies between use cases explicitly, as well as the trigger event that causes a use case to be initiated).

Use cases can be recorded with different degrees of formality and verbosity. For example, they can be as simple as a few sentences or they can be as formal as structured, cross-referenced specifications that conform to a particular template. They can even be described visually, such as with the Universal Modeling Language (UML) Use Case Diagram (Cockburn, 2000).

TIP: Use cases can be simple lists of short goal-oriented descriptions or they can be more formal structured specifications that follow a prescribed template.

In the more formal instance, there are many different template formats and styles for representing use cases textually. These templates tend to be very project-specific, and they can be as short or extensive as appropriate for that project. Don't get hung up on the details of your template. It's more important to communicate the requirements clearly than to conform to a rigid notation (Alexander, 2003). Nonetheless, a few common elements of a use case template include:

Name: *A unique identifier for the use case, often in verb–noun format such as Withdraw Cash or Buy Stamps.*
Version: *A number to differentiate different versions of the use case.*
Description: *A brief overview that summarizes the use case in one or two sentences.*
Goal: *A description of what the user wants to accomplish.*
Actors: *The actor roles that want to achieve the goal.*
Stakeholder: *The individual or organization that has a vested interest in the outcome of the use case* (e.g., *an ATM User or the bank).*
Basic Course: *A sequence of steps that describe the typical course of events. This should avoid conditional logic when possible.*
Extensions: *A list of conditions that cause alternative steps to be taken. This describes what to do if the goal fails* (e.g., *an invalid PIN number was entered).*
Precondition: *A list of conditions required for the trigger to execute successfully.*
Trigger: *The event that causes the use case to be initiated.*
Postcondition: *Describes the state of the system after the successful execution of the use case.*
Notes: *Additional information that doesn't fit well into any other category.*

Writing good use cases

Writing use cases should be an intuitive process. They are written in plain easy-to-read language to capture the user's perspective on how the API should be used. However, even supposedly intuitive tasks can benefit from general guidelines and words of advice.

- **Use domain terminology.** Use cases should be described in terms that are natural to the clients of an API. The terms that are used should be familiar to users and should come from the domain being targeted. In effect, users should be able to read use cases and easily understand the scenarios without their appearing too contrived.
- **Don't overspecify use cases.** Use cases should describe the black-box functionality of a system (i.e., you should avoid specifying implementation details). You should also avoid including too much detail in your use cases. Alistair Cockburn uses the example of inserting coins into a candy machine. Rather than trying to specify different combinations of inserting the correct quantity, such as "person inserts three quarters, or 15 nickels, or a quarter followed by 10 nickels," you just need to write "person inserts money."
- **Use cases don't define all requirements.** Use cases don't encompass all possible forms of requirements gathering. For example, they don't represent system design, lists of features, algorithm specifics, or any other parts of the system that are not user-oriented. Use cases concentrate on behavioral requirements for how the user should interact with the API. You may still wish to compile functional and nonfunctional requirements in addition to use cases.
- **Use cases don't define a design.** Although you can often create a high-level preliminary design from your use cases, you should not fall into the trap of believing that use cases directly define the best design. That they don't define all requirements is one reason. For example, they don't define performance, security, or network aspects of the API, which can greatly affect the most appropriate design. Also, use cases are written from the perspective of users. You may therefore need to reinterpret their feedback in light of conflicting or imprecise goals rather than treating them too literally (Meyer, 1997).
- **Don't specify design in use cases.** It's generally accepted that you should avoid describing user interfaces (UI) in use cases, because UI is a design not a requirement and because UI designs are more changeable (Cockburn, 2000). Although this axiom is not directly applicable to UI-less API design, it can be extrapolated to our circumstances by stating that you should keep API design specifics out of your use cases. Users may try to propose a particular solution for you to implement, but better solutions to the problem may exist. API design should therefore follow from your use case analysis. In other words, use cases define how a user wants to achieve a goal regardless of the actual design.

- **Use cases can direct testing.** Use cases are not test plans in themselves because they don't specify specific input and output values. However, they do specify the key workflows that your users expect to be able to achieve. As such, they are a great source to direct automated testing efforts for your API. Writing a suite of tests that verify these key workflows will give you the confidence that you've reached the needs of your users and that you don't break this functionality as you evolve the API in the future.
- **Expect to iterate.** Don't be too concerned about getting all of your use cases perfect the first time. Use case analysis is a process of discovery; it helps you learn more about the system you want to build. You should therefore look upon it as an iterative process in which you can refine existing use cases as you expand your knowledge of the entire system (Alexander, 2003). On the other hand, it's well-known that errors in requirements can significantly affect a project, causing major redesign and reimplementation efforts. This is why the first piece of advice I gave was to avoid making your use cases too detailed.
- **Don't insist on complete coverage.** For the same reasons that use cases do not encompass all forms of requirements, you should not expect your use cases to express all aspects of your API. However, you also don't need them to cover everything. Some parts of the system may already be well-understood or do not need a user-directed perspective. There's also the logistical concern that you will not have unlimited time and resources to compile exhaustive use cases, so you should focus the effort on the most important user-oriented goals and workflows (Alexander, 2003).

Putting all this information together, I will complete our ATM example by presenting a sample use case for entering a PIN number and use our template from earlier to format the use case:

Name: Enter PIN
Version: 1.0.
Description: User enters PIN number to validate her bank account information.
Goal: System validates User's PIN number.
Stakeholders.

1. User wants to use ATM services.
2. Bank wants to validate the User's account.

Basic Course.

1. System validates that ATM card is valid for use with the ATM machine.
2. System prompts the user to enter PIN number.
3. User enters PIN number.
4. System checks that the PIN number is correct.

—cont'd

Extensions.

a. System failure to recognize ATM card:
  a-1. System displays error message and aborts operation.
b. User enters invalid PIN:
  b-1. System displays error message and lets User retry.

Trigger: User inserts card into ATM.
Postcondition: User's PIN number is validated for financial transactions.

Requirements and agile development

Agile development is a general term for software development methods that align with the principles of the Agile Manifesto. Examples include extreme programming (XP), Scrum, and the Dynamic System Development Method. The Agile Manifesto (http://agilemanifesto.org/) was written in Feb. 2001 by 17 contributors who wanted to find more lightweight and nimble alternatives to the traditional development processes of the time. It states that the following qualities should be valued when developing software:

- **Individuals and interactions** over processes and tools
- **Working software** over comprehensive documentation
- **Customer collaboration** over contract negotiation
- **Responding to change** over following a plan

Agile methodologies therefore deemphasize document-centric processes, instead preferring to iterate on working code. However, this does not mean that they have no requirements. What it means is that the requirements are lightweight and easily changed. Maintaining a large, wordy, formal requirements document would not be considered agile. However, the general concept of use cases is very much a part of agile processes such as Scrum and XP, which emphasize the creation of user stories.

A user story is a high-level requirement that contains just enough information for a developer to provide a reasonable estimate of the effort required to implement it. It's conceptually similar to a use case, except that the goal is to keep them very short, normally just a single sentence. So the brief informal use case is more similar to a user story than the formal template-driven or UML use case. Another important distinction is that user stories are not all completed up-front. Many user stories will be added incrementally as the working code evolves. That is, you start writing code for your design early to avoid throwing away specifications that become invalid after you try to implement them.

> *TIP: User stories are a way to capture minimal requirements from users within an agile development process.*

Another important aspect of user stories is that they are written by project stakeholders, not developers: that is, the customers, vendors, business owners, or support personnel who are interested in the product being developed. Keeping the user story short allows stakeholders to write them in a few minutes. Mike Cohn suggests using a simple format to describe user stories (Cohn, 2004):

As a [*role*] I want [*something*] so that [*benefit*].

For instance, referring back to our ATM example, here's an example of five different user stories for interacting with a cash machine:

- As a customer I want to withdraw cash so that I can buy things.
- As a customer I want to transfer money from my savings account to my checking account so I can write checks.
- As a customer I want to deposit money into my account so I can increase my account balance.
- As a bank business owner I want the customer's identity to be verified securely so that the ATM can protect against fraudulent activities.
- As an ATM operator I want to restock the ATM with money so the ATM will have cash for customers to withdraw.

Given a set of well-written user stories, engineers can estimate the scale of the development effort involved, usually in terms of an abstract quantity like story points, and work on implementing these stories. Stakeholders will also often provide an indication of the priority of a user story, to help prioritize the order of work from the backlog of all stories. Stakeholders then assess the state of the software at regular intervals, such as during a sprint review, and can provide further user stories to focus the next iteration of development. In other words, this implies active user involvement and favors an iterative development style over the creation of large up-front requirements documents.

Cohn also presents an easy-to-remember acronym to help you create good user stories: INVEST, in which each letter stands for a quality of a well-written user story (Cohn, 2004):

Independent
Negotiable

Valuable
Estimable
Small
Testable

All of the advice that I offered earlier for writing good use cases applies equally well to user stories. For example, agile processes such as SCRUM and XP do not tell you how to design your API. So you must not forget that once you've built up your backlog of user stories, you still have to go through a separate design process to work out how best to implement those stories.

As I reflect on the process for producing 3D feature films at animation studios such as Pixar, I think there are some good analogies to be drawn. Animated movies are developed using a production pipeline where, generally speaking, the output of one department feeds into the next department. The first year or two of a production is driven by the story department, which defines the main characters and overall story arc for the movie. It can also produce physical assets such as storyboards and character studies. Only once these efforts are far enough along does any digital production work happen. I believe that a large part of the success of Pixar movies is due to this up-front emphasis on defining and refining the story elements before the expensive parts of a production are ramped up.

The analogy to software development is that a successful project should have a strong product team that defines the users and features of that product up-front. This might include developing assets such as persona modeling, requirements gathering, and use case definition. Only then should any engineers start writing production code. This is not to say that no code ever gets written while the product is being defined. Obviously, there can be prototypes and investigations to address known technical challenges or large risk areas. But there is little point in writing code when you don't even know who your users are and what problems they need to solve.

This production pipeline analogy may sound like I'm advocating for a traditional waterfall software development method, but I'm not. The process used for developing the software implementation can be whatever you want, once you've figured out what you have to build. The recommendation to think deeply about your users and product design up-front can still be combined with an agile sprint-based engineering process. And once the big design questions have been figured out, you can start using more agile processes for the design effort, too, such as designing smaller features during one sprint and then having engineers implement them during the next sprint.

Elements of API design

At last, we can talk about design! The secret to producing a good API design lies in coming up with an appropriate abstraction for the problem domain and then devising appropriate object and class hierarchies to represent that abstraction.

An abstraction is just a simplified description of something that can be understood with no knowledge of how it will be implemented programmatically. It tends to emphasize the important characteristics and responsibilities of that thing while ignoring details that are not important to understanding its basic nature. Furthermore, you often find that complex problems exhibit hierarchies, or layers, of abstractions (Henning, 2009).

For example, you could describe how a car works at a very high level with six basic components: a fuel system, engine, transmission, driveshaft, axle, and wheels. The fuel system provides the energy to turn the engine, which causes the transmission to rotate, whereas the driveshaft connects the transmission to the axle, allowing power to reach the wheels and ultimately cause the vehicle to move forward. This is one level of abstraction that's useful to understand the most general principles of how a car achieves forward motion. However, you could also offer another level of abstraction that provides more detail for one or more of these components. For example, an internal combustion engine could be described with several interconnected components, including a piston, crankshaft, camshaft, distributor, flywheel, and timing belt. Furthermore, an engine can be categorized as one of several different types, such as an internal combustion engine, an electric engine, a gas/electric hybrid, or a hydrogen fuel cell.

Similarly, most designs for complex software systems exhibit structure at multiple levels of detail, and those hierarchies can also be viewed in different ways. Grady Booch suggests that there are two important hierarchical views of any complex system (Booch et al., 2007):

1. **The object hierarchy**: Describes how different objects cooperate in the system. This represents a structural grouping based upon a part-of relationship between objects (e.g., a piston is part of an engine, which is part of a car).
2. **The class hierarchy**: Describes the common structure and behavior that is shared between related objects. It deals with the generalization and specialization of object properties. This can be thought of as an is-a hierarchy between objects (e.g., a hybrid engine is a type of car engine).

Both views are equally important when producing the design for a software system. Fig. 4.3 attempts to illustrate these two concepts, showing a hierarchy of related objects and a hierarchy of classes that inherit behavior and properties.

Related to this, it's generally agreed that the design phase of software construction consists of two major activities (Bourque et al., 2004):

1. **Architecture design.** Describes the top-level structure and organization of a piece of software.
2. **Detailed design.** Describes individual components of the design to a sufficient level at which they can be implemented.

Therefore, as a general approach, I suggest defining an object hierarchy to delineate the top-level conceptual structure (or architecture) of your system, and then refine this

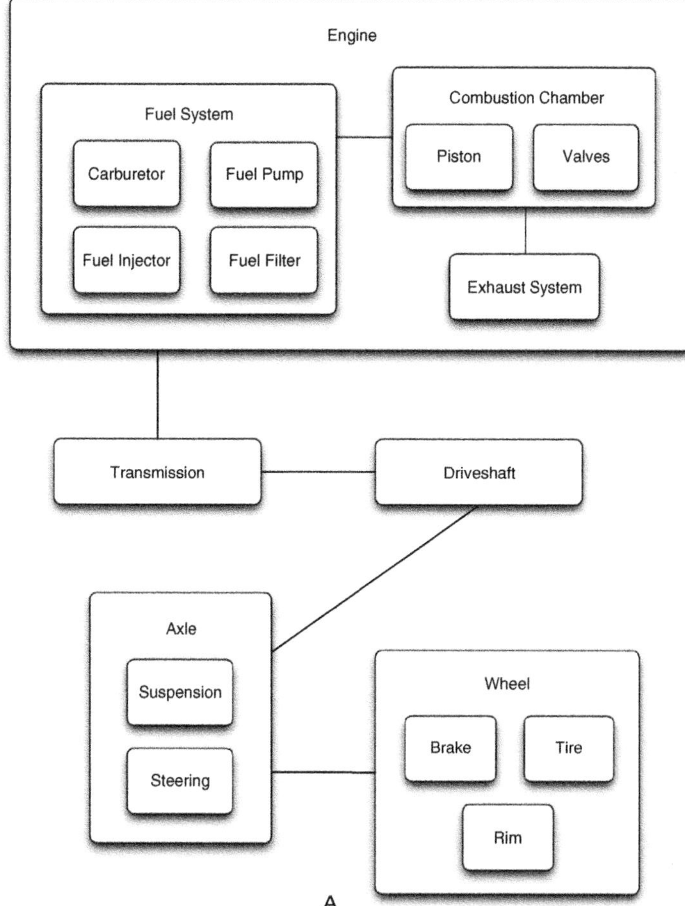

FIGURE 4.3 A car design shown as (A) a part-of object hierarchy.

with class hierarchies that specify concrete C++ classes for your clients to use. The latter process of defining the classes of your API also involves thinking about the individual functions and arguments that they provide. The rest of this chapter will therefore focus on each of these topics in turn:

1. Architecture design.
2. Class design.
3. Function design.

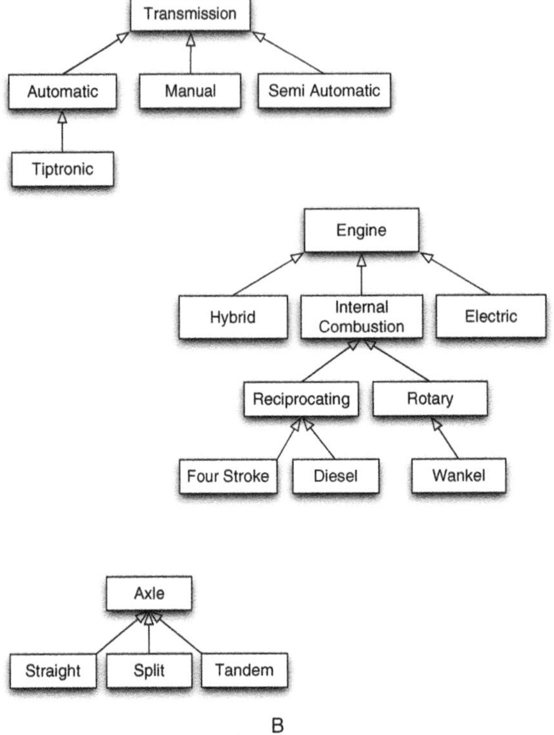

FIGURE 4.3 (B) an is-a class hierarchy. *Arrows* point from more specific to more general classes.

TIP: API design involves developing a top-level architecture and a detailed class hierarchy.

Architecture design

Software architecture describes the coarse structure of an entire system: the collection of top-level objects in the API and their relationships to each other. By developing an architecture, you gain an understanding of the different components of the system in the abstract, as well as how they communicate and cooperate with each other.

It's important to spend time thinking about the top-level architecture for your API because problems in your architecture can have far-reaching and extensive impact on your system. Consequently, in this Section I will detail the process of producing an architecture for your API and provide insight into how you can decompose a problem domain into an appropriate collection of abstract objects.

Developing an architecture

There's no right or wrong architecture for any given problem. If you give the same set of requirements to two different architects, you'll undoubtedly end up with two different solutions. The important aspect is to produce a well-thought-out purposeful design that delivers a framework to implement the system and resolves trade-offs between the various conflicting requirements and constraints (Bass et al., 2003). At a high level, the process of creating an architecture for an API resolves to four basic steps:

1. Analyze the functional requirements that affect the architecture.
2. Identify and account for the constraints on the architecture.
3. Invent the primary objects in the system and their relationships.
4. Communicate and document the architecture.

The first of these steps is fed by the earlier requirements gathering stage (refer back to Fig. 4.1), whether it is based upon a formal functional requirements document, a set of goal-oriented use cases, or a collection of informal user stories. The second step involves capturing and accounting for all factors that place a constraint on the architecture you design. The third step involves defining the high-level object model for the system: the key objects and how they relate to each other. Finally, the architecture should be communicated to the engineers who must implement it. Fig. 4.4 illustrates each of these steps.

It's important to stress that this sequence of steps is not a recipe that you perform only once and magically arrive at the perfect architecture. As I've stated, software design

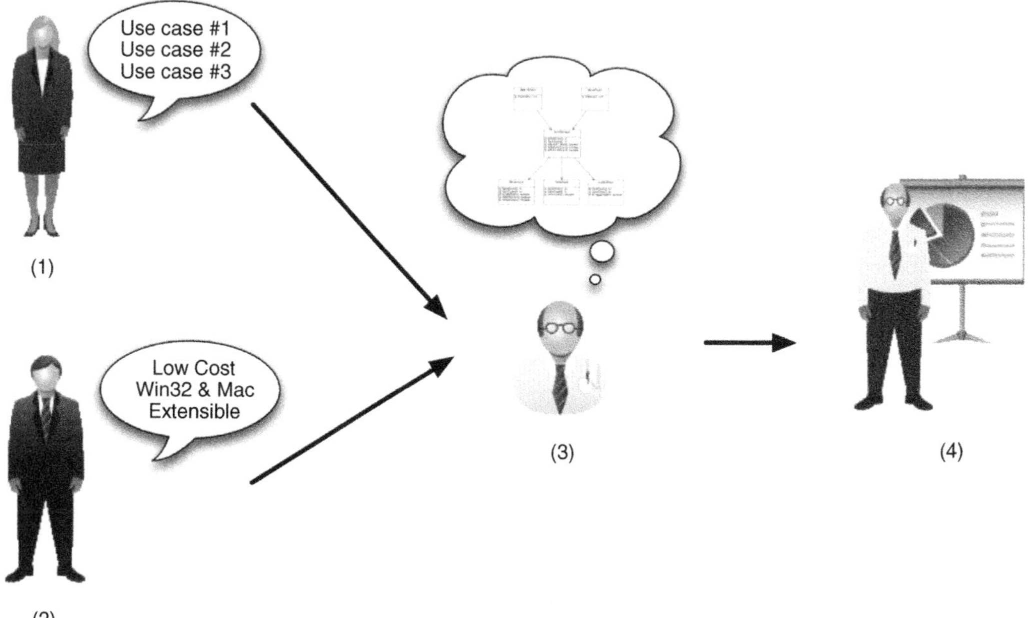

FIGURE 4.4 The steps to develop an application programming interface architecture: (1) gather user requirements, (2) identify constraints, (3) invent key objects, and (4) communicate design.

is an iterative process. You will rarely get each step right the first time. On the other hand, the first version of your API is critical because changes after that point will incur a higher cost. It's therefore important to try out your design early on and improve it incrementally before releasing it to clients who will then start to build upon it in their own programs.

Architecture constraints

APIs aren't designed in a vacuum. There will always be factors that influence and constrain the architecture. Before the design process can proceed in earnest, you must therefore identify and accommodate for these factors. Christine Hofmeister and co-authors refer to this phase as global analysis (Hofmeister et al., 2009). The term global in this respect means that the factors affect the system holistically and that as a group they are often interdependent and contradictory. These factors fall into three basic categories:

1. Organizational factors, such as:
   a. Budget
   b. Schedule
   c. Team size and expertise
   d. Software development process
   e. Build vs buy decision on subsystems
   f. Management focus (e.g., date vs feature vs quality)
2. Environmental factors, such as:
   a. Hardware (e.g., set-top box or mobile device)
   b. Platform (e.g., Windows, Mac, and Linux)
   c. Software constraints (e.g., use of other APIs)
   d. Client/server constraints (e.g., building a Web service)
   e. Protocol constraints (e.g., POP vs. IMAP for a mail client)
   f. File format constraints (e.g., must support GIF and JPEG images)
   g. Database dependencies (e.g., must connect to a remote database)
   h. Expose versus wrap decision on subsystems
   i. Development tools
3. Operational factors, such as:
   a. Performance
   b. Memory use
   c. Reliability
   d. Availability
   e. Concurrency
   f. Customizability
   g. Extensibility
   h. Scriptability
   i. Security
   j. Internationalization
   k. Network bandwidth

It's the job of the software architect to prioritize these factors, combined with the user constraints contained within the functional requirements, and to find the best compromises that produce a flexible and efficient design. Designing an API carefully for its intended audience can only serve to improve its usability and success. However, there's no such thing as a perfect design; it's all about trade-offs for the given set of organizational, environmental, and operational constraints. For example, if you are forced to deliver results under an aggressive schedule, then you may have to focus on a simpler design that leverages third-party APIs as much as possible and restricts the number of supported platforms.

TIP: Architecture design is constrained by a multitude of unique organizational, environmental, and operational factors.

Some constraints can be negotiated. For example, if one of the client's requirements places undue complexity on the system, the client may be willing to accept an alternative solution that costs less money or can be delivered sooner.

In addition to identifying the factors that will affect your initial architecture, you should assess which of these are susceptible to change during development. For example, the first version of the software may not be very extensible, but you know that you will eventually want to move to a plug-in model that lets users add their own functionality. Another common example is internationalization. You may not care about supporting more than one language at first, but later this may become a new requirement and one that can have a deep impact on the code. Your design should therefore anticipate the constraints that you reasonably expect to change in the future. You may be able to come up with a design that can support change, or if that's not feasible, then you may need to think about contingency plans. This is often referred to as design for change (Parnas, 1979).

TIP: Always design for change. Change is inevitable.

It's also worth thinking about how you can isolate your design from changes in any APIs on which your project will depend. If your use of another API is completely internal, then there's no problem. However, if you need to expose the concepts of a dependent API in your own public interface, then you should consider whether it's possible to limit the degree to which it's made visible. In some cases, this simply may not be practical. However, in other cases you may be able to provide wrappers for the dependent API so that you do not force your clients to depend upon that API directly. For example, the KDE API is built on top of the Qt library. However, KDE uses a thin wrapper over the Qt API so that users are not directly dependent on the Qt API. As a specific example, KDE offers classes such as `KApplication`, `KObject`, and `KPushButton` instead of exposing Qt's `QApplication`, `QObject`, and `QPushButton` classes directly. Wrapping dependent APIs in this way gives you an extra layer of indirection to protect against changes in a dependent API and to work around bugs or platform-specific limitations.

Finally, there are some common pitfalls to be aware of when developing an architecture. In particular, if you work at a large organization, you should be aware of Conway's law and its consequences. Conway originally defined his law as (Conway, 1968):

> *Any organization that designs a system (defined broadly) will produce a design whose structure is a copy of the organization's communication structure.*
>
> —*Melvin E. Conway.*

The implication of this law for software design is that if you work in an organization that has many separate groups that need to work together on an overall design, then the resulting design will likely be composed of parts that represent the focus areas of those separate groups. The positive side of this is that at least you have a team of engineers who will work on the problem; but the negative side is that this may not yield the best design to solve the problem. Eric S. Raymond gave an example of Conway's law, stating that "if you have four groups working on a compiler, you'll get a 4-pass compiler." So, when working on a software architecture for a large system, it's always good to question whether the design that's been developed is the best one to solve the problem, or just a repackaging of components that the constituent teams already own. A specific concern to be aware of is that if those components were developed separately and in isolation, they may not work well together as a coherent whole or fully solve the new problem being worked on.

Identifying the major abstractions

Once you've analyzed the requirements and constraints of the system, you are ready to start building a high-level object model. Essentially, this means identifying the major abstractions in the problem domain and decomposing these into a hierarchy of interconnected objects. Fig. 4.5 presents an example of this process. It shows a top-level architecture for the OpenSceneGraph API, an open source 3D graphics tool kit for visual simulation applications (http://www.openscenegraph.org/).

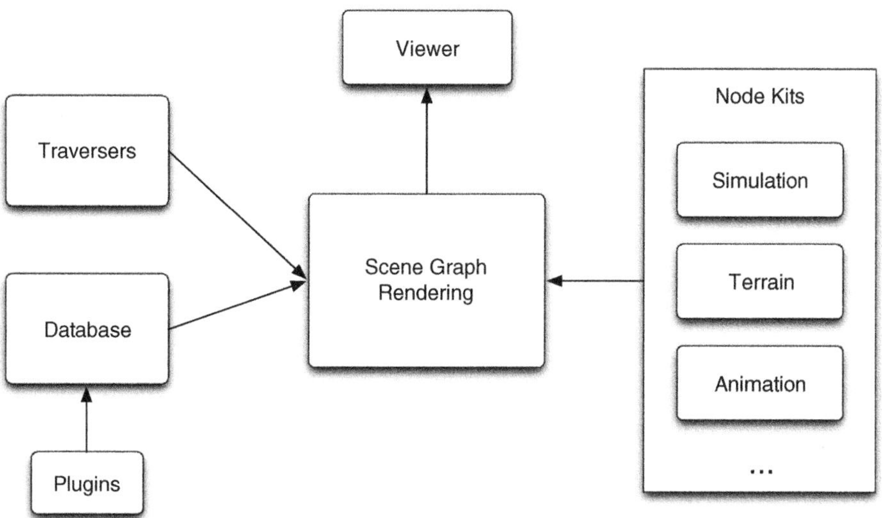

FIGURE 4.5 Example top-level architecture for the OpenSceneGraph application programming interface.

By basing the architecture on actual concepts in the problem domain, your design should remain general and robust to future changes in requirements. Recall that I listed this as the first API quality in Chapter 2: a good API should model the problem domain. However, decomposing a problem into a set of good abstractions is not an easy task. For well-understood problems such as writing a compiler or a Web server, you can take advantage of the collective knowledge that has been distilled and published by many other designers over time. However, for new problems that have had little or no previous research applied to them, the task of inventing a good classification can be far from obvious.

This is not a problem that's unique to computer science. The classification of the biology of our planet into a logical taxonomy has been an area of debate ever since the days of Aristotle. In the 18th century, Carolus Linnaeus proposed a two-kingdom model for life, composed of vegetables and animals. This was later refined in the 19th century to include microscopic life forms. Modern advances in electron microscopy have increased the number of kingdoms to five or six. However, research in the 21st century has contested the traditional view of kingdoms and proposed an alternative supergroup model. Additionally, the topic of deciding which characteristics should be used to create classifications has received much debate. Aristotle classified animals according to their method of reproduction, the binomial system groups organisms by their morphology (similar structure or appearance), whereas Darwinian-inspired taxonomies favor classification by common descent (whether organisms have a common ancestor).

Inventing the key objects

Despite the difficulty of classifying the major abstractions in a system, I can still offer some advice on how to tackle the problem. Accordingly, here are several techniques that you can draw upon to decompose a system into a set of key objects and identify their relationship to each other (Booch et al., 2007):

- **Natural language.** Using the analogy to natural language, it's been observed that (in general) nouns tend to represent objects, verbs represent functions, and adjectives and possessive nouns represent attributes (Bourque et al., 2004). I can illustrate this by returning to our address book API from Chapter 2. The real-world concepts of an address book and a person are both nouns and make sense to represent key objects in the API, whereas actions such as adding a person to the address book or adding a telephone number for a person are verbs and should be represented as function calls on the objects that they modify. However, a person's name is a possessive noun and makes more sense to be an attribute of the `Person` object rather than a high-level object in its own right.
- **Properties.** This technique involves grouping objects that have similar properties or qualities. This can be done using discrete categories in which each object is unambiguously either a member or not, such as red objects versus blue objects. Or it can involve a probabilistic grouping of objects that depends upon how closely each object matches some fuzzy criterion or concept, such as whether a film is categorized as an action or romance story.
- **Behaviors.** This method groups objects by the dynamic behaviors that they share. This involves determining the set of behaviors in the system and assigning these behaviors to different parts of the system. You can then derive the set of objects by identifying the initiators and participants of these behaviors.
- **Prototypes.** In this approach, you attempt to discover more general prototypes for the objects that were initially identified. For example, a beanbag, bar stool, and recliner are all types of chair, despite having very different forms and appearance. However, you can classify each of them based upon the degree to which they exhibit affordances of a prototypical chair.
- **Domains (Shlaer-Mellor).** The Shlaer-Mellor method first partitions a system horizontally to create generic domains and then partitions these vertically by applying a separate analysis to each domain (Shlaer and Mellor, 1988). One of the benefits of this divide-and-conquer approach is that the domains tend to form reusable concepts that can be applied to other design problems. For instance, using our earlier ATM example, a domain could be one of the following:
  - **Tangible domains**, such as an ATM machine or a bank note.
  - **Role domains**, such as an ATM user or a bank owner.
  - **Event domains**, such as a financial transaction.
  - **Security domains**, such as authentication and encryption.

- **Interaction domains**, such as PIN entry or a cash withdrawal.
- **Logging domains**, for the system to log information.

- **Domains (Neighbors).** James Neighbors coined the term domain analysis as the technique of uncovering classes and objects that are shared by all applications in the problem domain (Neighbors, 1980). This is done by analyzing related systems in the problem domain to discover areas of commonality and distinctiveness, such as identifying the common elements in all bug tracking systems or general features of all genealogy programs.
- **Domains (Evans).** A related issue to Neighbors' domain analysis is the term domain-driven design. This was introduced by Eric Evans and seeks to produce designs for complex systems using an evolving model of the core business concepts (Evans, 2003).

TIP: Identifying the key objects for an API is hard. Try looking at the problem from different perspectives and keep iterating and refining your model.

Most of these techniques work best when you have a well-organized and structured set of use cases from which to work. For example, use cases are normally constructed as sentences where a thing performs some action, often to or on another thing. You can therefore use these as input for a simple natural language analysis by taking the steps of each use case and identifying the subject or object nouns and use these to develop an initial candidate list of objects.

Each of these techniques can also involve different degrees of formal methods. For example, natural language analysis it not a very rigorous technique and is often discouraged by proponents of formal design methodologies. That's because natural language is intrinsically ambiguous and may express important concepts of the problem domain imprecisely or neglect significant architectural features. You should therefore be wary of naively translating all nouns in your use cases to key objects. At best, you should treat the result of this analysis as an initial candidate list from which to apply further careful analysis and refinement (Alexander, 2003). This refinement can involve identifying any gaps in the model, considering whether there are more general concepts that can be extracted from the list, and attempting to classify similar concepts.

In contrast, there are several formal techniques for producing a software design, including textual and graphical notations. One particularly widespread technique is UML (Booch et al., 2005). UML can be used to specify and maintain a software design visually, using a set of graphical diagrams. For instance, UML 2.3 includes 14 distinct types of diagrams to represent the various structural and behavioral aspects of a design (Fig. 4.6). As a specific example, UML sequence diagrams portray the sequence of

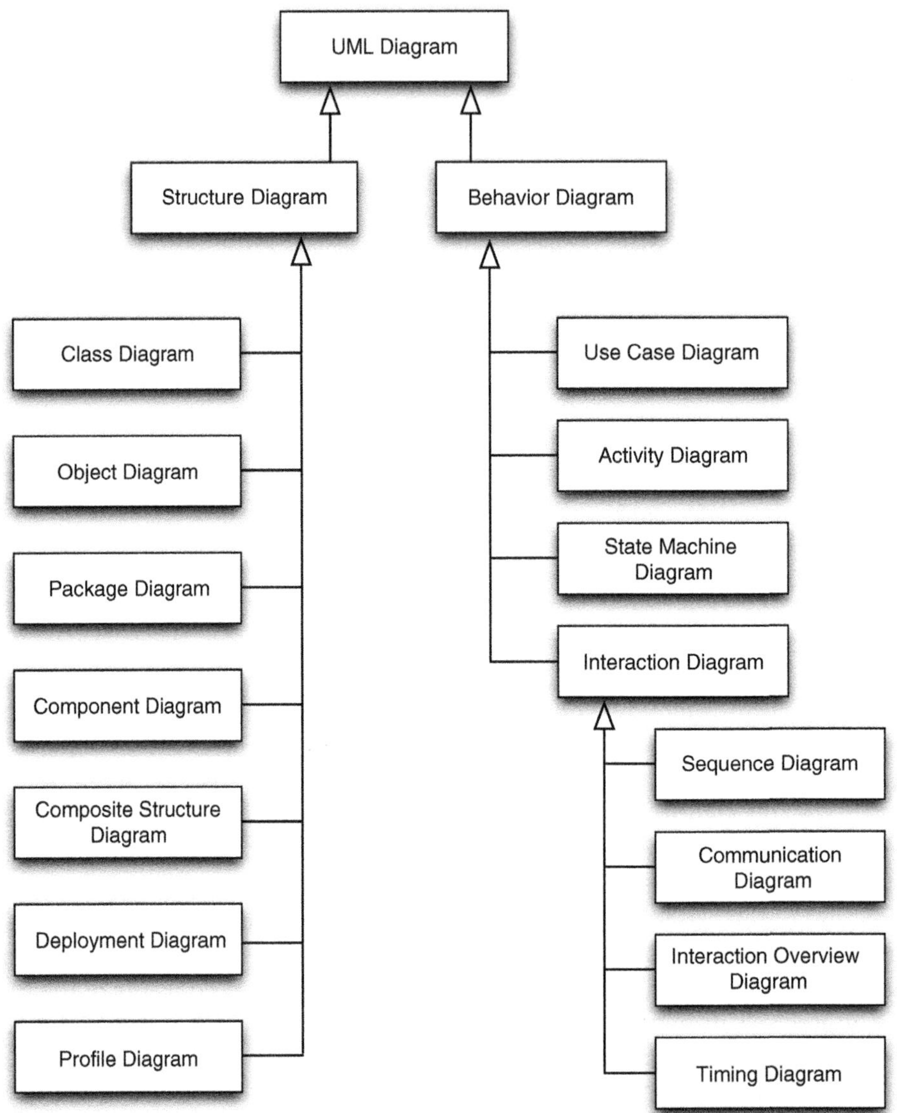

FIGURE 4.6 The 14 diagram types of Universal Modeling Language (UML) 2.3.

function calls between objects. These can be used during the analysis phase to represent use cases graphically. Then, during the design phase, the architect can use these formal diagrams to explore the object interactions within the system and flesh out the top-level object model.

Formal design notations can also be used to generate actual code. This ranges from the simple translation of class diagrams into their direct source code equivalents to the more comprehensive notion of an executable architecture. The latter is a sufficiently detailed description of an architecture that can be translated into executable software and run on a target platform. For example, Shlaer-Mellor notation was eventually evolved into a profile of UML called Executable UML (Mellor and Balcer, 2002), which itself became a cornerstone of Model Driven Architecture. The basic principle behind this approach is that a model compiler takes several executable UML models, each of which defines a different cross-cutting concern or domain, and combines these to produce high-level executable code. Proponents of executable architectures note the two-language problem that this entails: having a modeling language (e.g., UML) that gets translated into a separate programming language (e.g., C++, C#, or Java). Many of these proponents therefore posit the need for a single language that can bridge both concerns.

Architectural patterns

In Chapter 3, I covered various design patterns that be used to solve recurring problems in software design, such as Singleton, Factory Method, and Observer. These tend to provide solutions that can be implemented at the component level. However, there is a class of software patterns called architectural patterns that describe larger-scale structures and organizations for entire systems. As such, some of these solutions may be useful to you when you are building an API that maps well to a particular architectural pattern. The following list classifies several of the more popular architectural patterns (Bourque et al., 2004):

- **Structural patterns**: Layers, Pipes and Filters, and Blackboard
- **Interactive systems**: Model-View-Controller (MVC), Model-View-Presenter, and Presentation-Abstraction-Control
- **Distributed systems**: Client/Server, Three-Tier, Peer-to-Peer, and Broker
- **Adaptable systems**: Microkernel and Reflection

Many of these architectural patterns present elegant designs to avoid dependency problems between different parts of your system, such as the MVC pattern that I discussed in Chapter 3. At this point, it's worth noting that another important view of a system's architecture is the physical view of library files and their dependencies. I presented an example of this view back in Fig. 1.3, where I showed the layers of APIs that make up a complex end-user application. Even within a single API you'll likely have different layers of physical architecture, such as those listed subsequently and illustrated in Fig. 4.7:

1. API-neutral low-level routines, such as string manipulation routines, math functions, or your threading model
2. Core business logic that implements the primary function of your API
3. Plug-in or scripting APIs to allow users to extend the base functionality of the API
4. Convenience APIs built on top of the core API functionality
5. A presentation layer to provide a visual display of your API results

In this case, it's important to impose a strict dependency hierarchy on the different architectural layers of your system; otherwise you will end up with cyclic dependencies

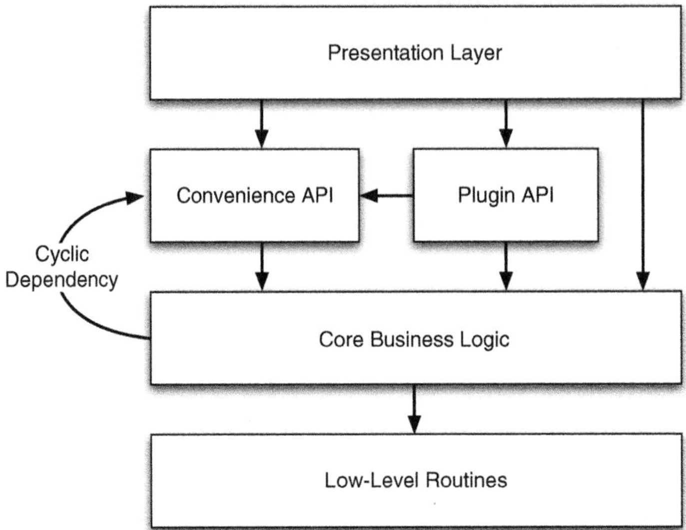

FIGURE 4.7 Example architectural layers of an application programming interface (API) showing a cyclic, or circular, dependency between two components.

between layers (Fig. 4.7). The same is true for individual components within those layers. The general observation is that lower-level components should not depend upon higher-level components of your architecture. For example, your core business logic cannot depend upon your convenience API because this would introduce a cycle between the two (if your convenience API also calls down into the core business logic). Referring back to the MVC architectural pattern, you will note that View depends upon Model, but not vice versa. David L. Parnas referred to this concept as loop-free hierarchies (Parnas, 1979).

SIDEBAR: Menvshared

During my early years at Pixar, we got into a situation where we had many cyclic dependencies between the core suite of animation libraries. These dependencies crept slowly into the system as we worked to meet many tight production deadlines.

In order to allow the system to continue to compile, we resorted to linking all of this interdependent code into a single huge shared library, called `libmenvshared.so` *(pronounced menvee-shared). This meant that any changes to one of the libraries in this shared library would require the entire shared library to be rebuilt, a process that could take at least 5–10 min.*

As you can imagine, this became a significant bottleneck to the development team's velocity. In fact, the massive size of the menvshared library even caused linker crashes when we tried porting our code to another platform.

Thankfully, a brave few engineers were eventually able to tackle this problem, and over many weeks we gradually teased apart the dependencies to finally get rid of menvshared.

Cyclic dependencies are bad for many reasons. For example, they mean that you cannot test each component independently and you cannot reuse one component without also pulling in the other. Basically, it's necessary to understand both components to understand either one (Lakos, 1996). This can also affect the speed of your development if you're forced to merge several components into one big über-component, as described in the accompanying sidebar. In Chapter 2 I presented various techniques to decouple dependencies, such as callbacks, observers, and notification systems. Fundamentally, an API should be an acyclic hierarchy of logically related components.

TIP: Avoid cyclic dependencies between components of your API.

Communicating the architecture

Once an architecture has been developed, it can be documented in various ways. This can range from simple drawings or wiki pages to various formal methods that provide modeling notations for architectures, such as UML or the set of Architecture Description Languages (Medvidovic and Taylor, 2000).

Whichever approach you adopt, documenting the architecture is an important factor in communicating the design to engineers. Doing so gives them more information to implement the system according to your vision and ensures that future changes continue to preserve the intent and integrity of the architecture. This is particularly important if the development team is large or geographically distributed.

One of the elements you should include in your architecture documentation is a rationale for the overall design: that is, which alternative designs and trade-offs were considered and why the final structure was judged to be superior. This design rationale can be very important for the long-term maintenance of the design and to save future designers from revisiting the same dead ends that you did. In fact, Martin Robillard notes that users of an API often find it difficult to learn and use an API if they don't understand its high-level architecture and design intents (Robillard, 2009).

TIP: Describe the high-level architecture and design rationale for your API in the accompanying user documentation.

Communication also allows for peer review of the design and for feedback and improvements to be received before the API is released. In fact, implementing design reviews early in the process to facilitate communication among architects, developers, and clients will help you to produce a more comprehensive and durable design. If the API architect also writes some of the code, then this can be an excellent way to communicate design principles through practical hands-on contribution.

Even though modern agile development methods deemphasize document-heavy processes because the design documents are frequently out of date, there is still a place for providing as much documentation about the system architecture as necessary, but no more. Augmenting any documentation with direct communication can be even more productive. This allows for a dialogue between the designer and implementer and can avoid misunderstandings that can happen when reading specifications. Ultimately, it's the jobs of the architect to be a passionate communicator, to be available to answer questions from engineers, and to ensure that the most efficient channels are used to keep architectural communication constantly stimulated (Faber, 2010).

SIDEBAR: Pitch your story

While working in the R&D department at Pixar, our task as software engineers and managers was to produce powerful yet easy-to-use programs for highly creative artists. We therefore endeavored to communicate our software plans and designs using analogies and terminology familiar to filmmakers.

For example, our design team was referred to as our Story department, a reference to the department in a film studio that is responsible for the initial planning and structuring of a movie. Our software schedules and system designs were then presented to our production users on standard storyboards. These are physical wood boards with a regular grid of 4×6-inch hand-drawn index cards pinned to them. These boards were presented to artists in a story pitch format, to use filmmaking terminology, thus allowing us to present our material in a meeting style and structure that was familiar to our users. Finally, we also produced several story reels, which were digital movies in which hand-drawn examples of the proposed applications were animated and narrated to show end-user workflows.

This specific approach worked well in our case to communicate our software plans with talented but nontechnical in-house users. Of course, this format will not be appropriate for all other software projects. However, the central tenet is that you should think deeply about how you can communicate your designs in the most natural and intuitive manner for your users.

Class design

With a high-level architecture in place, you can start refining the design to describe specific C++ classes and their relationship to other classes. This is the detailed design, in contrast to the top-level architecture design. It involves identifying actual classes that clients will use, how these classes relate to each other, and their major functions and attributes. For sufficiently large systems, this can also involve describing how classes are organized into subsystems.

Designing every single class in your API would be overkill for anything but the most trivial system. Instead, you should focus on the major classes that define the most important functionality. A good rule of thumb is the so-called 80/20 rule (i.e., you should concentrate on 20% of the classes that define 80% of your system's behavior) (McConnell, 2004).

TIP: Focus on designing 20% of the classes that define 80% of your API's functionality.

Object-oriented concepts

Before I talk more about the details of object-oriented design, let's take a moment to review some of the major object-oriented principles and their representation in C++. It is likely that you're already very familiar with these concepts, but let's briefly summarize them here in the interests of completeness and to ensure that we're on the same page:

- **Class**: A class is the abstract description, or specification, of an object. It defines the data members and member functions of the object.
- **Object.** An object is an entity that has state, behavior, and identity (Booch et al., 2007). It's an instance of a concrete class created at run time using the `new` operator in C++. A concrete class is one that is fully specified (e.g., it has no undefined pure virtual member functions).
- **Encapsulation**: This concept describes the compartmentalization of data and methods as a single object with access control specifications such as public, protected, and private to support hiding implementation details.
- **Inheritance**: This allows for objects to inherit attributes and behaviors from a parent class, and to introduce their own additional data members and methods. A class that's defined in this way is said to be a derived (or sub) class, of the parent (or base) class. A subclass can override any base class method, although normally you want to do this only when that base class method is declared to be virtual. A pure virtual method (indicated by appending its declaration with "= 0") is one where a subclass must provide an implementation of the method for it to be concrete (i.e., to allow instances of it to be created). C++ supports multiple inheritance, meaning that a subclass can inherit from more than one base class. Public inheritance is generally referred to as an is-a relationship between two objects, whereas private inheritance represents a was-a relationship (Lakos, 1996).
- **Composition**: This is an alternative technique to inheritance in which one or more simple objects are combined to create more complex ones. This is done by declaring the simpler objects as member variables inside the more complex object. The has-a relationship is used to describe the case in which a class holds an instance of another type. The holds-a relationship describes a class holding a pointer or reference to the other type.
- **Polymorphism**: This is the ability of one type to appear as, and to be used like, another type. This allows objects of different types to be used interchangeably if

they conform to the same interface. This is possible because the C++ compiler can delay checking the type of an object until run time, a technique known as late or dynamic binding. The use of templates in C++ can also be used to provide static (compile-time) polymorphism.

Class design options

For the creation of a class, there are many factors to be considered. As Scott Meyers notes, creating a new class involves defining a new type. You should therefore treat class design as type design and approach the task with the same thoughtfulness and attention that the designers of C++ put into the built-in types of the language (Meyers, 2005).

Here I list a few of the major questions that you should ask yourself when you embark upon designing a new class. This is not meant to be an exhaustive list, but it should provide a good starting point to help you define the major constraints on your design:

- **Use of inheritance.** Is it appropriate to add the class to an existing inheritance hierarchy? Should you use public or private inheritance? Should you support multiple inheritance? This affects whether you need to think about which member functions should be virtual.
- **Use of composition.** Is it more appropriate to hold a related object as a data member rather than inheriting from it directly?
- **Use of abstract interfaces.** Is the class meant to be an abstract base class, in which subclasses must override various pure virtual member functions?
- **Use of standard design patterns.** Can you employ a known design pattern to the class design? Doing so lets you benefit from well-thought-out and refined design methodologies and makes your design easier to use by other engineers.
- **Initialization and destruction model.** Will clients use `new` and `delete` or will you use a factory method? Will you override `new` and `delete` for your class to customize the memory allocation behavior? Will you use smart pointers?
- **Defining a copy constructor and assignment operator.** If the class allocates dynamic memory, you need both of these, as well as a destructor. This is called the Rule of Three, which I'll cover in the next chapter on C++ Usage. This will affect how your objects will be copied and passed by value.
- **Use of templates.** Does your class define a family of types rather than a single type? If so, then you may need to consider the use of templates to generalize your design.
- **Use of const and explicit.** Define arguments, return results, and methods as `const` wherever you can. Use the `explicit` keyword to avoid unexpected type conversions for single-parameter constructors.
- **Defining operators.** Define any operators that you need for your class, such as +, *=, `[]`, ==, or <<.

- **Defining type coercions.** Consider whether you want your class to be automatically coercible to different types and declare the appropriate conversion operators.
- **Use of friends.** Friends breach the encapsulation of your class and are generally an indication of bad design. Use them as a last resort.
- **Nonfunctional constraints.** Issues such as performance and memory use can place constraints on the design of your classes.

The SOLID principles

There is a popular acronym used in object-oriented design, called SOLID. This provides a good set of key design principles to consider when building an object model for any system (Feathers, 2004). I cover many of the aspects of SOLID throughout this book, but I'll summarize the core principles here:

- S = Single responsibility principle. This states that there should only ever be one reason for a class to change (i.e., an object should have only one responsibility). I mentioned this principle earlier in the Qualities chapter when talking about the Don't Repeat Yourself (DRY) principle.
- O = Open/closed principle. This states that objects should be open for extension but closed for modification (i.e., that it should be possible to modify the behavior of a class without changing its source code). I'll discuss this principle later in this chapter.
- L = Liskov substitution principle (LSP). This states that an object should be replaceable with a subclass of that object without changing any behavior. I'll also discuss this principle later in this chapter.
- I = Interface segregation principle. This states that clients should not have to depend on interfaces they don't use. Essentially, you should split large monolithic interfaces into smaller more specific ones. The goal of this principle is to reduce coupling, which I covered in detail in the earlier chapter on Qualities.
- D = Dependency inversion principle. This states that clients should depend on abstractions, not concrete implementations. The goal of this principle is also to reduce coupling, by having software components depend on abstract interfaces. I covered aspects of loose coupling in the chapter on Qualities and abstract interfaces in the chapter on Patterns.

Using inheritance

By far the biggest design decision that you'll face when designing your classes is when and how to use inheritance. For example, should you use public inheritance, private inheritance, or composition to associate related classes in your API? Because inheritance is such an important topic, and one that is often misused or overused, I will focus on this part of class design over the next few sections. Let's begin with some general design recommendations:

- **Design for inheritance or prohibit it.** The most important decision you can make is to decide whether a class should support subclasses. If it should, then you must think deeply about which methods should be declared as `virtual` and document their behavior. If the class should not support inheritance, a good way to convey this is to declare a nonvirtual destructor.
- **Use inheritance only where appropriate.** Deciding whether a class should inherit from another class is a difficult design task. In fact, this is perhaps the most difficult part of software design. I will present some guidance on this topic in the next section when I talk about the LSP.
- **Avoid deep inheritance trees.** Deep inheritance hierarchies increase complexity and invariably result in designs that are difficult to understand and software that's more prone to failure. The absolute limit of hierarchy depth is obviously subjective, but any more than two or three levels is already getting too complex (McConnell, 2004).
- **Use pure virtual member functions to force subclasses to provide an implementation.** A virtual member function can be used to define an interface that includes an optional implementation, whereas a pure virtual member function is used to define only an interface, with no implementation (although it is possible to provide a fallback implementation for a pure virtual method). Of course, a non-virtual method is used to provide behavior that cannot be changed by subclasses.
- **Don't add new pure virtual functions to an existing interface.** You should certainly design appropriate abstract interfaces with pure virtual member functions. However, be aware that after you release this interface to users, if you then add a new pure virtual method to the interface, you will break all of your clients' code. That's because client classes that inherit from the abstract interface will no longer be concrete until an implementation for the new pure virtual function is defined.
- **Don't overdesign.** In Chapter 2, I stated that a good API should be minimally complete. In other words, you should resist the temptation to add extra levels of abstraction that are currently unnecessary. For example, if you have a base class that's inherited by only a single class in your entire API, this is an indication that you've overdesigned the solution for the current needs of the system.

TIP: Avoid deep inheritance hierarchies.

Another important consideration is whether to use multiple inheritance (i.e., whether you will design classes that inherit from more than one base class). Bjarne

Stroustrup argued for the addition of multiple inheritance to C++ using the example of a `TemporarySecretary` class, where this inherits from both a `Secretary` and a `Temporary` class (Alexandrescu, 2001). However, opinion is divided in the C++ community on whether multiple inheritance is a good thing. On the one hand, it offers the flexibility to define composite relationships such as the `TemporarySecretary` example earlier. However, this can come at the cost of subtle semantics and ambiguities, such as the need to use virtual inheritance to deal with the diamond problem, in which a class inherits ambiguously from two or more base classes that themselves inherit from a single common base class.

Most languages that allow inheriting from only a single base class still support inheriting from multiple, more constrained types. For example, Java lets you inherit from multiple interface classes, and Ruby lets you inherit from multiple mixins. These are classes that let you inherit an interface (and implementation in the case of a mixin); however, they cannot be instantiated on their own. Similarly, the Swift programming language allows a class to inherit only from a single base class, but a class can conform to multiple protocols, similar to the concept of an interface.

Multiple inheritance can be a powerful tool if used correctly (see the C++ Standard Library iostreams classes for a good example). However, in the interest of robust and easy-to-use interfaces, I generally concur with Steve McConnell, who recommends that you should avoid the use of multiple inheritance, except to use abstract interfaces or mixin classes (McConnell, 2004).

TIP: Avoid multiple inheritance, except for interfaces and mixin classes.

As a point of interest, the C++11 specification introduced several improvements relating to inheritance. Of relevance here is the ability to explicitly specify your intent to override or hide a virtual method from a base class. This is done using the `[[override]]` and `[[hiding]]` attributes, respectively. This functionality is extremely helpful to avoid mistakes such as misspelling the name of a virtual method in a derived class.

Liskov substitution principle

This principle, introduced by Barbara Liskov in 1987, provides guidance on whether a class should be designed as a subclass of another class (Liskov and Zilles, 1974). The LSP states that if S is a subclass of T, then objects of type T can be replaced by objects of type S without any change in behavior.

At a first glance, this may seem to be a simple restatement of the is-a inheritance relationship, where a class S may be considered a subtype of T if S is a more specific kind of T. However, the LSP is a more restrictive definition than is-a.

Let's demonstrate this with the classic example of an ellipse shape type:

```
class Ellipse
{
public:
    Ellipse();
    Ellipse(float major, float minor);

    void SetMajorRadius(float major);
    void SetMinorRadius(float minor);
    float GetMajorRadius() const;
    float GetMinorRadius() const;

private:
    float mMajor;
    float mMinor;
};
```

You then decide to add support for a circle class. From a mathematical perspective, a circle is a more specific form of an ellipse, in which the two axes are constrained to be equal. It is therefore tempting to declare a `Sphere` class to be a subclass of `Ellipse`. For example,

```
class Sphere : public Ellipse
{
public:
    Sphere();
    explicit Sphere(float r);

    void SetRadius(float r);
    float GetRadius() const;
};
```

The implementation of `SetRadius()` can then set the major and minor radii of the underlying ellipse to the same value, to enforce the properties of a circle:

```
void Sphere::SetRadius(float r)
{
    SetMajorRadius(r);
    SetMinorRadius(r);
}

float Sphere::GetRadius() const
{
    return GetMajorRadius();
}
```

However, this poses a few problems. The most obvious is that `Sphere` will also inherit and expose the `SetMajorRadius()` and `SetMinorRadius()` methods of `Ellipse`. These could be used to break the self-consistency of our sphere by letting users change one radius without also changing the other. You could deal with this by overriding the `SetMajorRadius()` and `SetMinorRadius()` methods so that each sets both the major and minor radii. However, this poses several issues. First, you must go back and declare `Ellipse::SetMajorRadius()` and `Ellipse::SetMinorRadius()` to be virtual, so that you can override them in the `Circle` class. This should alert you that you're doing something wrong. Second, you have now created a nonorthogonal API: changing one property has the side effect of changing another property. Third, you have broken the LSP because you cannot replace uses of `Ellipse` with `Sphere` without breaking behavior, as the following code demonstrates:

```
void TestEllipse(Ellipse &e)
{
    e.SetMajorRadius(10.0);
    e.SetMinorRadius(20.0);
    assert(e.GetMajorRadius() == 10.0 &&
           e.GetMinorRadius() == 20.0);
}

...

Ellipse e;
Sphere s;
TestEllipse(e);
TestEllipse(s);   // fails!
```

The problem resolves to the fact that you've changed the behavior of functions inherited from the base class.

So, if you shouldn't use public inheritance to model a sphere as a kind of ellipse, how should you represent it? There are two main ways that you can correctly build your `Sphere` class upon the functionality of the `Ellipse` class: private inheritance and composition.

> *TIP: The LSP states that it should always be possible to substitute a base class for a derived class without any change in behavior.*

Private inheritance

Private inheritance lets you inherit the functionality, but not the public interface, of another class. In essence, all of the public members of the base class become private members of the derived class. This is referred to as a was-a relationship, in contrast to the is-a relationship of public inheritance. For example, you can redefine your `Sphere` class to inherit privately from `Ellipse` as:

```
class Sphere : private Ellipse
{
public:
    Sphere();
    explicit Sphere(float r);

    void SetRadius(float r);
    float GetRadius() const;
};
```

In this case, `Sphere` does not expose any of the member functions of `Ellipse` (i.e., there is no public `Sphere::SetMajorRadius()` method). This solution therefore does not suffer from the same problems as the public inheritance approach I discussed earlier. In fact, objects of type `Sphere` cannot be passed to code that accepts an `Ellipse` because the `Ellipse` base type is not publicly accessible.

Note that if you do want to expose a public or protected method of `Ellipse` in `Sphere`, then you can do this:

```
class Sphere : private Ellipse
{
public:
    Sphere();
    explicit Sphere(float r);

    // expose public methods of Ellipse
    using Ellipse::GetMajorRadius;
    using Ellipse::GetMinorRadius;

    void SetRadius(float r);
    float GetRadius() const;
};
```

Composition

Private inheritance is a quick way to fix an interface that violates the LSP if it already uses public inheritance. However, the preferred solution is to use composition. This simply means that instead of class S inheriting from T, S declares T as a private data member (has-a), or S declares a pointer or reference to T as a member variable (holds-a). For example:

```
class Sphere
{
public:
    Sphere();
    explicit Sphere(float r);

    void SetRadius(float r);
    float GetRadius() const;

private:
    Ellipse mEllipse;
};
```

Then the definition of the `SetRadius()` and `GetRadius()` methods might look like:

```
void Sphere::SetRadius(float r)
{
    mEllipse.SetMajorRadius(r);
    mEllipse.SetMinorRadius(r);
}

float Sphere::GetRadius() const
{
    return mEllipse.GetMajorRadius();
}
```

In this case, the interface for `Ellipse` is not exposed in the interface for `Sphere`. However, `Sphere` still builds upon the functionality of `Ellipse` by creating a private instance of `Ellipse`. Composition therefore provides the functional equivalent of private inheritance. However, there is wide agreement by object-oriented design experts that you should prefer composition over inheritance (Sutter and Alexandrescu, 2004).

TIP: Prefer composition to inheritance.

The main reason for this preference is that inheritance produces a more tightly coupled design. When a class inherits from another type, whether it is public, protected, or private inheritance, the subclass gains access to all public and protected members of the base class, whereas with composition, the class is coupled only to the public members of the other class. Furthermore, if you only hold a pointer to the other object, then your interface can use a forward declaration of the class rather than `#include` its full definition. This results in greater compile-time insulation and improves the time it takes to compile your code. Finally, you should not force an inheritance relationship when it's not appropriate. Our previous discussion tells us that a sphere should not be treated as an ellipse for the purposes of type inheritance. Note that there may still be a good case for a general `Shape` type from which all shapes, including `Circle` and `Ellipse`, inherit. However, a `Circle` should not inherit from `Ellipse` because it exhibits different behavior.

The open/closed principle

Bertrand Meyer introduced the open/closed principle (OCP) to state the goal that a class should be open for extension but closed for modification (Meyer, 1997). Essentially this means that the behavior of a class can be modified without changing its source code. This is a particularly relevant principle for API design because it focuses on the creation of stable interfaces that can last for the long term.

The principal idea behind the OCP is that once a class has been completed and released to users, it should be modified only to fix bugs. However, new features or changed functionality should be implemented by creating a new class. This is often achieved by extending the original class through either inheritance or composition. However, as I'll cover later in the book, you can also provide a plug-in system to allow users of your API to extend its basic functionality.

As an example of the OCP used to practical effect, the simple factory method I presented in Chapter 3 is not closed to modification or open for extensibility. That's because adding new types to the system requires changing the factory method implementation. As a reminder, here's the code for that simple renderer factory method:

```
IRenderer *RendererFactory::CreateRenderer(const std::string &type)
{
    if (type == "opengl") {
        return new OpenGLRenderer;
    }

    if (type == "directx") {
        return new DirectXRenderer;
    }

    if (type == "mesa") {
        return new MesaRenderer;
    }

    return nullptr;
}
```

By contrast, the extensible renderer factory that I presented later in Chapter 3 allows for the system to be extended without modifying the factory method. This is done by allowing clients to register new types with the system at run time. This second implementation therefore demonstrates the OCP: the original code doesn't need to be changed to extend its functionality.

However, when adhered to strictly, the OCP can be difficult to achieve in real-world software projects, and even contradicts some of the principles of good API design that I've advanced here. The constraint never to change the source code of a class after it's released is often impractical in large-scale complex systems, and the stipulation that any changes in behavior should trigger the creation of new classes can cause the original clean and minimal design to be diluted and fractured. In these cases, the OCP may be considered more of a guiding heuristic rather than a hard-and-fast rule. Also, although a good API should be as extensible as possible, there is a tension between the OCP and the specific advice in this book that you should declare member functions to be virtual in a judicious and restrained manner.

Nevertheless, if I restate the OCP to mean that the interface of a class should be closed to change, rather than considering the precise implementation behind that interface to be immutable, then you have a principle that aligns reasonably well with the focus of this book. That is, the maintenance of a stable interface gives you the flexibility to change the underlying implementation without unduly affecting your client's code. Furthermore, the use of extensive regression testing can allow you to make internal code changes without affecting existing behavior on which your users rely. And the use of an appropriate plug-in architecture (see Chapter 15) can provide your clients with a versatile point of extensibility.

> *TIP: Your API should be closed to incompatible changes in its interface, but open to extensibility of its functionality.*

The Law of Demeter

The Law of Demeter (LoD), also known as the principle of least knowledge, is a guideline for producing loosely coupled designs. The rule was proposed by Ian Holland based upon experiences developing the Demeter Project at Northeastern University in the late 1980s (Lieberherr and Holland, 1989). It states that each component should have only limited knowledge about other components, and even then only closely related components. This can be expressed more concisely as: only talk to your immediate friends.

When applied to object-oriented design, the LoD means that a function can:

- Call other functions in the same class
- Call functions on data members of the same class
- Call functions on any parameters that it accepts
- Call functions on any local objects that it creates
- Call functions on a global object (but you should never have globals)

By corollary, you should never call a function on an object that you obtained via another function call. For example, you should avoid chaining function calls such as:

```
void MyClass::MyFunction()
{
    mObjectA.GetObjectB().DoAction();
}
```

One way to avoid this practice involves refactoring object A so that it provides direct access to the functionality in object B, thus allowing you to do the following:

```
void MyClass::MyFunction()
{
    mObjectA.DoAction();
}
```

Alternatively, you could refactor the calling code so that it has an actual object B to invoke the required function directly. This can be done either by storing an instance or reference to object B in `MyClass` or by passing object B into the function that needs it. For example,

```
void MyClass::MyFunction(ObjectB &objectB)
{
    objectB.DoAction();
}
```

The downside of this technique is that you introduce lots of thin wrapper methods into your classes, increase the parameter count of your functions, or increase the size of your objects. However, the benefit is that you end up with more loosely coupled classes in which the dependencies on other objects are made explicit. This makes the code much easier to refactor or evolve in the future. In fact, the latter solution of explicitly passing an object into a function has clear parallels with the technique of dependency injection. Also, another application of the LoD involves creating a single method in object A that aggregates calls to multiple methods of object B, which resonates well with the Façade design pattern.

TIP: The LoD states that you should call functions only in your own class or on immediately related objects.

Class naming

While I've been largely concerned with the details of object-oriented design in these latest sections, once you've developed an appropriate collection of classes, an equally critical task is the development of expressive and consistent names for these classes. Accordingly, here are some guidelines for naming your classes:

- Simple class names should be powerful, descriptive, and self-explanatory. Moreover, they should make sense in the problem domain being modeled and they should be named after the thing they are modeling (e.g., `Customer`, `Bookmark`, or `Document`). As I've already noted, class names tend to form the nouns of your system: the principal objects of your design.
- Joshua Bloch states that good names drive good designs. Therefore, a class should do one thing and do it well, and a class name should instantly convey its purpose (Bloch, 2008). If a class is difficult to name, that's usually a sign that your design is lacking. Kent Beck offers an example in which he originally used the generic compound name `DrawingObject` for an object in a graphical drawing system, but later refined this to the more expressive term `Figure` by referring to the field of typography (Beck, 2007).
- Sometimes it's necessary to use a compound name to convey greater specificity and precision, such as `TextStyle`, `SelectionManager`, or `LevelEditor`. However, if you're using more than two or three words, then this can indicate that your design is too confusing or complex.
- Interfaces and abstract base classes tend to represent adjectives in your object model. They can therefore be named as such: `Renderable`, `Clonable`, or `Observable`. Alternatively, it's common to prefix interface classes with the uppercase letter I: for example, `IRenderer` and `IObserver`.

- Avoid cryptic abbreviations. Good class names should be obvious and consistent. Don't force your users to try and remember which names you've abbreviated and which you have not. I will revisit this point later when I discuss function naming.
- You should include some form of namespace for your top-level symbols, such as classes and free functions, so that your names don't clash with those in other APIs that your clients may be using. This can be done either via the C++ `namespace` keyword or through the use of a short prefix. For example, all OpenGL function calls start with gl and all Qt classes begin with Q.

Function design

The lowest granularity of API design is how you represent the individual function calls and their parameters. Although this may seem like an obvious exercise and not worth covering in much detail, there are many function-level issues that affect good API design. After all, function calls are the most used part of an API: they are how your clients access the API's behavior.

Function design options

There are many interface options you can control when designing a function call (Lakos, 1996). First, for free functions you should consider the following alternatives:

- Static vs. nonstatic function
- Pass arguments by value, reference, or pointer
- Pass arguments as const or nonconst
- Use of optional arguments with default values
- Return result by value, reference, or pointer
- Return result as `const`, `constexpr`, `consteval`, or nonconst
- Operator or nonoperator function
- Use of exception specifications
- Use of attributes such as `[[noreturn]]` or `[[deprecated]]`

For member functions, you should consider all those free function options as well as:

- Virtual vs. nonvirtual member function
- Pure virtual vs. nonpure virtual member function
- Const vs. nonconst member function
- Public, protected, or private member function
- Use of the explicit keyword for nondefault constructors

In addition to these options that control the logical interface of a function, there are a couple of organizational attributes that you can specify for a function, such as:

- Friend vs. nonfriend function
- Inline vs. noninline function

The proper application of these options can make a large impact on the quality of your API. For example, you should declare member functions as const wherever possible to advertise that they don't modify the object (see Chapter 6 on C++ Usage for more details). Passing objects as const references can reduce the amount of memory copying that your API introduces (see Chapter 8 on Performance). The use of the `explicit` keyword can avoid unexpected side effects for nondefault constructors (see Chapter 6), and inlining your functions can sometimes offer a performance advantage at the cost of exposing implementation details and breaking binary compatibility (see Chapters 8 and 10).

Function naming

Function names tend to form the verbs of your system, describing actions to be performed or values to be returned. Here are some guidelines for naming your free and member functions:

- Methods that are used to set or return some value should fully describe that quantity using standard prefixes such as `Get` and `Set`. For example, a function that returns the zoom factor for a Web view might be called `GetZoomFactor()` or, less expressively, just `ZoomFactor()`. If a getter function may perform nontrivial work to return its result, you may want to reflect that in its name (e.g., by using a prefix of `Find` or `Calculate` instead of `Get`).
- Functions that answer yes or no queries should use an appropriate prefix to indicate this behavior, such as `Is`, `Are`, or `Has`, and should return a bool result: for example, `IsEnabled()`, `ArePerpendicular()`, or `HasChildren()`. As an alternative, the C++ Standard Library tends to drop the initial verb, as can be seen in functions such as `empty()` instead of `IsEmpty()`. However, although terser, this naming style is ambiguous because it could also be interpreted as an operation that empties the container (unless you're astute enough to notice the const method decorator). The Standard Library scheme therefore fails the qualities of discoverability and difficulty to misuse.
- Methods that are used to perform some action should be named with a strong verb: for example, `Enable()`, `Print()`, or `Save()`. If you are naming a free function, rather than a method of a class, then you should include the name of the object to which the action will be applied: for example, `FileOpen()`, `FormatString()`, or `MakeVector3d()`.
- Method names should describe everything that the routine does. For example, if a routine in an image processing library performs a sharpening filter on an image and saves it to disk, the method should be called something like `SharpenAndSaveImage()` instead of just `SharpenImage()`. If this makes your method names too long, then this may indicate that they're performing too many tasks and should be split up (McConnell, 2004).

- Avoid abbreviations. Names should be self-explanatory and memorable, but the use of abbreviations can introduce confusing or obscure terminology. For example, the user has to remember if you are using `GetCurrentValue()`, `GetCurrValue()`, `GetCurValue()`, or `GetCurVal()`. Some software projects specify an explicit list of accepted abbreviations that must be followed, but in general it's simply easier for your users if they don't have to remember lists like this.
- Use consistent capitalization rules. You'll see that I favor `CamelCase` in this book, but you may prefer `camelCase` or `snake_case` for your own projects. You should also be consistent about how you include acronyms in your function names: for example, if a function that parses an XML file is called `ParseXMLFile()` or `ParseXmlFile()`. The former style preserves the familiar uppercase form of acronyms, but it can cause confusion when there are multiple acronyms (e.g., should `SendHTTPSXML()` be interpreted as "Send HTTPS XML" or "Send HTTP SXML"?). The capitalization approach makes these cases clearer: e.g., `SendHttpsXml()`.
- Functions should not begin with an underscore character (_). The C++ standard states that global symbols starting with an underscore are reserved for internal compiler use. The same is true for all symbols that contain a double underscore or begin with an underscore followed by an uppercase letter. Although you can find legal combinations of leading underscore names that navigate these rules, it's generally best to simply avoid this practice in your function names (some developers use this convention to indicate a private member).
- Methods that form natural pairs should use the correct complementary terminology. For example, `OpenWindow()` should be paired with `CloseWindow()`, not `DismissWindow()`. The use of precise opposite terms makes it clearer to the user that one method performs the opposite function of another method (McConnell, 2004). This list provides some common complementary terms:

Add/Remove	Begin/End	Create/Destroy
Enable/Disable	Insert/Delete	Lock/Unlock
Next/Previous	Open/Close	Push/Pop
Send/Receive	Show/Hide	Source/Target

Function parameters

The use of good parameter names can also have a big impact on the discoverability of your API. For example, compare these two signatures for the standard C function `strstr()`, which searches for the first occurrence of a substring within another string:

```
char *strstr(const char *s1, const char *s2);
```

and,

```
char *strstr(const char *haystack, const char *needle);
```

I think you'll agree that the second signature gives a much better indication of how to use the function, simply by using descriptive parameter names.

Another factor is to make sure that you use the right data type for your parameters. For example, when you have methods that perform linear algebra calculations, you should prefer using double-precision floats to avoid a loss of precision errors that are inherent in single-precision operations. Similarly, you should never use a floating-point data type to represent monetary values because of the potential for rounding errors (Beck, 2002).

When passing an array of items into a function, you should either try to use a dynamic container, such as an `std::vector`, or a static container of known size, such as an `std::array` or `double[3]`. Avoid having parameters that make assumptions about the size of an array. For example, the following function is a bad design because the user doesn't know how big the bounding box structure should be, which could lead to memory errors if the client provides an array that's too small:

```
void GetBoundingBox(double *bbox)   // bad design
{
    bbox[0] = mMinX;
    bbox[1] = mMaxX;
    bbox[2] = mMinY;
    bbox[3] = mMaxY;
    bbox[4] = mMinZ;
    bbox[5] = mMaxZ;
}
```

There's also a balance to be sought in terms of the number of parameters that you specify for each function. Too many parameters can make the call more difficult to understand and to maintain. It can also imply greater coupling and may suggest that it's time to refactor the function. Therefore, wherever possible, you should try to minimize the number of parameters in your public functions. In this regard, we have the often-cited research from the field of cognitive science, which states that the number of items we can hold in our short-term working memory is seven plus or minus 2 (Miller, 1956). This may suggest that you should not exceed around five to seven parameters; otherwise the user will find it difficult to remember all of the options. Indeed, Joshua Bloch suggests that five or more parameters are too many (Bloch, 2008).

TIP: Avoid long parameter lists.

For functions with many required parameters, you should consider whether an abstraction can be introduced to represent groupings of those parameters. For example, consider the following function that sets a 3D bounding box:

```
void SetBoundingBox(double x1, double y1, double z1, double x2,
double y2, double z3);
```

A better representation for this interface would be to introduce the concept of a 3D point, and then the bounding box can be defined with just two parameters, i.e.,

```
void SetBoundingBox(const Point &p1, const Point &p2);
```

For functions that accept many optional parameters, you may consider passing the arguments using a struct or map instead. For example,

```
struct OpenWindowParams
{
    OpenWindowParams();

    int mX;
    int mY;
    int mWidth;
    int mHeight;
    int mFlags;
    std::string mClassName;
    std::string mWindowName;
};

void OpenWindow(const OpenWindowParams &params);
```

This technique is also a good way to deal with argument lists that may change over the life of the API. A newer version of the API can simply add new fields to the end of the structure without changing the signature of the `OpenWindow()` function. You can also add a version field (set by the constructor) to allow binary compatible changes to the structure: the `OpenWindow()` function can then check the version field to determine what information is included in the structure. Other options include using a field that records the size of the structure in bytes, or simply using a different structure.

SIDEBAR: Reducing parameter lists

All the way back in the 1980s, the Commodore Amiga platform provided an extensive set of stable and well-designed APIs to build applications that run under AmigaOS. The original routine to open a new screen on the Amiga takes a single argument: a structure containing all of the necessary information to specify that screen:

```
struct Screen *OpenScreen(struct NewScreen *newscr);
```

Continued

SIDEBAR: Reducing parameter lists—cont'd

The `NewScreen` structure looks like:

```
struct NewScreen
{
    WORD LeftEdge, TopEdge, Width, Height, Depth;
    UBYTE DetailPen, BlockPen;
    UWORD ViewModes, Type;
    struct TextAttr *Font;
    UBYTE *DefaultTitle;
    struct Gadget *Gadgets;
    struct BitMap *CustomBitMap;
};
```

In Version 36 of the AmigaOS APIs, new functionality was added to this function. This was done by introducing the notion of tag lists, essentially an arbitrarily long list of keyword/value pairs. To support this new extensible scheme, a V36-only function was added to allow the explicit specification of these tag lists:

```
struct Screen *OpenScreenTagList(struct NewScreen *newscr,
                                 struct TagItem *taglist);
```

However, to maintain backward compatibility, it was also possible to pass a new `ExtNewScreen` structure to the `OpenScreen()` function:

```
struct Screen *OpenScreen(struct ExtNewScreen *newscr);
```

This extended structure looks like:

```
struct ExtNewScreen
{
    WORD LeftEdge, TopEdge, Width, Height, Depth;
    UBYTE DetailPen, BlockPen;
    UWORD ViewModes, Type;
    struct TextAttr *Font;
    UBYTE *DefaultTitle;
    struct Gadget *Gadgets;
    struct BitMap *CustomBitMap;
    struct TagItem *Extension;
};
```

When passing this new structure to `OpenScreen()` you had to set the `NS_EXTENDED` bit of the `Type` field to indicate that the structure included an `Extension` field at the end. In this way, you could pass either the old or the new form to newer versions of AmigaOS, but older versions of `amiga.lib` would safely ignore the new data.

Note that this is a plain C API, which cannot support function overloading, so the two versions of the `OpenScreen()` function were not specified in the same version of the API. Newer versions of the API would specify the `ExtNewScreen` signature, although code that tried to pass an older `NewScreen` structure would still compile fine under a C compiler (perhaps with a warning). In C++, this type mismatch would cause a compile error, but in that case you could simply provide two overloaded versions of `OpenScreen()`.

Taking this one step further, you can hide all of the public member variables and allow the values to be accessed only via getter/setter functions. The Qt API refers to this as a property-based API. It's also known more generally as the Named Parameter Idiom. For example:

```
QTimer timer;
timer.setInterval(1000);
timer.setSingleShot(true);
timer.start();
```

This lets you reduce the number of parameters required for functions; in this case, the `start()` function requires no parameters at all. The use of functions to set parameter values also offers the following benefits:

- Values can be specified in any order because the function calls are order-independent.
- The purpose of each value is more evident, because you must use a named function to set the value: e.g., `setInterval()`.
- Optional parameters are supported simply by not calling the appropriate function.
- The constructor can define reasonable default values for all settings.
- Adding new parameters is backward compatible because no existing functions need to change signature. Only new functions are added.

Error handling

A large amount of the code that application developers write is purely to handle error conditions. The actual amount of error handling code that's written will depend greatly upon the particular application. However, it's been estimated that up to 90% of an application's code is related to handling exceptional or error conditions (McConnell, 2004). This is therefore an important area of API design that will be used frequently by your clients. In fact, it's included in Ken Pugh's Three Laws of Interfaces (Pugh, 2006):

1. An interface's implementation shall do what its methods say it does.
2. An interface's implementation shall do no harm.
3. If an interface's implementation is unable to perform its responsibilities, it shall notify its caller.

Accordingly, the three main ways of dealing with error conditions in your API are:

1. Returning error codes
2. Throwing exceptions
3. Aborting the program

The last of these is an extreme course of action that should be avoided at all costs, and indeed it violates the third of Pugh's three laws, although there are far too many examples of libraries out there that call `abort()` or `exit()`. As for the first two cases, different engineers have various proclivities toward each of these techniques. I will not take a side on the exceptions versus error code debate here, but rather I'll attempt to present the arguments and drawbacks impartially for each option. Whichever technique you select for your API, the most important issues are that you use a consistent error reporting scheme and that it is well-documented.

TIP: Use a consistent and well-documented error handling mechanism.

The error codes approach involves returning a numeric code to indicate the success or failure of a function. Normally this error code is returned as the direct result of a function. For example, many Win32 functions return errors using the `HRESULT` data type. This is a single 32-bit value that encodes the severity of the failure, the subsystem responsible for the error, and an actual error code. The C Standard Library also provides examples of nonorthogonal error reporting design, such as the functions `read()`, `waitpid()`, and `ioctl()` that set the value of the `errno` global variable as a side effect. OpenGL provides a similar error reporting mechanism via an error checking function called `glGetError()`.

The use of error codes produces client code that looks like:

```
if (obj1.Function() == ERROR) {
    HandleError();
}

if (obj2.Function() == ERROR) {
    HandleError();
}

if (obj3.Function() == ERROR) {
    HandleError();
}
```

As an alternative, you can use C++'s exception capabilities to signal a failure in your implementation code. This is done by throwing an object for your clients to catch in their code. For example, several of the Boost libraries throw exceptions to communicate error conditions to the client, such as the `boost::iostreams` and `boost::program_options` libraries. The use of exceptions in your API results in client code such as:

```
try {
    obj1.Function();
    obj2.Function();
    obj3.Function();
} catch (const std::exception &e) {
    HandleError();
}
```

The error codes technique provides a simple, explicit, and robust way to report errors for individual function calls. It's also the only option if you're developing an API that must be accessible from plain C programs. The main dilemma comes when you wish to return a result as well as an error code. One way to deal with this is to return the error code as the function result and use an out parameter to fill in the result value. For example,

```
int FindName(std::string &name);

...

std::string name;
if (FindName(name) == OKAY) {
    std::cout << "Name: " << name << std::endl;
}
```

Dynamic scripting languages such as Python handle this more elegantly by making it easy to return multiple values as a tuple. However, this is still an option for modern C++ code. Since C++11, you can use `std::tuple` to return multiple results from a function. For example:

```
std::tuple<int, std::string> FindName();

...

std::tuple<int, std::string> result = FindName();
if (result.get<0>() == OKAY) {
    std::cout << "Name: " << result.get<1>() << std::endl;
}
```

And as of C++17, your clients can use structured binding to make this even easier to use: i.e.,

```
auto [num, name] = FindName();
if (num == OKAY) {
    std::cout << "Name: " << name << std::endl;
}
```

Even better, since C++23, you can use `std::expected` to return either a valid object or an error. You can refer to the section on Expected Values in the C++ Revisions chapter for more details about `std::expected`. In the meantime, here's an example of what the syntax looks like:

```
std::expected<std::string, int> FindName();

...

auto name = FindName();
if (name.has_value()) {
    std::cout << "Name: " << name.value() << std::endl;
}
```

By comparison, exceptions let your clients separate their error handling code from the normal flow of control, making for more readable code. They offer the benefit of being able to catch one or more errors in a sequence of several function calls without having to check every single return code, and they let you handle an error higher up in the call stack instead of at the exact point of failure. An exception can also carry more information than a simple error code. For example, C++ Standard Library exceptions include a human-readable description of the failure, accessible via a `what()` method. Also, most debuggers provide a way to break if an exception is thrown, making it easier to debug problems. Finally, exceptions are the only way to report failures in a constructor.

However, this flexibility does come with a cost. Although the use of exceptions should have no cost when an exception is not thrown, handling an exception when it is thrown can be an expensive operation owing to the run-time stack unwinding behavior. Also, an uncaught exception can cause your clients' programs to abort, resulting in data loss and frustration for their end users. Writing exception-safe code is difficult, and if not done correctly it can lead to resource leaks. Typically, the use of exceptions is an all-or-nothing proposition, meaning that if any part of an application uses exceptions, then the entire application must be prepared to handle exceptions correctly. This means that the use of exceptions in your API also requires your clients to write exception-safe code. It's noteworthy that Google forbids the use of exceptions in their C++ coding conventions because most of their existing code is not tolerant of exceptions. However, as a counterexample, the C++ Standard Library makes extensive use of exceptions to signal errors, so any modern C++ developer who uses the Standard Library must already be prepared to handle exceptions in their code.

If you do opt to use exceptions to signal unexpected situations in your code, here are some best practices to observe:

- Derive your own exceptions from `std::exception` and define a `what()` method to describe the failure.
- Consider the level of exception safety you will provide. There are four main types of exception guarantees (or Abrahams guarantees) you can offer (Stroustrup, 2001):
  - No-throw guarantee: Ensures that the function never throws an exception. If an error occurs within the function, it will be handled internally and not be exposed to clients. This is also known as failure transparency.
  - Strong exception safety: An operation that fails will have no side effects and any original values will be left intact. This is also known as commit or rollback semantics.
  - Basic exception safety: An operation that fails may have side effects, but all invariants will be preserved (i.e., the object should still contain valid values after the error is thrown and no resources should be leaked).
  - No exception safety: No guarantees are made in the case of an error (i.e., data may be corrupted, incorrect values may be returned, or memory may be leaked).

- Consider using RAII techniques to maintain exception safety (i.e., to ensure that resources get cleaned up correctly when an exception is thrown).
- Make sure that you document all the exceptions that can be thrown by a function in its comments.
- You might be tempted to use exception specifications, such as the `throw()` specifier, to document the exceptions that a function may throw. However, be aware that these constraints will be enforced by the compiler at run time, if at all, and that they can affect optimizations such as the ability to inline a function. Also, note that some of this functionality was removed in C++17. As a result, most C++ engineers steer clear of exception specifications such as:

```
void MyFunction1() throw();      // throws no exceptions
void MyFunction2() throw(A, B);  // throws either A or B
```

- Create exceptions for the set of logical errors that can be encountered, not a unique exception for every individual physical error that you raise.
- If you handle exceptions in your own code, then you should catch the exception by reference (as in the previous example) to avoid calling the copy constructor for the thrown object. Also, try to avoid the `catch(...)` syntax because some compilers also throw an exception when a programming error arises, such as an `assert()` or segmentation fault.
- If you have an exception that multiply inherits from more than one base exception class, you should use virtual inheritance to avoid ambiguities and subtle errors in your client's code where they attempt to catch your exceptions.
- Note that on Windows, exceptions cannot cross a DLL boundary with different run times. In that case, any use of exceptions should be contained within the DLL and any public API calls should catch all exceptions and return the error using a different technique, such as an error code or an output parameter.

TIP: Derive your own exceptions from `std::exception`.

In terms of error reporting best practices, your API should fail as fast as possible once an error occurs, and it should clean up any intermediate state, such as releasing resource that were allocated immediately before the error. However, you should also try to avoid returning an exceptional value such as `nullptr` when it's not necessary. Doing so causes your clients to write more code to check for these cases. For example, if you have a function that returns a list of items, consider returning an empty list instead of `nullptr` in exceptional cases. This requires your clients to write less code and reduces the chance that your clients will dereference a `nullptr`.

Also, any error code or exception description should represent the actual failure. Invent a new error code or exception if existing ones do not accurately describe the error. You'll infuriate your users if they waste time trying to debug the wrong problem because your error reporting was inaccurate or plain wrong. You should also give them as much information as possible to track down the error. For example, if a file cannot be opened, then include the filename in the error description and the cause of the failure (e.g., lack of permissions, file not found, or out of disk space).

TIP: Fail quickly and cleanly with accurate and thorough diagnostic details.

5 Styles

The previous chapters dealt with the qualities and approaches toward good application programming interface (API) design. Although I've illustrated these concepts with specific C++ examples, the abstract process of designing an API is language independent. However, in the next few chapters, I'll start to turn to more C++ specific aspects of producing a high-quality API.

In this chapter, I'll cover the topic of API style. Style in this context means how you decide to represent the capabilities of your API. That is, your API may provide access to internal state and routines to perform required functionality, but what is the form of invoking these actions? The answer to this question may seem obvious: you create classes to represent each key object in your API and provide methods on those classes. However, there are other styles you could adopt, and the object-oriented style may not be the best fit all of the time. In this chapter I'll present five very different API styles:

1. **Flat C APIs**: An API that can be compiled by a C compiler. This simply involves a set of free functions along with any supporting data structures and constants. Because this style of interface contains no objects or inheritance, it's often called flat or procedural.
2. **Object-Oriented C++ APIs**: As a C++ programmer, this is likely the style with which you're most familiar. It involves the use of objects with associated data and methods, and the application of concepts such as inheritance, encapsulation, and polymorphism.
3. **Template-Based APIs**: C++ also supports generic programming and metaprogramming via its template functionality. This allows functions and data structures to be written in terms of generic types that can be specialized later by instantiating them with concrete types.
4. **Functional APIs**: Functional programming is a style that relies on composing functions rather than executing sequences of imperative statements. It's a declarative programming style, meaning that it focuses more on what you want to do rather than how you want to do it.
5. **Data-Driven APIs**: This type of interface involves sending named commands to a handler, with arguments that are packaged within a flexible data structure, rather than invoking specific methods or free functions. This style maps well to supporting network protocols and file formats.

I'll now describe each of these API styles in turn and discuss the situations in which one style may be favored over another. Throughout the chapter, I'll use examples from the FMOD API to illustrate three of those styles. FMOD is a commercial library for

API Design for C++. https://doi.org/10.1016/B978-0-443-22219-1.00020-9

creating and playing back interactive audio that is used by many game companies such as Activision, Blizzard, Ubisoft, and Microsoft. It provides a flat C API, a C++ API, and a Data-Driven API to access its core audio functionality. As such it provides an instructive comparison for several of those API styles.

Flat C APIs

The term flat API is meant to convey the fact that the C language doesn't support the notion of encapsulated objects and inheritance hierarchies. Hence, an API that uses pure C syntax must be represented with a more restricted set of language features, such as typedefs, structs, and function calls that exist in the global namespace. Owing to the lack of the `namespace` keyword in C, APIs employing this style must use a common prefix for all public functions and data structures to avoid name collisions with other C libraries.

Of course, you can still use internal linkage (Lakos, 1996) to hide symbol names in your implementation, such as declaring them static at the file scope level of your `.cpp` files. In this way you can be assured that any such functions will not be exported externally and hence will not collide with the same symbol name in another library. (This applies equally to C++ programs as well, of course, although in C++ the use of anonymous namespaces is a preferred way to achieve the same result).

There are many examples of popular C APIs that are in use today, including:

- **The C Standard Library.** If you're writing a C program, then you must be familiar with the C Standard Library. This is composed of a collection of include files (such as `stdio.h`, `stdlib.h`, and `string.h`) and library routines for I/O, string handling, memory management, mathematical operations, etc. (such as `printf()`, `malloc()`, `floor()`, and `strcpy()`). Most C and many C++ programs are built using this library.
- **The Windows API.** Often referred to as the Win32 API, this is the core set of interfaces used to develop applications for the Microsoft Windows range of operating systems. It includes a group of APIs across various categories such as base services, the graphics device interface, the common dialogue box library, and network services. Another library, called Microsoft Foundation Class, provides a C++ wrapper to the Windows API.
- **The Linux Kernel API.** The entire Linux kernel is written in plain C. This includes the Linux Kernel API, which provides a stable interface for low-level software such as device drivers to access operating system functionality. The API includes driver functions, data types, basic C library functions, memory management operations, thread and process functions, and network functions, among many others.
- **GNOME GLib.** This is a general-purpose open source utility library containing many useful low-level routines for writing applications. This includes string utilities, file access, data structures such as trees, hashes, and lists, and a main loop abstraction. This library provides the foundation for the GNOME desktop environment and was originally part of the GIMP Toolkit (GTK+).

- **The Netscape Portable Runtime.** The Netscape Portable Runtime library provides a cross-platform API for low-level functionality such as threads, file I/O, network access, interval timing, memory management, and shared library linking. It is used as the core of the various Mozilla applications, including the Firefox Web browser and Thunderbird email client.
- **Image Libraries**: Most of the open source image libraries that help you add support for various image file formats to your applications are written entirely in C. For example, the libtiff, libpng, libz, libungif, and jpeg libraries are all plain C APIs.

ANSI C features

If you're used to writing C++ APIs, there will be many language features that you'll have to do without when writing a plain C API. For example, C does not support classes, references, smart pointers, templates, the C++ Standard Library, default arguments, access levels (public, private, or protected), or a bool type. Instead, C APIs are generally composed only of:

1. Built-in types such as `int`, `float`, `double`, `char`, and arrays and pointers to these.
2. Custom types created via the `typedef` and `enum` keywords.
3. Custom structures declared with the `struct` or `union` keywords.
4. Global free functions.
5. Preprocessor directives such as `#define`.

In fact, the complete set of C language keywords is quite short. The entire list is presented here as a reference:

- **auto**: defines a local variable as having a local lifetime.
- **break**: passes control out of a while, do, for, or switch statement.
- **case**: defines a specific branch point within a switch statement.
- **char**: the character data type.
- **const**: declares a variable value or pointer parameter to be unmodifiable.
- **continue**: passes control to the beginning of a while, do, or for statement.
- **default**: defines the fallback branch point for a switch statement.
- **do**: begins a do-while loop.
- **double**: the double-precision floating point data type.
- **else**: the statements to perform when an if statement resolves to false.
- **enum**: defines a set of constants of type int.
- **extern**: introduces the name of an identifier that is defined elsewhere.
- **float**: the single-precision floating point data type.
- **for**: defines a for loop.
- **goto**: transfers control to a labeled line of code.
- **if**: provides conditional execution of a sequence of statements.
- **int**: the integer data type.
- **long**: extends the size of certain built-in data types.
- **register**: instructs the compiler to store a variable in a CPU register.

- **return**: exits a function with an optional return value.
- **short**: reduces the size of certain built-in data types.
- **signed**: declares a data type to be able to handle negative values.
- **sizeof**: returns the size of a type or expression.
- **static**: preserves the value of a variable even after its scope ends.
- **struct**: allows multiple variables to be grouped into a single type.
- **switch**: causes control to branch to one of a list of possible statements.
- **typedef**: creates a new type in terms of existing types.
- **union**: groups multiple variables that share the same memory location.
- **unsigned**: declares a data type to handle only positive values.
- **void**: the empty data type.
- **volatile**: indicates that a variable can be changed by an external process.
- **while**: defines a loop that exists when the condition evaluates to false.

Although C is not strictly a subset of C++, well-written ANSI (American National Standards Institute) C programs will tend to be legal C++ programs, too. In general, a C++ compiler will impose greater type checking than a C compiler. When you're writing a plain C API, it's often a worthwhile task to try and compile your code with a C++ compiler and then fix any additional warnings or errors that are raised.

SIDEBAR: Stronger type checking saves brain cells

I recall one occasion early in my career when my manager at SRI International, Yvan Leclerc, had a crashing bug in a C program. He spent the best part of a day trying to track the problem down, and eventually the two of us stepped through his code together, line by line.

After much scratching of heads, we finally noticed that he was using the `calloc()` *function, but was passing only a single argument to it. As you may recall from* Chapter 2, *the* `malloc()` *function takes one parameter, whereas the* `calloc()` *function takes two parameters.*

He had switched from using `malloc()` *to* `calloc()` *to return an initialized block of memory but had forgotten to change the parameters to the function call. As a result, the returned block of memory was not the size he expected it to be. This is not an error in C (although these days most C compilers will at least give you a warning), but it is a compile error in C++. Using a C++ compiler to compile that C code would have turned up the problem immediately.*

TIP: Try compiling your C API with a C++ compiler for greater type checking, and to ensure that a C++ program can use your API.

Benefits of an ANSI C API

One of the main reasons to write an API in C is if it must integrate with an existing project that is written entirely in C. Situations such as these are becoming rarer as more projects are being written from the ground up in C++, but occasions may arise when your clients place this restriction on your API. Examples include any of the existing large C APIs that I listed earlier. For instance, if you're working on a Linux Kernel API, then you will need to write this interface in C.

Another reason to prefer the creation of a plain C API is binary compatibility. If you are required to maintain binary compatibility between releases of your API library, this is much easier to achieve with a plain C API than a C++ one. I will discuss the details of this in the chapter on Versioning, but suffice to say for now that seemingly minor changes to a C++ API can affect the binary representation for the resulting object and library files, thus breaking the ability for clients to simply drop in a replacement shared library and have it work without recompiling their code.

Of course, there's nothing to stop you producing an API that works under both C and C++. In fact, you may even decide that you wish to create a C++ API, to take advantage of the additional object-oriented features of C++, but also create a plain C wrapping of this interface for use within C-only projects, or to expose a simple and low–surface area version of your API for which binary compatibility is easier to enforce. The FMOD API is one such example of this, because it provides both a C++ and a C API.

Writing an API in ANSI C

The C language does not provide support for classes, so you cannot encapsulate data in objects along with the methods that act upon those data. Instead, you declare structs (or unions) that contain data, and then pass those as parameters to functions that operate on those data. For example, consider the following C++ class definition:

```
class Stack
{
public:
    void Push(int val);
    int Pop();
    bool IsEmpty() const;

private:
    int *mStack;
    int mCurSize;
};
```

This might look as follows in terms of a flat C API:

```
struct Stack
{
    int *mStack;
    int mCurSize;
};

void StackPush(Stack *stack, int val);
int StackPop(Stack *stack);
bool StackIsEmpty(const Stack *stack);
```

Each C function associated with the stack must accept the `Stack` data structure as a parameter, often as the first parameter. Also note that the name of the function must normally include some indication of the data on which it operates, because the name is not scoped within a class declaration as in C++. In this case I chose to prefix each function with the word Stack to make it clear that the functions operate on the `Stack` data structure. This example can be further improved by using an opaque pointer to hide the private data, such as:

```
typedef struct Stack *StackPtr;

void StackPush(StackPtr stack, int val);
int StackPop(StackPtr stack);
bool StackIsEmpty(const StackPtr stack);
```

Additionally, C does not support the notion of constructors and destructors. Therefore, any structs must be explicitly initialized and destroyed by the client. This is normally done by adding specific API calls to create and destroy a data structure:

```
StackPtr CreateStack();
void DestroyStack(StackPtr stack);
```

Now that I've compared what a C and C++ API might look like for the same task, let's look at the code that the client must write to use each API style. First, here's an example of using the C++ API:

```
Stack *stack = new Stack;
stack->Push(10);
stack->Push(3);
while (!Stack->IsEmpty()) {
    stack->Pop;
}
delete stack;
```

whereas the same operations performed with the C API might look like:

```
StackPtr stack = CreateStack();
StackPush(stack, 10);
StackPush(stack, 3);
while (!StackIsEmpty(stack)) {
    StackPop(stack);
}
DestroyStack(stack);
```

I'm ignoring the error handling in the previous case for simplicity. But it should be noted that the C++ `new` operator does not return `nullptr` in the case of an error and instead throws a `bad_alloc` exception. So the C++ example would require a `try/catch` block to check for memory allocation errors, whereas the C code would check for a `nullptr` returned from the `CreateStack()` function.

Calling C functions from C++

C++ compilers can also compile C code, and even though you're writing a C API you may want to allow C++ clients to use your API, too. This is a relatively easy task and one that I suggest you undertake as a matter of course when releasing a C API.

The first step is to make sure that your code compiles under a C++ compiler. As I've already noted, the C standard is more relaxed, and so a C compiler will let you get away with more sloppy code than a C++ compiler will.

As part of this process, you will, of course, want to make sure that you don't use any C++ reserved keywords in your code. For example, the following code is legal C, but will produce an error with a C++ compiler because `class` is a reserved word in C++:

```
int class = 0;
```

Finally, C functions have different linkage than C++ functions. That is, the same function is represented differently in object files produced by a C and C++ compiler. One reason for this is that C++ supports function overloading: declaring methods with the same name but different parameters or return values. As a result, C++ function names are mangled to encode additional information in the symbol name such as the number and type of each parameter. Because of this linkage difference, you cannot compile C++ code that uses a function, say `DoAction()`, and then link this against a library produced by a C compiler that defines the `DoAction()` function.

To get around this problem, you must wrap your C API in an `extern "C"` construct, which tells the C++ compiler that the contained functions should use C-style linkage. A C compiler will not be able to parse this statement, so it's best to compile it conditionally for C++ compilers only. This code snippet illustrates this best practice:

```
#ifdef __cplusplus
extern "C" {
#endif

// your C API declarations

#ifdef __cplusplus
}
#endif
```

> *TIP: Use an* `extern "C"` *scope in your C API headers so that C++ programs can compile and link against your API correctly.*

Case study: FMOD C API

The following source code presents a small program using the FMOD C API to play a single sound sample. This is provided to give you a real-world example of using a flat C API. Note the use of function naming conventions to create multiple layers of namespace, in which all functions begin with `FMOD_`, all system-level calls begin with `FMOD_System_`, and so on. Notice, in the interest of readability, that this example does not perform any error checking. Obviously, any real program would check that each function call completed without error:

```
#include "fmod.h"

int main(int argc, char *argv[])
{
    FMOD_SYSTEM *system;
    FMOD_SOUND *sound;
    FMOD_CHANNEL *channel = 0;
    unsigned int version;

    // Initialize FMOD
    FMOD_System_Create(&system);

    FMOD_System_GetVersion(system, &version);
    if (version < FMOD_VERSION) {
        printf("Error! FMOD version %08x required\n", FMOD_VERSION);
        exit(0);
    }

    FMOD_System_Init(system, 32, FMOD_INIT_NORMAL, NULL);

    // Load and play a sound sample
    FMOD_System_CreateSound(system, "sound.wav", FMOD_SOFTWARE,
                            0, &sound);
    FMOD_System_PlaySound(system, FMOD_CHANNEL_FREE, sound,
                          0, &channel);

    // Main loop
    while (!UserPressedEscKey()) {
        FMOD_System_Update(system);

        NanoSleep(10);
    }

    // Shut down
    FMOD_Sound_Release(sound);
    FMOD_System_Close(system);
    FMOD_System_Release(system);
    return 0;
}
```

Object-oriented C++ APIs

When you consider writing an API in C++, you probably think in terms of object-oriented design. Object-oriented programming (OOP) is a style of programming in which data and the functions that operate on those data are packaged together as an object. The origins of OOP date back to the 1960s with the development of the Simula and Smalltalk languages. However, it didn't take off as a dominant programming model until the 1990s, with the introduction of languages such as C++ and Java.

Because I have already covered many of the key techniques of OOP in the previous chapters, I will not spend too much time on this API style. I refer you to the Class Design section of the previous section, where I defined various OOP terms such as class, object, inheritance, composition, encapsulation, and polymorphism.

It should be noted that features such as method and operator overloading, default parameters, templates, exceptions, and namespaces are strictly not object-oriented concepts. However, they are new features that were included in the C++ language and are not a part of the original C language. C++ supports several programming models other than OOP, such as procedural programming (as seen in the previous section), generic programming, and functional programming (both of which I'll cover next).

Advantages of object-oriented APIs

The primary benefit of using an object-oriented API is the ability to use classes and inheritance: that is, the ability to model software in terms of interconnected data, rather than collections of procedures. This can provide both conceptual and technical advantages.

In terms of conceptual advantages, often the physical items and processes that you try to model in code can be described in terms of objects. For example, an address book is a physical item with which we are all familiar, and it contains descriptions for several people, which again is a conceptual unit to which anyone can relate. The core task of OOP is therefore to identify the key objects in a given problem space and determine how they relate to each other. Many engineers believe that this is a more logical way to approach software design than thinking in terms of the set of all actions that must be performed (Booch et al., 2007).

As for the technical advantages, using objects provides a way to encapsulate all of the data and methods for a single conceptual unit in one place. It essentially creates a unique namespace for all related methods and variables. For example, the methods of our earlier C++ Stack example all exist within the `Stack` namespace, such as `Stack::Push()` or `Stack::Pop()`. Objects also provide the notion of information hiding with public, protected, and private access levels, which is a critical concept for API design.

> *TIP: Object-oriented APIs offer the advantages of inheritance, encapsulation, and information hiding.*

Disadvantages of object-oriented APIs

However, there can be downsides to using object-oriented concepts. Many of these result from abuses of the power of object-oriented techniques. The first is that adding inheritance to your object model can introduce a degree of complexity and subtlety that not all engineers may fully understand, such as knowing that base class destructors must always be marked as virtual, or that an overridden method in a subclass will hide all overloaded methods with the same name in the base class.

Furthermore, deep inheritance hierarchies can make it challenging to figure out the complete interface offered by an object just by looking at header files, because the interface may be distributed across multiple headers (of course, good documentation tools such as Doxygen can help to abate this particular concern). Also, some engineers may abuse, or incorrectly use, the concepts of OOP, such as using inheritance in cases where it doesn't make sense (where the objects don't form an is-a relationship). This can cause strained and unclear designs that are difficult to work with.

Finally, creating a binary compatible API using object-oriented C++ concepts is an extremely difficult task. If binary compatibility is your goal, you may wish to choose one of the other API styles that I describe in this chapter, such as a flat C API or a data-driven API.

Case study: FMOD C++ API

The following source code presents the same program you saw in the earlier section, except that this example uses the FMOD C++ API instead of the C API. Namespacing is now achieved using the C++ namespace feature, so that all classes and functions exist within the FMOD namespace. Also the include file for the API has the same base name as the C API, except that it uses an `.hpp` extension to indicate that it's a C++ header. Once again, error checking has been omitted to make the code more legible:

```
#include "fmod.hpp"

int main(int argc, char *argv[])
{
    FMOD::System *system;
    FMOD::Sound *sound;
    FMOD::Channel *channel = 0;
    unsigned int version;

    // Initialize FMOD
    FMOD::System_Create(&system);

    system->getVersion(&version);
    if (version < FMOD_VERSION) {
        printf("Error! FMOD version %08x required\n", FMOD_VERSION);
        exit(0);
    }

    system->init(32, FMOD_INIT_NORMAL, NULL);

    // Load and play a sound sample
    system->createSound("sound.wav", FMOD_SOFTWARE, 0, &sound);
    system->playSound(FMOD_CHANNEL_FREE, sound, 0, &channel);

    // Main loop
    while (!UserPressedEscKey()) {
        system->update();

        NanoSleep(10);
    }

    // Shut down
    sound->release();
    system->close();
    system->release();
    return 0;
}
```

Template-based APIs

Templates are a feature of C++ that allow you to write functions or classes in terms of a generic yet-to-be-specified type. You can then specialize the template by instantiating it with a specific type or types. As a result, programming with templates is often called generic programming.

Templates are an extremely powerful and flexible tool. They can be used to write programs that generate code or that execute code at compile time (a technique known as metaprogramming). This can be used to achieve impressive results such as unrolling loops, precomputing certain values in a mathematical series, generating lookup tables at compile time, and expanding recursive functions that recurse a predetermined number of times. As such, templates can be used to perform work at compile time and thus improve run time performance.

However, it's not the focus of this book to provide a treatment of these aspects of template programming. There are many great books out there that already do this (Alexandrescu, 2001; Vandevoorde and Josuttis, 2002). Instead, our focus will be on the use of templates for API design. In this regard, there are several examples of well-designed template-based APIs that you can look to for reference and inspiration, such as:

- **The C++ Standard Library.** All of the Standard Library container classes with which you're familiar, such as `std::set`, `std::map`, and `std::vector`, are class templates. That's why they can be used to hold data of different types.
- **Boost.** The Boost libraries provide a suite of powerful and useful features, many of which have since been added to the C++ standard. Most of the Boost classes use templates such as `boost::shared_ptr`, `boost::function`, and `boost::static_pointer_cast`.
- **Loki.** This is a library of class templates written by Andrei Alexandrescu to support his book, *Modern C++ Design*. It provides implementations of various design patterns, such as Visitor, Singleton, and Abstract Factory. This elegant code provides an exemplar of good template-based API design.

Even though templates in C++ are often used in combination with object-oriented techniques, it's worth noting that the two are completely orthogonal concepts. Templates can be used equally well with free functions and with structs and unions (although of course, as you already know, structs are functionally equivalent to classes in C++, except for their default access level).

An example template-based API

Continuing our stack example, let's look at how you would create a generic stack declaration using templates, and then instantiate it for integers. You can define the template-based stack class in terms of a generic type `T` as:

```
#include <vector>

template <typename T>
class Stack
{
public:
    void Push(T val);
    T Pop();
    bool IsEmpty() const;

private:
    std::vector<T> mStack;
};
```

I have omitted the method definitions to keep the example clear. I will present some best practices for providing template definitions in the later chapter on C++ Usage. It's also worth noting that there's nothing special about the name `T`. It's common to use the name `T` for your generic type, but you could equally well call it `MyGenericType` if you prefer.

With this declaration for a generic stack, you can then instantiate the template for the type `int` by creating an object of type `Stack<int>`. This will cause the compiler to generate code for this specific type instance. You could also define a simple typedef to make it more convenient to access this instance of the template, such as:

```
using IntStack = Stack<int>;
```

Then this `IntStack` type can be used just as if you had written the class explicitly. For example:

```
IntStack *stack = new IntStack;
stack->Push(10);
stack->Push(3);
while (! stack->IsEmpty()) {
    stack->Pop;
}
delete stack;
```

Templates versus macros

An alternative to the templates approach would be to use the C preprocessor to define a block of text that you can stamp into the header multiple times, such as:

```
#include <vector>

#define DECLARE_STACK(Prefix, T) \
    class Prefix##Stack         \
    {                           \
    public:                     \
        void Push(T val);       \
        T Pop();                \
        bool IsEmpty() const;   \
                                \
    private:                    \
        std::vector<T> mStack;  \
    }

DECLARE_STACK(Int, int);
```

Aside from the ugliness of this code (e.g., having to end each line with a backslash and use preprocessor concatenation), the preprocessor has no notion of type checking or scoping. It's simply a text copying mechanism. This means that the declaration of the macro is not actually compiled, and so any errors in your macro will be reported on the single line where it's expanded, not where it's declared. Similarly, you can't step into your methods with a debugger because the whole code block is expanded in a single line of your source file. By contrast, templates provide a type-safe way to generate code at compile time, and you'll be able to debug into the actual lines of your class template.

Unless you're writing a plain C API and therefore don't have access to templates, you should avoid using the preprocessor to simulate templates.

Advantages of template-based APIs

The obvious power of templates is that they let you create (instantiate) many different classes from a single root declaration. In our previous stack example, you could add support for string-based and floating-point stack classes simply by adding these declarations:

```
using StringStack = Stack<std::string>;
using DoubleStack = Stack<double>;
```

As such, templates can help to remove duplication because you don't need to copy, paste, and tweak the implementation code. Without templates, you would have to create (and maintain) a lot of very similar-looking code to support `IntStack`, `StringStack`, and `DoubleStack` class.

Another important property of templates is that they can provide static (compile time) polymorphism, as opposed to the use of inheritance, which provides dynamic (run time) polymorphism. One element of this is that templates allow the creation of different classes that all exhibit the same interface. For example, every instance of our stack class, whether IntStack, DoubleStack, or StringStack, is guaranteed to provide the same set of methods. You can also use templates to create functions that accept any of these types, without the run time cost of using virtual methods. This is achieved by generating different type-specific versions of the function at compile time. The following template function demonstrates this ability: it can be used to pop the topmost element from any of our stack types. In this example, two different versions of the function will be generated at compile time:

```
template <typename T>
void PopAnyStack(T &stack)
{
    if (!stack.IsEmpty()) {
        stack.Pop();
    }
}

...

IntStack int_stack;
StringStack string_stack;

int_stack.Push(10);
string_stack.Push("Hello Static Polymorphism!");

PopAnySack(string_stack);
PopAnyStack(int_stack);
```

A further benefit of templates is that you can specialize certain methods of a class for a specific type instance. For instance, our generic stack template is defined as Stack<T>, but you could provide customized function implementations for certain types, such as for Stack<int>. This is very handy for optimizing the class for certain types or for adding customizations for certain types that behave uniquely. This can be done by providing a method definition with the syntax:

```
template <>
void Stack<int>::Push(int val)
{
    // integer specific push implementation
}
```

Disadvantages of template-based APIs

In terms of disadvantages of using templates, the most critical one for API design is that the definition of your class templates will normally have to appear in your public headers. This is because the compiler must have access to the entire definition of your template code to specialize it. This obviously exposes your internal details, which you know is a major sin of API development. It also means that the compiler will recompile the code each time the file is included, causing the generated code to be added to the object file for every module that uses the API. The result can be slower compilation times and code bloat (although C++11 introduced the notion of extern templates to address this). However, there are situations where you can in fact hide the implementation of a template in the `.cpp` file, using a technique called explicit instantiation. I will discuss this technique in more detail in the next chapter on C++ Usage.

The previous static polymorphism example demonstrates another potential source of code bloat. This is because the compiler must generate code for each different version of the `PopAnyStack()` function that's used. This is opposed to the virtual method flavor of polymorphism, which requires the compiler only to generate one such method, but then incurs a run time cost to know which class's `IsEmpty()` and `Pop()` methods to call. Therefore, if code size is more important to you than run time cost, you may decide to go with an object-oriented solution rather than use templates. Alternatively, if run time performance is critical for you, then templates may be the way to go.

Another commonly viewed disadvantage of templates is that most compilers can create verbose, long, or confusing messages for errors that occur in template code. It's common for simple errors in heavily templated code to produce dozens of lines of error output that cause you to scratch your head for a long time. In fact, there are even products on the market to simplify template error messages and make them easier to decipher, such as the STLFilt utility from BD Software. This is a concern not only for you as the developer of an API, but also for your clients because they will also be exposed to these voluble error messages if they use your API incorrectly. C++20 introduced a new feature called constraints to help alleviate this issue and provide more contextual template error messages.

Functional APIs

Functional programming is a programming style that relies on composing functions rather than executing sequences of imperative statements. It relies heavily on functions as a first-class concept, rather than objects. This approach is based on the formal system of computation called the lambda calculus, developed by Alonzo Church in 1936.

There are several programming languages that support only functional programming, such as Haskell, and there are others that support functional programming along with imperative style programming, such as C++. With the introduction of C++11, the language now has stronger support for functional programming, owing to additions like lambda functions and several new C++ Standard Library functions.

Functional programming concepts

Functional programming is a declarative programming style, meaning that it focuses more on what you want to do rather than the imperative approach of how you want to do it. That is, it focuses on the evaluation of expressions instead of the execution of statements (Čukić, 2018). For example, this imperative code shows how to calculate the average of a collection of integers:

```
auto nums = {1, 2, 4, 6, 1, 9, 3, 4};
int sum = 0;
for (int num : nums) {
    sum += num;
}
auto avg = sum / (double) nums.size();
```

whereas the following shows the same thing expressed in a functional style, using the C++20 `std::accumulate()` function:

```
auto nums = {1, 2, 4, 6, 1, 9, 3, 4};
auto avg = std::accumulate(nums.begin(), nums.end(), 0) / (double)
nums.size();
```

In the functional programming style, functions are a first-class concept, so they can be passed as arguments to other functions, returned as the result of a function, and assigned to variables. For example, the `std::accumulate` function also has an overloaded version in which you can provide a function to define an accumulation operator other than addition, such as:

```
auto nums = {1, 2, 3, 4, 5};
auto op = [](int a, int b) {
    return a * b;
};
int factorial = std::accumulate(nums.begin(), nums.end(), 1, op);
// returns 120
```

Another core aspect of functional programming is the use of pure functions. These are functions that have the attributes of:

- **Immutable arguments:** pure functions don't modify their arguments.
- **Stateless:** pure functions don't rely on any inputs other than their arguments.
- **No side effects:** pure functions don't mutate nonlocal variables or local static variables.

As a result, pure functions can be called multiple times with the same arguments and they'll always return the same result. For example, mathematical functions such as `sqrt()` and `max()` can be implemented as pure functions, whereas `rand()` and `time()` cannot. (Note that a pure function in this context of functional programming is very different from a pure virtual function that's used in object-oriented C++ programming.)

(It's interesting that there's been a proposal to add support for a [[pure]] attribute to the C++ standard so that developers could specify a function as being pure. Among other things, this would allow compilers to perform more aggressive optimizations. However, at the time of writing this book, this proposal has not been adopted by the C++ standards committee).

An example functional API

Continuing our example of defining an API for manipulating a stack, this behavior could be expressed in terms of a functional style by introducing various functions that operate on a stack data structure as:

```
namespace stack {

using Stack = std::vector<int>;

[[nodiscard]] Stack push(const Stack &stack, int element);
[[nodiscard]] Stack pop(const Stack &stack);
[[nodiscard]] Stack pop_if(const Stack &stack,
std::function<bool(int)> predicate);

}  // namespace stack
```

This design shows several pure functions that are contained within a stack namespace and that can be chained together to create composite expressions, such as:

```
auto stack = stack::push(stack::pop(stack::push(stack::push({}, 1),
2)), 3);
// returns {1, 3}
```

This API also shows the use of functions as a first-class concept by introducing a pop_if() function that accepts a function argument, in which elements will continue to be popped from the stack as long as the function returns true, such as:

```
auto isEven = [](int x) {
    return x % 2 == 0;
};
auto stack = stack::pop_if({1, 2, 3, 4, 6}, isEven);
// returns {1, 2, 3}
```

I've used the C++17 [[nodiscard]] attribute on each of the functions. That's because pure functions have no effect if their return value is not used, because they are defined to be stateless and without side effects. The use of [[nodiscard]] will therefore generate a compiler warning if the user ignores the return result. This can help if the user assumes, incorrectly, that the function will modify the data structure in place instead of returning its result as a new data structure.

There are several other ways that a functional style stack API could be implemented. Often, you'll find that you want to combine functional programming with template programming, so that your functions can work on different types. The example I provided earlier works only on a stack of integers, so that I could focus on functional concepts, but you could, of course, extend it to use function templates that work on any type, `T`. Also, many of the functional algorithms added to the C++ Standard Library, such as `std::accumulate`, `std::copy_if`, and `std::transform`, operate on iterators instead of containers. This makes them more general, but it also makes returning the results more convoluted and difficult to chain together as I did earlier.

Advantages of functional APIs

Functional programming has been a popular style of programming for many years, largely because of the many benefits it offers. These benefits include:

- **Declarative style:** Because functional programming focuses more on the how than the what, it can result in more self-explanatory code. For example, rather than writing `for` loops with custom logic to manipulate individual elements of a container, you can rely on higher-level functions such as `for_each`, `transform`, and `copy_if`.
- **Code readability:** Functional code can often be shorter than the equivalent imperative code because it tends to be specified at a higher level. This can help with the readability and maintainability of your clients' code.
- **Concurrency:** As I'll cover in the Concurrency chapter, the use of shared mutable state is the most common problem when trying to produce thread-safe code. However, pure functions are defined to have no shared state or side effects. They can therefore be trivially parallelized without the need to add synchronization primitives such as mutex locking.
- **Loose coupling:** The use of pure functions can also produce more robust code, and often fewer bugs as a result, because the lack of shared state and side effects removes any complex interactions with other parts of the system.
- **Testability:** A further benefit of pure functions is that they're easier to test. That's because they don't depend on external state, meaning that their behavior is fully expressed by the set of inputs that accept. Also, they will return the same result for the same set of inputs, making any automated tests more reproducible and robust.

Disadvantages of functional APIs

Every programming style has its disadvantages as well as its advantages. In terms of the functional style, some potential concerns to bear in mind include:

- **Memory usage:** Functional programming relies on making copies of data to avoid mutating existing variables. This style may therefore consume more memory than an equivalent imperative solution. This is a direct trade-off for all of the benefits of pure functions listed earlier. However, there may be optimizations that compilers can implement to offset these memory concerns, such as copy on write and return value optimization.
- **Performance:** Copying large amounts of memory can affect performance. Dedicated functional programming languages can often implement optimizations to offset these concerns, such as lazy evaluation and tail call recursion, but compilers for multiparadigm languages such as C++ may not be able to implement all of these optimizations as effectively.
- **Hardware compatibility:** Another aspect of performance is that imperative programming generally maps well to today's CPU and GPU architectures. However, functional programming does not map as well to modern hardware. The heavy use of recursion in traditional functional programming is one example, although it's fair to note that most functional implementations in C++ are implemented iteratively, not recursively.
- **Less support:** Objects and templates are the most common API styles for C++. As a result, there may be fewer functional programming experts that you can rely upon for your functional API designs and fewer resources such as tools and documentation available to you. Writing good functional APIs does require you to think differently, and the demands of producing a design for a general audience may require you to adopt the more mainstream methodologies.

Data-driven APIs

A data-driven API can be differentiated from a code-driven API:

- **Code-driven API:** an engineer writes code that's compiled into a binary and distributed to clients for them to run. If new behavior is needed, the engineer changes the code and distributes a new binary to the clients. All of the examples so far in this chapter are examples of code-driven APIs.
- **Data-driven API:** an engineer writes code for a general engine that can read and interpret data from a file, where that file defines much of the run time behavior of the system. If new behavior is needed, a nonengineer (potentially even the client) could change the data file without needing to recompile the software.

A data-driven program is therefore one that can perform different operations each time it's run by supplying it with different input data. For example, a data-driven program may simply accept the name of a file on disk that contains a list of commands to execute. This has an impact on the design of an API because it means that instead of relying on a collection of objects that provide various method calls, you provide more generic routines that accept named commands and a dictionary of named arguments. This is sometimes also called a message passing API or event-based API. The following function call formats illustrate how this API type differs from standard C and C++ calls:

- `func(obj, a, b, c)` = flat C-style function
- `obj.func(a, b, c)` = object-oriented C++ function
- `send("func", a, b, c)` = data-driven function with positional parameters
- `send("func", dict(arg1=a, arg2=b, arg2=c))` = data-driven function with a dictionary of named arguments (pseudocode).

> **SIDEBAR: A data-driven dialogue engine**
>
> *While at Apple, I developed a data-driven dialogue system for the Siri virtual assistant* (Rhoten and Reddy, 2020). *This was driven by a data file format called Conversation Authoring Template (CAT), which could contain localized dialogue with embedded parameter values and conditional logic.*
>
> *We developed an API that could load a CAT file for the current locale* (e.g., *en-US or fr-FR) and accept various parameters from the run time code. The parameters would be formatted, correctly inflected for the locale, and inserted into the dialogue template. Conditions defined in the file could be used to select among different templates based on the input parameter values. The resulting localized dialogue would be spoken by Siri and shown on the display device.*
>
> *Some of the benefits of this data-driven approach were that we could build rich authoring and localization GUI tools for nonengineers to use, we could employ extensive validation of the data files to catch errors early on, and we could push updates for Siri's dialogue to devices after an OS had shipped because the changes were expressed as data, not code.*

To provide a concrete example of what this looks like, let's see how you might redesign our Stack example using a more data-driven model:

```
class Stack
{
public:
    Stack();

    Result Command(const std::string &command,
                   const ArgList &args);
};
```

This simple class could then be used to perform multiple operations, such as:

```
s = new Stack;
s->Command("Push", ArgList().Add("value", 10));
s->Command("Push", ArgList().Add("value", 3));
auto r = s->Command("IsEmpty");
while (!r.convertToBool()) {
    s->Command("Pop");
    r = s->Command("IsEmpty");
}
delete s;
```

This is a more data-driven API because the individual methods have been replaced by a single method, `Command()`, which supports multiple possible inputs specified as string data. One could easily imagine writing a simple program that could parse the contents of an ASCII text file containing various commands and executing each command in order. The input file could look something like:

```
# Input file for data-driven Stack API
Push value:10
Push value:3
Pop
Pop
```

A program to consume this data file using the previous data-driven API would simply take the first whitespace-delimited string on each line (ignoring blank lines and lines that begin with # as a convenience). The program could then create an `ArgList` structure for any further whitespace-delimited strings that follow the initial command. It would then pass those to the `Stack::Command()` and continue processing the remainder of the file. This program could then perform vastly different stack operations by supplying a different text file, and notably without requiring the program to be recompiled.

Advantages of data-driven APIs

I've already pointed out one of the major benefits of data-driven APIs: that the business logic of a program can be abstracted out into a human-editable data file. In this way, the behavior of a program can be modified without the need to recompile the executable.

You may even decide to support a separate design tool to let users easily author the data file. Several commercial packages work this way, such as FMOD, which includes the FMOD Designer program that allows complex authoring of sound effects. The resulting `.fev` files can be loaded by the data-driven FMOD Event API. Also, the Qt UI toolkit includes the Qt Designer application that lets users create user interfaces in a visual and interactive fashion. The resulting `.ui` files can be loaded at run time by Qt's QUiLoader class.

Another major benefit of a data-driven API is that it tends to be far more tolerant of future API changes. That's because adding, removing, or changing a command can in many cases have no effect on the signatures of the public API methods. Often it will simply change the supported set of strings that can be passed to the command handler. In other words, passing an unsupported or obsolete command to the handler will not produce a compile-time error. Similarly, different versions of a command can be supported based upon the number and type of arguments that are provided, essentially mimicking C++'s method overloading.

Taking the earlier example of our data-driven Stack API, which simply provides a `Stack::Command()` method, a newer version of the API might add support for a `Top` command (to return the topmost element without popping it) and could also extend the `Push` command to accept multiple values, each of which are pushed onto the stack in turn. An example program using these new features might look like:

```
s = new Stack;
s->Command("Push", ArgList().Add("value1", 10).Add("value2", 3));
Result r = s->Command("Top");
int top = r.convertToInt();  // top == 3
```

Adding this new functionality involved no change whatsoever to the function signatures in the header file. It merely changed the supported strings, and the list of arguments, which can be passed to the `Command()` method. Because of this property, it's much easier to create backward compatible API changes when using a data-driven model, even when removing or changing existing commands. Similarly, it's much easier to create binary compatible changes because it's more likely that you'll not need to change the signature of any of your public methods.

One further benefit of data-driven APIs is that they more easily support data-driven testing techniques. This is an automated testing technique in which, instead of writing lots of individual test programs or routines to exercise an API, you can simply write a single data-driven program that reads a file containing a series of commands to perform and assertions to check. Then, writing multiple tests means simply creating multiple input data files. Test development iteration can therefore be faster because no compilation step is required to create a new test. Also, QA engineers who do not possess deep C++ development skills can still write tests for your API.

Remaining with our Stack example, you could create a test program that accepts input data files such as:

```
IsEmpty => True    # A newly created stack should be empty
Push value:10
Push value:3
IsEmpty => False   # A stack with two elements is nonempty
Pop => 3
IsEmpty => False   # A stack with one element is nonempty
Pop => 10
IsEmpty => True    # A stack with zero elements is empty
Pop => NULL        # Popping an empty stack is an error
```

This test program is very similar to the program I described earlier to read Stack commands from a data file. The main difference is that I've added support for a => symbol, which lets you check the result returned by the `Stack::Command()` method. With that small addition, you now have a flexible testing framework that allows you to create any number of data-driven tests for our API.

> *TIP: Data-driven APIs map well to Web services and other client/server APIs. They also support data-driven testing techniques.*

Disadvantages of data-driven APIs

As I've stated, the data-driven model is not appropriate for all interfaces. It may be useful for data communication interfaces such as Web services or for client/server message passing. However, it would not be an appropriate choice for a real-time 3D graphics API.

For one reason, the simplicity and stability of the API comes with a run time cost. This is because of the additional overhead of finding the correct internal routine to call given a command name string. The use of an internal hash table or dictionary that maps supported command names to callable functions can speed this process, but it will never be as fast as calling a function directly.

It's also worth noting that data-driven systems may require additional functionality to be built that you would not normally need for a code-driven solution. For example, with a code-driven solution you can rely on the compiler to emit warnings and errors if you make mistakes representing the logic of your program. However, with a data-driven approach you may need to write your own validation routines for the data formats that you support, so that you can detect syntactic and semantic errors in the data file inputs. You may also need to build a robust versioning system for your data format, so that your code can handle older or newer formats appropriately.

Furthermore, another downside of data-driven APIs is that your physical header files do not reflect your logical interface. This means that a user cannot simply look at your public header files and know what functionality and semantics are provided by the interface. However, recall in the first section of this book that I defined an API as a collection of header files … and associated documentation. So, as long as you provide good API documentation to specify the list of supported commands and their expected arguments, you can reasonably compensate for this disadvantage.

Supporting variant argument lists

Up to this point, I have glossed over the use of our `Result` and `ArgList` types in the various previous examples. These are meant to represent data values that can contain differently typed values that may not be known at compile time. For example, `ArgList` could be used to pass no arguments, a single integer argument, or two arguments in which one is a string and the other is a float. Weakly typed languages such as Python explicitly support this concept, but traditionally C++ has not: arrays and containers must contain elements that are all the same type, and where that type must be known at compile time.

However, C++17 introduced the `std::any` type, which provides a type-safe way to hold a value of different types. This can therefore be used to parse arbitrary data from a file. For example, the following types can be used to represent a dictionary of arbitrarily typed values, such as a JavaScript Object Notation (JSON) object, using the C++ `any` type:

```
using Arg = std::any;
using ArgList = std::map<std::string, Arg>;
```

This would allow us to create an interface with an optional number of named arguments that can be of type bool, int, double, or string. For example:

```
s = new Stack;
ArgList args;
args["NumberOfElements"] = Arg(2);
s->Command("Pop", args);
```

You can now support methods that accept a single parameter of type `ArgList`, which can be used to pass any combination of `bool`, `int`, `double`, or `std::string` arguments. As such, future changes to the API behavior (e.g., adding a new argument to the list of arguments supported by the method) can be made without changing the actual signature of the method.

Case study: FMOD data-driven API

Here, I present a simple program using the FMOD data-driven API, to give a real-world example of this API style. This is only one example of a data-driven interface, and it does not illustrate all of the concepts that I've discussed. However, it does illustrate the case in which much of the logic is stored in a data file that's loaded at run time. This is the `sound.fev` file, which is created by the FMOD Designer tool. The program then shows accessing a named parameter of an event in that file and changing that parameter's value:

```
#include "fmod_event.hpp"

int main(int argc, char *argv[])
{
    FMOD::EventSystem *eventsystem;
    FMOD::Event *event;
    FMOD::EventParameter *param;
    float param_val = 0.0f;
    float param_min, param_max, param_increment;

    // Initialize FMOD
    FMOD::EventSystem_Create(&eventsystem);
    eventsystem->init(64, FMOD_INIT_NORMAL,
                      0, FMOD_EVENT_INIT_NORMAL);

    // Load a file created with the FMOD Designer tool
    eventsystem->load("sound.fev", 0, 0);
    eventsystem->getEvent("EffectEnvelope", FMOD_EVENT_DEFAULT,
                          &event);

    // Get a named parameter from the loaded data file
    event->getParameter("param", &param);
    param->getRange(&param_min, &param_max);
    param->setValue(param_val);
    event->start();

    // Continually modulate the parameter until Esc pressed
    param_increment = (param_max - param_min) / 100.0f;
    bool increase_param = true;
    while (!UserPressedEscKey()) {
        if (increase_param) {
            param_val += param_increment;
            if (param_val > param_max) {
                param_val = param_max;
                increase_param = false;
            }
        } else {
            param_val -= param_increment;
            if (param_val < param_min) {
                param_val = param_min;
                increase_param = true;
            }
        }
        param->setValue(param_val);

        eventsystem->update();
        NanoSleep(10);
    }

    // Shut down
    eventsystem->release();
    return 0;
}
```

Data-driven Web services

Not all interfaces are appropriately represented with a data-driven style. However, this style is particularly suited to stateless communication channels such as client/server applications in which the API allows commands to be sent to a server and optionally to return results to the client. It's also useful for passing messages between loosely coupled components and for working with inherently data-driven entities such as file formats.

Web services in particular can be naturally represented using a data-driven API. A Web service is normally accessed by sending a URL with a set of query parameters to a given Web service, or by sending a message in some structured format such as JSON or XML. For instance, the Digg website supports an API to let users interact with the Digg.com Web service. As a specific example, the Digg API provides the `digg.getInfo` call to return extended information for a specific digg on a story. This is invoked by sending an HTTP GET request in the form:

```
http://services.digg.com/1.0/endpoint?method=digg.getInfo&digg_id=id
```

This maps well to the sort of data-driven APIs I've presented earlier, in which an HTTP request such as this could be invoked as:

```
d = new DiggWebService;
d->Request("digg.getInfo", ArgList().Add("digg_id", id));
```

This correlates very closely to the underlying protocol, although it still provides an abstraction from the details of that protocol. For example, the implementation can still decide whether it's more appropriate to send the request as a GET or POST, or even a JSON or XML description.

Idempotency

Because I'm discussing client/server applications, it's worth covering the topic of idempotency here, as well. An API is considered idempotent if calling a function more than once with the same set of parameters has no additional effect. For example, removing an element from an `std:set` or `std:map` is idempotent because if you attempt to remove the same element multiple times, the contents of the data structure will be the same as if you tried to remove the element only once.

Here's a simple example of an idempotent API, in which you can get or set a vector of strings, check if a string exists in the vector, and remove a string from the vector:

```
class Labels
{
public:
    std::vector<std::string> GetLabels() const;
    bool HasLabels(const std::string &label) const;
    void Setlabels(const std::vector<std::string> &labels);
    void DeleteLabel(const std::string &label);

private:
    std::vector<std::string> mLabels;
};
```

This API is idempotent because repeating any method call with the same input parameters will not change the underlying state of the object; for example,

```
Labels labels;
labels.SetLabels({"one", "two", "three"});
auto labels = labels.GetLabels();
auto hasLabel = labels.HasLabel("two");
labels.DeleteLabel("three");
```

will have the same effect as:

```
Labels labels;
labels.SetLabels({"one", "two", "three"});
labels.SetLabels({"one", "two", "three"});
auto labels = labels.GetLabels();
labels = labels.GetLabels();
auto hasLabel = labels.HasLabel("two");
hasLabel = labels.HasLabel("two");
labels.DeleteLabel("three");
labels.DeleteLabel("three");
```

However, we could extend this API with a new `AddLabel()` function to append a label to the vector of strings, in which the string argument is always appended to the vector even if it already exists. This new API might look like:

```
class Labels
{
public:
    std::vector<std::string> GetLabels() const;
    bool HasLabels(const std::string &label) const;
    void Setlabels(const std::vector<std::string> &labels);
    void AddLabel(const std::string &label);

private:
    std::vector<std::string> mLabels;
};
```

This new API call is not idempotent because calling `AddLabel()` multiple times will change the underlying data structure each time we call it:

```
Labels labels;
labels.AddLabel("one");
```

will produce a vector with one string in it, whereas:

```
Labels labels;
labels.AddLabel("one");
labels.AddLabel("one");
```

will produce a vector with two strings in it. So repeating the function call produced a different end state.

Enforcing an idempotent API is not something that you need to strive for in most cases: many well-designed APIs are not idempotent and that's fine. However, there are a few problem domains where having an idempotent interface is desirable because it offers additional fault tolerance guarantees. The most common example is when dealing with network protocols. As such, you often hear about idempotency with respect to RESTful Web APIs.

When sending information over a network, you can sometimes encounter time-outs or network outages. In the face of such errors, a common strategy is for the client to resend the original request to the server. With an idempotent interface, the client can safely retransmit the same request without being concerned that the data on the server will be corrupted by being applied multiple times. For example, imagine a banking application in which the client sends a money transfer request to the server. If the request times out and is resent by the client, then in a nonidempotent system there's a possibility that two transfers are executed instead of one.

Because idempotency is concerned primarily with the effect of an API call on the underlying state of the system, any read-only API calls can be considered idempotent. Similarly, any calls that replace the representation of an existing data entry, or any calls that delete a data entry, can also be considered idempotent because they can be repeated safely (as shown in the previous example). But calls that add new data to the system are generally not idempotent.

In terms of the HTTP protocol, Table 5.1 indicates which HTTP methods are considered idempotent and which are not (according to RFC 9110).

Table 5.1 HTTP methods and whether they are idempotent.

HTTP method	Is idempotent?
HEAD	Yes
GET	Yes
OPTIONS	Yes
TRACE	Yes
PUT	Yes
DELETE	Yes
POST	No
PATCH	No

It's possible to make an operation such as a POST HTTP call become idempotent. A common way to do this is to introduce a unique identifier that's generated for each data-modifying request. The server can keep a record of the recent transaction GUIDs, and if it encounters a GUID that it's already seen, it can ignore the request and return an appropriate success response without modifying the underlying data a second time.

It's not necessary for the duplicate call to return the same server response as the first one: idempotency is about ensuring the consistency of the data, so it could be valid for the second request to return a result indicating that it was ignored, and the client can choose how it wants to handle that information. It's also worth noting that the server may log a list of all requests that have been received, including any duplicate requests, so in that way the total underlying state will be different after a duplicate request, but the core business logic state should be unaffected.

So, in terms of API design guidelines, when a data-modifying operation is required to be fault tolerant, you can consider introducing a transaction GUID for each action so that the host can use this to handle duplicate requests correctly. Or, more generally, you could introduce the concept of beginning and ending an atomic transaction, in which the transaction is either guaranteed to be entirely executed or not at all, such as with a `BeginTransaction()` and `EndTranscation()` pair of functions in your API.

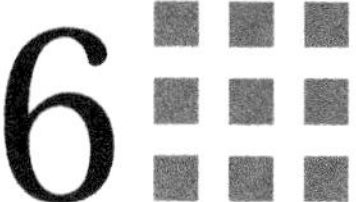

6 C++ usage

This chapter explores the qualities that make a good application programming interface (API) in C++. The generic API qualities that I covered in Chapter 2 could be applied to any programming language: the concepts of hiding private details, ease of use, loose coupling, and minimal completeness transcend the use of any programming language. Although I presented C++-specific examples for each of these topics, the concepts themselves are language agnostic.

However, there are many specific C++ style decisions that can affect the quality of an API, such as the use of namespaces, operators, friends, and const correctness. I will discuss each of these, and more, in the following sections.

This chapter will focus on issues that are applicable across all revisions of the C++ standard: in other words, those that relate only to C++98 and C++03. I will address newer features that relate to API design for C++11 and later revisions in the following chapter.

Note that I will defer some performance-related C++ topics, such as inlining and const references, until the later chapter on Performance. I will cover topics relating to multithreading in the chapter on Concurrency.

Namespaces

A namespace is a logical grouping of unique symbols. It provides a way to avoid naming collisions, so that two APIs don't try to define symbols with the same name. For example, if two APIs both define a class called `String`, then you cannot use the two APIs in the same program because only one definition can exist at any time. You should always use some form of namespacing in your APIs to ensure that they can interoperate with any other APIs that your clients may decide to use.

There are two common ways that you can add a namespace to your API. The first is to use a unique prefix for all of your public API symbols, which has the benefit that it can be used for plain C APIs as well. There are many examples of this type of namespacing:

- The OpenGL API uses "gl" as a prefix for all public symbols (e.g., `glBegin()`, `glVertex3f()`, and `GL_BLEND_COLOR`) (Shreiner, 2004).
- The Qt API uses the "Q" prefix for all public names (e.g., `QWidget`, `QApplication`, and `Q_FLAGS`).
- The libpng library uses the "png" prefix for all identifiers (e.g., `png_read_row()`, `png_create_write_struct()`, and `png_set_invalid()`).
- The GNU GTK + API uses the "gtk" prefix (e.g., `gtk_init()`, `gtk_style_new()`, and `GtkArrowType`).

API Design for C++. https://doi.org/10.1016/B978-0-443-22219-1.00010-6

- The Second Life source code uses the "LL" prefix (short for Linden Lab) for various classes, enums, and constants (e.g., `LLEvent`, `LLUUID`, and `LL_ACK_FLAG`).
- The Netscape Portable Runtime names all exported types with a "PR" prefix (e.g., `PR_WaitCondVar()`, `PR_fprintf()`, and `PRFileDesc`).

The second approach is to use the C++ `namespace` keyword. This essentially defines a scope where any names within that scope are given an additional prefix identifier. For example,

```
namespace MyAPI {

class String
{
public:
    ...
};

}  // namespace MyAPI
```

The `String` class in this case must now be referenced as `MyAPI::String`. The benefit of this style is that you don't need to meticulously ensure that every class, function, enum, or constant has a consistent prefix: the compiler does it for you. This method is used by the C++ Standard Library, in which all container classes, iterators, and algorithms are contained within the "std" namespace. You can also create nested namespaces, forming a namespace tree. For example, Intel's Threading Build Blocks API uses the "tbb" namespace for all of its public symbols and tbb:strict_ppl for internal code. The Boost libraries also use nested namespaces inside of the root "boost" namespace (e.g., the `boost::variant` namespace contains the public symbols for the Boost Variant API and `boost::detail::variant` contains the internal details for that API).

Using the namespace feature can produce verbose symbol names, particularly for symbols that are contained within several nested namespaces. However, C++ provides a way to make it more convenient to use symbols within a namespace with the `using` keyword:

```
using namespace std;
string str("Look, no std::");
```

or, preferably (because it limits the extent of symbols imported into the global namespace):

```
using std::string;
string str("Look, no std::");
```

However, you should never use the `using` keyword in the global scope of your public API headers! Doing so would cause all of the symbols in the referenced namespace to become visible in the global namespace. This would subvert the entire point of using namespaces in the first place (Stroustrup, 2000). If you wish to reference symbols in another namespace in your header files, then always use the fully qualified name (e.g., `std::string`).

> *TIP: Always provide a namespace for your API symbols, via either a consistent naming prefix or the C++ `namespace` keyword.*

Constructors and assignment

If you are creating objects that contain state and that may be copied or assigned by client programs (sometimes called value objects), you need to consider the correct design of your constructors and assignment operator. Your compiler will generate default versions of these methods for you if you don't define them yourself. However, if your class has any dynamically allocated objects, then you must explicitly define these methods yourself to ensure that the objects are copied correctly. Specifically, your C++03 compiler can generate default versions for the following four special methods (in the next chapter, we'll learn about two further special methods added in C++11):

- **Default constructor.** A constructor is used to initialize an object after it's been allocated by the `new` call. You can define multiple constructors with different arguments. The default constructor is defined as the constructor that can be called with no arguments (this could be a constructor with no arguments or with arguments that all specify default values). Your C++ compiler will automatically generate a default constructor if you do not explicitly define one. Note, however, that if you define a nondefault constructor, the compiler will no longer generate a default constructor for you.
- **Destructor.** The destructor is called in response to a `delete` call to release any resources the object is holding. There can be only one destructor for a class. If you don't specify a destructor, your C++ compiler will generate one automatically. The compiler will also generate code to call the destructors automatically for all your member variables in the reverse order they appear in the class declaration. The destructors for any base classes will be called automatically.
- **Copy constructor.** A copy constructor is a special constructor that creates a new object from an existing object. If you don't define a copy constructor, the compiler will generate one for you that performs a shallow copy of the existing object's member variables. If your object allocates any resources, you most likely need a copy constructor so you can perform a deep copy. The copy constructor gets called in the following situations:
  - An object is passed to a method by value or returned by value.
  - An object is initialized using the syntax `MyClass a = b;`
  - An object is placed in a brace-enclosed initializer list.

- An object is thrown or caught in an exception.

- **Assignment operator.** The assignment operator is used to assign the value of one object to another object (e.g., `a = b`). It differs from the copy constructor in that the object to which it is being assigned already exists. Some guidelines for implementing the assignment operator include:
  1. Use a const reference for the right-hand operand.
  2. Return `*this` as a reference to allow operator chaining.
  3. Destroy any existing state before setting the new state.
  4. Check for self-assignment (`a = a`) by comparing `this` with `&rhs`.

As a corollary to this, if you wish to create objects that should never be copied by your clients (also known as reference objects), then you should mark the object as noncopyable by declaring the copy constructor and assignment operator as private class members, or as of C++11, you can use the `= delete` specifier after the function's declaration.

Many novice C++ developers get into trouble because they have a class that allocates resources, and hence requires a destructor, but they don't also define a copy constructor and assignment operator. For example, consider the following simple integer array class where I show the implementation of the constructor and destructor inline to clarify the behavior:

```
class Array
{
public:
    explicit Array(int size) :
        mSize(size),
        mData(new int[size])
    {
    }

    ~Array()
    {
        delete[] mData;
    };

    int Get(int index) const;
    void Set(int index, int value);
    int Size() const;

private:
    int mSize;
    int *mData;
};
```

This class allocates memory but does not define either a copy constructor or an assignment operator. As a result, the following code will crash when the two variables go out of scope because the destructor of each will try to free the same memory:

```
{
    Array x(100);
    Array y = x;  // y now shares the same mData pointer as x
}
```

When creating a value object, it's therefore essential that you follow the rule of "the Big Three", also called "the Rule of Three". This term was introduced by Marshall Cline in the early 1990s and essentially states that there are three member functions that always go together: the destructor, the copy constructor, and the assignment operator (Cline et al., 1998). If you define one of these, you normally need to define the other two as well (declaring an empty virtual destructor to enable subclassing is one exception, because it doesn't perform an actual deallocation). James Coplien referred to this same concept as the orthodox canonical class form (Coplien, 1991).

TIP: If your class allocates resources, you should follow the Rule of Three and define a destructor, copy constructor, and assignment operator.

Note: C++11 introduced the move constructor and move assignment operator, resulting in the Rule of Three changing to the Rule of Five. However, following the Rule of Three is still valid because the new move functions will fall back to using your copy functions if not defined. I will discuss this further in the next chapter on C++ Revisions.

Defining constructors and assignment

Writing constructors and operators can be a tricky business, so here's an example that demonstrates the various combinations. It builds on the earlier array example and presents a class for storing an array of strings. The array is allocated dynamically, so you must define a copy constructor and assignment operator; otherwise the memory will be freed twice on destruction if you copy the array. Here's the declaration of the `Array` class in the header file:

```
#include <string>

class Array
{
public:
    // default constructor
    Array();
    // non-default constructor
    explicit Array(int size);
    // destructor
    ~Array();
    // copy constructor
    Array(const Array &in_array);
    // assignment operator
    Array &operator=(const Array &in_array);

    std::string Get(int index) const;
    bool Set(int index, const std::string &str);
    int Size() const;

private:
    int mSize;
    std::string *mArray;
};
```

And here are sample definitions for the constructors and assignment operator:

```
#include "array.h"
#include <algorithm>
#include <iostream>

// default constructor
Array::Array() :
    mSize(0),
    mArray(nullptr)
{
}

// non-default constructor
Array::Array(int size) :
    mSize(size),
    mArray(new std::string[size])
{
}

// destructor
Array::~Array()
{
    delete[] mArray;
}

// copy constructor
Array::Array(const Array &in_array) :
    mSize(in_array.mSize),
    mArray(new std::string[in_array.mSize])
{
    std::copy(in_array.mArray, in_array.mArray + mSize, mArray);
}

// assignment operator
Array &Array::operator=(const Array &in_array)
{
    if (this != &in_array) {  // check for self assignment
        delete[] mArray;      // delete current array first

        mSize = in_array.mSize;
        mArray = new std::string[in_array.mSize];
        std::copy(in_array.mArray, in_array.mArray + mSize, mArray);
    }

    return *this;
}
```

Given the previous `Array` class, the follow code demonstrates when the various methods will be called:

```
Array a;       // default constructor
Array a(10);   // non-default constructor
Array b(a);    // copy constructor
Array c = a;   // copy constructor (because c does not exist yet)
b = c;         // assignment operator
```

Note: there are certain cases in which your compiler may elide the call to your copy constructor: for example, if it performs some form of return value optimization (Meyers, 1998).

The explicit keyword

You may have noticed the use of the `explicit` keyword before the declaration of the nondefault constructor in the previous `Array` example. Adding `explicit` is a good general practice for any nondefault constructor that accepts a single argument. It's used to prevent a nondefault constructor from being called implicitly when constructing an object. For example, without the `explicit` keyword, this code is valid C++:

```
Array a = 10;
```

This will call the `Array` nondefault constructor with the integer argument of 10. However, this type of implicit behavior can be confusing, unintuitive, and in most cases unintended. As a further example of this kind of undesired implicit conversion, consider the function signature:

```
void CheckArraySize(const Array &array, int size);
```

Without declaring the single-argument constructor of `Array` as explicit, you could call this function as:

```
CheckArraySize(10, 10);
```

This weakens the type safety of your API because now the compiler will not enforce the type of the first argument to be an explicit `Array` object. As a result, there's the potential for the user to forget the correct order of arguments and pass them in the wrong order. This is why you should always use the `explicit` keyword for any single-argument constructors unless you know that you want to support implicit conversion.

You can also declare your copy constructor to be explicit, too. This will prevent implicit invocations of the copy constructor, such as passing an object to a function by value or returning an object by value. However, you will still be able to call the copy constructor explicitly using the "`Array a = b`" or "`Array a(b)`" syntax. As of C++11, you can use the `explicit` keyword in front of conversion operators as well as constructors. Doing so will prevent those conversion functions from being used for implicit conversions.

TIP: Consider using the `explicit` keyword before the declaration of any noncopy constructor with a single argument.

Const correctness

Const correctness refers to the use of the C++ `const` keyword to declare a variable or method as immutable. It's a compile time construct that you can use to maintain the correctness of code that shouldn't modify certain variables. In C++, you can define variables as const, meaning that they should not be modified, and you can also define methods as const, meaning that they should not modify any member variables of the class. Using const correctness is simply good programming practice. However, it can also provide documentation of the intent of your methods, and hence make them easier to use.

TIP: Ensure that your API is const correct.

Method const correctness

A const method cannot modify any of the member variables of the class. In essence, all of the member variables are treated as const variables inside a const method. This form of const correctness is indicated by appending the `const` keyword after the method's parameter list. There are two principal benefits of declaring a method as const:

1. To advertise the fact that the method will not change the state of the object. As I've just mentioned, this is helpful documentation for the users of your API.
2. To allow the method to be used on const versions of an object. A nonconst method cannot be called on a const version of an object.

Scott Meyers describes two camps of philosophy about what a const method represents. There's the bitwise constness camp, which believes that a const method should not change any of the member variables of a class, and then there's the logical constness camp, which says that a const method may change a member variable if that change cannot be detected by the user (Meyers, 2005). Your C++ compiler conforms to the bitwise approach. However, there are times when you really want it to behave in the logical constness manner. A classic example is if you want to cache some property of a class because it takes too long to compute. For example, consider a `HashTable` class that needs to return the number of elements in the hash table very efficiently. As a result, you decide to cache its size and compute the value lazily, on demand. Given the following class declaration:

```
class HashTable
{
public:
    void Insert(const std::string &str);
    int Remove(const std::string &str);
    bool Has(const std::string &str) const;
    int Size() const;
    ...

private:
    int  mCachedSize;
    bool mSizeIsDirty;
    ...
};
```

you may want to implement the `Size()` const method as:

```
int HashTable::Size() const
{
    if (mSizeIsDirty) {
        mCachedSize = CalculateSize();
        mSizeIsDirty = false;
    }

    return mCachedSize;
}
```

Unfortunately this is not legal C++, because the `Size()` method does actually modify member variables (`mSizeIsDirty` and `mCachedSize`). However, these are not part of our logical public interface: they are internal state that lets us offer a more efficient API. This is why there is the notion of logical constness. C++ provides a way around this problem with the `mutable` keyword. Declaring the `mCachedSize` and `mSizeIsDirty` variables as mutable means that they can be modified within a const method. Using `mutable` is a great way to maintain the logical constness of your API, rather than removing the `const` keyword on a member function that really should be declared const:

```
class HashTable
{
public:
    ...

private:
    mutable int mCachedSize;
    mutable bool mSizeIsDirty;
    ...
};
```

TIP: Declare methods and parameters as const as soon as you can. Trying to retrofit const correctness into an API later can be a time-consuming and frustrating activity.

Parameter const correctness

The `const` keyword can also be used to indicate whether you intend for a parameter to be an input or an output: that is, a parameter that's used to pass some value into a method or one that is used to receive some result. For example, consider a method such as:

```
std::string StringToLower(std::string &str);
```

It's not clear from this function signature whether this method will modify the string that you pass in. Clearly it returns a string result, but perhaps it also changes the parameter string. It certainly could do so if it wanted to. If the purpose of this method is to take the parameter and return a lowercase version without affecting the input string, then the simple addition of const can make this unequivocally clear:

```
std::string StringToLower(const std::string &str);
```

Now the compiler will enforce the fact that the function `StringToLower()` will not modify the string that the user passes in. As a result, it's clear and unambiguous what the intended use of this function is just by looking at the function signature.

Often, you'll find that if you have a const method, any reference or pointer parameters can also be declared const. Although this is not a hard and fast rule, it follows logically from the general promise that the const method doesn't modify any state. For example, in the following function the `root_node` parameter can be declared const because it's not necessary to modify this object to compute the result of the const method:

```
bool Node::IsVisible(const Node &root_node) const;
```

TIP: When passing a reference or pointer into a const method, think about whether that parameter can be declared const, too.

Return value const correctness

When returning the result of a function, the main reason to declare that result as const is if it references the internal state of the object. For example, if you're returning a result by value, then it makes little sense to specify it as const, because the returned object will be a copy and hence changing it will not affect any of your class's internal state.

Alternatively, if you return a pointer or reference to a private data member, then you should declare the result to be const; otherwise users will be able to modify your internal state without going through your public API. In this case, you must also think about whether the returned pointer or reference will survive longer than your class. If this is possible, you should consider returning a reference-counted pointer such as an `std::weak_ptr` or `std::shared_ptr`, as discussed earlier in Chapter 2.

Therefore, the most common decision you will have with respect to return value const correctness is whether to return the result by value or const reference, that is,

```
// return by value
std::string GetName() const
{
    return mName;
}

// return by const reference
const std::string &GetName() const
{
    return mName;
}
```

In general, I recommend that you return the result by value because it's safer. You may prefer the const reference method in a few cases where performance is critical, but it's likely that modern compiler optimizations will make most of those cases unnecessary. Returning by value is safer because you don't have to worry about clients holding on to references after your object has been destroyed, but also because returning a const reference can break encapsulation. It also makes multithreaded cases more complicated because the responsibility to perform locking is moved to the client instead of being kept behind the API call.

TIP: Prefer to return the result of a function by value, rather than const reference.

On the face of it, our earlier const reference `GetName()` method seems acceptable: the method is declared to be const to indicate that it doesn't modify the state of the object, and the returned reference to the object's internal state is also declared to be const so that clients can't modify it. However, a determined client can always cast away the constness of the reference and then directly modify the underlying private data member, such as in this example:

```
// get a const reference to an internal string
const std::string &const_name = object.GetName();

// cast away the constness
std::string &name = const_cast<std::string &>(const_name);

// and modify the object's internal data!
name.clear();
```

Templates

Templates provide a versatile and powerful ability to generate code at compile time. They are particularly useful for generating lots of code that looks similar but differs only by type. However, if you decide to provide class templates as part of your public API, there are several issues you should consider to ensure that you provide a well-insulated, efficient, and cross-platform interface. The following sections will address several of these factors.

Note that I will not cover all aspects of template programming here, only those features that affect good API design. For a more thorough and in-depth treatment of templates, there are several other good books on the market (Josuttis, 1999; Alexandrescu, 2001; Vandevoorde and Josuttis, 2002).

Template terminology

Templates are an often poorly understood part of the C++ specification, so let's begin by defining some terms so that we can proceed from a common base. I will use this template declaration as a reference for the definitions:

```
template <typename T>
class Stack
{
public:
    void Push(T val);
    T Pop();
    bool IsEmpty() const;

private:
    std::vector<T> mStack;
};
```

This class template describes a generic stack class where you can specify the type of the elements in the stack, `T`:

- **Template Parameters**: These are the names that are listed after the template keyword in a template declaration. For example, `T` is the single template parameter specified in our earlier `Stack` example.

- **Template Arguments**: These are the entities that are substituted for template parameters during specialization. For example, given a specialization `Stack<int>`, "int" is a template argument.
- **Instantiation**: This is when the compiler generates a regular class, method, or function by substituting each of the template's parameters with a concrete type. This can happen implicitly when you create an object based upon a template, or explicitly if you want to control when the code generation happens. For example, these lines of code create two specific stack instances and will normally cause the compiler to generate code for these two different types:

```
Stack<int>         myIntStack;
Stack<std::string> myStringStack;
```

- **Implicit Instantiation**: This is when the compiler decides when to generate code for your template instances. Leaving the decision to the compiler means that it must find an appropriate place to insert the code, and it must also make sure that only one instance of the code exists, to avoid duplicate symbol link errors. This is a nontrivial problem and can cause extra bloat in your object files or longer compile and link times to solve. Most important for API design, implicit instantiation means that you must include the template definitions in your header files so that the compiler has access to the definitions whenever it needs to generate the instantiation code.
- **Explicit Instantiation**: This is when the programmer determines when the compiler should generate the code for a specific specialization. This can make for much more efficient compilation and link times because the compiler no longer needs to maintain bookkeeping information for all of its implicit instantiations. However, the onus is then placed on the programmer to ensure that a particular specialization is explicitly instantiated once and only once. From an API perspective, explicit instantiation allows us to move the template implementation into the `.cpp` file, and so hide it from the user.
- **Lazy Instantiation**: This describes the standard implicit instantiation behavior of a C++ compiler in which it will generate code for only the parts of a template that are actually used. For example, given our two earlier instantiations, if you never called `IsEmpty()` on the `myStringStack` object, then the compiler would not generate code for the `std::string` specialization of that method. This means that you can instantiate a template with a type that can be used by some, but not all methods of a class template. For example, say one method uses the >= operator, but the type you want to instantiate does not define this operator. This is fine if you don't call the particular method that attempts to use the >= operator.

- **Specialization**: When a template is instantiated, the resulting class, method, or function is called a specialization. More specifically, this is an instantiated (or generated) specialization. However, the term specialization can also be used when you provide a custom implementation for a function by specifying concrete types for all the template parameters. I gave an example of this earlier in the API Styles chapter, where I presented this implementation of the `Stack::Push()` method, specialized for integer types. This is called an explicit specialization:

```
template <>
void Stack<int>::Push(int val)
{
    // integer specific push implementation
}
```

- **Partial Specialization**: This is when you provide a specialization of the template for a subset of all possible cases. That is, you specialize one feature of the template but still allow the user to specify other features. For example, if your template accepts multiple parameters, you could partially specialize it by defining a case in which you specify a concrete type for only one of the parameters. In our `Stack` example with a single template parameter, you could partially specialize this template specifically to handle pointers to any type T. This still lets users create a stack of any type, but it also lets you write specific logic to handle the case in which users create a stack of pointers. This partially specialized class declaration might look like:

```
template <typename T>
class Stack<T *>
{
public:
    void Push(T *val);
    T *Pop();
    bool IsEmpty() const;

private:
    std::vector<T *> mStack;
};
```

Implicit instantiation API design

If you want to allow your clients to instantiate your class templates with their own types, then you need to use implicit template instantiation. For example, if you provide a smart pointer class template, `smart_pointer<T>`, you don't know ahead of time what types your clients will want to instantiate it with. As a result, the compiler needs to be able to access the definition of the template when it's used. This essentially means that you must expose the template definition in your header files. This is the biggest disadvantage of the implicit instantiation approach in terms of robust API design. However, even if you can't hide the implementation details in this situation, you can at least try to isolate them.

Because you need to include the template definition in your header file, it's easy and therefore tempting simply to inline the definitions directly within the class definition. This is a practice that I have already classified as poor design, and that assertion is still true in the case of templates. Instead, I recommend that all template implementation details should be contained within a separate implementation header, which is then included by the main public header. Using the example of our `Stack` class template, you could provide the main public header:

```
// stack.h
#ifndef STACK_H
#define STACK_H

#include <vector>

template <typename T>
class Stack
{
public:
    void Push(T val);
    T Pop();
    bool IsEmpty() const;

private:
    std::vector<T> mStack;
};

// isolate all implementation details within a separate header
#include "stack_priv.h"

#endif
```

Then the implementation header, `stack_priv.h`, would look like:

```
// stack_priv.h
#ifndef STACK_PRIV_H
#define STACK_PRIV_H

template <typename T>
void Stack<T>::Push(T val)
{
    mStack.push_back(val);
}

template <typename T>
T Stack<T>::Pop()
{
    if (IsEmpty()) {
        return T();
    }
    T val = mStack.back();
    mStack.pop_back();
    return val;
}

template <typename T>
bool Stack<T>::IsEmpty() const
{
    return mStack.empty();
}

#endif
```

This is a technique used by many high-quality template-based APIs, such as the various Boost headers. It has the benefit of keeping the main public header uncluttered by implementation details while isolating the necessary exposure of internal details to a separate header that's clearly designated as containing private details. (The same technique can be used to isolate consciously inlined function details from their declarations.)

The technique of including template definitions in header files is referred to as the inclusion model (Vandevoorde and Josuttis, 2002). It's worth noting that there was an alternative to this style called the separation model. This was meant to allow the declaration of a class template in an `.h` file to be preceded with the `export` keyword. The implementation of the template methods could then appear in a `.cpp` file. From an API design perspective, this is a far preferable model because it would allow us to remove all implementation details from public headers. However, this part of the C++ specification was very poorly supported by most compilers. Ultimately the `export` keyword was deprecated in C++11 and then repurposed for different functionality in C++20.

Explicit instantiation API design

If you only want to provide a predetermined set of template specializations for your API and disallow your users from creating further ones, then you do in fact have the option of completely hiding your private code. For example, if you have created a 3D vector class template, `Vector3D<T>`, you may want to provide specializations of this template only for `int`, `short`, `float`, and `double`, and you may feel that it's not necessary to let your users create further specializations.

In this case, you can put your template definitions into a `.cpp` file and use explicit template instantiation to instantiate those specializations that you wish to export as part of your API. The `template` keyword can be used to create an explicit instantiation. For instance, using our previous `Stack` template example, you could create explicit instantiations for the type `int` with the statement:

```
template class Stack<int>;
```

This will cause the compiler to generate the code for the `int` specialization at this point in the code. As a result, it will subsequently no longer attempt to instantiate this specialization implicitly elsewhere in the code. Consequently, using explicit instantiation can also help to reduce build times.

Let's look at how you can organize your code to take advantage of this feature. Our `stack.h` header file looks almost the same as before, just without the `#include "stack_priv.h"` line:

```
// stack.h
#ifndef STACK_H
#define STACK_H

#include <vector>

template <typename T>
class Stack
{
public:
    void Push(T val);
    T Pop();
    bool IsEmpty() const;

private:
    std::vector<T> mStack;
};

#endif
```

Now you can contain all of the implementation details for this template in an associated .cpp file:

```
// stack.cpp
#include "stack.h"
#include <string>

template <typename T>
void Stack<T>::Push(T val)
{
    mStack.push_back(val);
}

template <typename T>
T Stack<T>::Pop()
{
    if (IsEmpty()) {
        return T();
    }
    T val = mStack.back();
    mStack.pop_back();
    return val;
}

template <typename T>
bool Stack<T>::IsEmpty() const
{
    return mStack.empty();
}

// explicit template instantiations
template class Stack<int>;
template class Stack<double>;
template class Stack<std::string>;
```

The important lines here are the last three, which create explicit instantiations of the Stack class template for the types int, double, and std::string. The user will not be able to create further specializations (and the compiler will not be able to create implicit instantiations for the user, either), because the implementation details are hidden in our .cpp file. On the other hand, our implementation details are now successfully hidden in our .cpp file.

To indicate to your users which template specializations they can use (i.e., which you have explicitly instantiated for them), you could add a few typedefs to the end of your public header, such as:

```
using IntStack = Stack<int>;
using DoubleStack = Stack<double>;
using StringStack = Stack<std::string>;
```

It's worth noting that by adopting this template style, not only do you (and your clients) get faster builds owing to the removal of the overhead of implicit instantiation, but by removing the template definitions from your header, you reduce the `#include` coupling of your API and the amount of extra code that your clients' programs must compile every time they `#include` your API headers.

TIP: Prefer explicit template instantiation if you need only a predetermined set of specializations. Doing so lets you hide private details and can reduce build times.

It's also worth noting that most compilers provide an option to turn off implicit instantiation completely, which may be a useful optimization if you plan to use only explicit instantiation in your code. This option is called `-fno-implicit-templates` in the GNU C++ and Intel ICC compilers.

As of C++11, support has been added for extern templates. That is, you will be able to use the `extern` keyword to prevent the compiler from instantiating a template in the current translation unit. With the use of extern templates, you can force the compiler to instantiate a template at a certain point and tell it not to instantiate the template at other points. For example:

```
// explicitly instantiate the template here
template class Stack<int>;

// do not instantiate the template here
extern template class Stack<int>;
```

Operator overloading

In addition to overloading functions, C++ allows you to overload many of the operators for your classes, such as `+`, `*=`, or `[]`. This can be very useful to make your classes look and behave more like built-in types and to provide a more compact and intuitive syntax for certain methods. For example, instead of having to use syntax such as:

```
add(add(mul(a,b), mul(c,d)), mul(a,c))
```

you could write classes that support the syntax:

```
a*b + c*d + a*c
```

Of course, you should only use operator overloading in cases where it makes sense: that is, where doing so would be considered natural to the user of your API and not violate the rule of least surprise. This generally means that you should preserve the natural semantics for operators, such as using the + operator to implement an operation analogous to addition or concatenation. You should also avoid overloading the operators `&&`, `||`, `&` (unary ampersand), and `,` (comma) because these exhibit behaviors that may surprise your users, such as short-circuited evaluation and undefined evaluation order (Meyers 1998; Sutter and Alexandrescu, 2004).

As I covered earlier in this chapter, a C++ compiler will generate a default assignment operator (=) for your class if you don't define one explicitly. However, if you wish to use any other operators with your objects, then you must explicitly define them; otherwise you'll end up with link errors.

Overloadable operators

Certain operators cannot be overloaded in C++, such as `.`, `.*`, `?:`, and `::`, the preprocessor symbols `#` and `##`, and the `sizeof` operator. Of the remaining operators that you can overload for your own classes, there are two main categories:

1. **Unary operators**: These are operators that act on a single operand. The list of unary operators includes:

Name	Example	Name	Example
Unary minus	−x	Unary plus	+x
Prefix decrement	−−x	Postfix decrement	x−−
Prefix increment	++x	Postfix increment	x++
Dereference	*x	Reference	&x
Logical NOT	!x	Bitwise NOT	~x
Function call	x()		

2. **Binary operators**: There are operators that act on two operands. The list of binary operators includes:

Name	Example	Name	Example
Addition	x + y	Subtraction	x − y
Assignment by addition	x += y	Assignment by subtraction	x −= y
Multiplication	x * y	Division	x / y
Assignment by multiplication	x *= y	Assignment by division	x /= y
Equality	x == y	Inequality	x != y
Assignment	x = y	Comma	x, y
Less than	x < y	Greater than	x > y

—cont'd

Name	Example	Name	Example
Less than or equal to	x <= y	Greater than or equal to	x >= y
Modulo	x % y	Bitwise XOR	x ^ y
Assignment by modulo	x %= y	Assignment by bitwise XOR	x ^= y
Bitwise AND	x & y	Bitwise OR	x | y
Assignment by bitwise AND	x &= y	Assignment by bitwise OR	x |= y
Logical AND	x && y	Logical OR	x || y
Bitwise left shift	x << y	Bitwise right shift	x >> y
Assignment by bitwise left shift	x <<= y	Assignment by bitwise right shift	x >>= y
Class member access	x -> y	Pointer-to-member selection	x ->* y
Array subscript	x[y]	C-style cast	(y) x

Free operators versus member operators

Operators can either be defined as members of your class or as free functions. Some operators must be defined as class members, but others can be defined either way. For example, this code illustrates the += operator defined as a class member:

```
class Complex
{
public:
    Complex(double real, double imag);

    // method form of operator+=
    Complex &operator+=(const Complex &other);

    double GetReal() const;
    double GetImaginary() const;

    ...
};
```

The following code shows an equivalent API using a free function version of the operator:

```
class Complex
{
public:
    Complex(double real, double imag);

    double GetReal() const;
    double GetImaginary() const;

    ...
};

// free function form of operator+=
Complex &operator+=(Complex &lhs, const Complex &rhs);
```

In this section I'll cover some best practices for whether you should make your operators free functions or methods.

To begin with, the C++ standard requires the following operators to be declared as member methods to ensure that they receive an lvalue (an expression that refers to an object) as their first operand:

- = Assignment
- [] Subscript
- -> Class member access
- ->* Pointer-to-member selection
- () Function call
- (T) Conversion, i.e., C-style cast
- new/delete

The remaining overloadable operators can be defined as either free functions or class methods. From the perspective of good API design, I recommend that you favor the free function version over the class method version of defining an operator. There are two specific reasons for this:

1. **Operator symmetry.** If a binary operator is defined as a class method, it must have an object to be applied to as the left-hand operand. Taking the * operator as an example, this means that your users would be able to write expressions such as "complexNumber * 2.0" (assuming that you have defined a constructor for Complex that accepts a single double argument), but not "2.0 * complexNumber" (because 2.0.operator +(complexNumber) does not make sense). This breaks the commutative property of the operator that your users will expect (i.e., that x * y should be the same as y * x). However, declaring the * operator as a free function lets you benefit from implicit type conversions for both the left- and right-hand operands.
2. **Reduced coupling.** A free function cannot access the private details of a class. It's therefore less coupled to the class because it can only access the public methods. This is a general API design statement that I covered in Chapter 2: turn a class method that does not need to access private or protected members into a free function to reduce the degree of coupling in your API (Meyers, 2000; Tulach, 2008).

Having stated this general preference toward free function operators, I will now present the exception to this rule: If your operator must access private or protected members of your class, then you should define the operator as a method of the class. I make this exception because otherwise you would have to declare the free operator to be a friend of your class. As I will discuss later in this chapter, adding friends to your classes is a greater evil. One specific reason I'll mention here is that your clients cannot change the friendship list of your classes, so they could not add new operators in this same way.

> *TIP: Prefer declaring operators as free functions, unless the operator must access protected or private members, or the operator is one of:* =, `[]`, `->`, `->*`, `()`, `(T)`, `new`, *or* `delete`.

Adding operators to a class

Let's develop the `Complex` class a little further to make these points more concrete. The += operator modifies the contents of an object, and we know that all member variables should be private; therefore you will most likely need to make the += operator be a member method. However, the + operator does not modify the left-hand operand. As such, it shouldn't need access to private members and can be made a free function. You also need to make it a free function to ensure that it benefits from symmetrical behavior, as described earlier. In fact, the + operator can be implemented in terms of the += operator, which allows us to reuse code and provide more consistent behavior. It also reduces the number of methods that might need to be overloaded in derived classes:

```
Complex operator+(const Complex &lhs, const Complex &rhs)
{
    return Complex(lhs) += rhs;
}
```

Obviously, the same technique applies to the other arithmetic operators, such as -, -=, *, *=, /, and /=. For example, *= can be implemented as a member function, whereas * can be implemented as a free function that uses the *= operator.

As for the relational operators ==, !=, <, <=, >, and >=, these must also be implemented as free functions to ensure symmetrical behavior. In the case of our `Complex` class, you can implement these using the public `GetReal()` and `GetImaginary()` methods. However, if these operators should need access to the private state of the object, there is a way to resolve this apparent dilemma. In this case, you can provide public methods that test for the equality and less than conditions. All of the relational operators could then be implemented in terms of these two primitive functions (Astrachan, 2000):

```
bool operator==(const Complex &lhs, const Complex &rhs)
{
    return lhs.IsEqualTo(rhs);
}

bool operator!=(const Complex &lhs, const Complex &rhs)
{
    return !(lhs == rhs);
}

bool operator<(const Complex &lhs, const Complex &rhs)
{
    return lhs.IsLessThan(rhs);
}

bool operator<=(const Complex &lhs, const Complex &rhs)
{
    return !(lhs > rhs);
}

bool operator>(const Complex &lhs, const Complex &rhs)
{
    return rhs < lhs;
}

bool operator>=(const Complex &lhs, const Complex &rhs)
{
    return rhs <= lhs;
}
```

The last operator I will consider here is <<, which I will use for stream output (as opposed to bit shifting). The stream operators must be declared as free functions because the first parameter is a stream object. Again, you can use the public `GetReal()` and `GetImaginary()` methods to make this possible. However, if the stream operator did need to access private members of your class, then you could create a public `ToString()` method for the << operator to call, as a way to avoid using friends.

Putting all these recommendations together, here's what the operators of our `Complex` class might look like:

```
#include <iostream>

class Complex
{
public:
    Complex(double real, double imaginary);
    Complex::~Complex();

    Complex(const Complex &obj);

    Complex &operator=(const Complex &rhs);

    Complex &operator+=(const Complex &rhs);
    Complex &operator-=(const Complex &rhs);
    Complex &operator*=(const Complex &rhs);
    Complex &operator/=(const Complex &rhs);

    double GetReal() const;
    double GetImaginary() const;

private:
    class Impl;
    Impl *mImpl;
};

Complex operator+(const Complex &lhs, const Complex &rhs);
Complex operator-(const Complex &lhs, const Complex &rhs);
Complex operator*(const Complex &lhs, const Complex &rhs);
Complex operator/(const Complex &lhs, const Complex &rhs);

bool operator==(const Complex &lhs, const Complex &rhs);
bool operator!=(const Complex &lhs, const Complex &rhs);
bool operator<(const Complex &lhs, const Complex &rhs);
bool operator>(const Complex &lhs, const Complex &rhs);
bool operator<=(const Complex &lhs, const Complex &rhs);
bool operator>=(const Complex &lhs, const Complex &rhs);

std::ostream &operator<<(std::ostream &os, const Complex &obj);
std::istream &operator>>(std::istream &is, Complex &obj);
```

Operator syntax

Table 6.1 provides a list of operators that you can overload in your classes and the recommended syntax for declaring each operator so that they have the same semantics as their built-in counterparts. This table omits operators that you can't overload, as well as those that I've previously stated you should not overload, such as `&&` and `||`. Where an operator can be defined as either a free function or a class method, I present both forms, but I list the free function form first because you should generally prefer this form unless the operator needs access to protected or private members.

Table 6.1 List of operators and the syntax for declaring these in your application programming interfaces.

Operator name	Syntax	Sample operator declarations
Assignment	x = y	`T1& T1::operator =(const T2& y);`
Dereference	*x	`T1& operator *(T1& x);` `T1& T1::operator *() const;`
Reference	&x	`T1* operator &(T1& x);` `T1* T1::operator &();`
Class member access	x->y	`T2* T1::operator ->();`
Pointer-to-member selection	x->*y	`T2 T1::operator->*(T2 T1::*);`
Array subscript	x[n]	`T2& T1::operator [](unsigned int n);` `T2& T1::operator [](const std::string &s);`
Function call	x()	`void T1::operator ()(T2& x);` `T2 T1::operator ()() const;`
C-style cast	(y) x	`T1::operator T2() const;`
Unary plus	+x	`T1 operator +(const T1& x);` `T1 T1::operator +() const;`
Addition	x + y	`T1 operator +(const T1& x, const T2& y);` `T1 T1::operator +(const T2& y) const;`
Assignment by addition	x += y	`T1& operator +=(T1& x, const T2& y);` `T1& T1::operator +=(const T2& y);`
Prefix increment	++x	`T1& operator ++(T1& x);` `T1& T1::operator ++();`
Postfix increment	x++	`T1 operator ++(T1& x, int);` `T1 T1::operator ++(int);`
Unary minus	-x	`T1 operator -(const T1& x);` `T1 T1::operator -() const;`
Subtraction	x - y	`T1 operator -(const T1& x, const T2& y);` `T1 T1::operator -(const T2& y) const;`
Assignment by subtraction	x -= y	`T1& operator -=(T1& x, const T2& y);` `T1& T1::operator -=(const T2& y);`
Prefix decrement	--x	`T1& operator --(T1& x);` `T1& T1::operator --();`
Postfix decrement	x--	`T1 operator --(T1& x, int);` `T1 T1::operator -(int);`
Multiplication	x * y	`T1 operator *(const T1& x, const T2& y);` `T1 T1::operator *(const T2& y) const;`
Assignment by multiplication	x *= y	`T1& operator *=(T1& x, const T2& y);` `T1& T1::operator *=(const T2& y);`
Division	x / y	`T1 operator /(const T1& x, const T2& y);` `T1 T1::operator /(const T2& y) const;`
Assignment of division	x /= y	`T1& operator /=(T1& x, const T2& y);` `T1& T1::operator /=(const T2& y);`
Modulo	x % y	`T1 operator %(const T1& x, const T2& y);` `T1 T1::operator %(const T2& y) const;`

Table 6.1 List of operators and the syntax for declaring these in your application programming interfaces.—cont'd

<table>
<tr><th>Operator name</th><th>Syntax</th><th>Sample operator declarations</th></tr>
<tr><td>Assignment of modulo</td><td>x %= y</td><td>T1& operator %=(T1& x, const T2& y);
T1& T1::operator %=(const T2& y);</td></tr>
<tr><td>Equality</td><td>x == y</td><td>bool operator ==(const T1& x, const T2& y);
bool T1::operator ==(const T2& y) const;</td></tr>
<tr><td>Inequality</td><td>x != y</td><td>bool operator !=(const T1& x, const T2& y);
bool T1::operator !=(const T2& y) const;</td></tr>
<tr><td>Less than</td><td>x < y</td><td>bool operator <(const T1& x, const T2& y);
bool T1::operator <(const T2& y) const;</td></tr>
<tr><td>Less than or equal to</td><td>x <= y</td><td>bool operator <=(const T1& x, const T2& y);
bool T1::operator <=(const T2& y) const;</td></tr>
<tr><td>Greater than</td><td>x > y</td><td>bool operator >(const T1& x, const T2& y);
bool T1::operator >(const T2& y) const;</td></tr>
<tr><td>Greater than or equal to</td><td>x >= y</td><td>bool operator >=(const T1& x, const T2& y);
bool T1::operator >=(const T2& y) const;</td></tr>
<tr><td>Logical NOT</td><td>!x</td><td>bool operator !(const T1& x);
bool T1::operator !() const;</td></tr>
<tr><td>Bitwise left shift (BLS)</td><td>x << y</td><td>T1 operator <<(const T1& x, const T2& y);
ostream& operator <<(ostream &, const T1 &x);
T1 T1::operator <<(const T2& y) const;</td></tr>
<tr><td>Assignment by BLS</td><td>x <<= y</td><td>T1& operator <<=(T1& x, const T2& y);
T1& T1::operator <<=(const T2& y);</td></tr>
<tr><td>Bitwise right shift (BRS)</td><td>x >> y</td><td>T1 operator >>(const T1& x, const T2& y);
istream& operator >>(istream &, T1 &x);
T1 T1::operator >>(const T2& y) const;</td></tr>
<tr><td>Assignment by BRS</td><td>x >>= y</td><td>T1& operator >>=(T1& x, const T2& y);
T1& T1::operator >>=(const T2& y);</td></tr>
<tr><td>Bitwise NOT</td><td>~x</td><td>T1 operator ~(const T1& x);
T1 T1::operator ~() const;</td></tr>
<tr><td>Bitwise AND</td><td>x & y</td><td>T1 operator &(const T1& x, const T2& y);
T1 T1::operator &(const T2& y) const;</td></tr>
<tr><td>Assignment by bitwise AND</td><td>x &= y</td><td>T1& operator &=(T1& x, const T2& y);
T1& T1::operator &=(const T2& y);</td></tr>
<tr><td>Bitwise OR</td><td>x | y</td><td>T1 operator |(const T1& x, const T2& y);
T1 T1::operator |(const T2& y) const;</td></tr>
<tr><td>Assignment by bitwise OR</td><td>x |= y</td><td>T1& operator |=(T1& x, const T2& y);
T1& T1::operator |=(const T2& y);</td></tr>
<tr><td>Bitwise XOR</td><td>x ^ y</td><td>T1 operator ^(const T1& x, const T2& y);
T1 T1::operator ^(const T2& y) const;</td></tr>
<tr><td>Assignment by bitwise XOR</td><td>x ^= y</td><td>T1& operator ^=(T1& x, const T2& y);
T1& T1::operator ^=(const T2& y);</td></tr>
<tr><td>Allocate object</td><td>new</td><td>void* T1::operator new(size_t n);</td></tr>
<tr><td>Allocate array</td><td>new []</td><td>void* T1::operator new[](size_t n);</td></tr>
<tr><td>Deallocate object</td><td>delete</td><td>void T1::operator delete(void* x);</td></tr>
<tr><td>Deallocate array</td><td>delete []</td><td>void T1::operator delete[](void* x);</td></tr>
</table>

Conversion operators

A conversion operator provides a way for you to define how an object can be automatically converted to a different type. A classic example is to define a custom string class that can be passed to functions that accept a `const char *` pointer, such as the standard C library functions `strcmp()` or `strlen()`:

```
class MyString
{
public:
    MyString(const char *string);

    // convert MyString to a C-style string
    operator const char *() { return mBuffer; }

private:
    char *mBuffer;
    int mLength;
};

// MyString objects get automatically converted to const char *
MyString mystr("Haggis");
int same = strcmp(mystr, "Edible");
int len = strlen(mystr);
```

Note that the conversion operator does not specify a return value type. That's because the type is inferred by the compiler based upon the operator's name. Also, note that conversion operators take no arguments. As of C++11, it's also possible to prefix a conversion operator with the `explicit` keyword to prevent its use in implicit conversations.

> *TIP: Add conversion operators to your classes to let them take advantage of automatic type coercion.*

Function parameters

In the following sections, I'll address a couple of C++ best practices relating to the use of function parameters. This includes when you should prefer to use pointers instead of references to pass objects into a function, and when you should use default arguments.

Pointer versus reference parameters

When specifying the parameters for your functions, you can chose to pass them as value parameters, pointers, or references. For example:

```
bool GetColor(int r, int g, int b);     // pass by value
bool GetColor(int &r, int &g, int &b);  // pass by reference
bool GetColor(int *r, int *g, int *b);  // pass by pointer
```

You pass a parameter as a reference or pointer when you want to receive a handle for the actual object rather than a copy of the object. This is done either for performance reasons (as I'll discuss in Chapter 7) or so that you can modify the client's object. C++ compilers normally implement references using pointers, so under the hood they are often the same thing. However, there are several practical differences, such as:

- References are used as if they were a value, such as `object.Function()` instead of `object->Function()`.
- A reference must be initialized to point to an object and does not support changing the referent object after initialization.
- You cannot take the address of a reference as you can with pointers. Using the & operator on a reference returns the address of the referent object.
- You can't create arrays of references.

The question of whether to use a pointer or a reference for a parameter is really a matter of personal taste. However, I will suggest that in general you should prefer the use of references over pointers for any input parameters. This is because the calling syntax for your clients is simpler, and you do not need to worry about checking for `nullptr` values (because references cannot be null). However, if you need to support passing `nullptr`, or you're writing a plain C API, then you must obviously use a pointer.

In terms of output parameters (parameters that your function may modify), some engineers dislike that references don't indicate to your clients that a parameter may be changed. For example, the reference and pointer versions of the previous `GetColor()` function can be called by clients as:

```
object.GetColor(red, green, blue);     // pass by reference
object.GetColor(&red, &green, &blue);  // pass by pointer
```

In both of these cases, the `GetColor()` function can modify the value of the `red`, `green`, and `blue` variables. However, the pointer version makes this fact explicit because of the required use of the & operator. For this reason, APIs such as the Qt framework prefer to represent output parameters using pointers instead of references. If you decide to follow this convention, too (which I recommend), then by implication all of your reference parameters should be const references.

> *TIP: Prefer the use of const references over pointers for input parameters where feasible. For output parameters, consider using pointers over nonconst references to indicate explicitly to the client that they may be modified.*

Default arguments

Default arguments are a very useful tool to reduce the number of methods in your API and provide implicit documentation on their use. They can also be used to extend an API call in a backward-compatible fashion, so that older client code will still compile, but newer code can optionally provide additional arguments (although it should be noted that this will break binary compatibility, because the mangled symbol name for the method will necessarily change if a new argument is added). As an example, consider this code fragment for a Sphere class:

```
class Sphere
{
public:
    Sphere(double x = 0, double y = 0, double radius = 10.0);
    ...
};
```

In this case, the user can construct a new `Sphere` object in several different ways, supplying as much detail as needed. For example:

```
Sphere s1();
Sphere s1(2.3);
Sphere s1(2.3, 5.6);
Sphere s1(2.3, 5.6, 1.5);
```

However, there are two issues to consider with this example. First, it supports combinations of arguments that don't make logical sense, such as supplying an `x` argument but no `y` argument. Also, the default values will be compiled into your client's programs. This means that your clients must recompile their code if you release a new version of the API with a different default radius. In essence, you are exposing the behavior of the API when you do not explicitly specify a radius value.

To illustrate why this might be bad, consider the possibility that you later add support for the notion of different default units, letting the user switch between values specified in meters, centimeters, or millimeters. In this case, a constant default radius of 10.0 would be inappropriate for all units.

An alternative approach is to provide multiple overloaded methods instead of using default arguments. For example:

```
class Sphere
{
public:
    Sphere();
    Sphere(double x, double y);
    Sphere(double x, double y, double radius);
    ...
};
```

Using this approach, the implementation of the first two constructors can use a default value for the attributes that are not specified. But importantly, these default values are specified in the `.cpp` file and are not exposed in the `.h` file. As a result, a later version of the API could change these values without any impact on the public interface.

> *TIP: Prefer overloaded functions to default arguments when the default value would expose an implementation constant.*

Not all instances of default arguments need to be converted to overloaded methods. If the default argument represents an invalid or empty value, such as defining `nullptr` as the default value for a pointer or `""` for a string argument, then this usage is unlikely to change between API versions. However, if you have cases where you are hardcoding specific constant values into your API that might change in future releases, then you should convert these cases to use the overloaded method technique instead.

As a performance note, you should also try to avoid defining default arguments that involve constructing a temporary object, because these will be passed into the method by value and can therefore be expensive.

Avoid #define for constants

The `#define` preprocessor directive is essentially used to substitute one string with another string in your source code. However, its use is generally frowned upon in the C++ community for several good reasons (Cline et al., 1998; DeLoura, 2001; Meyers, 2005). Many of these reasons are related to the subtle problems that can happen if you use `#define` to specify code macros that you wish to insert into multiple places. For example:

```
#define SETUP_NOISE(i, b0, b1, r0, r1) \
    t = vec[i] + 0x1000;                \
    b0 = (lltrunc(t)) & 0xff;           \
    b1 = (b0 + 1) & 0xff;               \
    r0 = t - lltrunc(t);                \
    r1 = r0 - 1.f;
```

However, you should never be using #define in this way for your public API headers, because of course it leaks implementation details. If you want to use this technique in your .cpp files, and you understand all the idiosyncrasies of #define, then go ahead, but never do this in your public headers.

That just leaves the use of #define to specify constants for your API, such as:

```
#define MORPH_FADEIN_TIME  0.3f
#define MORPH_IN_TIME      1.1f
#define MORPH_FADEOUT_TIME 1.4f
```

You should avoid even this use of #define (unless you're writing a pure C API, of course), for these reasons:

1. **No typing.** A #define does not involve any type checking for the constant you are defining. You must therefore make sure that you explicitly specify the type of the constant you're defining to avoid any ambiguities, such as the use of the "f" suffix on single-precision floating-point constants. If you defined a floating-point constant as simply 10, then it may be assumed to be an integer in certain cases and cause undesired math rounding errors.
2. **No scoping.** A #define statement is global and is not limited to a particular scope, such as within a single class. You can use the #undef preprocessor directive to undefine a previous #define, but this makes little sense for declaring a constant that you want your clients to be able to use.
3. **No access control.** You cannot mark a #define as public, protected, or private. It's essentially always public. You therefore can't use #define to specify a constant that should be accessed only by derived classes of a base class that you define.
4. **No symbols.** In the previous example, the symbolic names such as MORPH_IN_TIME may be stripped from your code by the preprocessor, and as such the compiler never sees this name and cannot enter it into the symbol table (Meyers, 2005). This can hide valuable information from your clients when they try to debug code using your API, because they will simply see the constant value used in the debugger, without any descriptive name.

The preferred alternative to using `#define` to declare API constants is to use a const variable. I will discuss some of the best practices of declaring constants in the later chapter on Performance, because it's possible to declare const variables in a way that adds bloat to your clients' programs. For now, I'll simply present a good conversion of the previous `#define` example:

```
class Morph
{
public:
    static const float FadeInTime;
    static const float InTime;
    static const float FadeOutTime;
    ...
};
```

where the actual values of these constants are specified in the associated `.cpp` file. If you really want your users to know what the values of these constants are, then you can tell them this information in the API documentation for the `Morph` class. Alternatively, as of C++17, you can use inline variables to define your constants only in your header files, which is a much more elegant solution. Note that this representation doesn't suffer from any of the problems I listed earlier: the constants are typed as floats, scoped to the `Morph` class, explicitly marked as publicly accessible, and will generate entries in the symbol table.

> *TIP: Use static const data members to represent class constants instead of #define (or use inline variables as of C++17).*

A further use of `#define` is to provide a list of possible values for a given variable. For example:

```
#define LEFT_JUSTIFIED   0
#define RIGHT_JUSTIFIED  1
#define CENTER_JUSTIFIED 2
#define FULL_JUSTIFIED   3
```

This is better expressed in C++ using enumerated types, via an `enum class`. Using enums gives you better type safety because the compiler will now ensure that you set any enum values with the symbolic name and not directly as an integer. This also makes it more difficult to pass illegal values, such as −1 or 23 in the previous example. You can turn the earlier `#define` lines into an enumerated type as:

```
enum class Justification {
    Left,
    Right,
    Center,
    Full
};
```

Avoid using friends

In C++, friendship is a way for your class to grant full access privileges to another class or function. The friend class or function can then access all protected and private members of your class. This can be useful when you need to split your class into two or more parts, but you still need each part to access private members of the other part. It's also useful when you need to employ an internal visitor or callback technique: that is, when some other internal class in your implementation code needs to call a private method in your class.

One alternative would be to expose the member data and function that need to be shared, converting them from private to public so that the other class can access them. However, this would mean that you're exposing implementation details to your clients, details that would not otherwise be part of your logical interface. From this point of view, friends are a good thing because they let you open up access to your class to only specific clients. However, friendship can be abused by your users, allowing them to gain full access to your class's internal details.

For example, consider the following class that specifies a single `Node`, as part of a `Graph` hierarchy. The `Graph` may need to perform various iterations over all nodes and therefore needs to keep track of whether a node has already been visited (to handle graph cycles). One way to implement this would be to have the `Node` object hold the state for whether it has already been visited, with accessors for this state. This is a purely implementation detail, so you don't want to expose this functionality in the public interface. Instead, you declare it as private, but explicitly give the `Graph` object access to the `Node` object by declaring it as a friend:

```
class Node
{
public:
    ...
    friend class Graph;

private:
    void ClearVisited();
    void SetVisited();
    bool IsVisited() const;
    ...
};
```

This seems okay on the face of it: you have kept the various *Visited() methods as private and only permitted the Graph class to access to our internal details. However, the problem with this is that the friendship offer is based upon the name of the other class only. It would therefore be possible for clients to create their own class called Graph, which would then be able to access all protected and private members of Node (Lakos, 1996). The subsequent client program demonstrates how easy it is to perform this kind of access control violation:

```
#include "node.h"

// define your own Graph class
class Graph
{
public:
    void ViolateAccess(Node &node)
    {
        // call a private method in Node
        // because Graph is a friend of Node
        node.SetVisited();
    }
};

...

Node node;
Graph local_graph;
local_graph.ViolateAccess(node);
```

So, by using friends you are leaving a gaping hole in your API that could be used to circumvent your public API boundary and break encapsulation.

In that example, a better solution that obviates the need to use friends would be for the Graph object to maintain its own list of nodes that it has already visited (e.g., by maintaining an std::set<Node *> container) rather than storing the visited state in the individual nodes themselves. This is also a better conceptual design because the information about whether another class has processed a Node is not inherently an attribute of the Node itself.

TIP: Avoid using friends. They tend to indicate a poor design and can allow users to gain access to all protected and private members of your API.

It's worth noting that there can be some acceptable uses of friends, such as to help with unit testing of private methods. I will show how to do this in the Testing Private Code section of the Testing chapter.

Exporting symbols

In addition to the language-level access control features (public, private, and protected), there are two related concepts that allow you to expose symbols in your API at the physical file level:

1. External linkage
2. Exported visibility

The term external linkage means that a symbol in one translation unit can be accessed from other translation units, whereas exporting refers to a symbol that is visible from a library file such as a DLL. Only external linkage symbols can be exported.

Let's look at external linkage first. This is the first stage that determines whether your clients can access symbols in your shared libraries. Specifically, global (file scope) free functions and variables in your `.cpp` file will have external linkage unless you take steps to prevent this. For example, consider this code that might appear in one of your `.cpp` files:

```
...

int INTERNAL_CONSTANT = 42;

std::string Filename = "file.txt";

void FreeFunction()
{
    std::cout << "Free function called" << std::endl;
}

...
```

Even though you have contained the use of these functions and variables inside a `.cpp` file, resourceful clients could easily gain access to these symbols from their own programs (ignoring symbol exporting issues for the moment). They could then call your global functions directly and modify your global state without going through your public API, thus breaking encapsulation. The following program fragment demonstrates how to achieve this:

```
extern void FreeFunction();
extern int INTERNAL_CONSTANT;
extern std::string Filename;

// call an internal function within your module
FreeFunction();

// access a constant defined within your module
std::cout << "Constant = " << INTERNAL_CONSTANT << std::endl;

// change global state within your module
Filename = "different.txt";
```

There are a couple of solutions to this kind of external linkage leakage problem:

1. **Static declaration.** Prepend the declaration of your functions and variables with the `static` keyword. This specifies that the function or variable should have internal linkage and hence will not be accessible outside the translation unit in which it appears.
2. **Anonymous namespace.** A more idiomatic C++ solution is to enclose your file-scope functions and variables inside an anonymous namespace. This is a better solution because it avoids polluting the global namespace. This can be done as:

```
...

namespace {

int INTERNAL_CONSTANT = 42;

std::string Filename = "file.txt";

void FreeFunction()
{
    std::cout << "Free function called" << std::endl;
}

} // namespace

...
```

TIP: Use internal linkage to hide file scope free functions and variables inside your .cpp files. This means using the static keyword or the anonymous namespace.

For symbols that have external linkage, there's the further concept of exporting symbols, which determines whether a symbol is visible from a shared library. Most compilers provide decorations for classes and functions that let you explicitly specify whether a symbol will appear in the exported symbol table for a library file. However, this tends to be compiler-specific behavior. For example:

1. **Microsoft Visual Studio.** Symbols in a DLL are not accessible by default. You must explicitly export functions, classes, and variables in a DLL to allow your clients to access them. You do this using the `__declspec` decorator before a symbol. For example, you specify `__declspec(dllexport)` to export a symbol when you're building a DLL. Clients must then specify `__declspec(dllimport)` to access the same symbol in their own programs.

2. **GNU/Clang C++ compiler.** Symbols with external linkage in a dynamic library are visible by default. However, you can use the visibility `__attribute__` decorator to explicitly hide a symbol. As an alternative to hiding individual symbols, the GNU C++ 4.0 compiler introduced the `-fvisibility=hidden` flag to set all declarations to hidden visibility by default. Individual symbols can then be explicitly exported using `__attribute__ ((visibility("default")))`. This is more like Windows behavior, in which all symbols are considered internal unless you explicitly export them. Using the `-fvisibility=hidden` flag can also have a dramatic impact on the load time performance of your dynamic library and produce smaller library files.

You can define various preprocessor macros to deal with these compiler differences in a cross-platform way. Here's an example of defining a `DLL_PUBLIC` macro to export symbols explicitly, and a `DLL_HIDDEN` macro to hide symbols when using the GNU C++ compiler. Note that you must specify an `_EXPORTING` define when you build the library file on Windows (i.e., `/D "_EXPORTING"`). This is an arbitrary define name; you can call it whatever you like (as long as you also update this code):

```
#if defined _WIN32 || defined __CYGWIN__
  #ifdef _EXPORTING // define this when generating DLL
    #ifdef __GNUC__
      #define DLL_PUBLIC __attribute__((dllexport))
    #else
      #define DLL_PUBLIC __declspec(dllexport)
    #endif
  #else
    #ifdef __GNUC__
      #define DLL_PUBLIC __attribute__((dllimport))
    #else
      #define DLL_PUBLIC __declspec(dllimport)
    #endif
  #endif
  #define DLL_HIDDEN
#else
  #if __GNUC__ >= 4
    #define DLL_PUBLIC __attribute__ ((visibility("default")))
    #define DLL_HIDDEN __attribute__ ((visibility("hidden")))
  #else
    #define DLL_PUBLIC
    #define DLL_HIDDEN
  #endif
#endif
```

For example, to export a class or function of your API you can do this:

```
DLL_PUBLIC void MyFunction();
class DLL_PUBLIC MyClass;
```

Many compilers also allow you to provide a simple ASCII file that defines the list of symbols that should be exported by a dynamic library, so that you don't need to decorate your code with macros such as the previous `DLL_PUBLIC`. Symbols that don't appear in this file will be hidden from client programs. For example, the Windows Visual Studio compiler supports `.def` files, whereas the GNU compiler supports export map files. See Appendix A for more details on these export files.

> *TIP: Explicitly export public API symbols to maintain direct control over the classes, functions, and variables that are accessible from your dynamic libraries. For GNU C++, this implies using the `-fvisibility=hidden` option.*

Note that with the introduction of modules in C++20, you can use the `export` keyword within a named module to indicate which symbols should be accessible from the module. This effectively standardizes the export behavior of symbols for translation units for all compilers, though only if you're using modules. See the Modules section in the C++ Revisions chapter for more details about this feature.

Coding conventions

C++ is a very complex language with many power features. The use of good coding conventions can partially help to manage this complexity by ensuring that all code follows certain style guidelines and avoids common pitfalls. It also contributes toward consistent code, which I identified as one of the important qualities of a good API in Chapter 2.

> *TIP: Specify coding standards for your API to help enforce consistency, define processes, and document common engineering pitfalls.*

Producing a coding standard document can be a lengthy and factious process, not only because of the complexity of the language but also because of the amount of personal taste and style that it aims to stipulate. Different engineers have different preferences for where to place brackets and spaces, or what style of comments to adopt, or whether to use lower or upper camelCase for function names. For example, in this book I have consistently formatted source code snippets with pointer or reference symbols next to variable names, not type names. For example:

```
char *a, *b;
```

instead of:

```
char* a, *b;
```

I favor the former style because from a language perspective the pointer is associated with the variable, not the type (in both of these cases, a and b are pointers). However, other software engineers prefer the latter style.

In this book, I do not advocate that you should adopt any particular style for your projects, but I do urge you to adopt some conventions for your API, whatever they are. The important point is to be consistent. Indeed, it's generally accepted among engineers that when editing a source file, you should adopt the conventions that are already in force in that file, rather than adding your own style into the mix and producing a file with a mixture of inconsistent styles (or you might be more antisocial and reformat the entire file to your own style).

Many large companies have already gone through the process of creating and publishing coding style documents, so you could always simply adopt one of these standards to make the decision easier. Doing a Web search for "C++ Coding Conventions" should return many hits for your consideration. In particular, the Google C++ Style Guide is a very extensive document used by many other groups (https://google.github.io/styleguide/). There are also some great books that provide even more depth and provide detail on numerous code constructs that should be used or avoided (Sutter and Alexandrescu, 2004).

Without making specific suggestions, I will enumerate some of the areas that a good coding standard should cover:

- **Naming conventions.** Whether to use `.cc`, `.c++`, or `.cpp`; capitalization of filenames; use of prefixes in filenames; capitalization of classes, functions, variables, differentiating private members, constants, typedefs, enums, macros, namespaces; use of namespaces; and so on.
- **Header files.** How to `#include` headers; use of double quotes versus angle brackets; ordering `#include` statements; using `#define` guards or `#pragma once`; use of forward declarations; inlining code policy; and so forth.
- **Comments.** Comment style; templates for commenting files, classes, functions, and so on; documentation requirements; commenting code with todo notes or highlighting known hacks; and so on.
- **Formatting.** Line length limit; spaces versus tabs; placement of braces; spacing between statements; how to break long lines; layout of constructor initialization lists; etc.
- **Classes.** Use of constructors; factories; inheritance; multiple inheritance; interfaces; composition, structs versus classes; access control; and so forth.

- **Best practices.** Use of templates; use of exceptions; use of enums; const correctness; use of pointers for output parameters, use of pimpl; initialization of all member variables; casting; operator overloading rules; virtual destructors; use of globals; and so on.
- **Portability.** Writing architecture-specific code; preprocessor macros for platforms; class member alignment; and so forth.
- **Process.** Compiling with warnings as errors; unit tests requirements; use of code reviews; use of use cases; SCM style check hooks; compile with extra checks such as `-Wextra` and `-Weffc++`; and so on.

The best kind of coding conventions are the ones that can be applied to your code automatically. There are several tools you can use to format your source files consistently; one that I like is clang-format. This comes with a few built-in styles such as LLVM, GNU, Google, Chromium, Microsoft, Mozilla, and WebKit. So if you want just to adopt one of these standard coding conventions, you can very easily conform all of your code with a single command line, such as:

```
clang-format -style=Google -i *.cpp *.h
```

And if there's some part of the style that you don't like, then clang-format supports an extensive list of settings that you can edit. To do this, you can dump out the configuration for one of the standard styles to a file, and then edit that file to suit your preferences. For example, here's how you could create a config file called `.clang-format` with all of the Microsoft style options, and then you can go in and edit that file to your own taste:

```
clang-format -style=Microsoft -dump-config > .clang-format
```

You can then format your code with your custom style by running the command:

```
clang-format -style=file -i *.cpp *.h
```

In this case, the tool will specifically look for a file called `.clang-format` in your source directory. If it doesn't find the file, it will look through each parent directory. A good practice is therefore to put this file in the root directory of your source repository so that it affects all of your code. You could also consider updating any build scripts to format your code automatically or define a hook for your revision control system to run it before pushing changes. That way, you can ensure that your code always follows your code conventions, and as an added bonus you can avoid code review feedback with nitpick comments about spacing or braces placement.

In fact, all of the code samples throughout this book have been formatted with `clang-format` to enforce a consistent style. You can find the configuration file that I used in the source code package on the book website, https://APIBook.com/.

7 C++ revisions

After the initial standardization of the C++ language in 1998 and a subsequent minor amendment in 2003, the language underwent a major revision in 2011 with C++11. Since that time, the ISO standard committee has committed to releasing updates to the language every 3 years. To date, we have seen C++14, C++17, C++20, and C++23. These have varied in scale, but generally they have sought to address problems or ambiguities in existing functionality as well as add new features to the core language and the Standard Library.

Rather than mix all of the features from all revisions together into a single discussion, I have chosen to break them out chronologically. The reason is that you will normally make a conscious choice about which revision of the standard you will use in each of your software projects, and so you only need to read up to the section that covers the revision you're using. Of course, you can always read beyond that point to learn about useful features that may inspire you to adopt a more recent version as well.

This chapter is not meant to be an exhaustive compendium of every feature introduced in each revision. Because this book is focused on application programming interface (API) design, I'll focus only on those features that have an impact on your interface designs. Although features such as the `auto` keyword, range-based `for` loops, lambda expressions, and new Standard Library algorithms are wonderful additions to the language, they have little impact on how you can deliver robust interfaces to your users, so I'll not spend time on those aspects.

Which C++ revision to use

Given the range of C++ revisions that have been published to date, you may reasonably ask which one you should target for your projects. Some factors to consider in your decision include:

1. **Dependencies:** The versions of C++ that your dependent libraries use can influence the choice of which version you can adopt.
2. **Clients:** The end users of your library may have limitations on the version of C++ they can support.
3. **Compilers:** Not all compilers on all platforms may fully support every new feature of the most recent versions of the language.
4. **Features:** You may want to use a more recent revision because it gives you access to a newer feature that you want to use.

An important question is whether you can mix and match different C++ versions within a single project. For example, can you use C++20 if your dependencies are all

API Design for C++. https://doi.org/10.1016/B978-0-443-22219-1.00006-4

using C++17? A key factor here is whether you're building your dependencies from source or binary. If you have the source for the other libraries, there should be no problem: a C++20 compiler should be able to compile all earlier versions.

The situation becomes more complex if you're depending on only the precompiled binaries for your dependencies. The C++ standard does not define a standardized application binary interface (ABI). So by default, you should assume that you can't mix and match code that was compiled with different versions of the compiler.

Having said that, some compiler implementations do try hard to provide a compatible ABI. For example, both gcc and clang generally let you link code that was compiled with different stable versions. For example, C++17 and C++11 are binary compatible for these compilers. By contrast, Microsoft Visual C++ does not try to maintain ABI compatibility between versions. The Microsoft C Runtime (MSVCRT) is versioned per release of the compiler, and apps cannot mix multiple run times.

As for a specific recommendation, C++11 was a huge leap forward for the language. It introduced a robust memory model, concurrency, smart pointers, and much more. Any modern C++ project should be using at least C++11. Beyond that, you should try to use the most recent version you can while considering the constraints of your clients, source and binary dependencies, and the platforms you need to support.

C++11 API features

C++11 was published in Sep. 2011 and introduced many changes to the language. These changes include improvements to the core language, built-in support for multithreading, templating enhancements, and significant changes to the C++ Standard Library. In the upcoming sections, I'll introduce the various features of C++11 that can further enhance your ability to deliver robust APIs, beyond the base functionality offered by C++03.

I'll not cover the new multithreading features introduced by C++11 here because I've dedicated an entire chapter to this important topic. Please refer to the Concurrency chapter for more details on this aspect of C++.

As some historical context, C++11 was referred to as C++0x during its development, because it was uncertain when the revision would be published. The x was meant to stand in for the year 2008 or 2009, although ultimately the update wasn't published until 2011.

Move constructors and the Rule of Five

The Rule of Three for C++03 code states that if a class defines one of the following special functions, then it must define all three:

- Destructor,
- Copy constructor,
- Assignment operator.

The reason for this rule is that if one of these functions is required, then it's most likely because the class dynamically allocates some resource. It's therefore important that the class explicitly defines how that resource should be freed, copied, and assigned to another object. With C++11, two special functions were added to implement move semantics for a class:

- Move constructor,
- Move assignment operator.

A move constructor is used to move the resources owned by one object to another. Whereas a copy constructor will leave the original object unchanged and create a copy of the object's data, a move constructor will transfer ownership of the data to the new object and update the original to stop referring to those data. Move constructors are therefore usually more efficient because they avoid unnecessary copying of an object's data. However, the use of a move constructor also implies that the original object will no longer be used.

A move assignment operator offers similar move semantics for assignment, such as to transfer a temporary object to an existing object (as opposed to a move constructor that creates a new object).

The Rule of Five for C++11 states that if you define a destructor, copy constructor, copy assignment operator, move constructor, or move assignment operator, then you should define all five.

(But if you don't define a move constructor and/or move assignment operator, your copy constructor and/or copy assignment operator will be used by default.)

The compiler can use a move constructor and move assignment operator to move data more efficiently between temporary variables that it creates. For example, when you store the result of a function call in a variable, such as `auto f = func()`, the compiler can use your move constructor to avoid having to copy the function return value into your variable. You can also allow the compiler to trigger move semantics in your own code using the `std::move` function. For example, this `swap` function template will create temporary copies of the provided objects:

```
template <class T>
void swap(T &a, T &b)
{
    T tmp(a);
    a = b;
    b = tmp;
}
```

However, this alternate implementation of the function uses the move constructor and move assignment operator to swap the two objects more efficiently:

```
template <class T>
void swap(T &a, T &b)
{
    T temp(std::move(a));
    a = std::move(b);
    b = std::move(temp);
}
```

(`std::move` doesn't directly call a move constructor. It just casts an object to an rvalue reference such as `MyClass &&` to indicate that it's a temporary value that won't persist beyond the expression where it's used. This lets the compiler decide to use the move constructor if it can.)

If you don't define a move constructor, the compiler will just use the copy constructor, which should be functionally identical albeit less efficient. Similarly, if you don't define a move assignment operator, the compiler will fall back to using the copy assignment operator. So, although the Rule of Three was expanded to the Rule of Five for C++11, if you only conform to the original Rule of Three, then your code will still behave correctly, just less efficiently than it could do in some circumstances.

The next example shows the use of a traditional destructor, copy constructor, and copy assignment operator, as well as how to define a move constructor and move assignment operator:

```
class MyClass
{
public:
    // Default constructor
    MyClass();

    // Non-default constructor
    explicit MyClass(size_t length);

    // Destructor
    ~MyClass();

    // Copy constructor
    MyClass(const MyClass &other);

    // Copy assignment operator
    MyClass &operator=(const MyClass &other);

    // Move constructor
    MyClass(MyClass &&other);

    // Move assignment operator
    MyClass &operator=(MyClass &&other);

private:
    std::vector<int> *mData;
};
```

The implementation of these functions looks like:

```
// Default constructor
MyClass::MyClass() :
    mData(nullptr)
{
}

// Non-default constructor
MyClass::MyClass(size_t length) :
    mData(new std::vector<int>())
{
    std::cout << "In constructor" << std::endl;
    mData->resize(length);
}

// Destructor
MyClass::~MyClass()
{
    std::cout << "In destructor" << std::endl;
    delete mData;
}

// Copy constructor
MyClass::MyClass(const MyClass &other) :
    mData(nullptr)
{
    std::cout << "In copy constructor" << std::endl;
    if (other.mData) {
        mData = new std::vector<int>();
        mData->resize(other.mData->size());
        std::copy(other.mData->begin(), other.mData->end(), mData->begin());
    }
}

// Copy assignment operator
MyClass &
MyClass::operator=(const MyClass &other)
{
    std::cout << "In copy assignment operator" << std::endl;
    if (this != &other) {
        delete mData;
        mData = new std::vector<int>();
        mData->resize(other.mData->size());
        std::copy(other.mData->begin(), other.mData->end(), mData->begin());
    }
    return *this;
}

// Move constructor
MyClass::MyClass(MyClass &&other) :
    mData(nullptr)
{
    std::cout << "In move constructor" << std::endl;
    if (other.mData) {
        mData = other.mData;
        other.mData = nullptr;
    }
}

// Move assignment operator
MyClass &
MyClass::operator=(MyClass &&other)
{
```

```
        std::cout << "In move assignment operator" << std::endl;
        if (this != &other) {
            delete mData;
            mData = other.mData;
            other.mData = nullptr;
        }
        return *this;
    }
```

If we pass two instances of this class to the first version of the previous swap function, it will produce the output:

```
In constructor
In constructor
In copy constructor
In copy assignment operator
In copy assignment operator
In destructor
In destructor
In destructor
```

whereas the second version of the swap function, using `std::move`, produces this output, showing the use of the move constructor and move assignment operator:

```
In constructor
In constructor
In move constructor
In move assignment operator
In move assignment operator
In destructor
In destructor
In destructor
```

Default and deleted functions

C++03 gave you little control over the compiler's behavior of automatically generating special functions such as constructors, destructors, and assignment operators. For example, if you don't declare a copy constructor, a C++03 compiler will always generate one for you. However, with C++11, you have explicit control over whether the compiler generates, or does not generate, these special functions.

For instance, you can now specifically indicate that you want the compiler to generate its default implementation for any special function using the new `= default` syntax in the function declaration, as in this example:

```
class MyClass
{
public:
    // default constructor
    MyClass() = default;

    // virtual destructor
    virtual ~MyClass() = default;

    // copy constructor
    MyClass(const MyClass &) = default;

    // move constructor
    MyClass(MyClass &&) = default;

    // copy assignment operator
    MyClass &operator=(const MyClass &) = default;

    // move assignment operator
    MyClass &operator=(MyClass &&) = default;
};
```

The resulting behavior should be the same as if you hadn't declared these functions at all: the compiler will generate its default implementation in both cases. However, the explicit default syntax makes your intention to use the compiler-generated implementation clearer.

A class that uses a compiler-generated default constructor is known as trivial. A trivial class offers certain properties such as being trivially copyable, so you can copy the object using `memcpy`. However, if you provide your own default constructor implementation, then your class becomes nontrivial, even if you are only defining an empty implementation such as:

```
class MyClass
{
public:
    MyClass() {}
    virtual ~MyClass() {}
};
```

Therefore, using = `default` to enforce the compiler-generated implementation is helpful to keep the object trivial, if you need that property for your object.

You can also tell the compiler to disable certain functions that would otherwise be generated for you, using the new = `delete` syntax. This is useful if you want to prevent a special function from being implemented. For example, this lets you make a class be noncopyable by deleting the copy constructor and copy assignment operator. Because this is C++11, you should include the move constructor and move assignment operator in that definition. For example:

```
struct NonCopyable
{
public:
    NonCopyable() = default;
    virtual ~NonCopyable() = default;
    NonCopyable(const NonCopyable &) = delete;
    NonCopyable &operator=(const NonCopyable &) = delete;
    NonCopyable(NonCopyable &&) = delete;
    NonCopyable &operator=(NonCopyable &&) = delete;
};
```

This is an alternative to the C++03 technique of declaring these functions as private, but the delete solution is better because in the private case you would still be able to access the copy constructor and assignment operator within the class where they're defined. Using the delete solution makes it certain that these functions can never be called from any context.

Moreover, you can put the = `default` and = `delete` specifiers in your `.cpp` files instead of your headers. In fact, that can be a better strategy from the point of view of good API design because then you're not exposing and inlining the implementation of the function in your header files. For example:

```
// MyClass.h
class MyClass
{
public:
    MyClass();
};

// MyClass.cpp
MyClass::MyClass() = default;
```

Object construction

C++11 introduced several features relating to constructors and object initialization. The first feature we'll cover is the ability for one constructor to call another. For example, consider a simple class that stores a floating-point range and caches the median value of the range:

```
class Range
{
public:
    Range();
    explicit Range(double max);
    Range(double min, double max);

    double GetMedian() const { return mMedian; }

private:
    double mMin;
    double mMax;
    double mMedian;
};
```

The implementation of these three constructors might look like:

```
Range::Range() :
    mMin(0),
    mMax(0),
    mMedian(0)
{}

Range::Range(double max) :
    mMin(0),
    mMax(max),
    mMedian(max / 2.0)
{}

Range::Range(double min, double max) :
    mMin(min),
    mMax(max),
    mMedian((max - min) / 2.0 + min)
{}
```

In this case, the logic to calculate the `mMedian` data member is duplicated in each constructor. It's possible to introduce a mathematical error in one constructor that would cause the object to behave differently, depending on which constructor was used.

Instead, it would be more robust to implement the first two constructors in terms of the third, more general constructor. This was not possible with C++03 without creating an additional member function that all three constructors can call.

However, starting with C++11, one constructor, known as a delegating constructor, can call another constructor, known as the target constructor. With this capability, the first two constructors in our previous example can now be rewritten more succinctly as:

```
Range::Range() :
    Range(0)
{}

Range::Range(double max) :
    Range(0, max)
{}
```

The first constructor calls the second constructor, which then calls the third constructor. If you're concerned about delegation cycles, your compiler should generate an error if, for example, you have one constructor that calls another constructor that then calls the first one again. If you use a delegating constructor, then you cannot include any other definitions in the constructor's initialization list.

This technique helps achieve the minimally complete API quality by defining the logic to calculate the median value in one place and reusing it across all constructors. As such, it's good practice to use delegating constructors where you can, to reduce redundancy.

The second feature related to object initialization we'll cover here is data member initializers. With C++11, you can now provide a default value for any data members as part of the class declaration. For example:

```
class MyObject
{
public:
    std::string GetName() const { return mName; }

private:
    std::string mName = "ObjectName";
};
```

In this case, the constructor will initialize the value of the `mName` data member to "ObjectName" when the object is created. As a result, you can omit defining the constructor if all it would do is set initial values for each data member. Instead, you can rely on the compiler-generated constructor to set the default values specified in the data member initializers.

This helps to achieve the so-called Rule of Zero, which states that you should avoid defining any special functions, such as constructors and destructors, if you can (i.e., you should rely upon the implicitly defined version of these functions that the compiler generates whenever possible).

It's still possible to define a value for each data member in the constructor. In that case, any initialization you implement in the constructor will override the value from the initializer. For example, we can add a constructor to the class as:

```
class MyObject
{
public:
    MyObject() {}
    MyObject(const std::string &name) :
        mName(name)
    {}

    std::string GetName() const { return mName; }

private:
    std::string mName = "ObjectName";
};
```

In this case, the zero-argument constructor will still initialize the `mName` data member with the string "ObjectName", but the second constructor that accepts a string argument will override the data member initializer. For example, `MyObject("OverrideName")` will set `mName` to "OverrideName".

TIP: For simple data member initialization cases, you can use C++11 data member initializers and avoid defining a custom constructor.

Initializer list constructors

C++11 introduced support for list initialization using curly bracket syntax, letting you initialize containers thus:

```
std::vector<int> intList = {1, 1, 2, 3, 5, 8};
```

or more simply,

```
std::vector<int> intList{1, 1, 2, 3, 5, 8};
```

You can also support initialization of your own classes using this syntax by defining a constructor that uses `std::initializer_list`. For example,

```
class MyClass
{
public:
    MyClass() = default;
    MyClass(int a);
    MyClass(std::initializer_list<int> intList);
};
```

Inside the `initializer_list` constructor, you can refer to `intList.size()` to get the number of values, and you can use `intList.begin()` and `intList.end()` to iterate over the elements.

With that class declaration, you can do the following:

```
MyClass obj1{1, 2, 3};  // calls initializer_list ctor
MyClass obj2{1};        // calls initializer_list ctor
MyClass obj3(1);        // calls single int ctor
MyClass obj4{};         // calls default ctor
MyClass obj5;           // calls default ctor
```

Calling `MyClass{1}` will call the `initializer_list` constructor, whereas the regular syntax `MyClass(1)` will call the single int constructor. Also, empty curly braces will generally prefer calling the default constructor instead of the `initializer_list` one, because an empty initializer list should behave like a default constructor, but the default constructor should be more efficient.

Smart pointers

A smart pointer is a class template that wraps a raw (or bare) pointer and provides management of the heap-allocated resource. From an implementation perspective, it's just a class that overloads the -> and * operators so that it behaves like a pointer, but with additional capabilities.

The key benefit of smart pointers is that they take care of deleting the memory associated with the pointer at some well-defined point in time. This can help avoid many common memory errors such as leaking memory, writing to unallocated memory, and double freeing memory.

C++11 introduced several different types of smart pointer in the `<memory>` header. We covered these in Chapter 2, but to recap them here they are:

- `std::shared_ptr`: a reference-counted pointer in which the pointer is deallocated when the reference count reaches zero.
- `std::weak_ptr`: provides a way to refer to a `shared_ptr` without influencing its reference count (e.g., to break reference cycles).
- `std::unique_ptr`: deallocates the pointer when the unique pointer goes out of scope.

To illustrate these concepts more concretely, here is an example of a raw pointer in which the pointer must be specifically deallocated using the `delete` keyword:

```
MyObject *obj = new MyObject();
...
delete obj;
```

For simple code, it can be easy to ensure that the object is deleted at the right time. But for more complex code, in which the pointer might be passed around through different functions and stored by different objects, it can be difficult to know when to delete the object safely.

By contrast, an `std::shared_ptr` will automatically delete the pointer when all references go out of scope, such as:

```
{
    std::shared_ptr<MyObject> obj1 = std::make_shared<MyObject>();
    {
        std::shared_ptr<MyObject> obj2 = obj1; // ref count == 2
        ...
    } // obj2 goes out of scope, ref count == 1
    ...
} // obj1 goes out of scope, ref count == 0, object deleted
```

If you're returning pointers from your API, then it's much safer to return them as shared pointers. I like to introduce a type definition for a shared pointer of any object that I intend to reference with a pointer, just to make the syntax easier to work with. For example,

```
#include <memory>

class MyObject;
using MyObjectPtr = std::shared_ptr<MyObject>;

class MyObject
{
    ...
};
```

In addition to returning pointers from your API, you can use a shared pointer to allocate a private data member object in your constructor and have it be automatically deallocated when the object is destroyed, without having to handle this explicitly in the destructor. For example:

```
class MyObjectHolder
{
public:
    MyObjectHolder() :
        obj(std::make_shared<MyObject>())
    {}
    ~MyObjectHolder() = default;

private:
    std::shared_ptr<MyObject> obj;
};
```

As for weak pointers, these are useful to return an object when you don't want to grant shared ownership of the underlying object to the client. That is, you don't want the client's act of holding on to the pointer to prevent the object from being cleaned up when all other references are released. A classic example of this is returning a pointer to an object in a cache, in which the object might get ejected from the cache and deleted, even if the client is still holding on to the weak pointer. This is better than returning a raw pointer because the client can check whether the underlying data have already been deleted by using the `expired()` member function.

The final smart pointer type in C++11 is the unique pointer, `std::unique_ptr`. This is a simpler version of `std::shared_ptr` with no reference counting, meaning that you cannot copy a unique pointer. This is useful when you want to tie the lifetime of the object to a block of code. You can also use a unique pointer with a data member to avoid having to delete the pointer manually in the destructor, in the same way that we saw earlier for `std::shared_ptr`. The difference in the `std::unique_ptr` case is that the containing object will no longer be copyable. For example:

```
class MyObjectHolder
{
public:
    MyObjectHolder() :
        obj(std::make_unique<MyObject>())
    {}
    ~MyObjectHolder() = default;

private:
    std::unique_ptr<MyObject> obj;
};

MyObjectHolder holder1;
MyObjectHolder holder2 = holder1;  // compiler error: non copyable
```

To summarize, you should use `std::shared_ptr` if ownership of a resource can be shared among several objects. You should use an `std::unique_ptr` if the resource should be owned only by a single object. And you should use an `std::weak_ptr` if you want to share temporary ownership of the resource.

Enum classes

Enumerated types are user-defined types with named values. They are useful for referring to a set of constants with meaningful names. For example:

```
enum Mode {
    READ,
    WRITE,
    APPEND,
};
```

However, C++03-style enums have several problems, including:

1. **Type safety:** Enums are essentially integers. A value from one enum can be compared with a value from another enum, even though they may have no semantic relation. Also, enums can be implicitly converted to integers, which can sometimes lead to confusion.
2. **Naming collisions:** All enum values exist at the same level of scope as the enum definition, which means that no two enums at the same scope can share the same value. To get around this, you must either use unique prefixes for each enum value (e.g., `MODE_READ`) or define the enum within a class or a different namespace.
3. **Undefined size:** The underlying integer type used to represent an enum is implementation-dependent and depends upon the definition of its values. Furthermore, you cannot directly specify the size of an enum.
4. **No forward declarations:** Because the size of an enum is unknown, it's not possible to forward declare an enum type. As a result, if you want to refer to an enum, then you must include the header file where it's defined.

To address these issues, C++11 introduced the concept of enum classes, sometimes also called scoped enums. The syntax for an enum class is similar to a traditional enum with the addition of the class (or struct) keyword:

```
enum class Mode {
    Read,
    Write,
    Append,
};
```

You must refer to the values of an enum class by prefixing them with the name of the enum (e.g., `Mode::Read` or `Mode::Write`). This means that you can define two enum classes at the same level of scope that have the same value names, such as:

```
enum class Permission {
    Read,
    Write,
    Execute,
};
```

In this case, `Mode::Read` and `Permission::Read` do not clash with each other because of the enum class namespace. They also cannot be compared with or assigned to each other because they are different types.

Enum classes are also not implicitly converted to integers, which offers stronger type safety. For example, imagine a function that accepts an integer as an argument, such as:

```
void SetSize(int size);
```

It would be possible to call this function with a traditional enum value, such as `SetSize(READ)`, which may not make a lot of sense. However, this is not possible with an enum class because it cannot be implicitly converted to an integer. For example, `SetSize(Mode::Read)` will produce a compiler error.

Although the size of a traditional enum is undefined, the size of a class enum is defined to be the size of an int. You can change this if you need to by defining the base integral type of the enum (you can also do this for traditional enums as of C++11). For example, we could redefine our `Mode` enum class to be the same size as a char:

```
enum class Mode : char {
    Read,
    Write,
    Append,
};
```

Finally, because the default size of an enum class is well-defined, it's now possible to forward declare these types. This is particularly useful in API design because it lets you refer to an enum type without having to include the header file directly where it's defined. For example,

```
enum class Mode;

FILE *OpenFile(const std::string &filename, Mode mode);
```

It should be clear that enum classes have many advantages over regular C++03 enums. They offer stronger type safety, scoping of value names, well-defined memory sizes, and the ability to forward declare them. For all of these reasons, you should prefer using enum classes over regular enums in your APIs whenever possible.

TIP: Always prefer using a C++11 enum class over a regular C++03 enum.

Override and final specifiers

C++ supports an inheritance model in which a derived class can override the implementation of a virtual function in a base class. However, there are several ways to introduce subtle errors with the original C++03 inheritance syntax. For example, if you intend to override a function in a derived class, you should define the function in the base class as `virtual`. However, if you don't do that, your code may still appear to work as expected in some cases. Consider these two classes:

```
class Base
{
public:
    virtual ~Base() = default;
    void Func() { std::cout << "Base::Func()" << std::endl; }
};

class Derived : public Base
{
public:
    void Func() { std::cout << "Derived::Func()" << std::endl; }
};
```

This code declares a `Func()` function in both the `Base` and `Derived` classes, but the function is not declared as `virtual` in the `Base` class. Nevertheless, this code still works as you would expect:

```
Base base;
Derived derived;
base.Func();
derived.Func();
```

produces the following output:

```
Base::Func()
Derived::Func()
```

The problem, however, is that the compiler will not be able to do dynamic binding at run time without the `virtual` keyword. For example:

```
void Process(Base_Error *obj)
{
    obj->Func();
}

...

Base base;
Derived derived;
Process(&base);
Process(&derived);
```

produces the unexpected output of:

```
Base::Func()
Base::Func()
```

That is, the `Base` class version of the function is called in the second case even though we pass in a pointer to a `Derived` object. The solution is to declare the `Base::Func()` function as `virtual` to ensure that we get dynamic function binding instead of static binding; that is,

```
class Base
{
public:
    virtual ~Base() = default;
    virtual void Func() { std::cout << "Base::Func()" << std::endl; }
};
```

Another possible error that can be encountered overriding a function is if you don't match the function names exactly, either because you spelled the name wrong or you renamed the function in one class but forgot to rename it in the other. For example,

```
class Base
{
public:
    virtual ~Base() = default;
    virtual void Func() { std::cout << "Base::Func()" << std::endl; }
};

class Derived : public Base
{
public:
    void func() { std::cout << "Derived::Func()" << std::endl; }
};
```

In this case, `Base::Func()` is declared as virtual. However, in the `Derived` class, although the intent is to override the function, an error was made by using a lowercase version of the function name, `Derived::func()`. This will compile, but it will not achieve the intended goal of overriding the base class function.

C++11 introduced syntax to solve both of these kinds of errors with a new `override` specifier that can be defined on the overridden function. This lets the compiler know that you intend to override a virtual function in a base class. The compiler can then generate an error if the function is not marked as `virtual` in the base class, or if the function does not exist in the base class. For example, this will not compile in C++11:

```
class Base
{
public:
    virtual ~Base() = default;
    void Func();
};

class Derived : public Base
{
public:
    void Func() override;  // compiler error: non-virtual base function
};
```

Similarly, this code will also not compile:

```
class Base
{
public:
    virtual ~Base() = default;
    void Func();
};

class Derived : public Base
{
public:
    void func() override;  // compiler error: no base virtual function
};
```

C++11 introduced another feature to provide more flexibility around object inheritance: the `final` specifier. This can be used on a class to prevent it from being derived further. It can also be used on a virtual function to indicate that it can't be overridden in derived classes. For example, the following code will fail to compile because the `Base` class is marked as `final` and therefore cannot be used as a base class for the `Derived` class.

```
class Base final
{
public:
    virtual ~Base() = default;
};

class Derived : public Base  // compiler error: overriding a final class
{
};
```

The next example illustrates using the `final` specifier to prevent a specific function from being overridden in a derived class:

```
class Base
{
public:
    virtual ~Base() = default;
    virtual void Func();
};

class Derived : public Base
{
public:
    void Func() final;
};

class FurtherDerived : public Derived
{
public:
    void Func() override;  // compiler error: overriding final function
};
```

The `final` specifier can therefore be used to prevent your clients from deriving further subclasses or overriding virtual functions in further subclasses. It can also be used to improve performance by allowing a virtual function to be transformed into a direct method call, a process called devirtualization.

Microsoft introduced a `sealed` keyword in Visual C++ before the `final` keyword was added to the standard. The two keywords are equivalent, but `final` is the one you should use to be standards compliant.

A good practice to adopt is to ensure that any virtual function you define uses only one of `virtual`, `override`, or `final`:

1. Use `virtual` to introduce a new virtual function.
2. Use `override` to declare a nonfinal overridden function.
3. Use `final` to declare a virtual function that cannot be overridden further.

TIP: Always use exactly one of `virtual`, `override`, or `final` for every virtual function.

The noexcept specifier

C++11 introduced the `noexcept` exception specifier, which you can add to a function signature to indicate that it can or won't throw an exception. There's also a `noexcept` operator that can be used to perform a compile-time check to see if an expression may or won't throw an exception. For example:

```
void Func1() noexcept;         // won't throw
void Func2() noexcept(true);   // won't throw
void Func3() noexcept(false);  // may throw
void Func4();                  // may throw

std::cout << noexcept(Func1()) << "\n"; // true
std::cout << noexcept(Func2()) << "\n"; // true
std::cout << noexcept(Func3()) << "\n"; // false
std::cout << noexcept(Func4()) << "\n"; // false
```

You should obviously not use `noexcept` on a function that throws an exception. Your compiler will likely tell you if you try to do this. Remember that allocating memory with the `new` operator can throw an `std::bad_alloc` exception if it runs out of memory, so you shouldn't use `noexcept` on functions where you perform memory allocation.

So when should you use `noexcept`? The original reason for adding this specifier was to allow faster move constructors to be used within Standard Library code. For example, resizing an `std::vector` can be implemented just by moving every element if it's known that the element's move constructor won't throw an exception. Otherwise, the slower copy constructor will have to be used if there's a chance that an exception can be thrown. So you can make your types behave more optimally in an `std::vector` by specifying `noexcept` for the move constructor and move assignment operator. This will happen by default for any implicitly defined move functions. Thus the `noexcept` specifier is really of use only when you are defining your own move functions, or for code that you know uses `std::move_if_noexcept` or `std::is_nothrow_constructible` to provide a faster code path. Other than those cases, you probably don't need to worry about this feature.

The C++ language originally included a `throw` exception specifier, which was deprecated in C++11 and removed in C++17. This would let you indicate which exceptions a function might throw, as well as that it will throw no exceptions. The first of these cases is generally considered bad practice, partly because the compiler can't enforce it, and so the `throw` specifier was removed from the standard. The `noexcept` specifier was added to cover the one reasonable use of `throw`: to state whether a function will or won't throw an exception. You should therefore not use the `throw` specifier and use `noexcept` only in cases where it makes sense, as described earlier.

Inline namespaces

Namespaces are an important feature for API design in C++ because they help to ensure that your names don't clash with symbols in other libraries. To recap, you can place your names within a namespace such as:

```
namespace MyLibrary {

void MyFunction();

}  // namespace MyLibrary
```

Then you would refer to your names using the namespace as a prefix, such as `MyLibrary::MyFunction()`. You can also nest namespaces, such as:

```
namespace Parent {
namespace Child {

void MyFunction();

}  // namespace Child
}  // namespace Parent
```

where you would then refer to your function as `Parent::Child::MyFunction()`.

C++11 extended the namespace functionality with a feature called inline namespaces, in which all members of the namespace are also automatically added to the enclosing namespace. For example,

```
namespace Parent {
inline namespace Child {

void MyFunction();

}  // namespace Child
}  // namespace Parent
```

This makes your function accessible as `Parent::MyFunction()` in addition to `Parent::Child::MyFunction()`. One reason to do this with respect to API design is to be able to offer different versions of your API and easily declare one of these as the current or default version. For example, consider these declarations:

```
namespace MyLibrary {
namespace v1 {

void MyFunction();

}  // namespace v1

inline namespace v2 {

void MyFunction();

}  // namespace v2
}  // namespace MyLibrary
```

This API offers two nested namespaces: `MyLibrary::v1` and `MyLibrary::v2`. We have declared the second one to be `inline`, so its symbols are added to the `MyLibrary` namespace. As a result, your clients can simply refer to `MyLibrary::MyFunction()` to get the latest definition of your function. But if they needed to, they could also refer directly to a specific version of the function, such as `MyLibrary::v1::MyFunction()` or `MyLibrary::v2::MyFunction()`.

This also gives you a way to prerelease new functionality and let clients try it out before you officially release it. For example, you could add a `MyLibrary::v3` namespace but not mark it as `inline` initially. That way, clients who wanted to try out the new function could explicitly call the `MyLibrary::v3::MyFunction()` version and give you any relevant feedback. Then, when you are ready, you could mark the `v3` namespace as `inline` to officially release the new version.

When you refer to the inlined symbol `MyLibrary::MyFunction()` the compiler will resolve this to the full namespace in the generated object file. For example, in the previous instance that would be the symbol `MyLibrary::v2::MyFunction()`. This means that if you release a new version, such as `MyLibrary::v3`, any modules that are not recompiled with the new API version will continue to link against the `v2` symbols. It's therefore important not to remove old version namespaces; otherwise you will break ABI backward compatibility.

I'll further discuss the use of inline namespaces to manage API versioning in the later chapter on Versioning.

Type aliases with using

C++ has always let you introduce an alias for a type name with the `typedef` keyword, such as:

```
typedef std::vector<int> IntVector;
```

This example introduces a type alias called `IntVector` that can be used anywhere in place of `std::vector<int>`. As of C++11, you can also express this same type alias with the `using` keyword, as:

```
using IntVector = std::vector<int>;
```

The readability benefits of this new syntax become more obvious when you consider how to define a type alias for a function pointer. The `typedef` syntax looks like:

```
typedef int (*FuncPtr)(int, int);
```

whereas the `using` syntax is much more easy to interpret:

```
using FuncPtr = int (*)(int, int);
```

The `using` syntax offers the same semantics as the `typedef` syntax, so they can be used interchangeably. In particular, neither of these options defines a new type, just an alias for the type. However, there are additional capabilities that the `using` syntax offers for templates. The traditional `typedef` syntax does not support templates, so the following is not valid C++:

```
template <class T>
typedef std::vector<T> Vec;
```

However, this is now possible with the new using syntax:

```
template <class T>
using Vec = std::vector<T>;
```

Overall, the `using` syntax offers the same functionality as `typedef`, with additional support for templated type aliases, and it's also easier to read and understand. So the `using` syntax should generally be preferred over typedef for modern C++ API development.

TIP: Prefer `using` instead of `typedef` for better readability and template support.

User-defined literals

As we discussed in the Qualities chapter, a good API should be difficult to misuse. One possible area for misuse is providing the wrong value for a function parameter, either because you have multiple parameters of the same type or because it's not clear which units you should use. This is an area in which the new feature of user-defined literals can help.

C++ supports several different built-in formats to define literals of different types, including:

```
10ul   // unsigned long
10.0f  // floating-point
0xfea9 // hexadecimal
0234   // octal
```

As of C++11, you can define your own literal formats. This can be used to define units for certain quantities and to convert between those units. For example, imagine you have an API that accepts a weight parameter in units of grams. You can allow your clients to be specific that they are passing values in grams and allow them also to provide values in kilograms or pounds. The syntax for this looks like:

```
long double operator"" _g(long double x)
{
    return x;
}

long double operator"" _kg(long double x)
{
    return x * 1000;
}

long double operator"" _lbs(long double x)
{
    return x * 453.592;
}
```

With these definitions, clients can now use literals such as:

```
100.0_g
10.0_kg
5.2_lbs
```

For example:

```
double weight = 1.0_kg + 2.2_lbs;
std::cout << weight << " grams\n";
// outputs "1997.9 grams"
```

We can also combine this feature with custom types for your quantities, so that you can support parameters with specific types rather than using generic primitive types. For example, here's the definition of a class that holds a temperature value in degrees Celsius:

```
class Temperature
{
public:
    explicit Temperature(double celsius) :
        mCelsius(celsius)
    {}

private:
    double mCelsius;
};

Temperature operator"" _degC(long double celsius)
{
    return Temperature(static_cast<double>(celsius));
}

Temperature operator"" _degF(long double fahrenheit)
{
    return Temperature(static_cast<double>(5.0 / 9.0 * (fahrenheit -
32.0)));
}
```

You'll see that I've added a couple of user-defined literals to allow clients to specify values for this class in either Celsius or Fahrenheit. Moreover, by making the constructor be `explicit`, this prevents implicit type conversions, essentially forcing clients to use your new literals to specify the units of any values they provide, such as:

```
void ProcessTemperature(const Temperature &temp);
...
ProcessTemperature(100.0_degC);
ProcessTemperature(100.0_degF);
ProcessTemperature(100.0); // compiler error due to use of explicit
```

This feature is therefore a great tool to help you build APIs that make it difficult for your users to misuse your design.

As a final note on this topic, you can define these custom literals for only a limited set of types, which is why I used `long double` in my earlier examples and I had to deal with casting to `double`. This list defines the set of supported parameters on literal operators in C++11:

```
(const char *)
(unsigned long long int)
(long double)
(char)
(wchar_t)
(char16_t)
(char32_t)
(const char *, std::size_t)
(const wchar_t *, std::size_t)
(const char16_t *, std::size_t)
(const char32_t *, std::size_t)
```

C++20 added a couple of further types to this list:

```
(char8_t)                       // C++20
(const char8_t *, std::size_t). // C++20
```

Alternate function style

C++11 introduced a new style for defining a function's return type. For example, consider a simple function declaration to return the sum of two integers:

```
int Add(int x, int y);
```

Starting with C++11, you can express this function signature using the trailing return type style:

```
auto Add(int x, int y) -> int;
```

In this style, you use the `auto` keyword in the normal return type position to indicate that you'll provide the return type later. Then you can provide the type information after all of the parameters have been declared. These two forms are functionally equivalent, so it's really a matter of personal taste which one you use in your APIs.

There was, however, a legitimate reason for adding this new syntax to the specification. It was introduced to solve a particular problem related to function templates, in which you need to define the return type based on the type of one or more function arguments. You might try to represent this using the traditional function style as:

```
template <class LeftT, class RightT>
decltype(lhs + rhs) Add(const LeftT &lhs, const RightT &rhs);
```

However, this is not valid C++ because the `lhs` and `rhs` variables are defined after the return type, so the compiler cannot deduce their types. (The `decltype` keyword was introduced in C++11 to let you specify a type based on another entity or expression.)

The new trailing return type function style lets you solve this by putting the return type at the end of the function signature, once the types of `lhs` and `rhs` are known to the compiler:

```
template<class LeftT, class RightT>
auto Add(const LeftT &lhs, const RightT &rhs) -> decltype(lhs + rhs);
```

(C++14 introduced automatic return type deduction to make this function syntax even simpler. Refer to the section on C++14 for more details on that update.)

Tuples

A tuple is an ordered collection of a fixed number of heterogeneous values. Support for tuples was introduced in C++11 with the `std::tuple` template. Tuples are a generalization of the `std::pair` type from C++98, which can only hold exactly two values. Tuples can be used to return multiple results from a function. For example, this function returns a tuple containing information to describe a user:

```
#include <tuple>

std::tuple<std::string, int, bool>
GetUserInfo()
{
    return {"Fred", 123, false};
}
```

This function can be used as follows to access the elements of the tuple:

```
auto user = GetUserInfo();
std::cout << "Name = " << std::get<0>(user) << "\n";
std::cout << "UserId = " << std::get<1>(user) << "\n";
std::cout << "Admin = " << std::get<2>(user) << "\n";
```

The downside is that you can only reference each of the elements of the tuple using a compile-time constant index, such as `std::get<1>(user)`. That is, you can't use a meaningful name to refer to the components of a tuple. In most cases when you're designing an API, you will want to return information that has some semantic meaning. As such, it will usually be better to use a named struct to hold the information so that the elements can be referred to by an appropriate name, such as:

```
struct UserInfo
{
    std::string name;
    int userId;
    bool isAdmin;
};

UserInfo
GetUserInfo()
{
    return {"Fred", 123, false};
}
```

Your clients can then access the return values of the function by name; for example,

```
auto user = GetUserInfo2();
std::cout << "Name = " << user.name << "\n";
std::cout << "UserId = " << user.userId << "\n";
std::cout << "Admin = " << user.isAdmin << "\n";
```

TIP: Use a tuple as a return type only when the values represent independent entities. Otherwise, prefer using a class or struct to assign meaningful names to each element.

Constant expressions

C++98 offered the ability to declare a variable as `const`, to indicate that its value should not be changed. The value of such a constant can be evaluated at run time, and the compiler then prevents the programmer from modifying it further.

C++11 introduced a new specifier called `constexpr`, which provides the further constraint that it must be possible to calculate the constant value statically: that is, at compile time, not run time. A trivial example is:

```
constexpr int THE_ANSWER = 42;
```

Although you could also use `const` in this case, the additional benefit of declaring a variable as `constexpr` is that it can be used at points where a compile-time constant is needed. A nonexhaustive list of examples includes:

- The value of a switch `case` statement must be a compile-time constant.
- Nontype template parameters must be constant expressions.
- The expression in a `static_assert()` call must be known at compile time.

For example, consider this simple function, which calculates the factorial of a number:

```
int Factorial(int n)
{
    return (n <= 1) ? 1 : n * Factorial(n - 1);
}
```

This function cannot be used in the following cases where a compile-time constant is required. For example, all of the following code examples will fail to compile:

```
// compile error: constexpr variable requires a compile-time constant
constexpr int THE_ANSWER = Factorial(5);

// switch statements require constant expressions
switch (value) {
case Factorial(5):  // compile error
    ...
}

// non-type template parameters require compile-time values
template <int n>
class constN
{
};
...
constN<Factorial(5)> value;  // compile error

// compile error: static assertions require constant expressions
static_assert(Factorial(5) == 120, "Factorial function is wrong");
```

To make that code work as of C++11, all you need to do is to declare the result of the `Factorial` function to be `constexpr` (i.e., to make it a constant expression function):

```
constexpr int Factorial(int n)
{
    return (n <= 1) ? 1 : n * Factorial(n - 1);
}
```

With that change, all of the earlier examples will now compile (i.e., the result of this function can be calculated at compile time). Not every function can be defined as a constant expression. There are several constraints, including (among others):

- The function must not be virtual (until C++20).

- The type of any return value or parameters must be a literal type, such as a scalar (e.g., `std::is_scalar<T>()` returns `true`) or reference type (e.g., `std::is_reference<T>()` returns `true`).
- The function should contain exactly one `return` statement (unless it's a constructor).
- The function body cannot contain a `try` block (until C++20), a `goto` statement, a `static` variable, or call a non`constexpr` function.

TIP: Declare a constant or function as `constexpr` *if it might need to be evaluated at compile time.*

The nullptr keyword

The C and C++98 programming languages use a macro called `NULL` to represent a pointer value that's not pointing at valid data. The definition of this macro is implementation dependent, but it's often defined as `(void *)0`. One issue with this is that `NULL` can be implicitly converted to integral types, which can cause ambiguity issues such as:

```
void Function(int value);
void Function(char const *ptr);
...
Function(NULL); // compile error: call is ambiguous
```

This is an error because `NULL` can be interpreted as both a pointer and an integer, so the compiler doesn't know which overloaded function to use. To address this issue, C++11 introduced a keyword called `nullptr` that's of type `std::nullptr_t` and is not implicitly convertible or comparable to integer types (i.e., `int x = nullptr` will not compile).

Now we can use `nullptr` in the previous example and the compiler will know to use the overloaded version of `Function` that uses a pointer parameter:

```
Function(nullptr); // compiles!
```

TIP: Always prefer `nullptr` *over* `NULL` *when working with pointer types to improve type safety.*

Variadic templates

Templates in C++03 could accept only a fixed number of arguments. However, C++11 introduced variadic templates, which are templates that can accept a variable number of arguments. This is achieved by introducing the concept of parameter packs that accept zero or more template arguments, in which each argument can be a type, a nontype, or a template. The basic syntax for a variadic function template is:

```
template <typename... T>
void Function(T... args);
```

Note the use of the ellipsis (...) after the `typename` keyword. This is referred to as a template parameter pack and specifies that the template can accept zero or more arguments. Each of the arguments can be of a different type. Also note the ellipsis in the function argument list. This is called a function parameter pack and indicates that the function can accept the same multiple arguments.

Often you want to enforce that one or more arguments are passed to the function, instead of only zero or more. The standard way to do that is to define a regular template parameter followed by a parameter pack, which looks like:

```
template <typename T1, typename... T2>
void Function(T1 required, T2... varArgs);
```

Let's look at a specific example. In this case, we will implement a variadic function template called `printLine` that will output all of its arguments to stdout followed by a newline:

```
void printLine()
{
    std::cout << std::endl;
}

template <typename T1, typename... T2>
void printLine(T1 var1, T2... var2)
{
    std::cout << var1;
    printLine(var2...);
}
```

We can call this function as:

```
printLine("Wrote ", 5, " file(s).");
```

which will produce the output:

```
Wrote 5 file(s).
```

Variadic templates are often implemented using recursion. That is, we process the first parameter and then call the function template again with the remaining parameters. This will bottom out with a call to an overloaded version of the function that ends the recursion, known as the base case. That's what the nontemplated `printLine` function is for in the previous example. Walking through this in more detail,

1. We start with the initial invocation:

```
printLine("Wrote ", 5, " file(s).");
```

2. This causes the first argument to be output and then recursively calls the function with the remaining arguments:

```
std::cout << "Wrote ";
printLine(5, " file(s).");
```

3. Calling the function template again, the next argument is output and we recursively call the function with the last remaining argument:

```
std::cout << 5;
printLine(" file(s).");
```

4. We then output the last parameter and call the function with an empty parameter list:

```
std::cout << " file(s).";
printLine();
```

5. This will call the nontemplated overload of `printLine` that will output a newline character and terminate the recursion:

```
std::cout << std::endl;
```

Moreover, we can implement this variadic function template without having to define a base version of the function. We can do this by referring to the size of the parameter pack within the template using the `sizeof...()` operator:

```
template <typename T1, typename... T2>
void printLine(T1 var1, T2... var2)
{
    if (sizeof...(T2) == 0) {
        std::cout << std::endl;
    } else {
        std::cout << var1;
        printLine(var2...);
    }
}
```

C/C++ supported a mechanism to implement variable parameters for nontemplate functions before the introduction of variadic templates (i.e., through `va_args` in the `cstdarg` header). However, variadic templates offer a more type-safe solution in which the parameter expansion happens at compile time and therefore allows more compiler optimizations to be performed. As such, the use of C++11 variadic templates should be preferred over the C-style `va_args` approach where feasible.

> *TIP: Prefer the use of variadic templates when you need to define a function with a variable number of arguments.*

Migrating to C++11

I have covered many C++11 features that can improve your API designs in this section. We'll conclude this discussion by talking about how you can migrate an existing C++03 API to C++11, and which features you may want to prioritize adopting over others. One of the core concerns you may have in this regard is how these new features will affect backward compatibility for your API, so I'll organize the section with that concern in mind:

- **Compatible Changes:** Although ABI compatibility is ultimately implementation dependent for your compiler, it should be possible to adopt the following C++11 features without breaking ABI (or source) compatibility:
  - **Delegating constructors:** Using a delegating constructor lets you change how the constructor is implemented, but it doesn't change the function signature.
  - **Data member initializers:** If you are simply removing initialization of a member variable from a constructor's initialization list and replacing it with a data member initializer, this will not change the signature of the constructor.
  - **Alternate function style:** The alternate function style, in which the return value is declared after the function parameters, is simply a syntactical difference. The binary representation of the function should be the same whether you're using the original function syntax or the new format.
  - **Using versus typedef:** Employing the `using` keyword as an alternative to `typedef` is again just syntactic sugar. More importantly, these define an alias for a new type, not an actual new type, so whether you use `using` or `typedef` should not make any functional difference. However, the `using` syntax is a little easier to understand, so it's worth adopting for clarity reasons.
  - **Override keyword:** This keyword lets the compiler perform additional compile-time checks to ensure that you're overriding a virtual member function. However, it should not affect the generated binary code for your interface.
  - **Constant expressions:** Declaring your constants with `constexpr`, where possible, should be backward compatible because it will just add the ability for your constants to be used in cases where a compile-time value is required.

- **Breaking Changes:** The following C++11 features are worth considering but could cause breaking changes for your clients:
  - **Smart pointers:** Switching from raw C pointers to smart pointers will involve representing your pointers with a different type. This will most likely require your clients to change their code (unless they are compiling from your source and have already changed their code to use the C++11 auto keyword to store your pointers). However, the safety benefits of using smart pointers are large, so it's worth strongly considering adopting them.
  - **Enum classes:** The size of an enum may be different from an enum class, so switching between them is not guaranteed to be binary compatible. Also, your clients will have to change their code to include the enum name when they refer to its values. It may not be worth breaking API compatibility to move to enum classes, but if you're adding new functionality, then enum classes should be preferred over traditional C++03 enums.
  - **Inline namespaces:** This feature is a great way to introduce robust versioning into your APIs. However, if you try to add this to an existing API, then it will change the name mangling of your symbols so it will break ABI compatibility. However, it should be possible to preserve source compatibility if you're diligent.
  - **Final keyword:** It's possible that adding a final specifier to an existing class will not break the ABI. However, if clients have already written code to override the final function, or inherit from the final class, then their code will now fail to compile because of your new constraint. This is therefore something you could chose to do for an existing API, but you should first check with your clients to see if their code would break, and to understand their use cases if it does.
  - **Variadic templates:** If you were previously using `va_arg` to implement support for variable function arguments, you could consider using variadic templates instead to provide a more efficient and type-safe solution. However, although the function calling syntax can be made to be the same, these two approaches are not ABI compatible, so making this change will break your clients' projects if you're distributing your library in binary form.

C++14 API features

C++14 was published in Dec. 2014 as a minor update to C++11. It largely consisted of a few bug fixes and incremental improvements, so the list of features I'll cover here is not as large as for C++11. However, there are a few notable new features for your API designs that we'll discuss.

As of 2012, the C++ Standards Committee committed to a regular 3-year schedule to release updates to the C++ international standard. Nonetheless, C++14 was sometimes referred to as C++1y during its development before its final publication date was officially known, in the same way that C++11 was referred to as C++0x early on.

The auto return type

Recall that C++11 introduced a trailing return type function style that looks like:

```
auto Add(int x, int y) -> int;
```

C++14 took this a step further by introducing function return type deduction. Essentially, you no longer need to define the function return type after the parameter list. Instead, the return type will be deduced based on the actual `return` statements defined within the function body. The previous function signature can therefore be simplified to just:

```
auto Add(int x, int y);
```

To support this syntax, all return values in the function body must be of the same type. For example, you could not have one `return` statement return `2` and another returning `6.0` because those are different types (int and double, respectively). This would not be an error in a regular function where the return type is explicitly declared because the compiler can perform implicit conversion to the known return type.

More critically, functions declared in this way can be forward declared, but the compiler must be able to see the implementation of the function in the translation unit that uses it to deduce the return type. For example, imagine that our `Add` function is declared in an `add.h` header, which is used by client code in `client.cpp`:

```
// add.h

auto Add(int x, int y);


// client.cpp

#include "add.h"

int OnePlusOne()
{
    Return Add(1, 1);  // compiler error: cannot deduce return type
}
```

This code will not compile, because the code in `client.cpp` sees only the forward declaration of `Add` and does not have access to the implementation body to deduce the return type. Nevertheless the client code will eventually link against the `add.cpp` file with the implementation code. The key point is that the implementation is not available at compile time when compiling the `client.cpp` translation unit.

In other words, in terms of API design, return type deduction forces you to expose your implementation details. You should therefore avoid using this feature in your APIs. An exception would be if you're using this for a function template in which you're forced to expose the implementation code in the header file anyway. Of course, it's also fine to use this feature within your implementation code, such as for static functions defined within your `.cpp` files.

> *TIP: Do not use function return type deduction in your APIs because it exposes implementation details (unless it's for already exposed details such as a function template).*

The deprecated attribute

An important part of the API life cycle is being able to deprecate older functionality (often giving your clients a new way to achieve the same thing) before finally removing the older functionality from a future version of the API. An important way to communicate that something is deprecated is to cause compiler warnings in your clients' code when they call the deprecated functionality. That way, they get a visible and annoying nudge that they need to update their use of your API.

Before C++14, there were compiler-specific ways to achieve this, but there was no standards-compliant solution. C++14 addressed this by introducing the `[[deprecated]]` attribute. You can use this before any function signature to indicate that the function is deprecated. You can also optionally provide a message that will be included in the compiler warning. For example:

```
[[deprecated]] void Initialize();

class MyObject
{
public:
    [[deprecated("Use GetValues() instead")]] int GetValue() const;
    std::vector<int> GetValues() const;
};

int main(int argc, const char *argv[])
{
    Initialize();

    MyObject obj;
    int value = obj.GetValue();

    return 0;
}
```

Compiling this code with clang produces the following compiler output:

```
main.cpp: warning: 'Initialize' is deprecated [-Wdeprecated-declarations]
    Initialize();
    ^
main.cpp: warning: 'GetValue' is deprecated: Use GetValues() instead
[-Wdeprecated-declarations]
    int value = obj.GetValue();
                    ^
```

You can use [[deprecated]] in front of a function signature, whether it is a free function, a function template, or a class method. You can also use it in front of a variable definition to deprecate any constants that your API may define. However, you cannot currently use it to deprecate an entire class, struct, or enum.

> *TIP: Use the [[deprecated(msg)]] attribute to indicate that a function may be removed in a future version of your API. Always include a useful message to describe how clients can update their code to avoid the deprecated functionality.*

I will go into more detail about deprecating and ultimately removing functionality from your APIs in the later chapter on Versioning.

Variable templates

In C++03, you can create templated versions of classes and functions. C++11 added support for templating of type aliases with the using keyword and C++14 added templating of variables. A variable template is a template that defines a family of variables or static data members based on one or more template parameters. The general syntax is:

```
template <typename T>
T variableName = value;
```

For example, we can provide a definition of the constant pi:

```
template <typename T>
constexpr T pi = 3.1415926535897932385;
```

We can then instantiate this variable template for different types, such as pi<float> and pi<double>. An interesting aspect of this technique is that we can use template specialization to provide a custom value of the variable for different types. For example:

```
template <>
constexpr float pi<float> = 3.1415;

template <>
const std::string pi<std::string> = "pi";
```

(I'm using constexpr in each of these templates, except for the std::string specialization that uses const. That's because you can't define a std::string constexpr, at least until C++20, or until C++17 if you use std::string_view.)

The effect of these definitions can be demonstrated as:

```
std::cout << std::setprecision(9) << pi<int> << "\n";
std::cout << std::setprecision(9) << pi<float> << "\n";
std::cout << std::setprecision(9) << pi<double> << "\n";
std::cout << pi<std::string> << "\n";
```

which will output the following results:

```
3
3.1415
3.14159265
pi
```

So if you refer to `pi<float>` you will get the specialized value 3.1415, and if you refer to `pi<std::string>` you will get the self-descriptive string "pi". Although this may be abstractly interesting, the main reason for introducing variable template support was because C++ didn't previously allow a template to declare a variable. One common workaround for this was to wrap a variable in a templated struct or class, such as:

```
template <typename T>
struct Pi
{
    static constexpr T value = 3.1415926535897932385;
};
```

However, this meant having to refer to variables such as `Pi<float>::value` instead of just `pi<float>`, as is now possible with variable templates.

Variable templates can be declared only at the global (or namespace) scope or as static data members of a class. They are most useful for defining constants, and as such they are often declared as `const` or `constexpr`, as you saw in the previous examples.

Const expression improvements

In the earlier C++11 section of this chapter, I introduced the use of `constexpr` to evaluate the value of a variable or the result of a function at compile time. To make this feasible, various constraints were enforced on what you could do in the body of a function to allow it to be evaluated at compile time. With C++14, this set of constraints was relaxed to allow more complexity within constant expression function. Specifically, these functions can now contain:

- Declarations with initializers, except `static` or `thread_local` variables.
- Conditional branching statements (e.g., `if` and `switch`).
- Looping statements (e.g., `for` and `while`).
- Expressions that change the value of an object defined within the function.

For example, this function will not compile with C++11 because of the use of looping and local variables. However it will compile in C++14:

```
constexpr int factorial_iterative(int n)
{
    int result = 1;
    for (int i = 2; i <= n; i++) {
        result *= i;
    }
    return result;
}
```

So if you are building your project in an environment that supports C++14 or later, you can have more complex constant expression functions that allow more opportunity for compile-time optimization.

Binary literals and digit separators

C++ has always supported ways to declare numeric constants in different bases. There's the default base 10, of course, as well as hexadecimal (base 16) constants that use the `0x` prefix (e.g., `0xdeadbeef`) and octal (base 8) constants that use a 0 prefix (e.g., `0172`). C++14 introduced the ability to define numeric literals in binary (base 2) using the `0b` or `0B` prefix. This might be useful when defining constants for your APIs that are better expressed in binary representation, such as bit masks. For example:

```
constexpr int lowerBitMask    = 0b00000001;
constexpr int lowerNibbleMask = 0b00001111;
constexpr int lowerByteMask   = 0b11111111;
```

Another C++14 change related to numeric literals is the introduction of the single quote character as a digit separator for both integer and floating-pointing literals. The quote character has no impact on the interpretation of the number, but it can be used to make it easier for humans to parse large numbers by breaking the value up into smaller sections, such as:

```
constexpr int oneMillion = 1'000'000;
constexpr long largest32bit = 0xffff'ffff'ffff'ffff;
```

Migrating to C++14

As I noted in the introduction to this section, C++14 was a much smaller update than C++11 and focused largely on minor enhancements and bug fixes.

The main new C++14 feature that affects API design is the standardized support for deprecating functions and constants. This is a really important feature to have at our disposal because most long-lived APIs will need to go through design changes at some point to handle new functionality, offer more general capabilities, or even break backward compatibility. The most elegant ways to introduce these kinds of API design changes are to:

1. Introduce a new way to perform some task in your API.
2. Mark the existing way of performing the task as deprecated.
3. Communicate the new way of performing the task to all of your clients so they can change their code accordingly.
4. Once all clients have moved to the new approach, you can safely remove the deprecated way from the API.

As such, having the ability to mark functions and constants as deprecated is an important part of the life cycle of an API. You should use the `[[deprecated]]` attribute whenever you need to communicate your intent to make future breaking changes to your API.

Other than that, we covered a few other more minor C++14 features that could affect your API designs, including:

- Function return type deduction: as I noted, you should generally avoid using this in your public APIs unless it's part of a function template.
- Variable templates: these can be used to produce more concise constants in your templated code, such as `pi<float>` instead of `pi<float>::value`, but they will change your API surface and break backward compatibility.
- Constant expressions, binary literals, and digit separators: these are all minor enhancements that you can employ if you know that all of your clients are using a C++14 or later compiler.

C++17 API features

The C++17 standard was published in Dec. 2017 after being referred to temporarily as C++1z during its development. This revision to the standard added many new features to the core language and the Standard Library, including several container classes such as `std::optional`, `std::any`, and `std:variant`. It also removed several older features, such as `std::auto_ptr` (replaced by `std::unique_ptr`), the compiler hint keyword `register` (deprecated in C++11), and dynamic exception specifications (such as `void myFunc() throw(int)`). In the following sections, I'll cover the key features of C++17 that you can use in your API designs.

Inline variables

With C++03, if you wanted to declare a global or namespace scope constant, you could try:

```
// api.h
const std::string VERSION = "1.0";
```

However, this would cause every translation unit that included the header to create a new internal (i.e., static) linkage copy of the variable, which will cause the final binary to hold multiple copies of the data and be larger than necessary. The normal way to avoid this was to declare the variable as `extern` and then define the value within a `.cpp` file. For example,

```
// api.h
extern const std::string VERSION;

// api.cpp
const std::string VERSION = "1.0";
```

This solves the binary size issue, but it has the unfortunate downsides that clients can no longer see the value of the constant simply by inspecting your header file, and you must split the declaration and definition across two files.

C++17 introduced a solution to this with support for inline variables. This allows a global or namespace scope variable to be declared as `inline` in a header file. The header can then be included in multiple translation units and the variable will have external linkage in each translation unit, so the linker will merge all instances of the variable into a single definition at link time. You can therefore avoid using `extern` for your API constants and simply define everything in your headers, such as:

```
// api.h
inline const std::string VERSION = "1.0";
```

TIP: Use inline variables to define constants in your header files.

You can also use inline variables for static class data members. For example, if you want to create a class scope constant in C++03, then you would have to do this:

```
// api.h
class MyObject
{
public:
    static const std::string VERSION;
};

// api.cpp
const std::string MyObject::VERSION = "1.0";
```

You can't provide the value of the static data member in the header file. It must be defined in the associated `.cpp` file. But as of C++17, you can declare the static data member as `inline`, allowing you to provide the variable's value directly in the header file, such as:

```
// api.h
class MyObject
{
public:
    inline static const std::string VERSION = "1.0";
};
```

Moreover, any `static constexpr` class data members are also implicitly `inline` variables. So you can also just write something like:

```
// api.h
class MyObject
{
public:
    static constexpr std::string_view VERSION = "1.0";
};
```

As a result of allowing global variables to be defined only in your header files, inline variables can also make it easier for you to create header-only libraries.

String views

The original C language supports strings as either null-terminated character pointers or character arrays: for example, `const char *` or `char[]`. These are efficient to pass around because they are just pointers, but you must explicitly manage the memory for these strings yourself, normally with `malloc()` and `free()`.

C++98 introduced the `std::string` object to hold a text string and offer various member functions to operate on the string. This object takes care of the memory management for you, but copying these objects by value can be more expensive, particularly for large strings. As a result, it's common practice to pass `std::string` objects into functions as const references to avoid the cost of copying the string data.

C++17 introduced the `std::string_view` class, which provides support for read-only strings that are cheaper to pass around. These strings views are normally implemented with just two quantities:

1. A pointer to an existing character array, such as a `const char *`.
2. The size of the array, such as a `size_t`.

The character array itself is not owned by the string view or copied by the string view, so you must be careful not to use the string view after the underlying character data has been deleted. Also, unlike `std::string`, an `std::string_view` can hold a null pointer. Moreover, because a `string_view` keeps a specific record of the size of the character array, the array does not need to hold a terminating null character.

One useful feature of std::string_view is that it offers implicit conversions from std::string and const char * types. So it's easy to pass in different kinds of strings to a function that uses string view arguments. For example:

```
void ProcessStringView(std::string_view str);
...
std::string str("Hello World");
std::string_view str_view("Hello World");
const char *c_str = "Hello World";

ProcessStringView(str);
ProcessStringView(str_view);
ProcessStringView(c_str);
```

(It's idiomatic to pass std::string_view objects by value because they're cheap to copy and it gives the compiler greater opportunities for optimization.)

In addition to using std::string_view for function arguments, you might think about returning an std::string_view for an std::string data member, such as:

```
class NameHolder
{
public:
    NameHolder(std::string_view name) :
        mName(name)
    {}
    std::string_view GetName() const { return mName; }

private:
    std::string mName;
};
```

However, the string view does not take a copy of the string data, so if a client were to keep a hold of the string view returned by GetName() after the object had been deleted, the behavior would be undefined and could lead to a crash. You can think of returning a string view as equivalent to returning a const reference or pointer, which I generally discourage because of the lifetime concerns I just mentioned and because it returns a handle to your private implementation details.

For these reasons, I would avoid returning std::string_view from a function. There are certainly ways to make it safe (e.g., returning statically allocated data that will exist until the program ends), but unless the performance gain is really important to you, it's safer just to return an std::string. It's also less surprising for your clients because string views don't offer the same richness of methods. For example, they do not have a c_str() function because the string data may not be null terminated.

Similarly, I would avoid storing an std::string_view in a class data member if you don't also own the underlying string data within your library or you don't otherwise know the lifetime of the data. This is for the same reason that it's dangerous to hold on to a raw C pointer when you don't know when the memory will be deleted.

The main takeaway is that string views are a performance optimization feature that you should fully understand before you commit to exposing them in your API designs. It's helpful to think of them as just a `const char *` pointer and make the same API design decisions you would for such a type.

However, there is another benefit of `std::string_view`, which is that it can be used to initialize constant expressions, which is something for which you can't use `std::string` (at least until C++20) because it's not a literal type. For example,

```
constexpr std::string CONST_STR = "Hello World"; // compiler error
constexpr std::string_view CONST_STR_VIEW = "Hello World";
```

In these cases, the source string data are statically allocated by the compiler and will persist until the program ends, so you don't need to worry about lifetime concerns.

> *TIP: Consider using `std::string_view` instead of `std::string` for a function argument if you don't need to own or modify the string data and if the associated performance gain is important for your use case. You can also use `std::string_view` to define compile-time string constants with `constexpr`.*

Although `std::string_view` is used to represent a read-only string, it does offer several methods that appear to modify the string but in fact just change the underlying pointer or size data members. For example, the `remove_prefix()` method simply increases the pointer by n and reduces the size of the string by the same n. This is possible because the string view doesn't own the underlying data array, so it can be updated to point to any subset of characters within the array. However, if you need to change the contents of a string, you should use an `std::string`.

Optional

There are times when you may want to differentiate between a variable holding a value and holding no value. For example, if you are reading a true/false setting from a configuration file, you may want to know if a value for the setting was defined in the file and, if so, what that value was. If you simply use a `bool` variable to represent the setting with a default of false, then you cannot differentiate between a configuration file that has no value for the setting and one that specifically sets the value to false. This might be important if you only wanted to change the current value of the setting if a value was explicitly provided.

One way to solve this is to use a `bool` for the setting value and another `bool` for whether the setting was defined in the file. However, this means representing every value with two variables. Another solution is to use a pointer, such as an `std::shared_ptr<bool>`, in which you can use `nullptr` to represent the absence of a value. However, in this case you have the additional overhead of dynamically allocating memory.

A common solution to this problem in many programming languages is the concept of an optional: essentially, the ability to state that a variable has no value. C++17 introduced functionality for this with the `std::optional` class. Here's an example:

```
#include <optional>

auto empty = std::optional<bool>();
auto nonempty = std::optional<bool>(false);

std::cout << empty.has_value() << std::endl;  // false
std::cout << empty.value() << std::endl;      // throws exception

std::cout << nonempty.has_value() << std::endl;  // true
std::cout << nonempty.value() << std::endl;      // false
```

The basic operations you can perform on an optional are:

1. **Creation:** You can create either an empty optional or one that holds a value. There are various ways to do this, such as:

```
// make an empty optional
std::optional<int> empty1 = std::nullopt;
auto empty2 = std::optional<int>();
auto empty3 = std::make_optional<int>();

// make an optional that holds a value of zero
std::optional<int> nonempty1 = 0;
auto nonempty2 = std::optional<int>(0);
auto nonempty3 = std::make_optional<int>(0);
```

2. **Setting/resetting:** You can set the value of an optional simply by assigning a value of the corresponding type. You can make the optional hold no value by assigning `std::nullopt` to it or calling its `reset()` method: that is:

```
std::optional<int> intOpt;

// assign a value to the optional
intOpt = 1;

// assign no value to the optional
intOpt = std::nullopt;
intOpt.reset();
```

3. **Checking for a value:** You can check to see if an optional is holding a value by comparing it with `std::nullopt` or by using the `has_value()` method. For example,

```
bool hasValue1 = intOpt.has_value();
bool hasValue2 = intOpt != std::nullopt;
```

4. **Accessing the value:** If you have checked that the optional holds a value, you can access the value using the `*` operator or the `value()` method. If the optional holds an object, then you can use the `->` operator to access members of the object, such as:

```
std::optional<std::string> strOpt = "";
std::string value1 = *strOpt;
std::string value2 = strOpt.value();
size_t strSize = strOpt->size();
```

An optional is normally implemented so that the value it contains is allocated as part of the optional object itself. That is, it does not contain a pointer that's dynamically allocated. However the interface for `std::optional` uses the `*` and `->` operators, which are normally associated with pointer types.

TIP: Use `std::optional` when you need a quantity that can hold a value or that can represent having no value.

To continue with the example of a configuration file, imagine that the settings from the file are returned in a `Settings` object with accessors for each setting value. We could use an `std::optional` as the return value for each accessor so that the client can differentiate between a value that appeared in the configuration file versus there being no value defined in the file. For example:

```
class Settings
{
public:
    std::optional<bool> GetDarkMode() const;
    std::optional<float> GetZoomLevel() const;
};
```

The use of an optional as a function's return value can also help you differentiate between a valid return value and the absence of one. For example, consider a function that returns the offset of some item within a larger collection:

```
class Collection
{
public:
    std::optional<size_t> GetIndex(const Item &item) const;
};
```

In this case, the `GetIndex()` function can return `std::nullopt` if the item does not exist in the collection. This avoids the need to introduce a special sentinel value that you have to define for your clients (such as `std::string::npos`, for example).

Some possible uses of `std::optional` in your API designs include:

1. Differentiating between a value and the absence of a value, such as knowing whether the user provided an overriding value for a setting.
2. Returning the result of some computation in which the lack of a result is not inherently an error, such as searching for an item in a collection and handling the case in which the item doesn't appear.
3. Allowing lazy loading of a potentially expensive resource, where you can pass around an optional and instantiate it with an actual value only when you need it.
4. Supporting another way to express optional function parameters, particularly for arguments in which the optional value is intrinsic to the object, not just to the function signature (e.g., you may want to pass it through multiple function calls and preserve the optional nature), or when you wish to calculate the default value for the argument within the function body, either because of complexity or to hide it from your API.

For cases in which you actually want to return both a valid result and some kind of error code or information, then a container such as `std::pair` or `std::tuple` would be more appropriate. For example:

```
enum class Error {
    None,
    FileNotExist,
    ReadFailed,
};

std::pair<std::string, Error> LoadFile(const std::string &filename);

...

auto [contents, error] = LoadFile("filename");
if (error == Error::None) {
    std::cout << "Contents = " << contents << std::endl;
}
```

Or, if you are using C++23, you can use the `std::expected` class template, which I'll cover in more detail later in this chapter.

Any

C++ is a statically typed language, so the type of all variables must be known at compile time. This allows the compiler to do more checks and avoid certain classes of errors happening at run time. However, there are some cases in which you may want a variable to hold values of different types or you don't know the type until run time.

As an example, the JSON file format supports heterogeneous arrays and dictionaries, so they can hold objects of different types. For example, this is a valid JSON dictionary:

```
{
  "name": "Ferris Bueller",
  "age": 17,
  "present": false,
}
```

This dictionary has three values that are of different types (string, int, and bool). However, in C++ you must use a specific type for a `std::vector` or `std::map` container. For example, `std::vector<int>` can hold only integer values and no other type. So if we wanted to read this JSON file into a C++ variable we would need some way to define an `std::map` that could hold any type of variable as its value. That's exactly what the new `std::any` type in C++17 does:

```
#include <any>

std::map<std::string, std::any> jsonDictionary;
```

Before C++17, you might have used a `void *` as a way to hold any type, but now `std::any` provides a type-safe way to achieve the same thing. This is useful for several applications, such as passing arbitrary message values, representing the value of settings of different types, implementing aspects of a scripting language, and the previous example of parsing data files that support heterogeneous containers.

> *TIP: Use `std::any` as a type-safe alternative to `void *`.*

The basic operations you can perform on an `std::any` type include the following, assuming an `std::any` variable named `anyVal`:

- `anyVal.has_value()`: check whether the object holds a value.
- `anyVal.reset()`: destroy any contained object producing an empty object.
- `anyVal.type()`: return the `typeid` of the contained object.
- `std::any_cast<T>(anyVal)`: type-safe access to the contained object. Throws an exception if the contained object does not match the requested type.

For example:

```
std::any anyVal;
assert(!anyVal.has_value());

anyVal = 5;
assert(anyVal.has_value());
assert(anyVal.type() == typeid(int));
assert(std::any_cast<int>(anyVal) == 5);

anyVal = "Hello";
assert(anyVal.type() == typeid(const char *));

anyVal.reset();
assert(!anyVal.has_value());
```

The main issue to be aware of when using `std::any` is that it's likely implemented with a `void *`, because the maximum size of the contained type cannot be known at compile time. It therefore involves dynamic memory allocation that could be costly depending on your use. If you have a small set of possible types that need to be supported, then it may be more efficient to use an `std::variant`, which I'll cover next.

Variant

The `std::variant` class template was introduced in C++17 as a type-safe alternative to the `union` class type, which is an object that can hold only one of its nonstatic data members at any time and in which the size of the object is determined by the size of its largest member. For example, in C++98 (and plain C) you could define a union to hold one of several different primitive types. The standard convention is to pair this with a variable to track the type of the variable that is currently being held in the `union`: that is,

```
enum class PrimitiveType {
    Int,
    Float,
    Bool,
};

struct PrimitiveValue
{
    PrimitiveType mType;
    union
    {
        int mIntValue;
        float mFloatValue;
        bool mBoolValue;
    };
};
```

The downside of using a `union` is that you can easily refer to a data member that's not the right type for the data value that's being stored, potentially returning a garbage value. For example, given the previous definition, you could set `mFloatValue` to 3.1415 but still access the `mIntValue` member. This would essentially return `*((int *)&mFloatValue)`,

which is likely not what you wanted. Instead, you can now use a variant to represent the same thing in a more type-safe fashion, as:

```
std::variant<int, float, bool>
```

This variant can hold an `int`, a `float`, or a `bool`. For example:

```
std::variant<int, float, bool> varVal;
varVal = -1;
varVal = 3.1415f;
varVal = true;
varVal = "a string"; // compile error: type is not in variant
```

Taking the example of the JSON file in the previous section on `std::any`, if we know that the dictionary in the data file will hold only a string, int, or bool, then we could replace the use of `std::any` for the dictionary value with an `std::variant`. One benefit of this is that `std::variant` does not dynamically allocate its value on the heap:

```
#include <variant>

using DictionaryValue = std::variant<std::string, int, bool>;
std::map<std::string, DictionaryValue> jsonDictionary;
```

TIP: Use `std::variant` as a type-safe alternative to `union`.

The basic operations that you can perform on a variant are, for a variant variable called `varVal`:

- `varVal.index()`: returns the zero-based index of the currently held alternative.
- `std::get<size_t>(varVal)`: returns the value of the alternative at the given offset, such as 0 for the first type specified in the variant. Throws an exception for an out of bounds index.
- `std::get<T>(varVal)`: returns the value of the alternative with the given type, such as `std::string`. Throws an exception if the type does not exist in the variant.
- `std::holds_alternative<T>(varVal)`: checks if the currently held alternative is of the given type, such as `std::string`.

You can also inherit from `std::variant`. This might be useful if you want to extend the functionality of the variant, such as adding some simple accessors or to add support for output streams by defining a << operator. For example:

```
class StringOrInt : std::variant<std::string, int>
{
public:
    bool isString() const
    {
        return std::holds_alternative<std::string>(*this);
    }

    bool isInt() const
    {
        return std::holds_alternative<int>(*this);
    }
};
```

Nested namespaces

As we've already seen, the use of namespaces is an important feature for your API designs to ensure that your symbols don't clash with those in other libraries. Namespaces can be nested, which is a useful feature to break up your symbol names into several categories, such as:

```
namespace Parent {
namespace Child {
    ...
}  // namespace Child
}  // namespace Parent
```

With C++17, you can now declare a nested namespace with a single `namespace` statement, such as:

```
namespace Parent::Child {
    ...
}  // namespace Parent::Child
```

In addition to being more compact, this syntax offers a satisfying symmetry with how you would refer to the nested namespace in a `using` statement, or if you were to refer to a symbol in that namespace with its fully qualified name, such as:

```
using namespace Parent::Child;
```

Making use of nested namespaces can often lead to long composite names, particularly in header files where you can't employ the `using` directive. However, you can use namespace aliases to produce shorter alternate names to help with readability and reduce the amount of necessary typing for your clients. For example,

```
namespace Parent::Child::Grandchild {

class MyObject
{
};

}  // namespace Parent::Child::Grandchild

namespace Grandchild = Parent::Child::Grandchild;
Grandchild::MyObject obj;
```

Fold expressions

You may recall from earlier in this chapter that C++11 introduced the notion of variadic templates. This provided type-safe support for functions that can accept variable numbers of arguments of different types. As I noted, these are often implemented recursively, requiring an overloaded base case function to be defined to terminate the recursion. For example, the next C++11 code lets you multiply any number of values:

```
template <typename T>
T multiply(T arg)
{
    return arg;
}

template <typename T, typename... Args>
T multiply(T first, Args... args)
{
    return first * multiply(args...);
}
```

C++17 introduced a feature called fold expressions, which can help us simplify this code even further and avoid the use of recursion. This lets you apply a binary operator to all of the parameters in a parameter pack. For example, here's a fold expression using the multiply operator (the surrounding parentheses are required):

```
(... * args)
```

For a parameter pack with four arguments, the compiler will expand this in place to become:

$$(((\texttt{args}_1 * \texttt{args}_2) * \texttt{args}_3) * \texttt{args}_4)$$

This is referred to as a unary left fold. There's also a unary right fold as well as binary left and right folds. I define the general form for each of these four fold expression types in Table 7.1, along with an expanded example assuming a parameter pack with four arguments and the multiply operator.

Table 7.1 Fold expression types introduced in C++17.

Name	Fold expression and expanded example
Unary left fold	(... op args) $(((args_1 * args_2) * args_3) * args_4)$
Unary right fold	(args op ...) $(args_1 * (args_2 * (args_3 * args_4)))$
Binary left fold	(init op ... op args) $((((init * args_1) * args_2) * args_3) * args_4)$
Binary right fold	(args op ... op init) $(args_1 * (args_2 * (args_3 * (args_4 * init))))$

Taking this C++11 variadic template example, we can express this using a fold expression, which allows us to remove the base case overloaded function and avoid the recursive implementation because the fold expression is expanded in place:

```
template <typename T, typename... Args>
T multiply(T first, Args... args)
{
    return first * (args * ...);
}
```

We can also simplify this further using the deduced return type functionality from C++14 and remove the need to declare an initial `first` argument. (However, recall that I generally advise avoiding using deduced return types.)

```
template <typename... Args>
auto multiply(Args... args)
{
    return (args * ...);
}
```

As a further example, here's a simplified version of the `printLine()` variadic template that I defined earlier in the C++11 section. Here I use a fold expression over the comma operator to generate a sequence of output stream operations, and then I add a newline at the end:

```
template <typename... Args>
void printLine(Args... args)
{
    ((std::cout << args), ...) << std::endl;
}
```

> *TIP: Consider using fold expressions with variadic templates as an alternative to a recursive implementation.*

While we're on this topic, it's worth noting a couple of alternative approaches to implementing variable parameters. For example, you can use multiple overloaded functions to implement different combinations of parameters, although of course, this works only if the number of possible parameter combinations is constrained, because you would have to define a separate function for each permutation.

Another approach would be to pass in a container object with multiple parameter values. If all of the parameters are of a known fixed type, such as `std::vector<std::string>`, then that makes things simpler. Otherwise, you can use a container of "any" objects, or objects that can hold a value of any type. As we just learned, C++17 introduced the `std::any` type to solve exactly this problem, letting us use `std::vector<std::any>` as a way to define a heterogeneous array (i.e., a list of parameters of different types). We could even combine this approach with variadic templates, as in this example in which we initialize an `std::any` vector with the parameter pack from a variadic function template and then iterate over the parameters:

```
template <typename... T>
void printTypes(T... args)
{
    std::vector<std::any> vec = {args...};
    std::cout << sizeof...(T) << " args:";
    for (auto elem : vec) {
        std::cout << " " << elem.type().name();
    }
    std::cout << std::endl;
}
```

In this example, calling `printTypes(1, true, 3.0f)` will output: `3 args: i b f`

Checking for header availability

Your library may have optional dependencies, other libraries that you can take advantage of if present, but which your code will ignore if not. For example, perhaps your library depends on libz to compress some data, but if libz is not available at compile time, then your library will still compile successfully but will not support the ability to compress data. The traditional way to do this is to have some define in your code, such as:

```
#ifdef HAVE_ZLIB
#include "zlib.h"
// code that uses libz functions
#endif
```

Then your clients would use a compiler option (e.g., -DHAVE_ZLIB) to set the define when the library is available. Or you may use some configuration scripts that attempt to locate the library and, if found, update your compile line to include the define. This makes it more complicated to compile your code because the client must either manually specify the right set of defines for the environment or run some configuration script to do this.

C++17 introduced the ability for your code to check directly whether a header file is available (i.e., whether the #include statement would succeed), which can greatly simplify such a situation. This is done with the __has_include preprocessor directive; for example:

```
#if __has_include("zlib.h")
#include "zlib.h"
#define HAVE_ZLIB
#else
#undef HAVE_ZLIB
#endif
```

(You can use both "header" and <header> syntax with __has_include.)

In this code snippet, the zlib.h header will be included only if the compiler is able to find it in its header search path. Based on whether it can be included, then the code will also set or unset a HAVE_ZLIB define. You can then use this to place #if guards around other parts of the code where you try to reference symbols from the library. With this approach, your clients no longer need to set a define on the compile line because your code can figure out whether it can access the dependency itself.

If you need to support clients who are not using a compiler that supports C++17 (or later), you can still use this feature by placing an additional preprocessor check around it. For example:

```
#if defined __has_include
#if __has_include("zlib.h")
#include "zlib.h"
#define HAVE_ZLIB
#else
#undef HAVE_ZLIB
#endif
#else
#undef HAVE_ZLIB
#endif
```

If this code were to be encountered by a C++11 compiler that doesn't support the __has_include directive, then it would fail to enter the first #if and then unset the HAVE_ZLIB define, whereas a C++17 (or later) compiler would enter the first #if and then check for the existence of the zlib.h header file. You could also make the non-C++17 code path attempt to include the header if HAVE_ZLIB is defined, to support the traditional approach of using a compiler flag for those older compilers.

Byte type

C++17 introduced a new `std::byte` type to represent the concept of a single byte of memory. Previously, it was common to use `char` or `unsigned char` to access raw memory, but these types imply that you're addressing character data, and they also allow arithmetic operations to be performed on them. By contrast, `std::byte` is meant only to represent a collection of bits and it only defines bitwise operators, such as bit shifting and bitwise AND, OR, and XOR. The C++ standard defines `std::byte` as an enum class with an underlying type of `unsigned char`.

You can use `std::byte` to pass around opaque blocks of memory. This might be useful for applications such as data serialization or networking calls in which you want to reference binary data as opposed to textual information. For example, you can use `std::vector<std::byte>` to hold a contiguous block of raw data, or `std::byte *` to represent a pointer to raw memory instead of `char *`, which implies a pointer to textual characters, such as:

```
// serialize an object to a vector of bytes
std::vector<std::byte> SerializeObject(const MyObject &obj);

// deserialize a vector of bytes to an object
MyObject DeserializeObject(const std::vector<std::byte> &data);
```

The following code illustrates various ways to work with bytes:

```
// initialize a byte with a value
std::byte b1{2};
std::byte b2{0b01110011};

// bitwise operation
b1 &= b2;

// vectors of bytes and pointers to bytes
std::vector<std::byte> vec = {std::byte{0x1}};
std::byte *ptr = vec.data();

// compiler errors
b1 += 2;             // error: no arithmetic operations on bytes
std::byte err = 0;   // error: cannot initialize enum class with int
```

(Although a byte is often composed of 8 bits on modern computers, this should not be assumed. Some older machines and specialized chips such as DSPs can use a different number of bits per byte, as defined by `CHAR_BIT` in the `limits.h` header.)

The maybe_unused attribute

A new attribute was introduced in C++17 to suppress compiler warnings on unused entities. The `[[maybe_unused]]` attribute can be applied to most types, such as classes, structs, unions, typedefs, and enums, although the most common use for a public API will likely be for function arguments, to indicate that an argument may not be used within the function body.

This attribute can appear on both a function declaration (in the `.h` file) and its definition (in the `.cpp` file). However, I would recommend putting this attribute only on the function definition because it pertains to the implementation details of the function and does not affect how clients should use it. For example,

```
// sum.h
int Sum(const std::vector<int> &v, ExecutionPolicy policy);

// sum.cpp
int Sum(const std::vector<int> &v, [[maybe_unused]] ExecutionPolicy policy)
{
    ...
}
```

In this example, if the body of the `Sum()` function does not refer to the `policy` parameter, the compiler will not emit a warning because of the `[[maybe_unused]]` attribute. There are, of course, a few other ways you could achieve the same result before C++17:

1. Add a flag to the compiler command line to turn off warnings for unused function parameters, such as `-Wunused-parameter` for clang and g++.
2. Omit the parameter name in the function definition, such as:

```
int Sum(const std::vector<int> &values, ExecutionPolicy)
```

3. Add a `(void)policy;` line to the function body to reference the parameter but do nothing with it.

The benefit of the `[[maybe_unused]]` approach is that it works for code where the parameter may be used in some cases but not others. One way this could happen is if you use a `#ifdef` preprocessor statement inside the function body and only reference the parameter inside the preprocessor block. For example,

```
int Sum(const std::vector<int> &values, ExecutionPolicy policy)
{
#ifdef HAVE_PARALLEL_SUPPORT
    if (policy == std::execution::parallel_policy) {
        ...
    }
#endif
    ...
}
```

In this example, a compiler could generate a warning if `HAVE_PARALLEL_SUPPORT` is undefined. However, it would not issue a warning if the flag is defined. Adding the `[[maybe_unused]]` attribute to the parameter would suppress the warning in all cases.

Migrating to C++17

If you have an existing API and are considering migrating to C++17, this section will suggest some features to help improve your API design. As with earlier sections, I'll split this up into changes that can be made without breaking backward compatibility and those that are worth contemplating, but which could introduce breaking changes for your clients. I don't list all of the C++17 features I covered earlier, only those that I feel can have a worthwhile impact on your existing APIs:

- **Compatible Changes**
  - **Inline Variables:** These produces simpler and compact code and are a good feature to adopt as you upgrade to C++17. If you don't change the name of your global variables, adopting inline variables will be source compatible and for most compilers should also be binary compatible: extern versus inline symbols may be represented differently in your library binary, but the linker should still be able to resolve them without error.
  - **Nested Namespaces:** This is a purely syntactic change that will be both source and binary compatible. I personally prefer the simpler syntax of not having to define multiple namespace statements and balance multiple curly braces.
  - **Header Checks:** Making use of `__has_include` can help make the compilation and use of your library more robust to missing optional dependencies and simplify the requirements for building your code.
- **Breaking Changes**
  - **Fold Expressions:** If you're not using fold expressions to simplify the signature of your variadic template functions, using them within your function implementation should be a source compatible change. However fold expressions allow you to remove the base case function that terminates recursive expansion, so there's a possibility that your clients' code could have a reliance on that base case function symbol and fail to link against your new library without recompiling.
  - **Optional, Any, Variant:** These are useful new utility classes, but if you change any existing API function signatures to use them, then you'll break backward compatibility. Of course, using them in new functions would be safe.

C++20 API features

The C++20 version of the standard was published in Dec. 2020, replacing the prior C++17 version. This revision introduced some major features to the language, including support for modules and coroutines, which I'll cover in depth here.

There were many other updates that don't directly affect API designs, so we won't cover them here, but they still offer some great additions to the language, including new lambda function features, a `contains()` method for associative containers, and the `using enum` syntax to refer to enum class identifiers more concisely.

This revision also removed and deprecated several existing features, such as removing the use of `throw()` as an exception specification and deprecating most uses of the `volatile` keyword.

Modules

Modules represent a fundamental shift in how interfaces can be defined in C++. As of the time of writing this book, the feature is not broadly supported across all compilers, but as support and tooling improves, modules will likely become more prevalent as a way to define APIs.

A core tenant of this book is the physical separation of interface and implementation. Until C++20, the universal way to express this has been by declaring your interface in `.h` header files and your implementation in `.cpp` source files. This organization goes all the way back to the C language, with `.h` and `.c` files. One of the problems with this approach is that a C++ compiler treats each source file separately and it uses the preprocessor to expand all included header files recursively. This causes a large amount of repeated work. Even with the use of header guards or `#pragma once` statements to ensure that a header is expanded only once, the work is repeated for every source file the compiler visits. Another solution is the use of precompiled headers, although these are not defined by the standard and therefore tend to be compiler specific.

Modules provide a standardized solution to this problem. There are two main types of modules that were introduced in C++20:

1. **Named Modules:** These remove the use of header files and provide ways to separate your interface and implementation within your `.cpp` files.
2. **Header Units:** These provide a bridge and migration strategy between the use of header files and a full named module approach.

Named Modules

A named module consists of one or more module units, in which a module unit is simply a `.cpp` file that starts with a `module` declaration in the form:

```
[export] module module_name[:partition_name];
```

One of the module units will export the module name with the `export` keyword. This is called the module interface unit. If you spread the implementation of the module over other `.cpp` files, these are called module implementation units. Symbols that you want to be accessible outside the module must be explicitly exported with the `export` keyword. Let's look at an example:

```
// MathModule.cpp
export module MathModule;

export namespace MathModule {

int Square(int x);
int Cube(int x);

}  // namespace MathModule

// MathModuleImpl.cpp
module MathModule;

namespace MathModule {

int Square(int x)
{
    return x * x;
}
int Cube(int x)
{
    return x * x * x;
}

}  // namespace MathModule

// main.cpp
import MathModule;

import std;

int main()
{
    std::cout << MathModule::Square(10) << std::endl;
    std::cout << MathModule::Cube(10) << std::endl;
}
```

In this example, I created a module called `MathModule`. I put the interface (what you would normally put in a header file) in the module interface unit, `MathModule.cpp`, and the implementation in the module implementation unit, `MathModuleImpl.cpp`. (clang requires you to name your interface unit files with a `.cppm` extension.) The module exports a namespace, also called `MathModule`, which exports two functions: `Square()` and `Cube()`.

Here are a few things to note about this example:

- You must use the `import` keyword to access a module, not `#include`.
- There are no header files. In fact, a module cannot include any headers, except in an optional global module fragment that, if present, must come before the module declaration. Modules don't emit macros. For example, the use of `import std` will not expose Standard Library macros such as `assert`, `errno`, `va_arg`, or `INT_MAX`.

- Modules and namespaces are orthogonal concepts. In this example, I created a namespace with the same name as the module, but this is not required. You can, of course, also export symbols from a module at the global level.
- I explicitly exported the symbols that I want to be accessible from the module. Any symbols not marked with the `export` keyword cannot be called outside the module. This standardizes the export behavior of symbols from translation units, which the standard previously left up to implementations (Microsoft compilers famously have a different default symbol exporting behavior to gcc/clang).
- I used `import std` to pull in the entire Standard Library as a module, allowing me to access `std::cout`. I also could have used `import <iostream>` or even `#include <iostream>`; in other words, application code can intermingle modules and headers.

There are several other features that let you manage the interface and implementation parts of your modules. For example, you can define a private module fragment within your module interface unit to provide definitions that are not exported from that file. There's also a concept of module partitions, which can be used to split up larger modules into smaller pieces. And you can choose to export only specific symbols selectively, rather than enclosing the entire file in a single `export` scope. However, the approach I presented earlier keeps things simple and maps conceptually to how we've organized interface and implementation code for all time up until C++20.

Some of the benefits of adopting modules include improved compile times (because the compiler no longer needs to expand a large tree of header files into each translation unit) and greater reliability because the order of importing modules does not affect their semantics (because modules aren't affected by macros from other imports). There are, however, several practical considerations to be aware of when adopting modules. I've collected these caveats in the later section on Migrating to C++20.

> *TIP: When adopting named modules, you should still try to separate your interface and implementation code across a module interface unit and module implementation unit(s).*

Header units

In addition to named modules, C++20 introduced a concept known as header units to help ease the transition to modules. This feature allows you to import certain header files as modules with the `import "header.h"` or `import <header.h>` syntax. For example:

```
// MathHeader.h
namespace MathModule {

int Square(int x);
int Cube(int x);

}  // namespace MathModule

// main.cpp
import "MathHeader.h";  // import header file as a header unit

int main()
{
    std::cout << MathModule::Square(10) << std::endl;
    std::cout << MathModule::Cube(10) << std::endl;
}
```

This code assumes that you've compiled the associated `MathHeader.cpp` file to a binary module interface file, so that the compiler can access it like a regular named module.

Just like named modules, header units aren't affected by macros in other headers, but unlike named modules they expose any macros defined in the header files. This means they are useful for cases in which you still need to depend on any macros defined in the header files. The key benefit is that you don't have to remove the header files from your API and restructure your project to adopt named modules fully.

TIP: Header units can be useful to get many of the benefits of modules without having to restructure your code as much.

The spaceship operator

C++ lets you define various comparison operators to describe the result of comparing two objects. These are the ==, !=, <, <=, >, and >= operators. This list of operators was expanded in C++20 to include the spaceship (or three-way) comparison operator. Unlike the other comparison operators that return a `bool` result, the spaceship operator returns a three-way result that can be negative, zero, or positive. You can think of this as being conceptually similar to how the plain C `strcmp` function works. The name spaceship operator comes from the fact that the characters used for this operator look like an ASCII art TIE fighter: <=>.

Let's start by reviewing the use of the original comparison operators. I'll use a simple object called `Magnitude` that holds a numeric value:

```
class Magnitude
{
public:
    Magnitude(int value) :
        mValue(value)
    {}

private:
    int mValue = 0;
};
```

Trying to compare two objects of this class will fail because no comparison operators have been defined; that is,

```
Magnitude a(1);
Magnitude b(2);

// compile error: no == operator defined
std::cout << (a == b) << std::endl;
```

To make this work, we need to define the == operator for the class. Here's what the code looks like to add support for all comparison operators to the class:

```
class Magnitude
{
public:
    Magnitude(int value) :
        mValue(value)
    {}

    bool operator==(const Magnitude &rhs) const
    {
        return mValue == rhs.mValue;
    }
    bool operator!=(const Magnitude &rhs) const
    {
        return mValue != rhs.mValue;
    }
    bool operator<(const Magnitude &rhs) const
    {
        return mValue < rhs.mValue;
    }
    bool operator<=(const Magnitude &rhs) const
    {
        return mValue <= rhs.mValue;
    }
    bool operator>(const Magnitude &rhs) const
    {
        return mValue > rhs.mValue;
    }
    bool operator>=(const Magnitude &rhs) const
    {
        return mValue >= rhs.mValue;
    }

private:
    int mValue = 0;
};
```

The benefit of the spaceship operator is that you can remove most or all these operator definitions just by providing a definition for the spaceship operator. The compiler will then use your spaceship operator to provide an implementation for all other operators. The compiler can even generate a default implementation of the spaceship operator for you, so that class definition can be simplified to:

```
class Magnitude
{
public:
    Magnitude(int value) :
        mValue(value)
    {}

    auto operator<=>(const Magnitude2 &) const = default;

private:
    int mValue = 0;
};
```

Now this code will compile and work correctly:

```
Magnitude a(1);
Magnitude b(2);

std::cout << "a == b: " << (a == b) << std::endl;
std::cout << "a != b: " << (a != b) << std::endl;
std::cout << "a < b: "  << (a < b)  << std::endl;
std::cout << "a <= b: " << (a <= b) << std::endl;
std::cout << "a > b: "  << (a > b)  << std::endl;
std::cout << "a >= b: " << (a >= b) << std::endl;
```

TIP: Use the spaceship operator, <=>, to avoid having to define all comparison operators for your classes.

The default spaceship operator implementation will do a member-wise comparison of each of the data members within your class. It does this by first comparing any base class members and then the nonstatic members in declaration order. For example, given the definition of a `Vector` class:

```
class Vector
{
public:
    Vector(int x, int y) :
        mX(x),
        mY(y)
    {}

    auto operator<=>(const Vector &) const = default;

private:
    int mX = 0;
    int mY = 0;
};
```

the generated implementation will be equivalent to the code:

```
auto operator<=>(const Vector &rhs) const
{
    if (mX < rhs.mX) {
        return -1;
    }
    if (mX > rhs.mX) {
        return 1;
    }
    if (mY < rhs.mY) {
        return -1;
    }
    if (mY > rhs.mY) {
        return 1;
    }
    return 0;
}
```

This default behavior may not always be what you want, such as if the data members need to be compared in a different order or a more complex comparison algorithm is required. In that case, you can provide a custom implementation for the spaceship operator. This will require you to choose among three different return types to control how the other two-way comparison operators are generated. Those are:

1. `std::strong_ordering`: This result type is used for cases in which it's possible to determine that two objects are exactly equal (i.e., one can be substituted for the other). This implies that every data member is compared by the operator. Possible values for this type are:
   a. `std::strong_ordering::less`
   b. `std::strong_ordering::equal`
   c. `std::strong_ordering::greater`
2. `std::weak_ordering`: This type is used for the weaker relationship in which you can determine only whether two objects are equivalent rather than equal. The difference is that equivalent objects may have irrelevant differences whereas equal objects are indistinguishable. Possible weak ordering values are:
   a. `std::weak_ordering::less`
   b. `std::weak_ordering::equivalent`
   c. `std::weak_ordering::greater`
3. `std::partial_ordering`: This type is similar to weak ordering but with the added feature that there may be values that cannot be compared with each other, such as a NaN float. Possible partial ordering values are:
   a. `std::partial_ordering::less`
   b. `std::partial_ordering::equivalent`
   c. `std::partial_ordering::greater`
   d. `std::partial_ordering::unordered`

For example, let's assume that for our `Vector` class we want the comparison operators to compare the magnitude of the vector, in which magnitude is the square root of the sum of each element squared: `sqrt(mX*mX + mY*mY)`. We can remove the square root operator for performance reasons because we just care about relative comparisons, not the absolute value. We could therefore implement our spaceship operator as:

```
class Vector
{
public:
    Vector(int x, int y) :
        mX(x),
        mY(y)
    {}

    std::strong_ordering operator<=>(const Vector &rhs) const
    {
        int thisMagnitude2 = (mX * mX) + (mY * mY);
        int rhsMagnitude2 = (rhs.mX * rhs.mX) + (rhs.mY * rhs.mY);

        // could just do: return thisMagnitude2 <=> rhsMagnitude2
        // but here we show the actual conditions and return values
        if (thisMagnitude2 < rhsMagnitude2) {
            return std::strong_ordering::less;
        }
        if (thisMagnitude2 > rhsMagnitude2) {
            return std::strong_ordering::greater;
        }
        return std::strong_ordering::equal;
    }

    bool operator==(const Vector &rhs) const
    {
        int thisMagnitude2 = (mX * mX) + (mY * mY);
        int rhsMagnitude2 = (rhs.mX * rhs.mX) + (rhs.mY * rhs.mY);
        return thisMagnitude2 == rhsMagnitude2;
    }

private:
    int mX = 0;
    int mY = 0;
};
```

If you provide a custom spaceship operator implementation (as opposed to the compiler default implementation), then the compiler will require you to implement the == operator, if you want the == and != operators to work. The reasoning for this is that if you need a custom implementation for <=>, then you most likely also need a custom implementation for ==, because using <=> to test for equality may be inefficient. For example, when comparing a vector or a string for equality, it would be more efficient first to check whether the lengths are the same before iterating over each element.

Constraints and concepts

One of the most common complaints about C++ templates is that compiler errors can be verbose and difficult to understand. This is because errors are often triggered deep within the template implementation, making it hard to figure out what the real problem is. C++20 introduced features called constraints and concepts to improve this situation. These let you add compile-time constraints to template parameters, allowing compilers to emit more useful errors if clients try to use the template in a manner that doesn't meet those constraints.

As an example, let's consider a simple function template to return the minimum of two values:

```
template <typename T>
T min(T x, T y)
{
    return (x < y) ? x : y;
}
```

This function has an implicit constraint that the type of the two values being compared must define a less than operator, <. If you pass in a type that doesn't meet that expectation, then you'll get a compile error, and that error will most likely be located within the template implementation:

```
// integer parameters work fine
auto a = min(1, 2);

// passing an object causes a compile error inside of min()
class MyClass
{
};
auto b = min(MyClass(), MyClass());
```

Using constraints, we can now augment the function template declaration to specify that it's valid to call the function only with integer or floating-point parameters. This is done with the new `requires` keyword, which takes a compile-time constant expression, such as:

```
template <typename T>
requires std::integral<T> || std::floating_point<T>
T min(T x, T y)
{
    return (x < y) ? x : y;
}
```

Now the compiler will generate an error message at the actual call site of the function, not within the function template implementation, if it's called with nonnumeric parameters. You can also place the `requires` expression after the function signature if you prefer.

> *TIP: Use constraints to define the expectations your code has for any template parameter types.*

The `std::integral` and `std::floating_point` symbols are called concepts. More specifically they are built-in concepts provided by the C++ Standard Library. A concept is essentially a named set of requirements that can be used with the `requires` keyword. Here's a list of all of the built-in concepts that were added in C++20. These are all included in the new <concepts> header, except where I indicate otherwise:

- `std::same_as`: the type is the same as another type.
- `std::derived_from`: the type is derived from another type.
- `std::convertible_to`: the type is implicitly convertible to another type.
- `std::common_reference_with`: two types share a common reference type.
- `std::common_with`: two types share a common type.
- `std::integral`: the type is an integral type.
- `std::signed_integral`: the type is a signed integral type.
- `std::unsigned_integral`: the type is an unsigned integral type.
- `std::floating_point`: the type is a floating-point type.
- `std::assignable_from`: the type is assignable from another type.
- `std::swappable`: the type can be swapped (e.g., with `std::swap`).
- `std::swappable_with`: two types can be swapped with each other.
- `std::destructible`: an object of the type can be destroyed.
- `std::constructible_from`: a variable of the type can be initialized with the given set of argument types.
- `std::default_initializable`: an object of the type can be default constructed.
- `std::move_constructible`: an object of the type can be move constructed.
- `std::copy_constructible`: an object of the type can be copy and move constructed.
- `std::equality_comparable`: operators == and != specify an equivalence relation.
- `std::equality_comparable_with`: operators == and != specify an equivalence relation for two types.
- `std::totally_ordered`: the type comparison operators yield a total order.
- `std::totally_ordered_with`: the comparison operators for two types yield a total order.
- `std::three_way_comparable`: operator <=> produces consistent comparison results (defined in <compare>).
- `std::three_way_comparable_with`: operator <=> produces consistent comparison results for two types (defined in <compare>).
- `std::movable`: an object of the type can be moved and swapped.
- `std::copyable`: an object of the type can be copied, moved, and swapped.

- `std::semiregular`: an object of the type can be copied, moved, swapped, and default constructed.
- `std::regular`: the type that is both `semiregular` and `equality_comparable`.
- `std::invocable`: the callable type that can be invoked with a given set of argument types using `std::invoke`.
- `std::regular_invocable`: the callable type that can be invoked with `std::invoke` where the invoke expression is equality preserving.
- `std::predicate`: the callable type is a Boolean predicate.
- `std::relation`: the callable type is a binary relation.
- `std::equivalence_relation`: the relation imposes an equivalence relation.
- `std::strict_weak_order`: the relation imposes a strict weak ordering.

In addition to these built-in concepts, you can define your own with the `concept` keyword. For example, continuing our earlier `min()` function template, we could introduce a new concept for whether the type defines the < operator, rather than constraining the function to only integral and floating-point types. This can be done as follows:

```
template <typename T>
concept less_than_comparable = requires(T a, T b)
{
    a < b;
};
```

We can now use our custom concept to redefine the `min()` function template:

```
template <typename T>
requires less_than_comparable<T>
T min(T x, T y)
{
    return (x < y) ? x : y;
}
```

Constraints are a great way to give your clients more information about how you expect your templates to be used. They also allow the compiler to provide more helpful error messages if your clients pass in parameter types that don't conform to those expectations. You can use built-in concepts in the Standard Library to define these template parameter constraints or you can write your own concepts to express the requirements for your constraints more directly.

Abbreviated function templates

The syntax for defining templates in C++ can be verbose and sometimes intimidating. To address this, C++20 introduced simpler syntax to express templates. Taking our `min()` function template from the previous section:

```
template <typename T>
T min(T x, T y)
{
    return (x < y) ? x : y;
}
```

This can now be expressed more tersely using the `auto` keyword as just:

```
auto min(auto x, auto y)
{
    return (x < y) ? x : y;
}
```

You can also use the abbreviated syntax to refer to a constraint directly within the parameter list. For example, our constrained version of the function template from the previous sections looks like:

```
template <typename T>
requires less_than_comparable<T>
T min(T x, T y)
{
    return (x < y) ? x : y;
}
```

This can be simplified to the abbreviated form:

```
auto min(less_than_comparable auto x, less_than_comparable auto y)
{
    return (x < y) ? x : y;
}
```

Although abbreviated templates are just a simplification of the existing template syntax, not all templating features can be expressed in abbreviated form. There are cases in which you will need to use the original nonabbreviated syntax. For example,

- In the previous abbreviated form of `min()`, each `auto` parameter could potentially be a different type. If you want to enforce that both parameters should be of the same type, T, then you should use the original nonabbreviated syntax.
- If you want to specify a template parameter with a default type, then you should use the nonabbreviated syntax (e.g., `template<typename T = void>`).
- If you want to use variadic template parameter packs, then these will also require the nonabbreviated form (e.g., `template<typename... T>`).

The consteval specifier

C++11 introduced the `constexpr` specifier to allow functions or variables to be used in constant expressions. With C++20, we now also have the `consteval` specifier, which is used to declare an "immediate" function (i.e., a function that must produce a constant at compile time).

The difference between `constexpr` and `consteval` is that the former is a hint to the compiler to evaluate the function at compile time if possible, whereas `consteval` requires the function to be evaluated at compile time. The compiler will generate an error if it cannot. Also, `constexpr` can be applied to functions and variables, whereas `consteval` can be applied only to functions. Here's an example:

```
consteval double Deg2Rad(double degrees)
{
    return degrees * M_PI / 180;
}
```

This `consteval` function can be evaluated at compile time, allowing the compiler to insert the constant result instead of making a function call. An implication of this is the compiled object code would not need to have a `Deg2Rad` symbol. You can therefore think about this as a way to get the functionality of C preprocessor function macros using pure C++ syntax and with full type checking. In other words, this `consteval` function could replace a macro such as:

```
#define DEG2RAD(degrees) (degrees * M_PI / 180)
```

TIP: Prefer using a `consteval` function instead of a preprocessor function macro.

The constinit specifier

A related specifier that was also introduced in C++20 is `constinit`. This is used to ensure that a static or thread storage variable is initialized at compile time to avoid the well-known problems with static initialization order at library load time.

The `constinit` specifier can only be applied to variables (you can use `consteval` to evaluate a function at compile time, as I just described). A compile error will be generated if you try to initialize a `constinit` variable with a value that can't be evaluated at compile time, or if you try to use it on a local variable within a function. Let's look at an example:

```
constexpr std::string_view GetProjectName()
{
    return "Phoenix";
}

constinit std::string_view PROJECT_NAME = GetProjectName();
```

In this example, I've declared `PROJECT_NAME` as `constinit`, so it'll be initialized at compile time, avoiding any potential static initialization issues at library load time. To do that, the `GetName()` function had to be declared `constexpr` to indicate that it can be evaluated at compile time.

> *TIP: Use* `constinit` *to initialize static or thread storage variables at compile time.*

Migrating to C++20

If you're looking to move your API to support C++20, then perhaps the biggest feature to consider using is modules. Modules offer several useful features for building large software projects, including the potential for improved compile times and the lack of side effects when importing modules. However, there are also a few caveats to be aware of:

- Right now, support for modules is limited. Microsoft has good support for modules in its MSVC compiler, but the latest clang compiler included with Xcode does not yet support modules (although you can use `brew install llvm` to install a version with partial modules support on macOS). In particular, the current clang documentation describes header unit support as highly experimental as of 2023.
- The format used to represent a module is left as a compiler implementation detail, meaning that they are not portable across different compilers and perhaps not even between different versions of the same compiler.
- One of the benefits of modules is the potential to improve compile time performance, but we also need to think about how modules can be used as a distributing mechanism for APIs. For binary distributions, today you will provide clients with a prebuilt library and header files. If you are using modules, then you will presumably also need to package the precompiled module binary files (although just the module interface units, not any implementation units).
- Modules must be compiled in a specific order to handle dependencies correctly. If module A imports module B, then module B must be compiled first so that its binary module interface file is accessible to module A. With traditional header files, the C++ compiler could parallelize compilation of translation units to speed up compilation times. This is no longer possible with modules. At a minimum, any parallelization effort would need to be constrained by the module dependency ordering.
- Distributing header files with library binaries has the benefit that they can be read by humans in a text editor. However a binary module file cannot be interpreted as easily. Therefore either we must rely on good documentation generation tools or we need to distribute the source code for our module interface units so that clients

know how to use your API (or in the case of header units, you can just share your header files as you would do normally). This provides another reason to split your interface and implementation across different module units: so you don't need to share your implementation code if you don't want to.

Presumably many of these caveats will be addressed over time as modules are used more widely and become more mature. I look forward to the opportunity to update this section in a future revision of this book to cover those learnings and improvements.

In terms of other C++20 features to consider adopting in your APIs, template constraints are a great addition that will improve template error messages for your clients and make it more clear what types they can use as template parameters. Because constraints are a compile-time check, you should be able to add them to your APIs without breaking binary compatibility. Although it's possible that adding a constraint will mean that client code will now fail to compile where it worked before, it's more likely that those cases were already invalid uses of your interface that failed to compile in earlier versions of your API as well.

The use of abbreviated function syntax is purely a stylistic choice. If you like the terser syntax, then feel free to use it. Just remember that there are some template features that can't be expressed in the abbreviated syntax.

The `consteval` specifier is a type-safe alternative to using preprocess function-like macros, and as such it is a good feature to consider using. Nevertheless if you convert an existing nonmacro function to a `consteval` function, then the symbol will no longer be defined in your library binary, so binary compatibility will be broken.

The `constinit` specifier can help to reduce the cost and dangers of static initialization when your library code is loaded, so it, too, is worth adopting where feasible. Not all static or thread storage variable definitions can be converted to use `constinit`, only those that can be calculated at compile time. However the use of this specifier should be backward compatible because it just optimizes the initialization of a variable before function `main()` is called.

C++23 API features

The most recent C++ revision that I'll cover in this edition of the book is C++23. At the time of writing, this revision has not yet been published; however, the set of proposed new features is well-established.

Although the early working group meetings for C++23 were affected by the COVID-19 pandemic, a large number of new features are being added to this revision of the standard, such as a `contains()` function for `std::string` and `std::string_view`, additional monadic members on `std::optional` like `and_then()` and `or_else()`, and a `#warning` preprocessor directive. Modules support for the Standard Library is also being added with the introduction of `std` and `std.compat` named modules.

There are also several new features that can affect API design, including the `std::expected` template, multidimensional subscript operators, and several new preprocessor directives. I'll cover these new features in these final sections of the chapter.

Expected values

The `std::expected` class template provides a way to return either a valid object or an error. This can be useful as a function return value when you want to be able to return a valid result but also provide additional information in the case of an error. In that case, it's common to use an integer or an enum to represent the error value. For example, here's an image loading function that can return either the image data or an error enum:

```
using Image = std::vector<std::byte>;

enum class Error {
    NotFound,
    BadFormat,
    EmptyFile,
};

std::expected<Image, Error> LoadImage(const std::string &filename);
```

> *TIP: Use `std::expected` when you want a function to return either a valid result or details about the error (such as an error code or enum) if there was a failure.*

Callers of this function can check the `has_value()` member of the `std::expected` return result to see whether it holds a valid value. If this is `true`, then the `value()` member function can be used to access the value; otherwise `error()` can be used to access to error. For example:

```
auto image = LoadImage("image.png");
if (image.has_value()) {
    std::cout << "Bytes = " << image.value().size() << std::endl;
} else if (image.error() == Error::NotFound) {
    std::cout << "Image not found" << std::endl;
} else if (image.error() == Error::BadFormat) {
    std::cout << "Image has a bad format" << std::endl;
} else if (image.error() == Error::EmptyFile) {
    std::cout << "Image not found" << std::endl;
}
```

You can also use simpler pointer-like syntax to check whether there is a valid value and to access members of the value object, as shown here:

```
if (image) {
    std::cout << "Bytes = " << image->size() << std::endl;
}
```

Clients can also use the `value_or()` member of `std::expected` to access the value or to use a default value if there was an error, e.g.,

```
std::cout << "Bytes = " << image.value_or(Image()).size() <<
std::endl;
```

In the past, you might have used an `std::optional` to return a value but also indicate if an error happened. However, in that case you cannot provide additional information about the error, just that something unexpected happened and there is no valid value. The `std::expected` template is therefore a better choice if you want to return a valid object or return some error code or enum if the operation failed.

Multidimensional subscript operator

C++ lets you to define a subscript operator, `[]`, to access the underlying data in your classes using an array subscript syntax, such as `object[0]`. For example:

```
class SparseArray
{
public:
    int operator[](size_t index) const;
    int &operator[](size_t index);
};
```

This lets you use `array[n]` syntax to access elements of `SparseArray`, and you can also assign values to the array using the overload with the reference return type: for example, `array[n] = 2`.

However, this support is limited to one-dimensional arrays. There are many problem domains where multidimensional arrays are useful, such as representing matrices in linear algebra, 2D or 3D coordinate systems for games, and images for graphics. Common workarounds to this limitation include:

- Using a regular function call, such as `array.at(x, y, z)`.
- Overloading the call operator for an object, such as `array(x, y, z)`.
- Using a chain of one-dimensional arrays, such as `array[x][y][z]`.

However, as of C++23, we now have direct support for multidimensional arrays. For example, here is an example of defining a 2D array using C++23:

```
class IntArray2D
{
public:
    int operator[](size_t x, size_t y) const;
    int &operator[](size_t x, size_t y);
};
```

Clients can now get and set elements of this array using `array[x, y]` syntax; for example:

```
IntArray2D array;
array[0, 0] = 2;
int originValue = array[0, 0];
```

TIP: Use the multidimensional subscript operator when you want to model 2D, 3D, or higher dimensionality concepts.

This new syntax is possible because the use of the comma operator in subscript expressions was deprecated in C++20. This paved the way for C++23 to reuse the syntax to represent multidimensional arrays.

You can also use variadic templates in combination with the new multidimensional subscript operator. For example:

```
template <typename T>
struct Vector3D
{
    template <std::convertible_to<std::size_t>... U>

    T operator[](U... indices) const;
};
```

This allows clients to use a variable number of indices for the array access. For example, you could enforce a default value for a 3D coordinate if no index is provided, such as:

```
Vector3D<int> vec3;
int a = vec3[1];
int b = vec3[1, 2];
int c = vec3[1, 2, 3];
```

Preprocessor directives

The C preprocessor provides several directives to support conditional compilation in header and source files. These are the `#if`, `#else`, `#elif`, and `#endif` directives. There's also the `#ifdef` directive, which is another way to write `#if defined()`, and `#ifndef`, which is alternate syntax for `#if !defined()`. The `defined()` operator is used to verify whether a macro has been defined.

However, there are no equivalent shorthand versions for the `#elif` directive. This is generally considered an oversight in the original C language and was addressed in C++23 with the introduction of:

- `#elifdef`: shorthand for `#elif defined()`.
- `#elifndef`: shorthand for `#elif !defined()`.

For example:

```
#ifdef USE_FEATURE_A
    std::cout << "Using Feature A" << std::endl;
#elifdef USE_FEATURE_B
    std::cout << "Using Feature B" << std::endl;
#elifndef DONT_USE_FEATURE_C
    std::cout << "Using Feature C" << std::endl;
#else
    std::cout << "No Feature Enabled" << std::endl;
#endif
```

If this code is compiled with a compiler that does not support the new directives, then they will be skipped and the logic will move from the first `#ifdef` to the `#else`. You can use this behavior to detect whether your compiler supports the new directives:

```
#if 0
#elifndef NON_EXISTING_MACRO
#define ELIFDEF_SUPPORTED
#else
#endif
```

C++23 also introduced another preprocessor directive: `#warning`, which provides a compile-time warning when clients compile against your API. Many compilers already provided this feature, but it was not officially part of the standard until C++23. This feature provides a complement to the `#error` directive, which will issue a compile-time error. For example:

```
// generate a compile error if library not available
#ifndef HAS_REQUIRED_LIBRARY
#error "Required library not available"
#endif

// generate a compile warning if library not available
#ifndef HAS_OPTIONAL_LIBRARY
#warning "Optional library not available"
#endif
```

One further preprocessor directive introduced in C++23 is `#embed`. This provides a standard way to include binary data in an executable. For example, this can be used to embed the binary data easily for an image, video, or sound file in your code. Although you may generally prefer to use this feature within your source files, you can combine this with features such as `constinit` and `inline` to define an inline variable that provides the raw binary data for certain resources provided by your API. For example:

```
constinit inline const char image[] = {
#embed <image.png>
};
```

The `#embed` directive will load the specified file at compile time and insert the contents of the file as a comma-separated list of integers. So that code will be expanded to something like the following once the preprocessor phase has completed:

```
constinit inline const char image[] = {
0x50, 0x89, 0x47, 0x4e, 0x0a, 0x1a, ...
};
```

Migrating to C++23

If you're planning to move your API's minimum supported C++ version to C++23, then adding support for `std::expected` can provide an elegant way to return a valid result from a function or an error with more context regarding what went wrong. However this will, of course, break source and binary compatibility.

If your API naturally deals with multidimensional quantities, then you may want to consider adding support for the new subscript operator. This would be a new capability that you add to your API, so it should not break existing code.

Finally, using `#warning` and `#error` can be useful to provide compile-time feedback to your clients if they're missing some required or optional dependency. You could also use the new conditional compilation directives, `#elifdef` and `#elifndef`, if you prefer this syntax over the more verbose alternative. Even though these two new directives were introduced in C++23, the standard encourages compiler implementors to backport these features to older language modes as well. As noted earlier, `#warning` was already implemented by many compilers before C++23. So you may be able to take advantage of these features in most compilers even before upgrading to C++23.

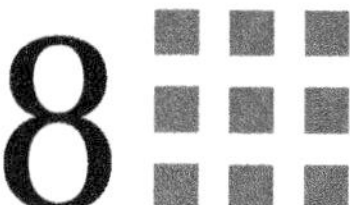

8 Performance

It's not the focus of this book to tell you how to optimize the performance of your implementation, or even to tell you whether it's necessary. Your implementation should be as fast as it needs to be: some application programming interfaces (APIs) are performance critical and must be called many times per second whereas other APIs are infrequently used and their speed is of less concern. However, it is the focus of this book to show you how certain API design decisions can affect performance, and therefore how you can optimize the performance of your interface.

Your implementation may not need to have high performance, but your interface should still be as optimal as possible so that it's not actively undermining performance. Requirements change, and you may find yourself needing to optimize your implementation after the first version of your API has been released. In this situation, you'll wish that you had considered the performance impact of your API beforehand, so that you're not forced to break backward compatibility to improve performance.

However, the most important point to make here is that you should strive never to warp your API for performance reasons. Good designs normally correspond with good performance (Tulach, 2008; Bloch 2008). Your API should continue to provide a clean and logical representation of the problem domain even after you've optimized its implementation. There are cases where this is simply not possible. For example, if you're writing an API that must communicate across a remote procedure call barrier, or if you're sending multiple HTTP requests to a Web-based API, you may hit the case that the overhead for performing many individual API calls is too slow and instead you feel the need to introduce a vectorized API, which batches up many calls into a single method. But other than these domain-specific examples, you will often find that a lot can be optimized behind your API without having to change the public interface. That's why you're writing an API after all!

> *TIP: Don't unduly warp your API to achieve high performance.*

There are several components to API performance. I'll consider each of these in the following subsections:

1. **Compile-time speed.** The impact of your API on the time it takes to compile a client program. This can affect the productivity of your users.

API Design for C++. https://doi.org/10.1016/B978-0-443-22219-1.00005-2

2. **Run-time speed.** The time overhead for calling your API methods. This is important if your methods must be called frequently.
3. **Run-time memory overhead.** The memory overhead for calling your API methods. This is important if you expect many of your objects to be created and kept in memory. It can also affect CPU cache performance.
4. **Library size**: The size of the object code for your implementation that clients must link into their applications. This affects the total disk and memory footprint of your clients' applications.
5. **Start-up time**: The time it takes to load and initialize a dynamic library that implements your API. Various factors can affect this, such as template resolution, binding unresolved symbols, calling static initializers, and library path searching.

In addition to specific API factors that affect these performance metrics, you should, of course, investigate your compiler's options to see whether there are any flags you can turn on or off to give a performance boost. For example, turning off run-time type information if you don't need to use the `dynamic_cast` operator is one common decision (`-fno-rtti` for the GNU C++ compiler).

One of the most important lessons to learn about performance optimization is that you should never trust your instincts about which parts of your implementation you think will be slow. You should always measure the actual performance profile of your API in real-world situations and then focus your optimization effort on the areas that give you the biggest impact. A corollary to this is that you don't need to start with the most efficient implementation: do it the easy way first and then figure out which parts of your implementation need to be optimized once everything is working.

At Pixar, we would regularly have various teams work on different components of a particular feature for one of our films. For example, the driving system in *Cars* involved work by the R&D team to implement a generic simulation plugin system, work by GUI engineers to provide direct manipulation controls for animators, and work by the production teams on the movie to integrate everything into their pipeline. Then, once all of the software was functional and integrated, we convened "speed team" meetings to assess where the bottlenecks were and assign work to the relevant engineers so that the overall system would meet specific performance criteria.

The important point is always to remember Amdahl's law. This states that the overall performance improvement gained by optimizing a single part of a system is limited by the fraction of time that the improved part is actually used. You may increase the performance of one part of your API by a factor of 10, but if a client's program spends only 1% of its time in that code, then the overall improvement is reduced to a factor of only 0.1 (10 * 0.01).

> *TIP: To optimize an API, instrument your code and collect performance data for real-world examples. Then target your optimization effort at the actual bottlenecks. Don't guess where the performance hot spots are.*

Pass input arguments by const reference

In Chapter 6, I recommended that you prefer const references over pointers to pass input parameters (i.e., parameters that are not changed by the function), but that you should prefer pointers over nonconst references for output parameters so that their mutability is clearly advertised to clients. This section now offers some additional performance reasons to support the use of const references to pass input arguments into a function.

By default, function arguments in C++ are passed "by value". This means that the object being passed into the function is copied, and then that copy is destroyed when the function returns. As a result, the original object that's passed into the method can never be modified. However, this involves the overhead of calling the object's copy constructor and then destructor. Instead, you should pass a const reference to the object. This has the effect of only passing the pointer to the object, but also ensuring that the object is not modified by the method:

```
void SetValue(std::string str);        // pass by value
void SetValue(std::string &str);       // pass by reference
void SetValue(const std::string &str); // pass by const reference
```

> *TIP: Always prefer passing a nonmutable object as a const reference rather than passing it by value. This will avoid the memory and performance costs to create and destroy a temporary copy of the object, and all of its member and inherited objects.*

This rule applies only to objects. It's not necessary for built-in types such as `int`, `bool`, `float`, `double`, or `char` because these are small enough to fit in a CPU register. In addition, C++ Standard Library iterators, smart pointers, and function objects are designed to be passed by value. However, for all other custom types, you should favor references or const references. Let's look at a specific example:

```
class MyObject
{
public:
    // constructor
    MyObject();
    // destructor
    ~MyObject();
    // copy constructor
    MyObject(const MyObject &obj);
    // assignment operator
    MyObject &operator=(const MyObject &obj);

private:
    std::string mName;
};

class MyObjectHolder
{
public:
    MyObjectHolder();
    ~MyObjectHolder();

    MyObject GetObject() const;
    void SetObjectByValue(MyObject obj);
    void SetObjectByConstReference(const MyObject &obj);

private:
    MyObject mObject;
};
```

If you assume that the `SetObjectByValue()` and `SetObjectByConstReference()` methods both simply assign their argument to the `mObject` member variable, then the sequence of operations that's performed when each of these methods is called is:

- **SetObjectByValue(object)**
  - `std::string` constructor
  - `MyObject` copy constructor
  - `MyObject` assignment operator
  - `MyObject` destructor
  - `std::string` destructor
- **SetObjectByConstReference(object)**
  - `MyObject` assignment operator

The situation becomes worse if `MyObject` is derived from some base class, because then the copy constructor and destructor of each base class in the object hierarchy would also have to be called for the pass by value case.

There's another reason to avoid passing arguments by value, and that's the "slicing problem" (Meyers, 2005). This is the problem in which, if a method accepts a base class argument (by value) and you pass a derived class, then any extra fields of the derived class will be sliced off. This is because the size of the object to be passed by value is determined, at compile time, to be the size of the base class. Passing arguments as const references instead of passing by value avoids the slicing problem.

Minimize #include dependencies

The time it takes to compile a large project can depend greatly on the number and depth of `#include` files. As such, one of the common techniques for decreasing build times is to try and reduce the number of `#include` statements in header files. As you know, it's good practice to put include guards around your header files to avoid their being expanded multiple times during the complication of a single translation unit. But there are other factors you can consider to improve build times.

Avoid "Winnebago" headers

Some APIs provide a single large header file that pulls in all of the classes and global definitions for the interface. This can seem like a convenient affordance for your clients. However, it only serves to increase the compile-time coupling between your clients' code and your API, and it means that even the most minimal use of your API must pull in every public symbol.

For example, the standard Win32 header `windows.h` pulls in well over 200,000 lines of code (under Visual Studio 9.0). Every `.cpp` file that includes this header effectively adds over 4 MB of extra code that needs to be loaded from around 90 separate files and compiled for every source file. Similarly, the macOS header `Cocoa/Cocoa.h` expands to over 100,000 lines of code at over 3 MB.

Precompiled headers can help to alleviate this burden by preprocessing these large common include files to a more optimal form, such as a `.pch` or `.gch` file. However, a more modular and loosely coupled solution would involve providing a collection of smaller individual headers for each component of your API. Clients can then choose to `#include` only the declarations for the subset of your API that they're using. This can make for longer lists of `#include` statements in your clients' code, but the result is an overall reduction in the amount of your API that their code must pull in.

Forward declarations

A header file, A, includes another header file, B, to pull in the declaration of a class, function, struct, enum, or other entity that's used in header A. The most common situation in an object-oriented program is that header A wants to pull in the declaration of one or more classes from header B. However, in many situations, header A doesn't actually need to include header B and can instead simply provide a forward declaration for the classes needed. A forward declaration can be used when:

1. The size of the class is not required. If you include the class as a member variable or subclass from it, then the compiler will need to know the size of the class.
2. You don't reference any member methods of the class. Doing so would require knowing the method prototype: its argument and return types.

3. You don't reference any member variables of the class; but you already know never to make those public (or protected).

For example, you can use forward declarations if header A refers only to the name of classes from header B via pointers or references:

```
class B;   // forward declaration

class A
{
public:
    void SetObject(const &B obj);

private:
    B *mObj;
};
```

However, if you were to change the definition of class A so that the compiler needs to know the actual size of class B, then you must include the actual declaration of class B (i.e., you must `#include` its header). For example, if you store an actual copy of B inside A:

```
#include "B.h"

class A
{
public:
    void SetObject(const &B obj);

private:
    B mObj;
};
```

As noted in the chapter on Patterns, you can use the pimpl idiom to avoid having to define object B as a data member in your header file, and thereby avoid the need to `#include` the `B.h` header in the previous case.

Obviously, you will need to `#include` the full header in any `.cpp` file that uses the classes in that header (e.g., `A.cpp` must include `B.h`). A forward declare simply tells the compiler to add the name to its symbol table with a promise that you'll provide the full declaration when it actually needs it.

One interesting point is that a forward declaration is sufficient if you pass a variable to a method by value or return it by value (although you should prefer a const reference over passing a parameter by value). You might think that the compiler needs to know the size of the class at this point, but it's needed only by the code that implements the method and any client code that calls it. So the following example is quite legal:

```
class B;

class A
{
public:
    void SetObject(B obj);
    B GetObject() const;
};
```

> *TIP: As a rule of thumb, you should need to #include the header file for a class only if you use an object of that class as a member variable in your own class or if you inherit from that class.*

Note: it's generally safer to use forward declarations only for your own code. By using forward declarations, you are embedding knowledge of how symbols are declared in the header that you're eliding. For example, if you use forward declares for a foreign header and your API is used in environments with different versions of that header, the declaration of a class within that header could be changed to a typedef, or the class could be changed to a templated class, which would break your forward declaration. This is one reason why you should specifically never try to forward declare C++ Standard Library objects. For example, always do `#include <string>`, never attempt to forward declare `std::string`.

> *TIP: Only forward declare symbols from your own API.*

Of note here is the `<iosfwd>` header offered by the Standard Library. This header contains forward declarations for the input/output library, so that you don't have to pull the full definitions into your own header files.

Finally, a header file should declare all of its dependencies as either forward declarations or explicit `#include` lines. Otherwise, the inclusion of the header in other files may become order dependent. This is at odds with the statement to minimize the number of `#include` statements, but it is an important exception for the sake of API robustness and elegance. A good way to test this is to ensure that an empty `.cpp` file that includes only your public header can compile without error.

TIP: A header file should #include or forward declare all of its dependencies.

Redundant #include guards

Another way to reduce the overhead of parsing too many include files is to add redundant preprocessor guards at the point of inclusion. For example, if you have an include file, `bigfile.h`, which looks like this:

```
#ifndef BIGFILE_H
#define BIGFILE_H

// lots and lots of code

#endif
```

Then you might include this file from another header by doing this:

```
#ifndef BIGFILE_H
#include "bigfile.h"
#endif
```

This saves the cost of pointlessly opening and parsing the entire include file if you've already included it. This may seem like a trivial optimization, and indeed it can be for small include hierarchies. However, if you have a large code base with many include files, this optimization can make a significant difference. Back in 1996, John Lakos performed several experiments to demonstrate the degree of performance improvements that this optimization can elicit on a large project. The results were striking (Lakos, 1996). However, because these results are from the mid-1990s, I designed a similar experiment to test this effect on a modern compiler, and the results correlate well with those of Lakos.

For a given number *N*, I generated *N* include files that each included the *N*-1 other include files. Each include file also contained around 100 lines of class declarations. I also generated *N* `.cpp` files, in which each `.cpp` file included only one header. I then timed how long it took to compile every `.cpp` file. This experiment therefore chooses a worst-case $O(n^2)$ include structure, although it also includes the time to run the compiler *N* times. The experiment was performed for a set of include files that used redundant include guards and a set that did not. Table 8.1 shows the averaged results of this experiment, using the GNU C++ compiler, version 4.2.1, on an Intel Core 2 Duo processor running macOS 10.6 with 2 GB of RAM.

This behavior will, of course, vary by compiler and platform, so in the interest of good experimental technique, I repeated the experiment with the Microsoft C++ compiler,

Table 8.1 Compilation time speedup from using redundant include guards for a worst-case include hierarchy containing *N* include files.

N	Without guards (s)	With guards (s)	Speedup
2	0.07	0.07	1.00 ×
4	0.15	0.14	1.07 ×
8	0.35	0.31	1.13 ×
16	0.98	0.76	1.29 ×
32	4.07	2.12	1.92 ×
64	25.90	6.82	3.80 ×
128	226.83	24.70	9.18 ×

version 14.0, on an Intel Core 2 Quad CPU running Windows XP with 3.25 GB of RAM. The results in this case were even more pronounced, with a speedup of around 18 times for the $N = 128$ case.

TIP: Consider adding redundant #include guards to your headers to optimize compile time for your clients.

By comparison, I found that the experiment showed almost no effect under Linux (1.03× speedup for $N = 128$), in which presumably the combination of the GNU compiler and the Linux disk cache produces a more efficient environment. However, the users of your API may be using a broad range of platforms, so this optimization could have a large impact on many of them. Even a speedup of only 1.29× could make a big difference to the amount of time they spend waiting for a build to finish. The code to run the experiment is included in the full source code package on the accompanying website for this book, so you can try it out for your own platforms.

Herb Sutter recommends against using redundant `#include` guards, partly because they break encapsulation (Sutter and Alexandrescu, 2004). I also do not support breaking encapsulation. You should never use this technique across libraries or to include headers that you do not maintain. Use this only within your own APIs. As with all optimizations, you should measure performance for your compilers first. Modern compilers are improving in this area, although I was still able to measure significant differences across platforms as of 2024.

This technique has been used to practical benefit by many large-scale APIs. To give a retro example, the Commodore Amiga platform used this technique to improve the performance of the AmigaOS APIs. For instance, here's what the top of the `intuition/screens.h` header file looked like for the Amiga in the early 1990s:

```
#ifndef INTUITION_SCREENS_H
#define INTUITION_SCREENS_H TRUE
/*
**   $Filename: intuition/screens.h $
**   $Release: 2.04 Includes, V37.4 $
**   $Revision: 36.36 $
**   $Date: 91/10/07 $
**
**   The Screen and NewScreen structures and attributes
**
**   (C) Copyright 1985-1999 Amiga, Inc.
**          All Rights Reserved
*/

#ifndef EXEC_TYPES_H
#include <exec/types.h>
#endif

#ifndef GRAPHICS_GFX_H
#include <graphics/gfx.h>
#endif

#ifndef GRAPHICS_CLIP_H
#include <graphics/clip.h>
#endif

#ifndef GRAPHICS_VIEW_H
#include <graphics/view.h>
#endif

#ifndef GRAPHICS_RASTPORT_H
#include <graphics/rastport.h>
#endif

#ifndef GRAPHICS_LAYERS_H
#include <graphics/layers.h>
#endif

#ifndef UTILITY_TAGITEM_H
#include <utility/tagitem.h>
#endif
```

Declaring constants

Often you want to define several public constants for your API. This is a great technique for avoiding the proliferation of hardcoded values throughout your client's code, such as maximum values or default strings. For example, you might declare several constants in the global scope of your header in this way:

```
const int MAX_NAME_LENGTH = 128;
const float LOG_2E = log2(2.71828183f);
const std::string LOG_FILENAME = "filename.log";
```

The issue to be aware of here is that only very simple constants for built-in types will be inlined by your C++ compiler. By default, any variable that you define in this way will cause your compiler to store space for the variable in every module that includes your header. In the previous case, this will happen for both the float and the string constant. If you declare a lot of constants, and your API headers are included in many `.cpp` files, then this can cause bloat in the client's `.o` object files and the final binary. The solution is to declare the constants as extern:

```
extern const int MAX_NAME_LENGTH;
extern const float LOG_2E;
extern const std::string LOG_FILENAME;
```

Then define the value of each constant in the accompanying `.cpp` file. In this way, the space for the variables is allocated only once. This has the additional benefit of hiding the actual constant values from the header file: they are implementation details, after all.

A better way to do this is if you can declare the constants within a class. Then you can declare them as static const (so they will not count toward the per-object memory size):

```
// myapi.h
class MyAPI
{
public:
    static const int MAX_NAME_LENGTH;
    static const int MAX_RECORDS;
    static const std::string LOG_FILENAME;
};
```

You can then define the value for these constants in the associated `.cpp` file:

```
// myapi.cpp
const int MyAPI::MAX_NAME_LENGTH = 128;
const int MyAPI::MAX_RECORDS = 65525;
const std::string MyAPI::LOG_FILENAME = "filename.log";
```

You may think that this is a trivial optimization and will save an insignificant amount of space. However, at Linden Lab, we decided to clean up all of these instances in the Second Life Viewer source code. The end effect was an appreciable 10% reduction in the size of the generated files.

If you're using a C++17 (or later) compiler, then you should instead take advantage of inline variables to solve this issue for you, allowing you simply to put your constants, and their values, in your header files.

Another option to avoid the bloat issue in certain cases is to use enums as an alternative to variables. Or you could also use getter methods to return the constant values, such as:

```
// myapi.h
class MyAPI
{
public:
    static int GetMaxNameLength();
    static int GetMaxRecords();
    static std::string GetLogFilename();
};
```

TIP: Declare global scope constants with extern, or declare constants in classes as static const. Then define the value of the constant in the .cpp file. This reduces the size of object files for modules that include your headers. Even better, hide these constants behind a function call or use C++17 inline variables.

The constexpr, consteval, and constinit keywords

The problem with most of the previous options is that the compiler can no longer evaluate the const expression at compile time, because the actual value is hidden in the `.cpp` file. For example, clients would not be able to do either of these:

```
char name1[MyAPI::MAX_NAME_LENGTH];        // illegal
char name2[MyAPI::GetMaxNameLength()];     // illegal
```

However, the `constexpr` keyword was added as of C++11 to allow for more aggressive compile-time optimizations. This can be used to indicate functions or variables that you know to be constant, so that the compiler can perform greater optimization. For example, consider this code:

```
int GetTableSize(int elems) { return elems * 2; }
double myTable[GetTableSize(2)];  // illegal in C++03
```

This is illegal according to the C++03 standard, because the compiler has no way of knowing that the value returned by `GetTableSize()` is a compile-time constant. However, under C++11, you can tell the compiler that this is in fact the case:

```
constexpr int GetTableSize(int elems) { return elems * 2; }
double myTable[GetTableSize(2)];  // legal in C++11
```

The `constexpr` keyword can also be applied to variables. However, marking the result of a function as a compile-time constant opens the door to letting us define constants using a function call while allowing clients to use the constant value at compile time. For example:

```
// myapi.h
class MyAPI
{
public:
    constexpr int GetMaxNameLength() { return 128; }
    constexpr int GetMaxRecords() { return 65525; }
    constexpr std::string GetLogFilename() { return "log.txt"; }
};
```

As I covered in the C++ Revisions chapter, C++20 added further const specifiers to the standard with the introduction of `consteval` and `constinit`. The `consteval` specifier allows functions to be used in constant expressions and requires the function to be evaluated at compile time. The `constinit` specifier ensures that a static or thread storage variable is initialized at compile time, to avoid static initialization problems at library load time. Here's a quick summary of the differences between each of these similar-sounding keywords:

- `const` declares that an object or variable is immutable (e.g., that a method does not change any data members or that a variable cannot be modified).
- `constexpr` can be used on functions and variables to produce a constant expression that can be evaluated at compile time or at run-time; `constexpr` functions and static data members are also implicitly `inline`.
- `consteval` can be used only on functions and forces compile-time evaluation of the function; `consteval` implies `inline`. It cannot be applied to destructors or allocation/deallocation functions, and you cannot use `consteval` and `constexpr` together.
- `constinit` can be used only on variables that can be initialized at compile time. You cannot use `constinit` and `constexpr` together.

Initialization lists

C++ provides constructor initialization lists to let you easily initialize all member variables in your class. Using this feature can afford a slight performance increase over simply initializing each member variable in the body of the constructor. For example, instead of writing:

```
// avatar.h
class Avatar
{
public:
    Avatar(const std::string &first, const std::string &last)
    {
        mFirstName = first;
        mLastName = last;
    }

private:
    std::string mFirstName;
    std::string mLastName;
};
```

you could write:

```
// avatar.h
class Avatar
{
public:
    Avatar(const std::string &first, const std::string &last) :
        mFirstName(first),
        mLastName(last)
    {
    }

private:
    std::string mFirstName;
    std::string mLastName;
};
```

Member variables are constructed before the body of the constructor is called, so in the first example the default `std::string` constructor will be called to initialize the `two` member variables, and then inside the constructor the assignment operator is called (DeLoura, 2001). However, in the second example, only the assignment operator is invoked. Using an initialization list therefore avoids the cost of calling the default constructor for each member variable that you include in the list.

Of course, in terms of good API design, you should hide as much implementation detail from your header files as possible. The best approach is therefore to define the class in your header file as:

```
// avatar.h
class Avatar
{
public:
    Avatar(const std::string &first, const std::string &last);

private:
    std::string mFirstName;
    std::string mLastName;
};
```

Then provide the constructor and initialization list in the associated `.cpp` file:

```
// avatar.cpp
Avatar::Avatar(const std::string &first, const std::string &last) :
    mFirstName(first),
    mLastName(last)
{
}
```

> *TIP: Use constructor initialization lists to avoid the cost of a constructor call for each data member, but declare these in the .cpp file to hide implementation details.*

Here are a few things to be aware of when using initialization lists:

1. The order of variables in the initialization list must match the order in which the variables are specified in the class.
2. You can't specify arrays in an initialization list. However, you can specify an `std::vector`, which may be a better data structure choice anyway.
3. If you're declaring a derived class, the default constructor for any base classes will be called implicitly. You can use the initialization list to call a nondefault constructor instead. If specified, a call to a base class constructor must appear before any member variables.
4. If you've declared any of your member variables as references or as const, then you must initialize them via the initialization list (to avoid the default constructor defining their initial, and only, value).

In C++03, constructors cannot call other constructors. However, this constraint was loosened in C++11, allowing constructors to call other constructors in the same class. This lets you avoid copying code between constructors by delegating the implementation of one constructor to another, such as in this example:

```
class MyClass
{
public:
    MyClass(int answer) :
        mAnswer(answer)
    {}
    MyClass() :
        MyClass(42)   // legal in C++11
    {}

private:
    int mAnswer;
};
```

Although the same effect could be achieved by using a single constructor with a default argument, that would bake the default value into your client's code. The C++11 syntax lets you hide this value because you can (and should) define the initialization list in the `.cpp` file.

Memory optimization

On modern CPUs, memory latency can be one of the largest performance concerns for a large application. That's because while processor speeds have been improving at a rate

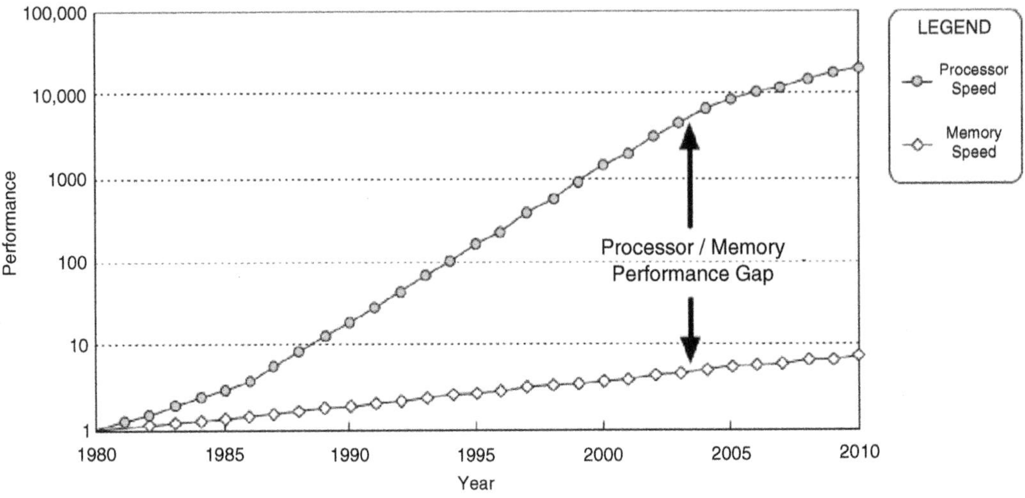

FIGURE 8.1 The widening gap in CPU performance improvements versus memory improvements. The vertical axis is on a logarithmic scale. *Adapted from Hennessy, J.L. and Patterson, D.A., 2006. Computer Architecture: A Quantitative Approach, Fourth ed. Morgan Kaufmann, ISBN 0123704901. Copyright Morgan Kaufmann Publishers.*

of roughly 55% per year, access times for DRAM have been improving at around 7% per year (Hennessy and Patterson, 2006). This has resulted in the so-called processor–memory performance gap, as shown in Fig. 8.1.

Because of this trend, memory overhead is now a principal factor in the execution time of most programs. This is further exacerbated by the fact that the cost for a cache miss (i.e., the cost to access main memory) has increased from a few CPU cycles 30 years ago to over 400 cycles on modern architectures. The effect of this is that a seemingly elegant and demonstrably correct algorithm can behave poorly in real-world situations because of unanticipated cache behavior (Albrecht, 2009). As a result, cache-miss optimization has become an extremely important element of performance optimization activities in recent years.

Although it's not the focus of this book to provide techniques to optimize your implementation details in the presence of caches, there are some API-related efforts that you can undertake to improve data cache efficiency. One key technique is to reduce the size of your objects: the smaller your objects are, the more of them can potentially fit into a cache. There are several ways that you can reduce the size of your objects:

1. **Cluster member variables by type.** Modern computers access memory a single word at a time. Your C++ compiler will therefore align certain data members so that their memory addresses fall on word boundaries. A number of unused padding bytes may be added to a structure to make this happen. By clustering all member variables of the same type next to each other, you can minimize the amount of memory lost to these padding bytes. Table 8.2 provides example alignment figures for member variable on the Windows platform.
2. **Use bit fields.** A bit field is a decorator for a member variable that specifies how many bits the variable should occupy (e.g., `int tinyInt:4`). This is particularly useful

Table 8.2 Typical alignment of member variables of different types under Windows on x86 CPUs.

Type	Size (bytes)	Alignment (bytes)
bool	1	1
char	1	1
short int	2	2
int	4	4
float	4	4
double	8	8
pointer/reference (32-bit)	4	4
pointer/reference (64-bit)	8	8

These sizes may vary by platform and processor.

for packing several bools into a single byte or squeezing two or more numbers into the space of a single int. The downside is that there's normally a performance penalty for using bit field sizes that are not a multiple of 8, but if memory is your biggest concern, then this may be an acceptable cost. When implementing performance optimizations, you often have to trade speed for size, or vice versa. Remember, when in doubt about the impact of a feature, measure the real-world performance.

3. **Use unions.** A union (or `std::variant` in C++17) is a structure in which data members share the same memory space. This can be used to allow multiple values that are never used at the same time to share the same area of memory, thus saving memory. The size of a union is the size of the largest type in the union. For example:

```
union
{
    float floatValue;
    int intValue;
} FloatOrIntValue;
```

4. **Don't add virtual methods until you need them.** I recommended this as a way to keep an API minimally complete back in Chapter 2, but there are also performance reasons to do this. Once you add one virtual method to a class, that class needs to have a vtable. Only one copy of the vtable needs to be allocated per class type, but a pointer to the vtable is stored in every instance of your object. This adds the size of one pointer to your overall object size (normally 4-bytes for a 32-bit application or 8-bytes for a 64-bit application).
5. **Use explicit size-based types.** The size of various types can differ by platform, compiler, and whether you're building on a 32-bit or a 64-bit architecture. If you want to specify the exact size of a member variable, then you should use a type

that specifically enforces this, rather than assuming that types such as `bool`, `short`, or `int` will be a specific size. As of C++11, you can use the `cstdint` header to access various fixed-width integer types, such as `int8_t`, `uint32_t`, and `int64_t`.

Let's look at an example. The following structure defines a collection of variables to describe a fireworks effect. It contains information about the color and color variance of the firework particles, some flags such as whether the effect is currently active, and a screen location for the effect to begin. A real fireworks class would have a lot more state, but this is sufficient for illustration purposes:

```
class Fireworks_A
{
public:
    bool mIsActive;
    int mOriginX;
    int mOriginY;
    bool mVaryColor;
    char mRed;
    int mRedVariance;
    char mGreen;
    int mGreenVariance;
    char mBlue;
    int mBlueVariance;
    bool mRepeatCycle;
    int mTotalParticles;
    bool mFadeParticles;
};
```

The variables in this class are ordered roughly in terms of their logical function, without any consideration to how efficiently they're packed in terms of memory. Most member variables are ordered in this way, or sometimes even more randomly simply by adding new variables to the end of the class. For this particular example, the total size for the structure on a 32-bit computer is 48 bytes: that is, `sizeof(Fireworks_A)` == 48.

If you simply cluster the member variables based upon their type and sort them based upon the size of each type (bools, chars, then ints), then the size of the structure can be reduced to 32 bytes, a 33% reduction:

```
class Fireworks_B
{
public:
    bool mIsActive;
    bool mVaryColor;
    bool mRepeatCycle;
    bool mFadeParticles;
    char mRed;
    char mGreen;
    char mBlue;
    int mRedVariance;
    int mGreenVariance;
    int mBlueVariance;
    int mTotalParticles;
    int mOriginX;
    int mOriginY;
};
```

TIP: Cluster member variables by their type to optimize object size.

You can still squeeze a few more bytes out of the structure, however. By using bit fields you can make each of the bool flags occupy a single bit instead of an entire byte. Doing this lets you get the structure size down to 28 bytes, a 42% reduction:

```
class Fireworks_C
{
public:
    bool mIsActive : 1;
    bool mVaryColor : 1;
    bool mRepeatCycle : 1;
    bool mFadeParticles : 1;
    char mRed;
    char mGreen;
    char mBlue;
    int mRedVariance;
    int mGreenVariance;
    int mBlueVariance;
    int mTotalParticles;
    int mOriginX;
    int mOriginY;
};
```

TIP: Consider the use of bit fields to compress an object further, but be aware of their performance impact.

Finally, you can step back and consider the actual size requirements for each variable rather than simply using ints for all integer values. Doing so, you might decide that the RGB variance values need to be only 1 byte, and the screen space coordinates can be 2 bytes each. I'll use the types of `char` and `short`, respectively, to simplify the subsequent figure, but in reality you could use size-specific types such as `int8_t` and `uint16_t`:

```
class Fireworks_D
{
public:
    bool mIsActive : 1;
    bool mVaryColor : 1;
    bool mRepeatCycle : 1;
    bool mFadeParticles : 1;
    char mRed;
    char mGreen;
    char mBlue;
    char mRedVariance;
    char mGreenVariance;
    char mBlueVariance;
    int mTotalParticles;
    short mOriginX;
    short mOriginY;
};
```

TIP: Use size-specific types, such as `int32_t` or `uint16_t`, to specify the maximum number of required bits for a variable.

This version of our structure occupies only 16 bytes, a reduction of 66% from the original unoptimized size of 48 bytes. That's quite a massive memory savings, just by rearranging the member variables and thinking a little more about how large they need to be. Fig. 8.2 shows you exactly where all of your memory is going in each of these four configurations.

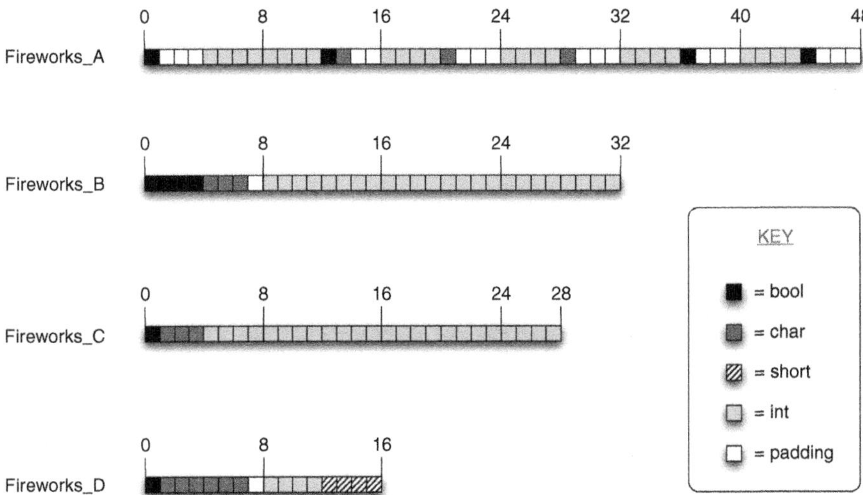

FIGURE 8.2 The memory layout for four different configurations of the member variables in a class. Fireworks_A is the original unoptimized version, Fireworks_B employs type clustering, Fireworks_C uses bit fields to compress the bool variables, and Fireworks_D uses smaller integer types.

Don't inline functions until you need to

This may seem like strange advice to give in a chapter on performance: don't inline code! However, this is first and foremost a book about APIs, and inlining code in your header files breaks one of the cardinal rules of API design: don't expose implementation details. That's not to say that you should never inline code. Sometimes your performance requirements demand it, but you should do so with your eyes wide open and with a full understanding of the implications, such as:

1. **Exposing implementation details.** As I've just covered, the primary reason for avoiding inlining in public API headers is that it causes you to expose the

implementation of your API methods directly in the header. I spent an entire section in Chapter 2 detailing why you shouldn't do that.

2. **Code embedded in client applications.** Inlined code in your API headers is compiled directly into your clients' applications. This means that clients must recompile their code whenever you release a new version of the API with any changes to inlined code. They cannot simply drop a new version of your shared library into their installation and expect their application just to work. In other words, inlining breaks binary compatibility.
3. **Code bloat.** Excessive inlining can significantly grow the size of your object files and resulting binary. This is, of course, because each call to an inlined method is replaced by all of the operations of that method. This larger code size can end up reducing performance by causing more disk access and virtual-memory page faults.
4. **Debugging complications.** Many debuggers have problems dealing with inlined code. This is perfectly understandable: it's difficult to put a breakpoint in a function that doesn't actually exist! The common way to circumvent these problems is to turn off inlining for debug code.

As Donald Knuth famously stated: "Premature optimization is the root of all evil" (Knuth, 1974).

Despite these downsides, there still may be cases in which you need to put inlined code into your API's public headers. The two main reasons for doing this are:

1. **Performance.** Using getter and setter methods to wrap access to member variables can cause a performance impact on your code if those methods are called many times a second. Inlining can recover those performance loses while allowing you to keep the getter/setter methods (the accompanying source code for this book has a simple program to let you test this for yourself). However, marking a function as inline may not necessarily give you performance gains. For one, this is simply a hint to the compiler that can be ignored. Some situations in which the request is likely to be ignored are using loops in the function, calling another inline function, or recursion. Even when the compiler does inline the method, the resulting code could be larger or smaller, and it could be faster or slower, depending upon the original size of the method, your CPU's instruction cache, and your virtual memory system (Cline et al., 1998). Inlining tends to work best for small, simple, frequently called functions.
2. **Templates.** You may also be using templates in your header and so are forced to inline the template implementation. However, as I covered in the previous chapter on C++ Usage, you can sometimes use explicit template instantiation to avoid this.

TIP: Avoid using inlined code in your public headers until you have proven that your code is causing a performance problem and confirmed that inlining will fix that problem.

For those cases in which you need to use inlined code, I will discuss the best way to do it. One way to inline code is simply to include the implementation of a method in the class body, such as:

```
class Vector
{
public:
    double GetX() const { return mX; }
    double GetY() const { return mY; }
    double GetZ() const { return mZ; }
    void SetX(double x) { mX = x; }
    void SetY(double y) { mY = y; }
    void SetZ(double z) { mZ = z; }

private:
    double mX, mY, mZ;
};
```

This perfectly demonstrates the concern of including implementation details in your header files: a user of your API can look at your header and see exactly how an inlined method is implemented. In this example, the implementation is simple, but it could easily expose a lot more complexity.

Another way to inline code is by using the C++ `inline` keyword. This approach offers at least one improvement over the previous syntax in that it lets you define the code for a method outside the class body. Although the code is still in the header, at least you don't obfuscate the class declaration with code:

```
class Vector
{
public:
    double GetX() const;
    double GetY() const;
    double GetZ() const;
    void SetX(double x);
    void SetY(double y);
    void SetZ(double z);

private:
    double mX, mY, mZ;
};

inline void   Vector::SetX(double x) { mX = x; }
inline void   Vector::SetY(double y) { mY = y; }
inline void   Vector::SetZ(double z) { mZ = z; }
inline double Vector::GetX() const { return mX; }
inline double Vector::GetY() const { return mY; }
inline double Vector::GetZ() const { return mZ; }
```

An even better style would be to hide the inline statements in a separate header, where the filename of that header indicates that it contains implementation details. This is the same technique that I suggested earlier for dealing with templates, and it's used by several industry-strength APIs such as the Boost headers. Boost uses the convention of a "detail" subdirectory to hold all private details that must be exposed in header files and then `#include` those from the public header files. For example:

```
class Vector
{
public:
    double GetX() const;
    double GetY() const;
    double GetZ() const;
    void SetX(double x);
    void SetY(double y);
    void SetZ(double z);

private:
    double mX, mY, mZ;
};

#include "detail/Vector.h"
```

The Boost headers also often use the convention of using a "detail" subnamespace to contain all private implementation code, such as `boost::tuples::detail`. This is a good practice to segment necessary private code further in public headers.

Finally, I'll make the point that this section was about avoiding inlining functions. With C++17, you can also inline variable definitions with external linkage. This is a great feature that you should use where possible to define constants in your header files.

Copy on write

One of the best ways to save memory is not to allocate any until you need to. This is the essential goal of copy-on-write (COW) techniques. These work by allowing all clients to share a single resource until one of them needs to modify the resource. Only at that point is a copy made. Hence the name: copy on write. The advantage is that if the resource is never modified, then it can be shared for all clients. This is related to the Flyweight design pattern, which describes objects that share as much memory as possible to minimize memory consumption (Gamma et al., 1994).

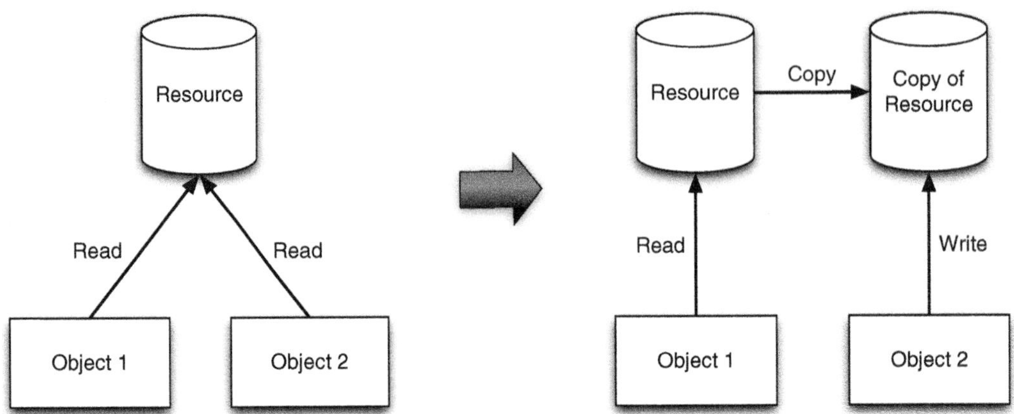

FIGURE 8.3 Illustrating copy on write: Object1 and Object2 can share the same resource until Object2 wants to change it, at which point a copy must be made so that the changes do not affect Object1's state.

> *TIP: Use copy-on-write semantics to reduce the memory cost for many copies of your objects.*

For example, several string objects that store the same text could all share the same memory buffer. Then when one of the strings must modify the text, it creates a copy of the memory buffer so that the modification doesn't affect the other strings. Fig. 8.3 illustrates this concept. Most C++ Standard Library string implementations use copy on write so that passing them by value is fairly cheap (Stroustrup, 2000).

There are several ways to implement copy on write. One popular way is to declare a class template that lets you create pointers to objects managed with copy-on-write semantics, in the same way that you would create a shared or weak pointer template. The class normally contains a standard shared pointer, to track the reference count of the underlying object, and provides a private `Detach()` method for operations that modify the object and hence require detaching from the shared object and creating a new copy. An implementation of this is shown next using shared pointers. I show this with inline methods within the class declaration for reasons of clarity. In real-world applications, you should, of course, hide these inline definitions in a separate header as they obscure the interface declarations:

```
#include <memory>

template <class T>
class CowPtr
{
public:
    using RefPtr = std::shared_ptr<T>;

    inline CowPtr() :
        mPtr(0)
    {}
    inline ~CowPtr() {}
    inline explicit CowPtr(T *other) :
        mPtr(other)
    {}
    inline CowPtr(const CowPtr<T> &other) :
        mPtr(other.mPtr)
    {}

    inline T &operator*()
    {
        Detach();
        return *mPtr.get();
    }
    inline const T &operator*() const
    {
        return *mPtr.get();
    }
    inline T *operator->()
    {
        Detach();
        return mPtr.get();
    }
    inline const T *operator->() const
    {
        return mPtr.get();
    }
    inline operator T *()
    {
        Detach();
        return mPtr.get();
    }
    inline operator const T *() const
    {
        return mPtr.get();
    }
    inline T *data()
    {
        Detach();
        return mPtr.get();
    }
    inline const T *data() const
    {
        return mPtr.get();
    }
    inline const T *constData() const
    {
        return mPtr.get();
    }
    inline bool operator==(const CowPtr<T> &other) const
    {
        return mPtr.get() == other.mPtr.get();
```

```
    }
    inline bool operator!=(const CowPtr<T> &other) const
    {
        return mPtr.get() != other.mPtr.get();
    }
    inline bool operator!() const
    {
        return !mPtr.get();
    }
    inline CowPtr<T> &operator=(const CowPtr<T> &other)
    {
        if (other.mPtr != mPtr) {
            mPtr = other.mPtr;
        }
        return *this;
    }
    inline CowPtr &operator=(T *other)
    {
        mPtr = RefPtr(other);
        return *this;
    }

private:
    inline void Detach()
    {
        T *temp = mPtr.get();
        if (temp && !mPtr.unique()) {
            mPtr = RefPtr(new T(*temp));
        }
    }

    RefPtr mPtr;
};
```

This class can then be used as:

```
CowPtr<std::string> string1(new std::string("Share Me"));
CowPtr<std::string> string2(string1);
CowPtr<std::string> string3(string1);
string3->append("!");
```

In this example, `string2` points to the same object as `string1`, whereas `string3` points to a copy of the object because it needed to modify it. As I've mentioned, many implementations of `std::string` will use copy-on-write semantics anyway. I'm simply using this as a convenient example.

There's a loophole that can be exploited in the previous `CowPtr` implementation. It's possible for users to dig into the copy-on-write pointer and access the underlying object to hold on to references to its data. They could then modify the data directly, thus affecting all `CowPtr` variables that share that object. For example:

```
CowPtr<std::string> string1(new std::string("Share Me"));
char &char_ref = string1->operator[](1);
CowPtr<std::string> string2(string1);
char_ref = 'p';
```

In this code, the user takes a reference to a character in the underlying `std::string` of `string1`. After `string2` is created, which shares the same memory as `string1`, the user then directly changes the second character in the shared string, causing both `string1` and `string2` now to equal "Spare Me".

The best way to avoid this sort of misuse is simply not to expose `CowPtr` to your clients. It most cases, you don't need your clients to know you are using a copy-on-write optimization: it's an implementation detail, after all. Instead, you could use `CowPtr` to declare member variables in your objects and not change your public API in any way. This is called implicit sharing by the Qt library. For example:

```
// myobject.h
class MyObject
{
public:
    MyObject();

    std::string GetValue() const;
    void SetValue(const std::string &value);

private:
    CowPtr<std::string> mData;
};
```

where the implementation of `MyObject` might look like:

```
// myobject.cpp
MyObject::MyObject() :
    mData(0)
{}

std::string MyObject::GetValue() const
{
    return (mData) ? *mData : "";
}

void MyObject::SetValue(const std::string &value)
{
    mData = new std::string(value);
}
```

In this way, your clients can use your `MyObject` API without any knowledge that it uses copy on write, but underneath the covers the object is sharing memory whenever possible and enabling more efficient copy and assignment operations:

```
MyObject obj1;
obj1.SetValue("Hello");

MyObject obj2 = obj1;
std::string val = obj2.GetValue();

MyObject obj3 = obj1;
obj3.SetValue("There");
```

In this example, `obj1` and `obj2` will share the same underlying string object, whereas `obj3` will contain its own copy because it modified the string.

Iterating over elements

Iterating over a collection of objects is an extremely common task for client code, so it's worth spending some time looking at alternative strategies that offer different strengths and weaknesses. That way you can choose the best solution for your API requirements.

Iterators

The C++ Standard Library approach to this problem is to use iterators. These are objects that can traverse over some or all elements in a container class (Josuttis, 1999). An iterator points to a single element in a container, with various operators available, such as `operator*` to return the current element, `operator->` to access the members of the container element directly, and `operator++` to step forward to the next element. This design intentionally mimics the interface of plain pointer manipulation in C/C++.

Clients can then use the `begin()` and `end()` methods on each container class to return iterators that bound all elements in the container, or they can use various Standard Library algorithms that return iterators within the set of all elements, such as `std::find()`, `std::lower_bound()`, and `std::upper_bound()`. The next code segment provides a simple example of using a Standard Library iterator to sum all of the values in an `std::vector`:

```
float sum = 0.0f;
std::vector<float>::const_iterator it;
for (it = values.begin(); it != values.end(); ++it) {
    sum += *it;
}
```

This is purely an illustrative example. If you really wanted to calculate the sum of all elements in a container, you should prefer the use of the Standard Library algorithm `std::accumulate`, as described in the Functional APIs section of the Styles chapter.

In terms of your own API designs, here are some reasons why you may want to adopt an iterator model to allow your clients to iterate over data:

- Iterators are a well-known pattern with which most engineers are already familiar. As such, using an iterator model in your own APIs will minimize the learning curve for users. This addresses the ease of use quality I introduced in Chapter 2. For example, most engineers will already be aware of any performance issues, such as knowing that they should prefer the preincrement operator for iterators (`++it`) as opposed to postincrement (`it++`), to avoid the construction and destruction of temporary variables.
- The iterator abstraction can be applied to simple sequential data structures, such as arrays or lists, as well as more complicated data structures, such as sets and maps, which are often implemented as self-balancing binary search trees such as red-black trees (Josuttis, 1999).
- Iterators can be implemented quite efficiently, even as simply as a pointer in some cases. In fact, `std::vector` iterators were actually implemented this way in Visual C++ 6 and GNU C++ 3 (although most modern Standard Library implementations now use dedicated iterator classes).
- Iterators can be used to traverse massive datasets that may not even fit entirely into memory. For example, the iterator could be implemented to page in blocks of data from disk as needed, and free previously processed blocks. Of course, the client can also stop traversing at any point without having to visit every element in the container.
- Clients can create multiple iterators to traverse the same data and use these iterators simultaneously. If clients wish to insert or delete elements while traversing the container, there are established patterns for doing this while maintaining the integrity of the iterators.

Random access

An iterator allows clients to traverse linearly through each element in a container. However, you may have cases where you wish to support random access to any element, such as accessing a specific element in an array or vector container. C++ Standard Library container classes that support random accesses provide this in a couple of ways:

1. **The `[]` operator.** This is meant to simulate the array indexing syntax of C/C++. Normally this operator is implemented without any bounds checking, so it can be made very efficient.
2. **The `at()` method.** This method is required to check if the supplied index is out of range and throw an exception in this case. As a result, this approach can be slower than the `[]` operator.

To illustrate these concepts, the iterator source example in the previous section can be recast in terms of the `[]` operator as:

```
float sum = 0.0f;
size_t len = values.size();
for (size_t it = 0; it < len; ++it) {
    sum += values[it];
}
```

In terms of performance, these two methods are essentially equivalent. Obviously one approach may prove to be marginally more efficient than the other for a given platform and compiler, but in general they should involve an equivalent degree of overhead.

If you plan to add random access functionality to your API, you should strive to adopt the previous design to capitalize on consistency with the Standard Library. However, if your API doesn't need to provide random access to the underlying data, you should prefer using the iterator model over the `operator[]` approach, simply because an iterator expresses the user's intent more clearly and results in client code that is more obvious and consistent.

Array references

As an alternative to iterators, some APIs use an approach in which the user passes in an array data structure by reference. The API then fills the array with the requested elements and returns it to the user. The Maya API uses this pattern extensively. Autodesk Maya is a high-end 3D modeling and animation system used extensively in the film and game industry. The package includes a C++ and Python API that provides programmatic access to the underlying 2D and 3D data in a scene.

As an example of this pattern, the `MfnDagNode::getAllPaths()` method is used to return a sequence of node paths in the Maya scene graph. This is achieved by passing in an `MDagPathArray` object by reference, which is then populated with `MDagPath` references. Some reasons for this design, and hence the reasons why you may prefer this approach for your own APIs, are:

- The primary purpose of this method is performance. In essence, it's a way to collapse a series of connected nodes of a graph data structure into a sequential array data structure. This provides a data structure that can be very efficiently iterated over, but also locates elements adjacent to each other in memory. The result is a data structure that can take better advantage of CPU caching strategies, as opposed to a tree structure in which individual nodes in the tree may be fragmented across the process's address space.
- This technique is particularly efficient if the client keeps the same array around to service multiple calls to `getAllPaths()`. Also, any initial performance overhead to fill the array can be compensated for if the array is kept around to support multiple iterations over its elements.
- This technique also offers a specific feature that the iterator model does not: support for noncontiguous elements. That is, a traditional iterator cannot handle different orderings of elements or omit certain elements from a sequence, whereas using the array reference technique, you can fill the array with any subset of elements in any order.

This concept can be seen in other languages, too, such as the `iterator_to_array()` function in PHP, which can be used to convert an iterator into an array for faster traversal in certain cases.

As an alternative to consuming a user supplied array, you could also return a const container of objects and rely on the compiler's return value optimization to avoid copying data.

> *TIP: Adopt an iterator model for traversing simple linear data structures. If you have a linked list or tree data structure, then consider using array references if iteration performance is critical.*

Extern templates

One issue with the use of templates in C++03 is that compilers must generate code in each translation unit where a template is instantiated. Most modern compilers will discard any duplicate code when assembling the final binary at link time, but this can still cause increases in compilation time and intermediate object code size.

C++11 introduced the concept of extern templates to improve this situation. This lets you define a template without instantiating it in a translation unit. Ultimately, this feature does not affect your public interface, but it can help with compilation and link performance for your clients.

For example, consider this class template defined in a `header.h` file:

```
// header.h

template <class T>
class Range
{
public:
    Range(T min, T max);
    T GetMin() const;
    T GetMax() const;

private:
    T mMin;
    T mMax;
};
```

This template can be instantiated in multiple translation units. For example, these two modules both instantiate a specialization of the template for type `int`:

```
// module1.cpp

#include "header.h"

void Module1Function()
{
    Range<int> v1(1, 2);
    ...
}
```

```
// module2.cpp

#include "header.h"

void Module2Function()
{
    Range<int> v1(2, 4);
    ...
}
```

In this case, the compiler will generate code for the template instantiation in both `module1.o` and `module2.o` object files. To avoid this, you can prevent the template instantiation in one of the modules by using an extern template. For example,

```
// module2.cpp

#include "header.h"

// tell the compiler to not instantiate the template in this module
extern template class Range<int>;

void Module2Function()
{
    Range<int> v1(1, 2);
    ...
}
```

With this change, the compiler will no longer generate code for the template instantiation in `module2.o`. The resulting library or executable binary will be the same in each case, but the compiler can do less work and produce smaller intermediate products.

When using extern templates, it's incumbent on you to ensure that the template is instantiated in at least one translation unit; otherwise you'll get an undefined symbol linker error.

Performance analysis

In the final section in this chapter on performance, I'll look at some tools and techniques to help you measure the performance of your system. Most of these are aimed at analyzing the performance of your implementation code and as such are not directly related to how you design your API. However, this is obviously still a very important part of producing an efficient API and therefore worthy of focus.

I'll consider several different aspects of performance: time-based performance, memory overhead, and multithreading contention. Also, whereas the preceding sections in this chapter dealt with stable features of C++, the subsequent text presents software products that may change over time. Products come, go, change ownership, and change focus. However, I have endeavored to make this list as up-to-date as possible at the time of publication.

Time-based analysis

The most traditional interpretation of performance is how long it takes your code to perform various operations. For example, if you are developing an image processing library, how long does your `Sharpen()` or `RedEyeReduction()` method take to run on a given image? The implication here is that you must write some sample or test programs that use your API so that you can then time your API's performance under different real-world scenarios. If you have written such programs, there are several forms of performance analysis you could consider using:

1. **In-house instrumentation.** The most targeted and efficient profiling you can perform is the kind that you write yourself. Every piece of software is different, and so the performance-critical sections of your code will be specific to your API. It's therefore extremely beneficial to have access to a fast timer class that can be inserted into your code at key points to gather accurate timing information. The results can be output to a file and analyzed offline, or your clients could integrate a visual display of the timer results into their end-user applications.

   The Second Life Viewer provides this capability via its `LLFastTimer` class. This works by inserting `LLFastTimer()` calls into critical sections of the code using an extensible label that identifies the area being analyzed, such as `LLFastTimer(RENDER_WATER)`. The Second Life Viewer itself then provides a debugging overlay display to view the cumulative result of the timers in real time. See Fig. 8.4 for an example of this debugging view.

2. **Binary instrumentation.** This technique involves instrumenting a program or shared library by adding code that records details for each function call. Running the instrumented binary then creates an exact trace of the function calls for that session. Processing the resulting data can then determine the top call stacks where the program spent most of its time.

   One drawback of this approach is that the extra instrumentation overhead can slow the program execution significantly, sometimes by as much as 10-100 times, although relative performance should be preserved. Finally, this technique will obviously not be able to time functions that do not appear as symbols in the binary file, such as inline functions.

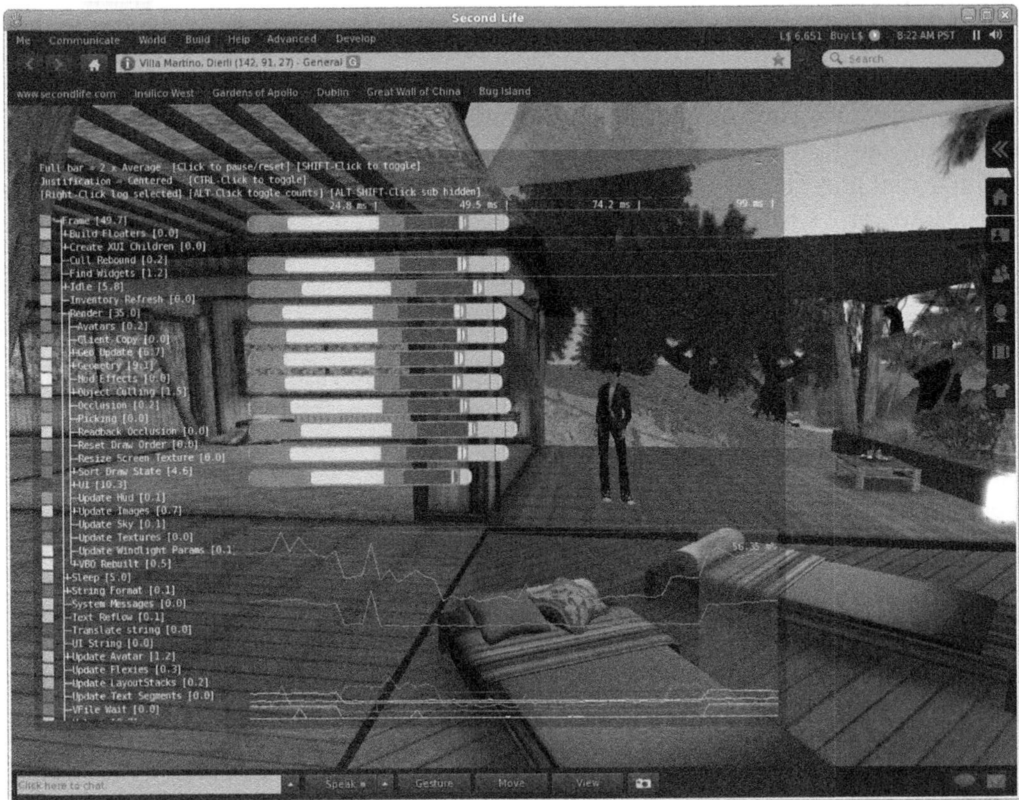

FIGURE 8.4 Screenshot of the Second Life Viewer showing its built-in view to display the results of various timers embedded in the code.

3. **Sampling.** This involves the use of a separate program that continually samples your test application to determine its program counter. This is a low-overhead statistical technique, meaning that it may not log every single function call made by your application, but with a sufficiently high sample rate, it can still be useful in telling you where your application is spending most of its time.

   Sampling can be performed at a system level (e.g., to see if your application spends a lot of time in system calls, such as whether it is I/O bound), or it can be isolated to the functions in your application. In addition to recording time samples, this technique can refer to sampling processor events, such as cache misses, mispredicted branches, and CPU stalls.

4. **Counter monitoring.** Many commercially available operating systems provide performance counters that report how well various subsystems are performing, such as the processor, memory, network, and disk. For example, Microsoft provides the Performance Counters API to access counter data on the Windows platform. By monitoring these counters while your application is running you can determine system bottlenecks and evaluate how inadequate system resources can affect your API's performance.

Given this categorization of performance analyzing techniques, the next list provides a cross-section of profiling tools that were on the market at the time of this book's publication:

- **Intel VTune**: This is a commercial performance analysis suite available for both Microsoft Windows and Linux platforms. It includes a binary instrumentation feature (called call graph), time- and event-based sampling, and a counter monitor, among various other tools. It comes with extensive and powerful graphical tools to visualize the resulting performance data.
- **gprof**: GProf is the GNU profiler. It uses binary instrumentation to record the number of calls, and time spent within each function. It is integrated with the GNU C++ compiler and activated via the -pg command line option. Running an instrumented binary creates a data file in the current directory that can be analyzed with the gprof program (or the Saturn application on macOS).
- **OProfile**: This is an open source performance tool for Linux. It is a system-wide sampling profiler that can also leverage hardware performance counters. Profile data can be produced at the function or instruction level and the software includes support for annotating source trees with profile information.
- **AMD CodeXL**: A freely available profiler from AMD that runs on Windows and Linux. It is based upon OProfile with specific support for analyzing and visualizing the pipeline stages of AMD processors.
- **Open SpeedShop**: An open source performance measurement tool for Linux based upon SGI's IRIX SpeedShop and currently supported by the Krell Institute. Open SpeedShop uses a sampling technique with support for hardware performance counters. It provides support for parallel and multithreaded programs and also includes a Python scripting API.
- **Sysprof**: This is an open source performance profiler for Linux. It uses a system-wide sampling technique to profile the entire Linux system while your application is running. A simple user interface is provided to browse the resulting data.
- **Callgrind**: Callgrind is a part of the valgrind instrumentation framework for Linux and macOS. It uses a binary instrumentation technique to collect call graph and instruction data for a given program run. The separate KCachegrind tool can be used to visualize the profile data. An optional cache simulator can be used to profile memory access behavior.

- **Apple Shark**: Shark is a system-wide sampling profiler written by Apple and provided for free as part of their developer tools. It can also profile hardware and software events such as cache misses and virtual memory activity. Shark includes an intuitive and easy-to-use interface to browse the hot spots of your Apple applications.
- **DTrace**: DTrace is a unique and powerful tracing framework that can be used to monitor applications in real time. This is done by writing custom tracing programs that can define a list of probes and actions to be performed when a probe fires. Probes include opening a file, starting a process, and executing a specific line of code; actions can analyze the run-time context such as the call stack. Apple added DTrace to macOS 10.5 with an accompanying GUI called Instruments. It is also available as ktrace on FreeBSD.

Memory-based analysis

As I stated earlier in this chapter, memory performance can be just as important as time-based performance. Algorithms that allocate and deallocate memory frequently, or whose memory allocation profiles don't map well to modern processor caches, can end up performing a lot slower than expected. Also, memory bugs such as doubling freeing or accessing unallocated memory can corrupt data or cause crashes, and memory leaks can build up over time to the point that they consume all available memory and reduce the performance of your clients' applications to a crawl or cause them to crash.

These tools can be used to profile your API's memory performance and to detect memory bugs:

- **UNICOM Systems PurifyPlus**: This is a commercial memory debugger that uses binary instrumentation to detect memory access errors in C/C++ programs. After a program run, Purify outputs a report file that can be browsed via a graphical interface. It includes an API that you can access within your programs. Purify is available for Solaris, Linux, AIX, and Windows.
- **Valgrind**: Valgrind is an open source instrumentation framework for Linux and macOS that began life as a memory profiling tool. However, it has since matured into a more general performance analysis tool. It works by performing binary instrumentation of your executable file and outputs a textual report when your program ends. Several front-end GUIs are available to browse the output file, such as Valkyrie and Alleyoop.
- **Perforce TotalView MemoryScape**: A commercial memory analysis tool available for Unix and macOS platforms that works without binary instrumentation. It provides a real-time graphical view of your heap memory, including memory use, allocation bounds violations, and leaks. It handles parallel and multithreaded programs and also incorporates a scripting language to perform batch testing.
- **Parasoft Insure++**: A commercial memory debugger that is available for Windows, Linux, Solaris, AIX, and HP-UX. Insure++ performs instrumentation at

the source code level by prepending your compile line with the insure program. You can even set your debugger to stop whenever it detects an error, by adding a breakpoint in `__Insure_trap_error()`. Of course, there is a GUI tool to let you browse the detected memory errors.

- **Synopsys Coverity**: Coverity is a tool different from the others I have listed. It's a static analysis tool, which means that it checks your source code without executing your programs. It records all potential coding errors in a database using a unique ID for each error that is stable across multiple analysis runs. A Web interface is provided to view the results of the static analysis.

Multithreading analysis

The final aspect of performance that I'll cover here is multithreaded performance. Writing efficient multithreaded code is a very difficult task, but luckily there are various tools out there to help you find logical threading errors in your code, such as race conditions or deadlocks, as well as profile the performance of your threaded code to find concurrency bottlenecks:

- **Intel Inspector**: This is a commercial threading analysis tool available for 32-bit and 64-bit Windows and Linux. It can be used to discover logical threading errors, such as potential deadlocks. You can use it as a command-line tool that outputs a textual report, or you can use the accompanying visual GUI that maps potential errors to source code lines.
- **Intel VTune**: The thread profiler component of VTune lets you visualize your threaded application's behavior by displaying a time line that shows what your threads are doing and how they interact. This lets you determine if you're getting the maximum concurrency from your code. It runs on Windows and Linux.
- **Intel Parallel Studio XE**: Intel's Parallel Studio provides a suite of tools to support parallel applications on multicore systems, including a utility to identify candidate functions for parallelizing, the Intel Threading Building Blocks, an inspector tool to detect threading and memory errors, and a performance analysis tool for parallel applications.
- **ParaTools ThreadSpotter**: Lets you find performance problems in multithreaded and OpenMPI applications on Solaris, Linux, and Windows. It can identify areas where a program is using processor cache memory, and in some cases it can suggest ways to restructure the code to make more effective use of cache memory.
- **Helgrind and DRD**: Helgrind and DRD are both modules of the open source Valgrind instrumentation framework. They can be used to detect synchronization errors in pthreads-based applications, including misuses of the pthreads API, deadlocks, and race conditions. They can be used on Linux and macOS.

Concurrency

Concurrency is the act of breaking down a problem into independently executing elements and then composing those together to solve some larger problem. This is distinct from the concept of parallelism, which is concerned with executing multiple operations simultaneously. The difference is that a concurrent algorithm might run correctly on a single processor or 100 of them, whereas a parallel program is about employing multiple cores to execute different elements of a task at the same time.

Concurrency is often achieved through multithreading, in which a single process can have multiple threads of execution. Some of the complexities introduced by multithreaded programming include the need to communicate between threads, such as through message passing or shared memory, and managing access to shared resources, such as global data. Done incorrectly, your code can exhibit severe faults such as crashes, deadlocking, or resource starvation.

In this chapter, I'll cover some of the core multithreading features offered by C++. I'll then explain some of the related terminology, such as race conditions, thread safety, and reentrancy. I'll also look at ways to initialize and synchronize data so that your code is safe to use from multiple threads simultaneously. And finally, I'll introduce a collection of best practices for writing concurrent code and discuss the Thread-Safe Interface design pattern as one way to write thread-safe object-oriented code.

Because this is a book focused on the design of software interfaces, I'll not dive deeply into all of the aspects of how to write robust multithreaded code in C++. There are many other good books on the market with that focus (e.g., Williams, 2012). Instead, I'll look at how adopting a multithreading model can influence your application programming interface (API) design decisions. Having said that, this topic does require delving one level down into your implementation code because that's where many of the decisions are made that ultimately determine the thread safety of your API, such as mutex locking.

Multithreading with C++

The C++98 and C++03 standards did not specify how code should behave in the presence of multiple threads. As a result, it was not possible to write portable multithreaded code using compilers based on these early standards. Some of the problems that could be faced writing C++03 code include optimizing compilers that may reorder statements and multiple processor caches that might reorder shared memory writes. As a result, developers had to rely on external libraries, such as pthreads or Boost, to make their multithreaded programs work correctly.

API Design for C++. https://doi.org/10.1016/B978-0-443-22219-1.00007-6

Fortunately, the C++11 standard introduced direct support for multithreading, including a memory model that defines how a compiler can access memory. This involved the introduction of atomic load and store operations and memory access settings that specify how operations on different threads should be synchronized. The C++20 revision added some incremental improvements to this memory model, in particular for Power and ARM CPUs and NVIDIA GPUs.

> *TIP: Building a modern robust multithreaded library in C++ will generally require the use of C++11 or later.*

At the language level, the C++11 standard also enforced guarantees around the behavior of the `static` keyword. This is a typical way to implement shared data in C++, but it wasn't until C++11 that the standard required the initialization of a static object to be thread-safe. Now the standard states that only one thread can enter the initialization of a static and that the compiler must not introduce any deadlocks around its initialization.

Additionally, C++11 introduced the specific concept of a thread and various new primitives for working with threads, such as semaphores, mutex locks, and condition variables. The writers of the specification even added higher-level concepts in the form of asynchronous tasks that abstract away many of the details of managing threads.

The standard also introduced minimum requirements for C++ Standard Library implementations when it comes to thread safety. For example, data containers can't use a static object for internal purposes without synchronization. However, in general, if you expect any threads to write to Standard Library data containers, you should add your own synchronization mechanisms to keep them thread-safe.

Because of the large number of threading features that were added to C++11 (and later revisions), several different categories of solutions operate at different levels of abstraction. You may not need to worry about all of the different tiers for most projects. This list gives an overview of these levels of abstraction, from the lowest to the highest level:

1. **Memory model:** The standard introduced the `std::atomic` template, which can be used to access data atomically across multiple threads, along with the `std::memory_order` enum, which specifies how atomic operation memory accesses are ordered. It also enforced the thread safety of the `static` keyword and introduced per-thread static storage with the `thread_local` keyword.
2. **Threading model:** The `std::thread` class was introduced to represent a single thread of execution along with related synchronization primitives to help you

coordinate work across multiple threads such as `std::mutex`, `std::lock_guard`, `std::condition_variable`, and `std::init_once`.
3. **Task execution:** The `std::async` function template can be used to execute a function asynchronously without having to worry about the details of coordinating multiple threads. The result of an asynchronous function is returned in an `std::future`, which offers the ability to wait until the future has a valid result.
4. **Standard Library containers and algorithms:** The C++ standard defines the basic thread safety guarantees for Standard Library container classes, and as of C++17 the set of algorithms includes parallel variants that can take advantage of multiple processors, including functions such as `std::copy`, `std::reverse`, and `std::transform`.

Terminology

Before I dive in deeper, I'll explain some of the terms you may encounter when trying to build an interface that can take advantage of concurrency and parallelism. In the remaining sections of this chapter, we will look at how multithreaded programming can affect your API designs and how to build robust thread-safe and concurrent interfaces.

Data races and race conditions

A data race is when one thread accesses a memory location while another thread is writing to the same address, causing undefined or unpredictable behavior. This can happen when code tries to access some shared data without any synchronization. The standard way to fix these kinds of faults is therefore to add appropriate synchronization around the data accesses, or to use an operation that's guaranteed to execute atomically, so that only one thread can access the shared data at any time.

In contrast, a race condition is a semantic flaw that happens when the timing or order of events affects the correctness of a section of code. The classic example of a race condition is a bank account transfer operation, such as:

```
bool TransferBalance(const Amount &amount, Account *from, Account *to)
{
    if (from->GetBalance() < amount) {
        return false;
    }
    to->AddToBalance(amount);
    from->AddToBalance(-amount);
    return true;
}
```

In this example, it's possible for two threads to get past the initial balance check, but then both threads remove money from the account, potentially resulting in a negative balance. This is an example of a check-then-act race condition. The fix in this case is to make the entire function operate atomically by introducing synchronization around all of the code inside the function.

Race conditions and data races are very different types of concurrency errors, and neither is a subset of the other. Data races are concerned with access to the same memory address whereas race conditions are concerned with the correctness of blocks of code. Many race conditions can be caused by data races, but it's also possible to have race conditions in code that has no data races.

Thread safety

A thread-safe API is one in which its functions can be called by multiple threads simultaneously without producing unexpected behavior or faults (i.e., without exhibiting any data races or race conditions). Another way of describing this is an API that only accesses shared resources in a way that ensures safety and correctness when executed by multiple threads at the same time.

There's not generally just one solution to ensuring thread safety because achieving this will depend upon the specific characteristics of your code and the usage pattern of clients exercising your code. Race conditions in particular will depend upon the semantics of your code and whether certain blocks of code need to be executed atomically. It can therefore sometimes be better to think of three basic levels of guarantees that your code offers:

1. **Not Thread-Safe.** The API doesn't provide any specific correctness guarantees when accessed by multiple threads simultaneously. Most code written using C++98/C++03 should be considered not thread-safe unless it specifically employs a threading library such as pthreads or Boost threads.
2. **Partially Thread-Safe.** The API provides some guarantees of thread safety when used in specific ways. For example, most of the C++ Standard Library containers offer the guarantee that simultaneous reads of the same object are okay and simultaneous writes of different objects are okay. But simultaneous writes of the same object are not directly supported, requiring clients to implement their own synchronization to handle these cases. We'll see an example of this later in the chapter.
3. **Thread-Safe.** The API commits to being free of data races and race conditions when accessed by multiple threads at the same time. This requires a concerted effort on the part of the developer to ensure that all access to shared resources is done safely and that operations that need to be executed atomically are done so.

Even if your API offers full thread safety, clients of your interface could still use it in ways that introduce race conditions into their code. For example, consider this simple code fragment to pop an element from the end of a vector:

```
if (!vec.empty()) {
    vec.pop_back();
}
```

The `pop_back()` method of `std::vector` produces undefined behavior if the vector is empty (using the clang 14 compiler, it causes a crash). However, in the previous code fragment, it would be possible for two threads to pass the condition check, which would then cause undefined behavior if the vector had only one element and both threads tried to call `pop_back()`. Ultimately, the authors of this code fragment need to perform their own locking around this entire code block to ensure thread safety. This is why the Standard Library container classes aren't required to implement locking of their underlying data: because the users of the container class will often need to implement locking at the level of their own code anyway.

One possible way to address this at the API level might be for the container class to offer a `pop_back_if_empty()` function, in which the need to introduce locking can then be hidden entirely within the API implementation and not exposed to the client. However, this could mean adding many of these composite operations, which would quickly pollute the API and violate the principle of being minimally complete that we introduced in the Qualities chapter.

Reentrancy

An API call is considered reentrant if it can be safely called again before the previous invocation has finished: for example, if a function can be interrupted at any point during its execution and then safely reentered before that earlier invocation is done.

This sounds similar to the definition of thread safety, but the key distinction is that reentrancy is a term that comes from a time before the introduction of modern multitasking operating systems and therefore refers to the ability to reenter a function within a single-threaded environment: for example, if a system interrupt is triggered and the interrupt handler calls the reentrant function while it's already in the process of being executed.

A function that's reentrant can often also be thread-safe, but this is not always guaranteed: a reentrant function might not be thread-safe, and a thread-safe function might not be reentrant. For example, the use of mutex locking can make a function thread-safe in the presence of multiple threads, but if that function locked its mutex and was then called again within the same thread before releasing that lock, it would be deadlocked waiting for the mutex to be released.

Some best practices for writing reentrant functions:

1. The function should work only on the data provided by the caller.
2. It should not hold static or global nonconst data unless atomic access can be guaranteed.
3. It should not return the memory address to any static or global nonconstant data.
4. It must not call other nonreentrant functions.
5. Locking any shared resource should be avoided, unless you can use a reentrant lock, such as a recursive mutex lock.
6. Self-modifying code should be avoided unless this can be done in its own thread-specific memory.

For the purposes of most modern projects, thread safety is probably the more important issue to be concerned about. But if you are developing an API that may be used in an interrupt or signal handler, then you will want to consider whether your function calls are reentrant within the same thread.

Asynchronous tasks

C++11 introduced a number of primitives to help developers directly create and manage threads, but it also introduced some higher-level concepts for performing work asynchronously in the form of the `std::async` and `std::future` templates.

The `std::async` function template is used to run any C++ functor asynchronously, and `std::future` is used to access the result of that function once the work has completed. This makes it a lot easier to perform background processing tasks without having to worry about managing threads and semaphores. For example, the next code snippet demonstrates using a lambda function to perform some work asynchronously using `std::async`. The main thread can then do some other work and then call `future.get()` to block on the result from `std::async`:

```
auto future = std::async(std::launch::async, []() {
    int result = 0;
    // perform work to calculate the result value
    return result;
});

// Do something else while the async task is running

auto result = future.get();
```

Asynchronous tasks make it a lot easier to write multithreaded code in C++. However, in terms of implications for your API designs, there's perhaps not a lot to talk about here. That's because you can use any function or object method with `std::async` (i.e., there's nothing special you need to add to your class definitions to make them callable with `std::async`). This is an elegant feature of the C++ implementation, particularly compared with other languages such as Swift, in which functions must be decorated with an `async` keyword and callers of those async functions must use the `await` keyword at the call site, which can cause a cascade of changes to happen to add asynchronous support to a Swift API.

Parallelism

As noted at the start of this chapter, concurrency and parallelism are related but distinct concepts. For the purposes of the discussion here, I'll define parallelism as splitting a task into multiple subtasks that can then be executed simultaneously on multiple CPUs. For example, consider an algorithm that runs a function on every value in a vector and returns another vector containing all of the results. This input vector could be split into several smaller vectors and each of those could be computed on a different CPU to parallelize the work.

The C++17 standard introduced parallel support for many of the algorithms in the C++ Standard Library. These are provided as overloaded versions of functions such as `std::sort`, `std::find`, `std::reverse`, and `std::count`. These new parallelized versions accept a new execution policy argument, which can be one of the following values:

1. `std::execution::seq`: Runs the algorithm sequentially. This is the default if you don't specify an execution policy.
2. `std::execution::par`: The implementation can choose to run the algorithm in parallel on multiple processors. You're responsible for ensuring no data races happen within the function.
3. `std::execution::par_unseq`: The implementation can choose to run the algorithm in parallel on multiple processors. Additionally, this policy enables the use of vectorization/single instruction, multiple data technology where available.
4. `std::execution::unseq` (introduced in C++20): Indicates that the algorithm can be vectorized (e.g., it can execute on a single thread but take advantage of CPU vectorization technology such as SSE on ×86 chips or Altivec on PPC chips).

For instance, the previous example of executing a function over every element in a vector is performed by the `std::transform` function. You can use the C++17 parallelized version of this algorithm as follows. In this example, we use a lambda function to double the value of every element in the vector. The key part in this example is the use of the `std::execution::par` policy to enable the parallel implementation:

```
std::vector<double> inVec;
std::vector<double> outVec;
...
std::transform(std::execution::par,
               inVec.begin(), inVec.end(),
               outVec.begin(),
               [](double value) { return value * 2.0; });
```

Like the previous asynchronous task section, this functionality likely has little impact on the actual design of your APIs. The new parallel algorithms are features that your clients could use in their code or that you can use within your API implementation, but they don't have much of an impact on the surface area of your API itself. One exception might be if you want to provide your own parallel algorithms, in which case you might consider being consistent with the functions in the C++ Standard Library and handle an execution policy parameter to configure the parallel behavior of your calls.

Accessing shared data

Stateless APIs

Threading errors are generally caused when multiple threads try to manipulate a shared resource at the same time, such as a global variable. Therefore, objects that don't maintain or rely on any shared state can be safely accessed by multiple threads. Any function parameters and local variables within a function are allocated on the stack, and each thread has its own stack, so these are safe to use within your functions.

There are certain categories of problem that lend themselves more toward this kind of approach, such as mathematical calculations or unit conversion routines. For example:

```
class MathUtils
{
public:
    static size_t Factorial(size_t num)
    {
        size_t result = 1;
        for (size_t i = 2; i <= num; i++) {
            result *= i;
        }
        return result;
    }
};
```

In this case, the `Factorial()` function uses only the value parameter `num` and a local variable `result`, so this can be considered stateless and safe to call from multiple threads. The function is declared `static`, indicating that it does not refer to any object instance state. We are not passing in any pointers or references to shared state.

Stateless functions such as this will often be side effect free and deterministic, meaning that for the same set of inputs they'll produce the same result every time they're called unless, of course, they depend upon some other nondeterministic function, such as a random number generator. This is essentially the same as the concept of pure functions, which I described earlier in the Functional APIs section of the Styles chapter. The functional programming style often adapts well to concurrency without the need for additional synchronization.

It's sometimes possible to recast an otherwise stateful API as a stateless one. This can be done by having any state that it may produce be returned as the result of the function. For example, consider this class that holds a vector of numbers with a routine to reverse the vector:

```
class NumberSequence
{
public:
    explicit NumberSequence(const std::vector<int> &seq);
    std::vector<int> Get() const;
    void Reverse();

private:
    std::vector<int> mSequence;
};
```

This object maintains the state for the number sequence in an instance variable. Without appropriate synchronization, calling the `Reverse()` method at the same time from multiple threads on the same object could have unpredictable results. Instead, we can develop a stateless version of this API:

```
class NumberSequence
{
public:
    static std::vector<int> Reverse(const std::vector<int> &seq);
};
```

In this case, the number sequence is passed into the API and the modified version of the sequence is returned as the function result. If the implementation of this function uses only temporary local variables, then this should be safe to call from multiple threads. However if the caller needs to store this result in a shared data structure, we've simply moved the burden of correctly synchronizing the code onto the caller.

An example of this in the C++ Standard Library is reversing a string. You can use the `std::reverse()` function to reverse a string in place (i.e., to change the state of the string). Or you can use the `std::string` constructor to return a new string that is the reverse of the input string, such as:

```
std::string str = "Hello World";

// reverse the string in place
std::reverse(str.begin(), str.end());

// return a reversed version of the string
std::string rev = std::string(str.rbegin(), str.rend());
```

Initializing shared data

If you're unable to avoid shared data in your projects, then you must make sure that all access to any shared data is appropriately synchronized to make it safe in the presence of multiple threads. The first part of achieving this is to ensure that the data are safely initialized. If multiple threads try to initialize your data at the same time, then it can leave your data in an inconsistent state.

There are multiple ways to declare global data, with different levels of scoping. This code provides examples of a few of these:

```
// 1: global variables are accessible across multiple translation units
int myGlobal = 1;

// 2: file statics are accessible only in the current translation unit
static int myFileStatic = 2;

namespace myNamespace {

// 3: namespace statics are scoped within the namespace
static int myNamespaceStatic = 3;

class MyClass
{
public:
    int GetSharedValue() const
    {
        // 4: function statics are scoped to within a single function
        static int myFunctionStatic = 4;
        return myFunctionStatic;
    }

private:
    // 5: class statics are accessible to all instances of a class
    static int myClassStatic = 5;
};

} // namespace myNamespace
```

Data defined at the file, global, namespace, or class scope (i.e., 1, 2, 3, and 5) should generally be avoided because they introduce several problems. First, they are initialized when a library is first loaded into a process and they use memory in every linked process, even when those processes don't use the library. Second, if you call any functions to initialize the data, then you can run into complex initialization dependency problems that in the worst case can cause your library to crash on load.

A static variable is initialized when it's first encountered, so using a function-scoped static (i.e., 4) is far preferable because it avoids library load time dependency issues and it creates the shared data only if that function is executed. This is described in more detail in the section on Singletons in the Patterns chapter.

As we noted earlier in this chapter, before C++11 it was simply not possible to implement thread-safe initialization of data using the `static` keyword. However, since C++11, static data initialization must now be implemented by compilers in a thread-safe manner. So you should use static data without further synchronization only if you're using C++11 or a later revision.

Nevertheless, there may be times when you need to perform several instructions to initialize your shared data fully and a single static declaration is insufficient. In those cases, you can rely on another new feature of C++11: `std::call_once`. This will execute a piece of code exactly once even if called concurrently from multiple threads. Here's an example:

```
MyClass &GetInstance()
{
    static MyClass sData;
    static std::once_flag sFlag;
    std::call_once(sFlag, []() {
        // perform initialization of sData
    });
    return &sData;
}
```

Because we're talking about static data and threading, we should mention the `thread_local` keyword that was also introduced with C++11. This is a way to define a static variable with a separate copy of the data per thread. A thread-local variable is created at thread creation time and destroyed when the thread finishes. A typical example of this is a random number generator in which each thread has its own random seed, as shown in this example:

```
#include <random>

double randomDouble()
{
    // return a random number from 0 to 1 with a per-thread seed
    thread_local static std::random_device rd;
    thread_local static std::mt19937 gen(rd());
    thread_local std::uniform_real_distribution<double> dist(0.0, 1.0);
    return dist(gen);
}
```

Finally, another way to ensure that your shared data are initialized in a thread-safe manner is to perform the initialization while the app is still single threaded. This is something that only the client of your library can fully determine, so this approach means providing an initialization function that your clients must call at an appropriately early point in their programs.

Synchronized data access

Now that we've covered how to initialize shared data safely, we can move on to discuss how to modify those data safely without causing a data race. A data race is when two or more threads try to access the same memory address and at least one of the threads is writing to the data.

The way to avoid data races is to introduce locking around all accesses to the data, such as with a mutually exclusive lock, also known as a mutex. This ensures that only a single thread can access the shared data at any point in time. You generally must acquire the mutex lock before you access the shared data and then release it afterward. If you don't balance your lock and unlock calls, perhaps due to an early function exit, then no other attempt to acquire the lock will succeed and your program will deadlock.

To avoid this, it's better to apply RAII principles and have the lock be automatically released when your code goes out of scope. This is also referred to as the scoped locking idiom (Schmidt, 1999). C++11 introduced new objects for this purpose, such as `std::lock_guard` and `std::unique_lock`. For example, when an `std::lock_guard` object is created it locks the mutex, and when it goes out of scope the mutex is released. Here's an example:

```
class NumberSequence
{
public:
    NumberSequence(const std::vector<int> &seq) :
        mSequence(seq)
    {}

    std::vector<int> Get() const
    {
        std::lock_guard<std::mutex> lock(sMutex);
        return mSequence;
    }
    void Reverse()
    {
        std::lock_guard<std::mutex> lock(sMutex);
        std::reverse(mSequence.begin(), mSequence.end());
    }

private:
    std::mutex mMutex;
    std::vector<int> mSequence;
};
```

We don't need to add locking code to the constructor. Objects are constructed by a single thread and can be accessed by other threads only once they've been initialized. So it's safe to eschew the lock in this case. Of course, if the constructor were to try to access any shared or global resources, then locking would be required.

In this example I've added the mutex as a class data member, so the mutex will be initialized when the object is created and we will create a separate mutex for each instance of the object. This is generally preferrable to the alternative of having a single mutex for all instances because the way it's expressed in the previous example allows different threads to access different instances of the object at the same time. However, if you did want to use a single mutex for all instances, you could declare the mutex to be a static class data member. You would want to do this if the shared data you want to protect are also a static data member (i.e., the data are shared across all instances of the object).

TIP: Any time you create global or static data for your APIs, you should think about the thread safety implications of accessing those shared data from multiple threads.

Concurrent API design

In terms of API design, it's important to bear in mind that building an API that handles concurrency can be different from building a single-threaded one. Nonetheless, many of the design principles we've already covered will help you in both cases. For example, designing loosely coupled, minimally complete interfaces in which you hide implementation details and don't duplicate concepts will help you as you design APIs that support concurrency. And if you chose to adopt a functional programming style, then your use of pure functions is likely already thread-safe.

Concurrency best practices

In his book *Clean Code*, Robert Martin offers a series of great tips for writing good concurrent code (Martin, 2008). Also, Bjarne Stroustrup and Herb Sutter provide several recommendations related to concurrency on their C++ Core Guidelines website. I've summarized several of these best practices here as they relate to API design:

- **Limit Access to Shared Data**: As we've already covered, most multithreaded problems come down to accessing shared resources, such as shared mutable data. Code that doesn't share writable data won't have data races. So, limiting the amount of shared data you have in your design and ensuring that all access goes through an API where you can enforce synchronization will help to avoid data races.
- **Return Copies of Data**: If your clients can work on copies of your data, then this will be much simpler than trying to synchronize all access to a single shared instance. For example, you can have an API that returns a read-only copy of the data if clients don't always need the ability to change it. Or you can have an API that returns a copy of your data for clients to work on and then provide a single synchronized API call for them to return the modified version.

- **Independent Threads**: You should try to partition your data into independent objects so that threaded code doesn't need to share data from other threads. By allowing threads to be as independent as possible, you minimize the need to perform synchronization and avoid data races. Similarly, you should be aware of dependencies between API calls that require synchronization because these can introduce subtle bugs.
- **Minimize Synchronized Sections**: When you add locking to some shared resource, you're ensuring that only one thread can execute that critical section of code at any time. If you perform a lot of work within this critical section, you can drastically affect the overall performance of your system. So try to limit locks to just the code that needs to be synchronized and don't include other unrelated calls.
- **Think in Terms of High-Level Tasks**: Threads are a low-level implementation concept. But usually your clients are focused on achieving higher-level tasks. Thinking about your API design in terms of tasks, rather than threads, can therefore make it easier for your clients to reason about your design. C++11 introduced the concept of `std::async` to run a function asynchronously. Consider using this task-level feature to introduce support for concurrency in your designs where feasible.
- **Don't Rely on the Volatile Keyword**: The `volatile` keyword is used to tell the compiler that a variable can be changed by some external event. This can be used to avoid compiler optimizations such as caching the value in a register. This may sound like it could also be useful for multithreading because multiple threads may try to change the same variable, but it's important to appreciate that `volatile` has nothing to do with concurrency (i.e., it doesn't perform any synchronization or prevent instruction reordering). Adding this keyword will not fix any data races in your code.
- **Test Concurrent Code**: When writing concurrent code, you should not ignore seemingly one-off failures because they can indicate that you have an underlying threading problem. Make sure you thoroughly test concurrent code by running it on different platforms, running it multiple times with different timings and orderings, and running with more threads than processors. Also, take advantage of tools that can help you find potential threading errors in your code. You can refer to the Multithreading Analysis section of the Performance chapter for a review of several tools that can check your multithreaded code.

Thread-Safe Interface pattern

If you have multiple functions within your API that call each other, there can be the potential for self-deadlocking (i.e., when one function locks a resource and then calls another function that also tries to lock the same resource). One way to solve this is with a recursive mutex, but that can introduce additional locking overhead. A better solution is to use the Thread-Safe Interface design pattern. The basic principles of this pattern are (Schmidt, 1999):

1. Only public methods should acquire and release locks. These functions can then call into private or protected implementation methods to perform the actual work.
2. Private and protected methods should perform work only when called by the public methods. These functions should never call public methods because that could introduce self-deadlocking.

   This is demonstrated in this code:

```
class MyObject
{
public:
    void Foo()
    {
        std::lock_guard<std::mutex> lockGuard(mLock);
        DoFoo();
    }

private:
    void DoFoo()
    {
        // perform the actual work
    }

    std::mutex mLock;
};
```

The use of inheritance can also introduce subtle concurrency bugs around locking. For example, you may have an API that does locking to synchronize access to a shared resource, but a client could override a virtual function and provide a new implementation that doesn't do any locking. The Thread-Safe Interface pattern can help here, too. You just need to ensure that any virtual functions are defined as protected or private (i.e., that clients can provide an alternative implementation for a virtual function, but the functionality is only ever accessed by calling a public function that performs all necessary locking). For example:

```
class MyObject
{
public:
    void Foo()
    {
        std::lock_guard<std::mutex> lockGuard(mLock);
        DoFoo();
    }

protected:
    virtual void DoFoo() {
        // Base class functionality, no locking required
    }

private:
    std::mutex mLock;
};

class MyDerivedObject : public MyObject
{
protected:
    void DoFoo() override
    {
        // Overridden functionality, no locking required
    }
};
```

In this example, you can see that the client created a new `MyDerivedObject` class that overrides the protected virtual function `DoFoo()`. The client doesn't need to worry about locking any shared resources in this override because the only correct way to call `DoFoo()` is by calling the public `Foo()` function, which performs all necessary locking before forwarding to the `DoFoo()` implementation.

10

Versioning

Up to this point, I have largely considered the design of an application programming interface (API) as a discrete task that is finished once the API is fully specified and released to users. Of course, this is simply the beginning of a continuous and complex process. After an API has been released, that's when the real work begins and when your API development process is put to the test.

Very few if any APIs stop development after the 1.0 product is released. There will always be bugs to fix, new features to integrate, workflows to refine, architecture to improve, other platforms to support, and so on.

The primary objective for all releases after the initial release of an API must be to cause zero impact on existing clients, or as close to zero as practically possible. Breaking the interface, or the behavior of the interface, between releases will force your clients to update their code to take advantage of your new API. The more you can minimize the need for this manual intervention on their part, the more likely your users are to upgrade to your new API, or even to keep using your API at all. If your API has a reputation for introducing major incompatible changes with each new release, you are giving your clients incentive to look for an alternative solution. On the other hand, an API with a reputation for stability and robustness can be the largest factor in the success of your product.

To this end, this chapter will cover the details of API versioning, explaining the different types of backward compatibility and describing how you can achieve backward compatibility for your API.

Version numbers

Each release of your API should be accompanied with a unique identifier so that the latest incarnation of the API can be differentiated from previous offerings. The standard way to do this is to use a version number.

Version number significance

There are many different schemes in use to provide versioning information for a software product. Most of these schemes attempt to impart some degree of the scale of change in a release by using a series of numbers, normally separated by a period (.) symbol. Most commonly, either two or three separate integers are used, for example

API Design for C++. https://doi.org/10.1016/B978-0-443-22219-1.00002-7

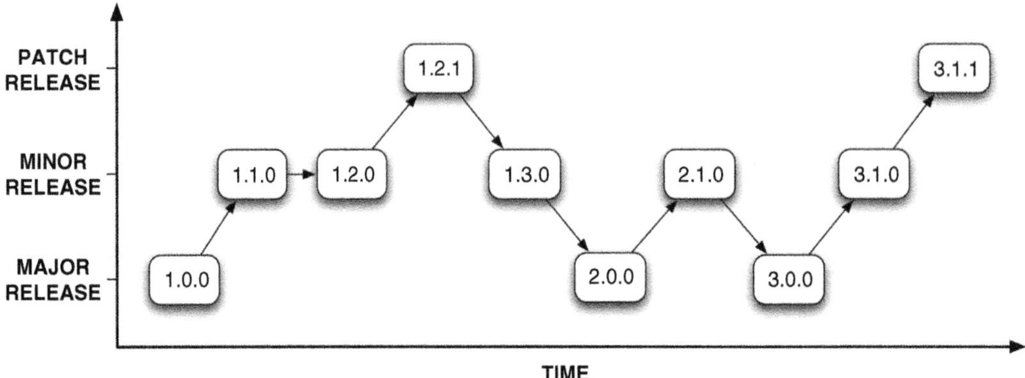

FIGURE 10.1 Illustrating the progression of version numbers using a standard MAJOR.MINOR.PATCH numbering scheme.

"1.2" or "1.2.3". The subsequent list, along with Fig. 10.1, explains the significance of each of these integers:

1. **Major version.** This is the first integer in a version number, such as **1**.0.0. It's normally set to 1 for the initial release and is increased whenever significant changes are made. In terms of API change, a major version change can signal the backward-compatible addition of substantial new features, or it can signify that backward compatibility has been broken. In general, a bump of the major version of an API should signal to your users to expect significant API changes.
2. **Minor version.** This is the second integer in a compound version number (e.g., 1.**0**.0). This is normally set to 0 after each major release and increased whenever smaller features or significant bug fixes have been added. Changes in minor version number should not normally involve incompatible API changes. Users should expect to be able to upgrade to a new minor release without making any changes to their own software. However, some new features may be added to the API, which, if used, would mean that users could not revert to an earlier minor version without changing their code.
3. **Patch version.** The (optional) third integer is the patch number, sometimes also called the revision number (e.g., 1.0.**0**). This is normally set to 0 after each minor release and increased whenever important bug or security fixes are released. Changes in patch number should imply no change to the actual API interface (i.e., only changes to the behavior of the API). In other words, patch version changes should be backward and forward compatible. That is, users should be able to revert to an earlier patch version and then switch back to a more recent patch version without changing their code (Rooney, 2005).

Some software products employ additional numbers or symbols to further describe a release. For example, an automated build number might be used so that every single build of the software can be differentiated from previous builds. This build number could be derived from the revision number of the last change checked into the revision control system, it may be derived from the current date, or it could just be a monotonically increasing number.

Software is often provided to users before the final release to get feedback and valuable field testing. In these cases, it's common to add a symbol to the version string to indicate the phase of the development process to which the software relates. For example, "1.0.0a" might refer to an alpha release, "1.0.0b" might refer to a beta release, and "1.0.0rc" might refer to a release candidate. However, once you start deviating from a purely numeric identification system, doing comparisons of version numbers starts to become more complicated (see Python PEP 0386 at http://www.python.org/dev/peps/pep-0386/ for an example of this complexity).

> *TIP: It is good practice to include your API's major version number in your library names, particularly if you've made non–backward compatible changes, such as* `libFoo.so`, `libFoo2.so`, *or* `libFoo3.so`.

Esoteric numbering schemes

I've also decided to list some nonstandard or imaginative versioning schemes that have been used by software projects in the past. This section is more for fun than actual practical advice, although each scheme obviously offers advantages for certain situations. For API development, however, I recommend sticking with the widely understood major, minor, patch scheme.

The TeX document processing system, originally designed by Donald Knuth, produces new version numbers by including additional digits of precision to the value pi, π. The first TeX version number was 3, then came 3.1, then 3.14, and so on. The current version as of 2021 was 3.141592653. Similarly, the version numbers for Knuth's related METAFONT program asymptotically approach the value e, 2.71828182.

Although this may seem at first to be simply the wry sense of humor of a mathematician, this numbering scheme does convey an important quality about the software. Even though Knuth himself recognizes that some areas of TeX could be improved, he has stated that no new fundamental changes should be made to the system and any new versions should contain only bug fixes. As such, the use of a versioning scheme that introduces increasingly smaller floating-point digits is insightful. In fact, Knuth's recognition of the importance of feature stability and backward compatibility, to the extent that he encoded this importance in the versioning scheme for his software, is food for thought for any API designer.

Another interesting versioning scheme is the use of dates as version numbers. This is commonly used for large end-user software releases such as Microsoft's Visual Studio 2022 and games such as EA's FIFA 23. However, a more subtle system is used by the Ubuntu flavor of the Linux operating system. This uses the year and month of a release as the major and minor version number, respectively. The first Ubuntu release, 4.10, appeared in Oct. 2004 whereas 9.04 was released during Apr. 2009. Ubuntu releases are also assigned a code name, consisting of an adjective and an animal name with the first same letter, such as Breezy Badger and Lucid Lynx. With the exception of the first two releases, the first letter of these code names increases alphabetically for each release. These schemes have the benefit of imparting how recent an Ubuntu release is, but they do not convey any notion of the degree of change in a release. This may be fine for a continually evolving operating system, although you should prefer a more traditional number scheme for your API to give your users an indication of the degree of API change to expect in a release.

The Linux kernel currently uses an even/odd numbering scheme to differentiate between stable releases (even) and development releases (odd). For example, Linux 2.4 and 2.6 are stable releases, whereas 2.3 and 2.5 are development releases. This numbering scheme is also used by the Second Life Server releases.

Creating a version API

The version information for your API should be accessible from code, to allow your clients to write programs that are conditional on your API's version number: for example, to call a new method that exists only in recent versions of your API, or to work around a bug in the implementation of a known release of your API.

To offer maximum flexibility, users should be able to query your API's version at compile time as well as runtime. The compile-time requirement is necessary so that the user can use `#if` preprocessor directives to conditionally compile against newer classes and methods that would cause undefined reference errors if linking against older versions of your API. The runtime requirement allows clients to choose between different API calls dynamically or to provide logging output with your API version number included. These requirements suggest the creation of a version API. I present a simple generic API for this purpose:

```
// version .h
#include <string>

#define API_MAJOR 1
#define API_MINOR 2
#define API_PATCH 0

class Version
{
public:
    static int GetMajor();
    static int GetMinor();
    static int GetPatch();

    static std::string GetVersion();

    static bool IsAtLeast(int major, int minor, int patch);

    static bool HasFeature(const std::string &name);
};
```

There are a few features of note in this `Version` class. First, I provide accessors to return the individual major, minor, and patch numbers that comprise the current version. These simply return the values of the respective `#define` statements, `API_MAJOR`, `API_MINOR`, and `API_PATCH`. Although I stated in the C++ Usage chapter that you should avoid `#define` for constants, this is an exception to that rule because you need users to be able to access this information from the preprocessor.

The `GetVersion()` method returns the version information as a user-friendly string, such as 1.2.0. This is useful for the client to display in an About dialog or to write to a debug log in the application.

Next, I provide a method to let users perform version comparisons. This lets them do checks in their code, such as checking that they are compiling against an API that is greater than or equal to the specified (major, minor, patch) triple. Obviously you could add other version math routines here, but `IsAtLeast()` provides the most common use case.

TIP: Provide version information for your API.

Finally, I provide a `HasFeature()` method. Normally when users want to compare version numbers, they don't really care about the version number itself, but instead they're using this designator as a way to determine whether a feature they want to use is present in the API. Instead of making your users be aware of which features were introduced in which versions of your API, the `HasFeature()` method lets them test for the availability of the feature directly. For example, in version 2.0.0 of your API, perhaps you made the API thread-safe. You could therefore add a feature tag called THREADSAFE, so that users could do a check such as:

```
if (Version::HasFeature("THREADSAFE")) {
    ...
}
```

Although you probably don't need to define any feature tags for your 1.0 release, you should include this method in your version API so that it's possible for a client to call it in any release of your API. The method can simply return false for 1.0, but for future releases, you can add tags for new features or major bug fixes. These strings can be stored in an `std::set` and lazily initialized on the first call so that it's efficient to determine whether a feature tag is defined. The source code that accompanies this book provides an implementation of this concept.

Feature tags are particularly useful if you have an open source project in which clients may fork your source code, or an open specification project in which vendors can produce different implementations of your specification. In these cases, there could be multiple versions of your API that offer different feature sets in releases with the same version number. This concept is employed by the OpenGL API, in which the same version of the OpenGL API may be implemented by different vendors but with different extensions available. For example, the OpenGL API provides the `glGetStringi(GL_EXTENSION, n)` call to return the name of the n'th `extension`. Also, C++11 introduced a number of feature testing macros to detect the presence of new language features, such as `__cpp_consteval`.

Software branching strategies

Before I talk in more depth about API versioning, let's cover some basics about the related topic of software branching strategies. Although small projects with one or two engineers can normally get by with a single code line, larger software projects normally involve some form of branching strategy to enable simultaneous development, stabilization, and maintenance of different releases of the software. In the next couple of sections I'll cover some of things to consider when choosing a branching strategy and policy for your project.

Branching strategies

Every software project needs a trunk or main code line, which is the enduring repository of the project's source code. Branches are made from this trunk code line for individual releases or for development work that must be isolated from the next release. This model supports parallel development in which new features can be added to the project while imminent releases can lock down changes and stabilize the existing feature set.

There are many different branching schemes that can be devised. Each engineering team will normally adapt a strategy for its own individual needs, process, and workflow. However, Fig. 10.2 provides one example branching strategy that's frequently seen. In this case, major releases are branched off the trunk and minor releases occur along that code line. If an emergency patch is required while work is happening for the next minor release, then a new branch may be created for that specific hotfix. Longer-term development work that needs to skip the next release because it won't be ready in time is often done in its own branch and then landed at the appropriate time. Note the resemblance between Figs. 10.1 and 10.2. This similarity is, of course, not accidental.

Branching policies

This basic structure is used by many projects to support parallel development and release management. However, there are many policy decisions that can be used to customize the actual workflow, such as in which branches developers work, how many development branches are in flight at once, at what point in the process are release branches created, how often changes are merged between branches, and whether automatic merges between branches are attempted.

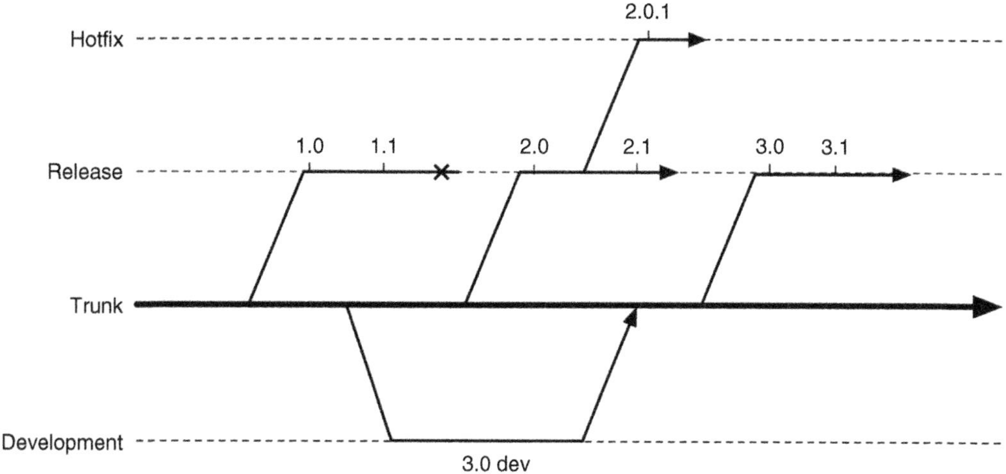

FIGURE 10.2 An example branching diagram for multiple releases of a software product.

Although different branching policies make sense for different situations, I'll comment that in my experience, ongoing development work should happen in the trunk code line, with development branches used where necessary for longer-term work. Quality assurance (QA) should also be focused on trunk, particularly during periods of development between releases. The trunk is where your project's crown jewels live: this is the code that will live on past individual releases. If no developers or QA engineers are working in trunk, and merges are allowed to take place into trunk unattended, the trunk will soon become unstable and buggy. Another arrangement is to have only release branches, although even in that case the advice is the same, that most of your active work should be happening on the most recent release branch.

Your choice of revision control system also has an impact on your branching policy, because different source control management (SCM) products make certain branching strategies easier than others. For example, supporting and merging between branches in Subversion can be a painful endeavor, whereas in distributed SCM systems such as Git or Mercurial, branching is designed into the core of the system. For example, using an SCM such as Git it's possible to consider merging between branches daily because this is an easy and low-impact operation. The more often you merge between branches, the less code divergence occurs and the easier it will be eventually to land the branch into trunk, if that's the end goal. Release branches will normally just be end of lifed when the release is done, as represented by the X symbol after the 1.1 release in Fig. 10.2. Another decision factor that relates to your SCM system is whether all of your engineers are on-site, in which case a server-based solution such as Perforce is acceptable, or whether you have open source engineers working out of their homes, in which case a distributed solution such as Git will be more appropriate.

> *TIP: Branch only when necessary, and branch late. Prefer branching to freezing a code line. Merge between branches early and often. (From the Perforce High-Level Best Practices white paper).*

APIs and parallel branches

Once an API has been released, changes to that API should appear (at least externally) to follow a serialized process. That is, you do not release incompatible nonlinear versions of your API: the functionality in version *N* should be a strict superset of the functionality in version *N*-1. Although this may seem obvious, large software projects tend to involve developers working in several parallel branches of the code, and there can be several concurrently supported releases of an API. It's therefore important that the teams working in different parallel branches do not introduce incompatible features. There are several policy approaches to deal with this potential problem, such as:

- **Target development branches.** Your project will generally have development branches and release branches. By enforcing that no API changes occur directly in release branches, you minimize the chance that an API change is made in one release but is lost in the next release because it was never merged down to trunk. If an API change is needed in a release branch, it should be committed to trunk and then merged up from there. This is generally true of any change in a release branch, but interface changes have a higher cost if they're lost between releases.
- **Merge to trunk often.** Any changes to a public API should be developed in the common trunk code line or merged into trunk as early as possible. This also assumes that teams are regularly syncing their development branches to the trunk code, which is good practice anyway. This avoids surprises further down the line when two teams try to merge development branches with conflicting APIs.
- **Review process.** A single API review committee should oversee and vet all changes to public APIs before they're released. It's the job of this committee to ensure that no conflicting or non–backward compatible changes have been made to APIs. They are the gatekeepers and the last line of defense. This group should be sufficiently empowered to slip release deadlines if necessary to address API problems. I'll discuss how to run an API review process later in this chapter.

These solutions attempt to keep a one true definition of the API in the trunk code line rather than fracture changes across multiple branches. This may not always be possible, but if you strive for this goal, you will make your life easier later.

The problems become more difficult if you have an open source product in which users may create many forks of your source code and make changes to your APIs that are beyond your control. You obviously cannot do too much about this situation. However, if these changes are to be merged back into your source repository, then you can and should apply the same thoughtful review process to community patches as you would to internally developed changes. It can be difficult or awkward to deny changes from open source developers, but you can minimize any hurt feelings by clearly documenting the review process and expectations, offering advice on technical direction early on, and providing constructive feedback on how a patch can be changed to make it more acceptable.

File formats and parallel products

A colleague once described to me a project that he worked on in which a decision was made to support two different variants of the product: a Basic version and an Advanced version. Until that point, there was a single variant of the product and a single file format. The team had a policy of increasing the file format major version number when an incompatible change was introduced into the format, with the last single-variant version being 3.0. The file format was XML based and included a version tag, so it was known which version of the product generated the file. The file format reader would ignore tags that it didn't understand in versions that differed only by minor version number, so that it could still read files that were generated from newer but compatible versions of the product. Both the Basic and Advanced variants could read all files from 3.0 and earlier.

This all seems reasonable so far.

It wasn't long before the Advanced variant introduced new features that required non–backward compatible additions to the file format. So the team decided to increment the major version number to 4.x. However, then there was a need to evolve the entire file format in an incompatible way (i.e., to require a major version bump for Basic and Advanced files). To deal with this, the Basic variant format was updated to 5.x and the Advanced variant was bumped to 6.x. This meant that:

- 3.x builds couldn't read any of 4.x through 6.x formats, which is fine.
- 4.x builds (old Advanced) couldn't read 5.x files (new Basic) or 6.x files (new Advanced).
- 5.x builds (new Basic) couldn't read 4.x files (old Advanced).
- 6.x builds (new Advanced) could read any existing format, which is also fine.

Then, of course, eventually another major version bump was required, introducing a 7.x (newer Basic) and 8.x (newer Advanced). Things then started to get really messy.

With the benefit of hindsight, we talked about how this situation could've been avoided. The key observation is that in this case, the information about which variant had created the file was being conflated with the file format version. One solution would have been to tease apart those two concepts and to write both into the file (i.e., a version number, such as 3.2 and a variant name such as Basic). In this way, the Basic variant could easily know whether it could read a format: it could read any file with an empty or Basic variant name. This essentially creates two version number spaces, in which the version numbers for the two variants can advance independently of each other. A product first checks the variant name for compatibility, then version number compatibility works in the usual linear fashion.

Learning from this experience, I offer this advice: when supporting different variants of a product, store the variant's name in any files that should be shared between the variants in addition to the version number of the variant that wrote the file.

> *TIP: When creating Basic versus Advanced versions of the same APIs, accompany the version number with a Basic or Advanced string in any generated files. Don't try to use the version number solely to glean whether the file was generated by the Basic or Advanced API.*

Life cycle of an API

In this section I'll examine the life of an API and the various phases it goes through, from conception to end of life.

Maintaining an API is not necessarily the same as maintaining a normal software product. That's because of the extra constraints that are placed on API development to not break existing clients. In a normal end user software product, if you change the name of a method or class in your code, this doesn't affect the user-visible features of the application. However, if you change the name of a class or method in an API, you may break the code of all of your existing clients. An API is a contract, and you must make sure that you uphold your end of the contract.

Fig. 10.3 provides an overview of the life span of a typical API. The most important event in this life span is the initial release, marked by the thick vertical bar in the figure. Before this pivotal point, it's fine to make major changes to the design and interface. However, after the initial release, once your users can write code using your API, you have committed to providing backward compatibility, and the extent of the changes you can make is greatly limited. Looking at the life span as a whole, there are four general stages of API development (Tulach, 2008):

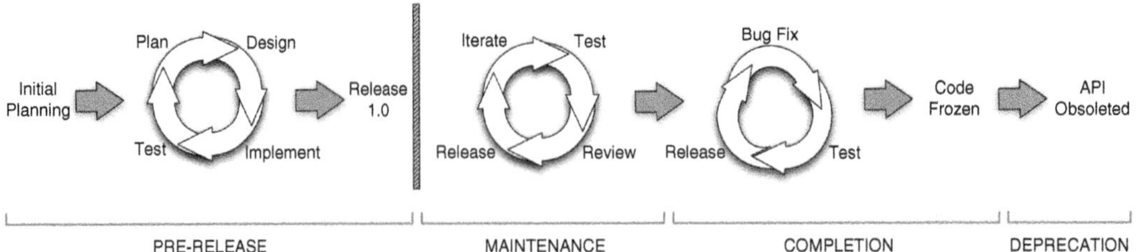

FIGURE 10.3 The life cycle of an application programming interface (API). Before the initial release, extensive redesign of the API can be performed. After the initial release, only incremental changes can be tolerated.

1. **Prerelease**: Before the initial release, an API can progress through a standard software cycle, including requirements gathering, planning, design, implementation, and testing. Most notably, as I've stated, the interface can go through major changes and redesigns during this period. You may release these early versions of your API to your users to get their feedback and suggestions. You should use a version number of 0.x for these prerelease versions to make it clear to users that the API is still under active development and may change radically before 1.0 is delivered.
2. **Maintenance**: An API can still be modified after it has been released, but to maintain backward compatibility any changes must be restricted to adding new methods and classes, as well as fixing bugs in the implementation of existing methods. In other words, during the maintenance phase you should seek to evolve an API, not change it incompatibly. To ensure that changes do not break backward compatibility, it is good practice to conduct API reviews before a new version is released.
3. **Completion**: At some point, the project leads will decide that the API has reached maturity and that no further changes should be made to the interface. This may be because the API solves the problems it was designed to solve, or it may be because team members have moved on to other projects and can no longer support the API. Stability is the most important quality at this point in the life span. As such, only bug fixes will generally be considered. API reviews could still be run at this stage, but if changes are indeed restricted to implementation code and not public headers, then they may not be necessary. Ultimately, the API will reach the point where it's considered to be complete and no further changes will be made.
4. **Deprecation**: Some APIs eventually reach an end of life state where they're deprecated and then removed from circulation. Deprecation means that an API should not be used for any new development and that existing clients should migrate away from the API. This can happen if the API no longer serves a useful purpose or if a newer, incompatible API has been developed to take its place.

TIP: After release, you can evolve an API but not change it.

Levels of compatibility

Up to this point I've talked only vaguely about what backward compatibility means. It's now time to get concrete and define our terms more precisely. Accordingly, the next few sections will detail what is meant by the specific terms backward compatibility, forward compatibility, functional compatibility, source (or API) compatibility, and application binary interface (ABI) compatibility.

Often you'll provide different levels of compatibility promises for major, minor, and patch releases of your API. For example, you may promise that patch releases will be both backward and forward compatible (Subversion promises this), or you may promise only to break binary compatibility for major releases (KDE promises this for core libraries).

Backward compatibility

Backward compatibility can be defined simply as an API that provides the same functionality as a previous version of the API. In other words, an API is backward compatible if it can fully take the place of a previous version of the API without requiring the user to make changes.

This implies that the newer API is a superset of the older API. It can add new functionality, but it cannot incompatibly change functionality that is already defined by the older API. The cardinal rule of API maintenance is never to remove anything from your interface.

There are different types of API backward compatibility, including:

1. Functional compatibility,
2. Source compatibility, and
3. Binary compatibility.

I will define each of these in more detail in the following sections. In addition, there are also data-oriented backward compatibility issues, such as:

1. Client/server compatibility, and
2. File format compatibility.

For example, if your API involves communication over a network, then you also need to consider the compatibility of the client/server protocol you use. This means that a client using an older release of the API will still be able to communicate with a newer version of the server. Also a client using a newer release of the API will still be able to communicate with an older version of the server (Rooney, 2005).

Additionally, if your API stores data in a file or database, then you will need to consider the compatibility of that file format or database schema. For example, more recent versions of the API need to be able to read files generated by older versions of the API.

These data-oriented features should also have version numbers, and these don't have to be the same as your API version number (e.g., you may make a change to your API but not make a change to the underlying file format in that release, so the API version number changes while the file format version number stays the same).

> *TIP: Backward compatibility means that client code that uses version N of your API can be upgraded without change to version N+1.*

Functional compatibility

Functional compatibility is concerned with the runtime behavior of an implementation. An API is functionally compatible if it behaves exactly the same as a previous version of the API. However, as Jaroslav Tulach notes, an API will hardly ever be 100% backward compatible in this respect. Even a release that fixes only bugs in implementation code will have changed the behavior of the API, behavior that some clients may be depending on.

For example, if your API provides the function:

```
void SetImage(Image *img);
```

this function may have a bug in version 1.0 of your API, causing it to crash if you pass it a `nullptr`. In version 1.1, you fix this bug so that your code no longer crashes in this case. This has changed the behavior of the API, so it's not strictly functionally compatible. However, it has changed the behavior in a good way: it's fixed a crashing bug. So, although this metric is useful as a basic measure of change in the runtime behavior of an API, that functional change may not necessarily be a bad thing. Most API updates will intentionally break functional compatibility.

As an example of a case in which functional compatibility is useful, consider a new version of an API that focused solely on performance. In this case, the behavior of the API is not changed at all. However, the algorithms behind the interface are improved to deliver the same results in less time. In this respect, the new API could be considered 100% functionally compatible.

> *TIP: Functional compatibility means that version N+1 of your API behaves the same as version N.*

Source compatibility

Source compatibility is a looser definition of backward compatibility. It basically states that users can recompile their programs using a newer version of the API without making any change to their code. This says nothing about the behavior of the resulting program, only that it can be successfully compiled and linked. Source compatibility is also sometimes referred to as API compatibility.

For example, these two functions are source compatible, even though their function signatures are different:

```
// version 1.0
void SetImage(Image *img);

// version 1.1
void SetImage(Image *img, bool keep_aspect = true);
```

This is because any user code that was written to call the 1.0 version of the function will also compile against version 1.1 (the new argument is optional). In contrast, the next two functions are not source compatible because users will be forced to go through their code to find all instances of the `SetImage()` method and add the required second parameter:

```
// version 1.0
void SetImage(Image *img);

// version 1.1
void SetImage(Image *img, bool keep_aspect);
```

Any changes that are completely restricted to implementation code, and therefore do not involve changes to public headers, will obviously be 100% source compatible, because the interfaces are the same in both cases.

> *TIP: Source compatibility means that a user who wrote code against version N of your API can also compile that code against version N+1 without changing their source.*

Binary/application binary interface compatibility

Binary compatibility implies that clients only need to relink their programs with a newer version of a static library or simply drop a new shared library into the install directory of their end user application. This contrasts with source compatibility, in which users must recompile their programs whenever any new version of your API is released.

This implies that any changes to the API must not affect the representation of any classes, methods, or functions in the library file. The binary representation of all API elements must remain the same, including the type, size, and alignment of structures and the signatures of all functions. This is also often called ABI compatibility.

Binary compatibility can be difficult to attain using C++. Most changes that you make to an interface in C++ will cause changes to its binary representation. For example, here are the mangled names of two different functions (i.e., the symbol names that are used to identify a function in an object or library file):

```
// version 1.0
void SetImage(Image *img)
-> _Z8SetImageP5Image

// version 1.1
void SetImage(Image *img, bool keep_aspect = false)
-> _Z8SetImageP5Imageb
```

These two methods are source compatible but they are not binary compatible, as evidenced by the different mangled names that each produce. This means that code compiled against version 1.0 cannot simply use the version 1.1 libraries because the `_Z8SetImageP5Image` symbol is no longer defined.

The binary representation of an API can also change if you use different compile flags. It tends to be compiler specific, too. One reason for this is that the C++ standards committee decided not to dictate the specifics of name mangling. As a result, the mangling scheme used by one compiler may differ from that of another compiler, even on the same platform. (The mangled names presented earlier were produced by GNU C++ 4.3.)

> *TIP: Binary compatibility means that an application written against version N of your API can be upgraded to version N+1 simply by replacing or relinking against the new dynamic library for your API.*

Next I provide two lists of specific API changes, detailing those that will require users to recompile their code and those that should be safe to perform without breaking binary compatibility.

Binary incompatible API changes

- Removing a class, method, or function.
- Adding, removing, or reordering member variables for a class.
- Adding or removing base classes from a class.
- Changing the type of any member variable.
- Changing the signature of an existing method in any way.
- Adding, removing, or reordering template arguments.
- Changing a noninlined method to be inlined.

- Changing a nonvirtual method to be virtual, and vice versa.
- Changing the order of virtual methods
- Adding a virtual method to a class with no existing virtual methods.
- Adding new virtual methods (some compilers may preserve binary compatibility if you add only new virtual methods after existing ones).
- Overriding an existing virtual method (this may be possible in some cases, but is best avoided).

Binary compatible API changes

- Adding new classes, nonvirtual methods, or free functions.
- Adding new static variables to a class.
- Removing private static variables (if they are never referenced from an inline method).
- Removing nonvirtual private methods (if they are never called from an inline method).
- Changing the implementation of an inline method.
- Changing an inline method to be noninline.
- Changing the default arguments of a method (however, this requires recompilation to use the new default argument).
- Adding or removing friend declarations from a class.
- Adding a new enum to a class.
- Appending new enumerations to an existing enum
- Using unclaimed remaining bits of a bit field.

Restricting any API changes to only those listed in this second list should allow you to maintain binary compatibility between your API releases. Some further tips to help you achieve binary compatibility include:

- Rather than adding parameters to an existing method, you can define a new overloaded version of the method. This ensures that the original symbol continues to exist but provides the newer calling convention, too. Inside your `.cpp` file, the older method may be implemented simply by calling the new overloaded method:

```
// version 1.0
void SetImage(Image *img)

// version 1.1
void SetImage(Image *img)
void SetImage(Image *img, bool keep_aspect)
```

(This technique may affect source compatibility if the method is not already overloaded, because client code can no longer reference the function pointer `&SetImage` without an explicit cast).

- The pimpl idom can be used to help preserve binary compatibility of your interfaces, because it moves all implementation details—those elements that are most likely to change in the future—into the `.cpp` file where they don't affect the public `.h` files.
- Adopting a flat C style API can make it much easier to attain binary compatibility (and also cross-compiler compatibility), simply because C does not offer you features such as inheritance, optional parameters, overloading, exceptions, and templates. To get the best of both worlds, you may decide to develop your API using an object-oriented C++ style and then provide a flat C style wrapping of the C++ API.
- If you do need to make a binary incompatible change, then you might consider naming the new library differently so that you don't break existing applications. This approach was taken by the libz library. Builds before version 1.1.4 were called ZLIB.DLL on Windows. However, a binary incompatible compiler setting was used to build later versions of the library, and so the library was renamed to ZLIB1.DLL, in which the 1 indicates the API major version number.

Binary compatibility is transitive, so the ABI compatibility of your library dependencies can affect the compatibility of your library. For example, the clang compiler lets you choose between different C++ runtimes: libc++ (the LLVM runtime) or libstdc++ (the GNU runtime). However, these are not binary compatible: a library compiled against one will not link against the other. In particular, libc++ uses an inlined namespace of `std::__1` for all of its symbols but libstdc++ does not. So, even if you used a plain C API for your library, if you have a C++ library dependency that was linked against libc++, then your clients will not be able to link against libstdc++ because it's not binary compatible with your dependencies.

Forward compatibility

An API is forward compatible if client code written using a future version of the API can be compiled without modification using an older version of the API. Forward compatibility therefore means that users can downgrade to a previous release and still have their code work without modification.

Adding new functionality to an API breaks forward compatibility because client code written to take advantage of these new features will not compile against the older release where those changes are not present.

For example, these two versions of a function are forward compatible:

```
// version 1.0
void SetImage(Image *img, bool unused = false);

// version 1.1
void SetImage(Image *img, bool keep_aspect);
```

This is because code written using the 1.1 version of the function, in which the second argument is required, can compile successfully in the 1.0 version. However, these two versions are not forward compatible:

```
// version 1.0
void SetImage(Image *img);

// version 1.1
void SetImage(Image *img, bool keep_aspect = false);
```

This is because code written using the 1.1 version can provide an optional second argument, which if specified will not compile against the 1.0 version of the function.

Forward compatibility is obviously a difficult quality to provide any guarantees about because you can't predict what will happen to the API in the future. You can, however, give this your best effort before the 1.0 version of your API is released. In fact, this is an excellent activity to engage in before your first public release, to try to make your API as future-proof as possible.

This means that you must give thought to the question of how the API could evolve in the future. What new functionality might your users request? How would performance optimizations affect the API? How might the API be misused? Is there a more general concept that you may want to expose in the future? Are you aware of functionality that you plan to implement in the future that will affect the API?

Here are some ways in which you can make an API forward compatible:

- If you know that you'll need to add a parameter to a method in the future, then you can use the technique shown in the first example mentioned earlier: that is, you can add the parameter even before the functionality is implemented and simply document (and name) this parameter as unused.
- You can use an opaque pointer or typedef instead of using a built-in type directly if you anticipate switching to a different built-in type in the future. For example, create a typedef for the `float` type called `Real`, so that you can change the typedef to double in a future version of the API without causing the API to change.
- The data-driven style of API design, described in the Styles chapter, is inherently forward compatible. A method that simply accepts an `ArgList` variant container essentially allows any collection of arguments to be passed to it at runtime. The implementation can therefore add support for new named arguments without requiring changes to the function signature.

TIP: Forward compatibility means that client code that uses version N of your API can be downgraded without change to version N-1.

How to maintain backward compatibility

Now that I've defined the various types of compatibility, I'll describe some strategies for maintaining backward compatibility when releasing newer versions of your APIs.

Adding functionality

In terms of source compatibility, adding new functionality to an API is generally a safe thing to do. Adding new classes, new methods, or new free functions does not change the interface for preexisting API elements and so will not break existing code.

As an exception to this rule, adding new pure virtual member functions to an abstract base class is not backward compatible: that is,

```
class ABC
{
public:
    virtual ~ABC();
    virtual void ExistingCall() = 0;
    virtual void NewCall() = 0;  // added in new release of API
};
```

This is because all existing clients must now define an implementation for this new method; otherwise their derived classes will not be concrete and their code will not compile. The workaround for this is simply to provide a default implementation for any new methods that you add to an abstract base class (i.e., to make them virtual but not pure virtual). For example:

```
class ABC
{
public:
    virtual ~ABC();
    virtual void ExistingCall() = 0;
    virtual void NewCall();  // added in new release of API
};
```

TIP: Do not add new pure virtual member functions to an abstract base class after the initial release of your API.

In terms of binary (ABI) compatibility, the set of elements that you can add to the API without breaking compatibility is more restricted. For example, adding the first virtual method to a class will cause the size of the class to increase, normally by the size of one pointer, to include a pointer to the vtable for that class. Similarly, adding new base classes, adding template parameters, or adding new member variables will break binary compatibility. Some compilers will let you add virtual methods to a class that already has virtual methods without breaking binary compatibility, as long as you add the new virtual method after all other virtual methods in the class.

Refer to the list in the Binary Compatible section for a more detailed breakdown of API changes that will break binary compatibility.

Changing functionality

Changing functionality without breaking existing clients is a trickier proposition. If you only care about source compatibility, then it's possible to add new parameters to a method as long as you order them after all previous parameters and declare them as optional. This means that users are not forced to update all existing calls to add the extra parameter. I gave an example of this earlier, which I will replicate here for convenience:

```
// version 1.0
void SetImage(Image *img);

// version 1.1
void SetImage(Image *img, bool keep_aspect = true);
```

Also, changing the return type of an existing method, in which the method previously had a void return type, is a source compatible change because no existing code should be checking that return value:

```
// version 1.0
void SetImage(Image *img);

// version 1.1
bool SetImage(Image *img);
```

If you wish to add a parameter that doesn't appear after all of the existing parameters, or if you are writing a flat C API in which optional parameters are not available, then you can introduce a differently named function and perhaps refactor the implementation of the old method to call the new method. As an example, the Win32 API uses this technique extensively by creating functions that have an Ex suffix to represent extended functionality. For example:

```
HWND CreateWindow(
    LPCTSTR lpClassName,
    LPCTSTR lpWindowName,
    DWORD dwStyle,
    int x,
    int y,
    int nWidth,
    int nHeight,
    HWND hWndParent,
    HMENU hMenu,
    HINSTANCE hInstance,
    LPVOID lpParam
);
```

```
HWND CreateWindowEx(
    DWORD dwExStyle,
    LPCTSTR lpClassName,
    LPCTSTR lpWindowName,
    DWORD dwStyle,
    int x,
    int y,
    int nWidth,
    int nHeight,
    HWND hWndParent,
    HMENU hMenu,
    HINSTANCE hInstance,
    LPVOID lpParam
);
```

The Win32 API also provides examples of deprecating older functions and introducing an alternative name for the newer functions, instead of simply appending Ex to the end of the name. For example, the `OpenFile()` method is deprecated, and instead the `CreateFile()` function should be used for all modern applications.

In terms of template use, adding new explicit template instantiations to your API can potentially break backward compatibility because your clients may already have added an explicit instantiation for that type. If this is the case, those clients will receive a duplicate explicit instantiation error when trying to compile their code.

In terms of maintaining binary compatibility, any changes you make to an existing function signature will break binary compatibility, such as changing the order, type, number, or constness of parameters, or changing the return type. If you need to change the behavior of an existing method and maintain binary compatibility, then you must resort to creating a new method for that purpose, potentially overloading the name of the existing function. This technique was shown earlier in this chapter:

```
// version 1.0
void SetImage(Image *img);

// version 1.1
void SetImage(Image *img);
void SetImage(Image *img, bool keep_aspect);
```

Finally, it will be common to change the behavior of an API without changing the signature of any of its methods. This could be done to fix a bug in the implementation or to change the valid values or error conditions that a method supports. These kinds of changes will be source and binary compatible, but they will break functional compatibility for your API. Often, these will be desired changes that your affected clients will find agreeable. However, in cases where the change in behavior may not be desirable to all clients, you can make the new behavior opt-in. For example, if you have added multithreaded locking to your API, you could allow clients to opt-in to this new behavior by calling a `SetLocking()` method to turn on this functionality (Tulach, 2008). Alternatively, you could integrate the ability to turn on/off features with the `HasFeature()` method I introduced earlier for the Version class. For example:

```
// version.h
class Version
{
public:
    ...

    static bool HasFeature(const std::string &name);
    static void EnableFeature(const std::string &name, bool);
    static bool IsFeatureEnabled(const std::string &name);
};
```

With this capability, your clients could explicitly enable new functionality while the original behavior is maintained for existing clients, thus preserving functional compatibility. For example,

```
Version::EnableFeature("LOCKING", true);
```

Deprecating functionality

A deprecated feature is one that clients are actively discouraged from using, normally because it's been superseded by newer, preferred functionality. A deprecated feature remains in the API, so users can still call it, although doing so may generate some kind of warning. The expectation is that deprecated functionality may be completely removed from a future version of the API.

Deprecation is a way to start the process of removing a feature while giving your clients time to update their code to remove their dependency on the feature.

There are various reasons to deprecate functionality, including addressing security flaws, introducing a more powerful feature, simplifying the API, or supporting a refactoring of the API's functionality. For example, the standard C function `tmpnam()` has been deprecated in preference to more secure implementations such as `tmpnam_s()` or `mkstemp()`.

When you deprecate an existing method you should mark this fact in the documentation for the method, along with a note on any newer functionality that should be used instead. In addition to this documentation task, there are ways to produce warning messages if the function is ever used. As of C++14, there's a standard way to do this with the `[[deprecated]]` attribute. For example:

```
class MyObject
{
public:
    [[deprecated("Use GetValues() instead")]] int GetValue() const;
    std::vector<int> GetValues() const;
};
```

You can use `[[deprecated]]` in front of a function signature, whether it is a free function, a function template, or a class method. You can also use it in front of a variable definition to deprecate any of your API constants. However, you cannot currently use it to deprecate an entire class, struct, or enum.

If you're unable to use C++14, most compilers provide their own way to decorate a class, method, or variable as being deprecated and will output a compile-time warning if a program tries to access a symbol decorated in this fashion. In Visual Studio C++, you prefix a method declaration with `__declspec(deprecated)`, whereas in the GNU C++ compiler you use `__attribute__ ((deprecated))`. This code defines a `DEPRECATED` macro that will work for either compiler:

```
// deprecated.h
#ifdef __GNUC__
#define DEPRECATED __attribute__((deprecated))
#elif defined(_MSC_VER)
#define DEPRECATED __declspec(deprecated) func
#else
#define DEPRECATED
#pragma message("DEPRECATED is not defined for this compiler")
#endif
```

Using this definition, you can mark certain methods as being deprecated:

```
#include "deprecated.h"

#include <string>

class MyClass
{
public:
    DEPRECATED std::string GetName();
    std::string GetFullName();
};
```

If users try to call the `GetName()` method, their compiler will output a warning message indicating that the method is deprecated. For example, this warning is emitted by the GNU C++ 4.3 compiler:

```
In function 'int main(int, char**)':
warning: 'GetName' is deprecated (declared at myclass.h:21)
```

As an alternative to providing a compile-time warning, you could write code to issue a deprecation warning at runtime. One reason to do this is so you can provide more information in the warning message, such as an indication of an alternative method to use. For example, you could declare a function that you call as the first statement of each function you wish to deprecate, such as:

```
void Deprecated(const std::string oldfunc,
                const std::string newfunc = "");

...

std::string MyClass::GetName()
{
    Deprecated("MyClass::GetName", "MyClass::GetFullName");
    ....
}
```

The implementation of `Deprecated()` could maintain an `std::set` with the name of each function for which a warning has already been emitted. This would allow you to output a warning on only the first invocation of the deprecated method, to avoid spewage to the terminal if the method gets called a lot. Noel Llopis describes a similar technique in his Game Gem, except that his solution also keeps track of the number of unique call sites and batches up the warnings to output a single report at the end of the program's execution (DeLoura, 2001).

Removing functionality

Some functionality may eventually be removed from an API after it's gone through at least one release of being deprecated. Removing a feature will break all existing clients that depend upon that feature, which is why it's important to give users a warning about your intention to remove the functionality, by first marking it as deprecated.

Removing functionality from an API is a drastic step, but it's sometimes warranted when the methods should no longer be called for security reasons, if that functionality is simply no longer supported, or if it's restricting the ability of the API to evolve.

One way to remove functionality and yet still allow legacy users to access the old functionality is to bump the major version number and declare that the new version is not backward compatible. Then you can completely remove the functionality from the latest version of the API but still provide old versions of the API for download, with the understanding that they are deprecated and unsupported and should be used only by legacy applications. You may even consider storing the API headers in a different directory and renaming the libraries so that the two APIs do not conflict with each other. This is a big deal, so don't do it often. Once in the lifetime of the API is best. Never is even better.

This technique was used by the Qt library when it transitioned from version 3.x to 4.x. Qt 4 introduced several new features at the cost of source and binary compatibility with Qt 3. Many functions and enums were renamed to be more consistent, some functionality was simply removed from the API, whereas other features were isolated into a new Qt3Support library. A thorough porting guide was also provided to help clients transition to the new release. This allowed Qt to make a radical step forward and improve the consistency of the API while providing support for certain Qt 3 features to legacy applications.

Another option when deciding to remove some functionality is to leave it in the code so that legacy code continues to compile, but remove it from the associated API documentation so that its existence is no longer advertised to clients. The Qt library provides another good example of this, with the use of a preprocessor define that can compile out parts of a header during documentation generation, such as:

```
QGraphicsItem(QGraphicsItem *parent = 0
#ifndef Q_QDOC
              , QGraphicsScene *scene = 0  // obsolete argument
#endif
);
```

You can also achieve this with the Doxygen tool by surrounding the function declaration with the `\cond` and `\endcond` commands, such as:

```
/// \cond TestFunctions
void MyDeprecatedFunction();
/// \endcond
```

Inline namespaces for versioning

C++11 introduced the concept of inline namespaces, in which all members of an inlined namespace are added to the enclosing namespace. This feature can be used to help with versioning of an API by enclosing all of its symbols within a version namespace, such as:

```
namespace MyLibrary {
inline namespace v1 {

void MyFunction();

}  // namespace v1
}  // namespace MyLibrary
```

The function in this example is available as `MyLibrary::v1::MyFunction()` and also simply `MyLibrary::MyFunction()`. Now, if we release a new version of the API, we can preserve the original implementation as well as introduce the newer version. For example:

```
namespace MyLibrary {
namespace v1 {

void MyFunction();

}  // namespace v1
inline namespace v2 {

void MyFunction();

}  // namespace v2
}  // namespace MyLibrary
```

Here I've introduced a new `v2` namespace and made it be the inlined namespace instead of `v1`. As a result, the `MyLibrary::v2::MyFunction()` is now the version that is available as `MyLibrary::MyFunction()`. However, your clients can still access the previous version by referring explicitly to `MyLibrary::v1::MyFunction()`, or they could use `using namespace MyLibrary::v1` to use only the `v1` symbols consistently, even though `v2` is the default inlined version.

When your clients refer to `MyLibrary::MyFunction()` in their code, the compiler will use the full namespace with the version number in the object file. This means that if you were to break the binary compatibility of your API in a newer version, you wouldn't actually break your clients' projects because they would be explicitly linking against the previous version symbols. This approach therefore gives you the ability to insulate your clients from binary breaking changes, although at the cost of having to maintain an entire set of symbols for every version that you distribute. You could reduce this cost by employing this technique only for major version number changes of your API, or only for breaking version changes.

However, if you plan to use inline namespaces for versioning, you should use them from the very first version of your API. Otherwise, you could break ABI compatibility for your clients. For example, if they were originally linking against `MyLibrary::MyFunction()`, but you moved to using inline namespaces, then that symbol would no longer exist and instead there would be `MyLibrary::v1::MyFunction()` and `MyLibrary::v2::MyFunction()`.

API reviews

Backward compatibility doesn't just happen. It requires dedicated and diligent effort to ensure that no new changes to an API have silently broken existing code. This is best achieved by adding API reviews to your development process. In this section, I'll present the argument for performing API reviews, discuss how to implement these successfully, and describe a number of tools that can be used to make the job of reviewing API changes more manageable.

There are two models for performing API reviews. One is to hold a single prerelease meeting to review all changes since the previous release. The other model is to enforce a precommit change request process in which changes to the API must be requested and approved before being checked in. You can do both, of course.

The purpose of API reviews

You wouldn't release source code to your users without at least compiling it. In the same way, you shouldn't release API changes to your clients without checking that they don't break their applications. API reviews are a critical and essential step for anyone who is serious about API development. In case you need further encouragement, here are a few reasons to enforce explicit checks of your API before you release it to clients:

- **Maintain backward compatibility.** The primary reason to review your API before it's released is to ensure that you've not unknowingly changed the API in a way that breaks backward compatibility. As I mentioned, if you have many engineers working on fixing bugs and adding new features, it's possible that some will not understand the vital importance of preserving the public interface.
- **Maintain design consistency.** It's crucial that the architecture and design plans that you obsessed about for the version 1.0 release are maintained throughout subsequent releases. There are two issues here. The first is that changes that don't fit into the API design should be caught and recast; otherwise the original design will become diluted and deformed until you eventually end up with a system that has no cohesion or consistency. The second issue is that change is inevitable, and if the structure of the API must change, then this requires revisiting the architecture to update it for the new functional requirements and use cases. As a caution, John Lakos points out that if you implement one new feature for 10 clients, then every client gets nine features they didn't ask for, and you must implement, test, and support 10 features that you did not originally design for (Lakos, 1996).
- **Change control.** Sometimes a change may simply be too risky. For example, an engineer may try to add a major new feature for a release that's focused on bug fixing and stability. Changes may also be too extensive or poorly tested, appear too late in the release process, violate API principles such as exposing implementation details, or not conform to the coding standards for the API. The maintainers of an API should be able to reject changes that they feel are inappropriate for the current API release and target them for a future release, perhaps with feedback on ways to improve the change.
- **Allow future evolution.** A single change to the source code can often be implemented in several different ways. Some of those ways may be better than others in that they consider future evolution and put in place a more general mechanism

that will allow future improvements to be added without breaking the API. The API maintainers should be able to demand that a change be reimplemented in more future-proof fashion. Tulach calls this being evolution ready (Tulach, 2008).
- **Revisit solutions.** Once your API has been used in real situations and you've received feedback from your clients on its usability, you may come up with better solutions. API reviews can also be a place where you revisit previous decisions to see if they're still valid and appropriate.

If your API is so successful that it becomes a standard, then other vendors may write alternative implementations. This makes the need for API change control even more critical. For example, the creators of OpenGL recognized this need, and so they formed the OpenGL Architecture Review Board (ARB) to govern the standard. The ARB was responsible for changes to the OpenGL API, advancing the standard, and defining conformance tests. This ran from 1992 until 2006, at which point control of the API specification was passed to the Khronos Group.

> *TIP: Introduce an API review process to check all changes before releasing new version of the API.*

Prerelease API reviews

A prerelease API review is a meeting that you hold just before the API is finally released. It's often a good activity to perform right before or during the alpha release stage. The best process for these reviews for your organization will obviously differ from what works best elsewhere. However, here are some general suggestions to help guide you. First, you should identify who the critical attendees are for such a review meeting. For example:

1. **Product owner.** This is the person who has overall responsibility for product planning and representing the needs of clients. Obviously, this person must be technical, because your users are technical. In Scrum terminology, the product owner represents the stakeholders and the business.
2. **Technical lead.** The review will generate questions about why a particular change was added and whether it was done in the best way. This requires deep technical knowledge of the code base and strong software design skills. This person should be intimately familiar with the architectural and design rationale for the API, if they didn't develop this themselves.
3. **Documentation lead.** An API is more than code. It's also documentation. Not all issues that are raised in the review meeting need to be fixed by changing the code. Many will require documentation fixes. Strong documentation is a critical

component of any API, and so the presence of a person with technical writing skills at the review is very important.

4. **Testing lead.** One way to feel confident that you're not breaking backward compatibility is to ensure that all existing tests have been run and continue to pass. It is also important to be sure that new tests have been written that cover any new functionality. The developer will normally be responsible for maintaining unit tests, but a separate QA engineer may also be involved to deal with higher-level automated testing such as integration or performance tests.

You may decide to add further optional attendees to the review meeting, to try and cast a wider net in your effort to ensure that the new API release is the best it can be. At Pixar, we performed API reviews for all of the animation system public APIs we delivered to our film production users. In addition to those attendees, we would include a couple of senior engineers who work daily in the source code. You'll obviously also want someone from project management present to capture all of the notes for the meeting so that nothing is forgotten.

Depending upon the size of your API and the extent of the changes, these meetings can be long and grueling. We would regularly have several 2-hour meetings to cover all API changes for a release. You shouldn't rush these meetings or continue them when everyone is getting tired. Assume that they'll take several days to complete, and ensure that developers and technical writers have time to implement any required changes coming out of the meeting before the API is released.

For larger organizations, it's particularly important to have a good process for engineers to submit proposals for new public APIs or changes to existing public APIs. It's also vital to have a dedicated team focused on reviewing and enforcing the organization's best practices for API design. One large company where I worked has a particularly good API review process. They have an internal website where proposals for new or changed APIs can be submitted. The submissions are managed through GitHub, allowing reviewers and stakeholders to collaborate through comments, to enforce rules such as requiring two reviewer approvals, and to have an entire revision history for all API decisions. The format of these submissions is also well-defined and is composed of sections such as:

- **Metadata**: Each submission has a unique ID, a list of authors and sponsors, the target software release, any related bug tracking system IDs, and so on.
- **Introduction**: Provides a short elevator pitch for the API change.
- **Motivation**: Goes into detail about the problem being solved.
- **Proposed Solution**: Provides an overview of the solution, ideally with examples.
- **Detailed Design**: Declarations of new or changed public symbols, lifted verbatim from the code.
- **Impacts on Quality**: Any concerns on the usability or testing of the API.
- **Alternatives Considered**: Summarizes decisions that were made to get to the current proposal.

In terms of activities that should be performed during an API review, the most important thing is to focus on the interface being delivered, not on the specifics of the implementation. In other words, the primary focus should be on the public symbols and the documentation, such as the automatically generated API documentation (see Chapter 11 on Documentation for specifics on using Doxygen for this purpose). More specifically, you are most interested in changes to the API since the last release. This may involve the creation of a tool to report the differences between the current API and the previous version. For each change, the review committee should ask various questions, such as:

- Does this change break backward compatibility?
- Does this change break binary compatibility (if this is a goal)?
- Does this change have sufficient documentation?
- Could this change have been implemented in a more future-proof manner?
- Does this change have negative performance implications?
- Does this change break the architecture model?
- Does this change conform to the API coding standards?
- Does this change introduce cyclic dependencies into the code?
- Should the API change include upgrade scripts to help clients update their code or data files?
- Are there existing automated tests for the changed code to verify that functional compatibility has not been affected?
- Does the change need automated tests, and have they been written?
- Is this a change we want to release to clients?

Precommit API reviews

Of course, you don't have to wait until right before release to catch these problems. The prerelease API review meeting is the last line of defense to ensure that undesirable changes aren't released to users. However, the work of the API review can be greatly decreased if the API owners are constantly vigilant during the development process, watching checkins to the source code and flagging problems early on so that they can be addressed before they reach the API review meeting.

Many organizations or projects will therefore institute precommit API reviews. That is, they will put in place a change request process, in which engineers wishing to make a change to the public API must formally request permission for the change from the API review committee. Implementation-only changes that do not affect the public API do not normally need to go through this additional process. This is particularly useful for open source software projects, in which patches can be submitted from many engineers with differing backgrounds and skills.

For example, the open-source Symbian mobile device OS imposes a change control process for all Symbian platform public API changes. The stated goal of this process is to ensure that the public API evolves in a controlled manner. The process is started by submitting a change request with the following information:

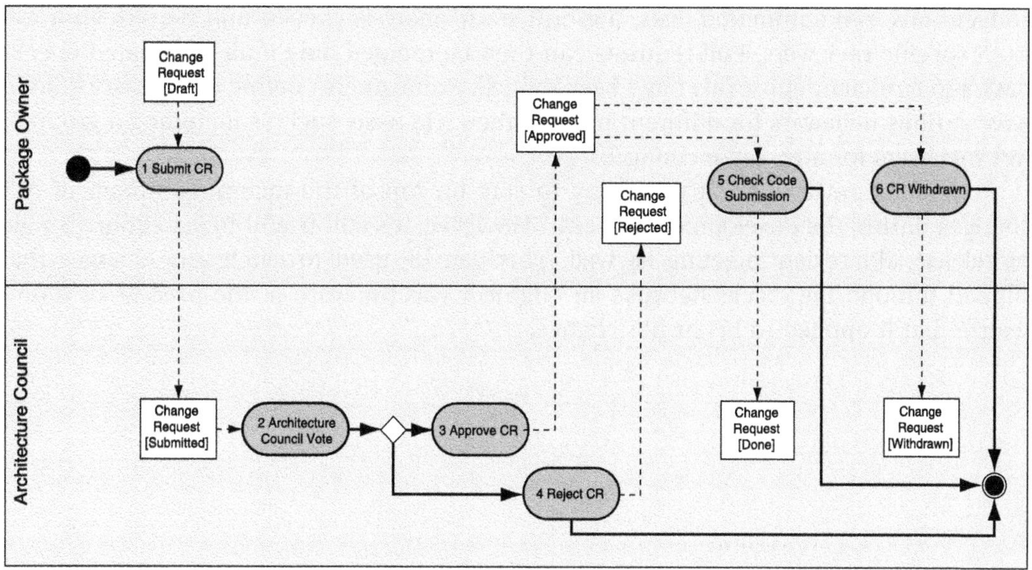

FIGURE 10.4 The public interface change request (CR) process for the Symbian platform. Copyright (c) 2009 Symbian Foundation limited. Licensed under the Creative commons license 2.0 by Stichbury.

- A description of the change and why it's necessary.
- An impact analysis for any clients of the API.
- A porting guide to the new version of the API.
- Updates to the backward compatibility test cases.

This is then reviewed by the architecture council, who will either approve or reject the request and provide the rationale for the decision. Once approved, the developer can submit the code, documentation, and test updates. Fig. 10.4 provides an overview of this process.

As another example, the Java-based NetBeans project defines an API review process for accepting patches from developers. This is done to supervise the architecture of the NetBeans IDE and related products. Changes to existing APIs, or requests for new APIs, are required to be reviewed before they are accepted into the trunk code line. This process is managed through the NetBeans bug tracking system, in which requests for review are marked with the keyword API_REVIEW or API_REVIEW_FAST. The review process will result in a change being either accepted or rejected. In the case of rejection, the developer is normally given direction on improvements to the design, implementation, or documentation that would make the change more acceptable. Of course, similar feedback may still be provided for accepted changes. For details on this process, see https://netbeans.apache.org/wiki/main/wiki/APIReviews/.

You can also rely on automation to make precommit checks easier. For example, using an SCM such as GitHub you can set up various commit hooks to enforce coding conventions, run automated tests, perform static analysis checks, and require approval from specific reviewers. Pull requests can then be merged only if all automated checks pass and sufficient approvals have been gained. You can also define a `CODEOWNERS` file to have various reviewers for different parts of the code base, such as including a group of API reviewers for any `.h` file changes.

Precommit reviews are a good way to stay on top of the incoming stream of API changes during the development process. However, it's still useful to schedule a single prerelease API review meeting as well. This can be used to catch any changes that slipped through the cracks because an engineer was unaware of the process or didn't realize that it applied to his or her change.

Documentation

In the very first chapter of this book, I defined an application programming interface (API) as one or more header files plus supporting documentation. In fact, I claim that an API is incompletely specified unless it includes accompanying documentation. This is because header files do not specify the behavior of the API, only the calling conventions for the various functions and methods. David L. Parnas stated this well (Parnas, 1994):

> *Reuse is something that is far easier to say than to do. Doing it requires both good design and very good documentation. Even when we see good design, which is still infrequently, we won't see the components reused without good documentation.*

Well-written documentation is therefore a critical element of your API. Accordingly, I am dedicating an entire chapter to this topic. I will start by covering some of the reasons why you should care about documentation, and then describe the various types of documentation that you could provide along with several tools that can be used to aid the task of writing documentation.

One of the easiest ways to document your API is to use a tool that automatically extracts comments from your header files and builds API documentation for you. One of the most popular and full featured of these tools is Doxygen, from Dimitri van Heesch. I will therefore spend some time looking at how you can use Doxygen in your projects and provide some sample templates for documenting various elements of your API.

Documenting your implementation code is good practice, too, but that documentation is for your own internal use, whereas the need to document your API header files is more vital because that documentation is for your users.

I feel obliged to admit at this point that most of the source code examples in this book have little or no comments or documentation. This was done simply to keep these examples as minimal and focused as possible, both to make them easier to digest by you, the reader, and so the book doesn't grow to twice its current size. As a compromise, I've ensured that the source code examples that accompany this book are all well-documented, as should any API headers that you write, too.

Reasons to write documentation

I hope I don't need to convince you that providing your users with details on how to use your API is a good thing. It's certainly true that if you follow the core principles from the Qualities chapter, then your interfaces should already be consistent, discoverable, and easy to use. However, that's not a replacement for good documentation. Professionally

API Design for C++. https://doi.org/10.1016/B978-0-443-22219-1.00015-5

written documentation is an equally important component of delivering high-quality, world-class APIs. In fact, good documentation can make the difference between a user adopting your API or looking for an alternative.

Defining behavior

An API is a functional specification. That is, it should define how to use an interface as well as how the interface should behave. Simply looking at a method in a header file will tell you the number and types of its arguments and the type of its return value, but it says nothing about how that method will behave. To illustrate this point, consider this definition of a class that represents an RGB color:

```
class RGBColor
{
public:
    RGBColor(float, float, float);
    ~RGBColor();

    float Red() const;
    float Green() const;
    float Blue() const;

    void Set(float, float, float);
};
```

This class definition fully specifies the arguments and return types for each method, so you can write code that creates instances of this class and call each method. However, several things are unknown about the behavior of this class. For example:

- Are the red, green, and blue floating-point values represented by a 0.0 to 1.0 range, as a percentage from 0% to 100%, as a floating-point range from 0.0 to 255.0 (so that the midvalue of 127.5 can be accurately represented), or from 0 to 65,535?
- What happens if values outside this range are passed to the constructor or `Set()` method? Will it leave them as is, clamp them, take the modulus of the values, or attempt to scale them into the valid range?
- What's the order of the parameters to the constructor and the `Set()` method? Presumably it's red, green, and then blue, but this hasn't been explicitly documented in the function signatures.

The subsequent updated version of this class provides some documentation in the form of comments, to help clarify these points. The triple slash comment style will be explained later in the chapter when I talk more about Doxygen:

```
/// Represents a single RGB (Red, Green, Blue) color
class RGBColor
{
public:
    /// Create an RGB color from three floats in the range 0..1
    /// Out of range values will be clamped to this range
    RGBColor(float red, float green, float blue);
    ~RGBColor();

    /// Return the red component of the color, in the range 0..1
    float Red() const;
    /// Return the green component of the color, in the range 0..1
    float Green() const;
    /// Return the blue component of the color, in the range 0..1
    float Blue() const;

    /// Set the RGB color with three floats in the range 0..1
    /// Out of range values will be clamped to lie between 0..1
    void Set(float red, float green, float blue);
};
```

These comments are quite minimal, but nonetheless they specify the behavior of the API to a sufficient degree for a client to use it. Documentation doesn't have to be long-winded to be thorough. Although more details will be appreciated by your users, minimal documentation is better than none at all.

TIP: Good documentation describes how to use your API and how it will behave under different inputs.

It's good practice to write these kinds of comments as you're writing the header file itself. You can always go back and add more content or revise the details once the API is more stable and is approaching release quality. But if you do run out of time and need to get the API out the door quickly, you will at least have some basic level of documentation. Even better, you should explicitly schedule a task to write the documentation for your API before it is released.

You may even consider creating commit hooks (also known as triggers) for your revision control system to reject new public API code that doesn't include documentation. If that's too draconian a policy, you can still perform the code checks but treat them as advisory feedback rather than roadblocks to checking in new code.

TIP: Write API documentation as you implement each component. Then revise it once the API is complete.

Documenting the interface's contract

The term design by contract was introduced by Bertrand Meyer as an approach to defining formal specifications for software components (Meyer, 1987). Meyer later trademarked the term in the United States, so many developers now refer to it as contract programming instead. The central notion is that a software component provides a contract, or obligation, for the services that it will provide, and that by using the component a client agrees to the terms of that contract. This concept is founded upon Hoare logic, a formal system developed by Tony Hoare for reasoning about the correctness of programs (Hoare, 1969). The main principles of this model are threefold:

1. **Preconditions**: the client is obligated to meet a function's required preconditions before calling a function. If the preconditions are not met, then the function may not operate correctly.
2. **Postconditions**: the function guarantees that certain conditions will be met after it's finished its work. If a postcondition is not met, then the function did not complete its work correctly.
3. **Class invariant**: these are constraints that every instance of the class must satisfy. This defines the state that must hold true for the class to operate according to its design.

The best way to specify the details of a contract is in the class and function documentation for your API. In other words, your functions should specify any expected preconditions that must be in force before calling them, as well as any resulting postconditions that will be met after they've completed. Also, your class documentation should specify any invariants that will hold true during the observable lifetime of each class (i.e., after construction, before destruction, and before and after each public method call). Let's look at a couple of examples to demonstrate this concept more concretely.

A square root function will have the precondition that the input number must be a positive number, or zero. It will have a postcondition that the result of the function squared should equal the input number (within an appropriate tolerance). These conditions can be documented using Doxygen syntax, as follows (in the next chapter on testing, I'll talk about how you can enforce these conditions in code):

```
///
/// \brief Calculate the square root of a floating-point number.
/// \pre value >= 0
/// \post fabs((result * result) - value) < 0.001
///
double SquareRoot(double value);
```

To give a more object-oriented example, consider a string container class. This has a class invariant that the length of the string, as returned by the `size()` method, must always be zero or greater. Also, the pointers returned by `c_str()` and `data()` must always be nonnull. For the class's `append()` method, the length of the string being changed must grow by the length of the input string. These contract terms are expressed in the following documentation comments (the term `@pre.size()` is meant to represent the length of the string before the method is called):

```
///
/// \brief A container that operates on sequences of characters.
/// \invariant size() >=0 && c_str() != NULL && data() != NULL
/// \invariant empty() implies c_str()[0] == '\0'
///
class String
{
public:
    ...

    ///
    /// \brief Append str to this string object.
    /// \post size() == @pre.size() + str.size()
    ///
    void append(const std::string &str);

    ...
};
```

TIP: Contract programming implies documenting the preconditions and postconditions for your functions and the invariants for your classes.

In terms of inheritance relationships between classes, Ken Pugh notes that the preconditions for derived classes can be weaker, but not stronger, than those in its base classes (Pugh, 2006). That is, a derived class should handle all cases that its base classes handle, but it may decide to handle additional cases by requiring fewer preconditions.

By contrast, a derived class should inherit all postconditions from its base classes. That is, a function in a derived class must meet all of its base class postconditions as well as the further postconditions that it defines itself.

Communicating behavioral changes

There are many cases in which an API change doesn't involve changes to any class definitions or function signatures. This happens when you change the underlying implementation for a function without affecting its interface. From a code perspective, this change will be invisible to your clients because it doesn't affect the function signatures in your header files. However, the API has indeed changed because its behavior has changed. In other words, the update will be source and binary compatible, but not functionally compatible.

In this case, you can communicate the change by modifying the documentation for the API call. For example, consider this function that returns a list of children in a hierarchy of nodes:

```
/// Return a list of children in the hierarchy.
///
/// \return A nullptr-terminated list of child nodes,
///         or nullptr if no children exist.
///
const Node *GetChildren() const;
```

According to the documentation, this function returns a `nullptr` if there are no children in the hierarchy. This behavior forces clients to guard against a `nullptr` return value, such as:

```
const Node *children = hierarchy.GetChildren();
if (children) {
    while (*children != nullptr) {
        // process node
        children++;
    }
}
```

Now let's say that in a future version of this API you realize that you can make the lives of your clients easier, and improve the stability of their code, if you instead return a valid `Node *` pointer that points to a `nullptr` value when there are no children. This obviates the need for special `nullptr` checks, so clients could instead write code such as:

```
const Node *children = hierarchy.GetChildren();
while (*children != nullptr) {
    // process node
    children++;
}
```

This change involved no modification to the function signature for the `GetChildren()` method. However, the change can still be communicated to the client by updating the documentation for the method. For example,

```
/// Return a list of children in the hierarchy.
///
/// \return A non-null pointer to a list of child nodes.
///         The list is terminated with a nullptr.
///
const Node *GetChildren() const;
```

What to document

You should document every public element of your API: every class, function, enum, constant, and typedef. If your clients can access it, you should tell them what it is and how they can use it (Bloch, 2008). In particular, you should focus on any important aspects that will help them to use your classes and methods productively. This includes describing the behavior of methods as well as their relationship to the rest of the API.

Specifying the units of parameters and return values is another particularly important element of good documentation. When doing so, you should consider whether you need to define the nature, accuracy, and precision of the values, too. A particularly pertinent example is that of a timer class that calls client code every *n* seconds. You may document the units of the timeout to be seconds, but it would still be reasonable for a client to ask:

- Does time refer to real-world (wall clock) time or process time?
- What's the accuracy of the timer?
- Will other API operations affect the accuracy of the timer or block it?
- Will the timer drift over time or will it always fire relative to the start time?

Defining these additional characteristics will help your users work out whether the class is appropriate for their tasks. For example, an idle task that just needs to perform some trivial work roughly every second doesn't care whether it gets woken up after 1.0 s or 1.1 s. However, an analog clock that increments its second hand on every invocation would soon show the wrong time under the same conditions.

As an aid to working out what you should document for your APIs, this list provides a checklist of questions to ask yourself when you're producing the documentation for your classes and functions:

- What is the abstraction that a class represents?
- What are the valid inputs (e.g., can you pass a `nullptr`)?

- What are the valid return types (e.g., when does it return true versus false)?
- What error conditions does it check for (e.g., does it check for file existence)?
- Are there any preconditions, postconditions, or side effects?
- Is there any undefined behavior, such as `sqrt(-1.0)`?
- Does it throw any exceptions?
- Is it thread-safe?
- What are the units or any parameters?
- What is the space or time complexity, such as O(log n) or $O(n^2)$?
- What is the memory ownership model (e.g., is the caller responsible for deleting any returned objects)?
- Does a virtual method call any other method in the class (i.e., what methods should clients call when they override the method in a derived class)?
- Are there any related functions that should be cross-referenced?
- In which version of the API was this feature added?
- Is this method deprecated, and if so what's the alternative?
- Are there any known bugs in this feature?
- Do you wish to share any planned future enhancements?
- Are there any code examples you can provide?

In addition to this list of API behaviors, Diomidis Spinellis developed a short list of qualities for which all good code documentation should strive (Spinellis, 2010). He recommends that documentation should be:

1. Complete.
2. Consistent.
3. Effortlessly accessible.
4. Nonrepetitive.

TIP: Document every public element of your API.

It's also good practice to have another person review your documentation. As the developer who wrote the code, you are often too close to the problem space and may assume knowledge that the normal user will not have. As Michi Henning puts it, the developer of the API is "mentally contaminated by the implementation" (Henning, 2009). Therefore, have a fellow developer, quality assurance (QA) engineer, or technical writer look over your API documentation and be open to the feedback from that fresh perspective.

Types of documentation

There are various types of documentation you can provide for your API. Not all of these will be appropriate for every project, but it's worth being aware of the various options that your users may expect.

Moreover, many software products provide documentation that can accept contributions from users. This may be via the use of a wiki, in which any user can create and edit documentation, such as the massive World of Warcraft wiki at https://wowpedia.fandom.com/ or Web pages that support user comment addendums, such as the PHP manual at http://www.php.net/manual/. This is obviously a very powerful capability that can build a community around your API and create collective documentation that goes beyond what you could achieve on your own. Even if you don't employ an automated system for user contributions, you should provide a way for them to send feedback to you so that it can be incorporated into the documentation manually.

Much of the documentation that you (or your users) write will apply to specific versions of the API. One trap that's easy to fall into with wiki-based solutions is that the wiki content will eventually refer to a range of versions of the API. It's worth considering how you can create distinct versioned sets of documentation, perhaps generating a separate copy of the documentation for each major release. This will allow clients who are using older versions to view the documentation for that specific release, while letting you keep the latest documentation focused on the current version of the API.

Fig. 11.1 shows a diagram of the Qt reference documentation index. This is included as an example of extensive and well-written documentation that demonstrates all of the various types of documentation that I'll cover next. Also, all of the Qt documentation is versioned along with the API.

Automated API documentation

The most obvious kind of documentation for an API is the type that's automatically generated from comments in your source code. This allows you to keep the documentation for classes and methods right next to their declaration in the header files. The result is that you're more likely to keep the documentation up-to-date when you change the API. It's also easy to generate the latest documentation for a new release, and this documentation can have cross-references between related functionalities of the API.

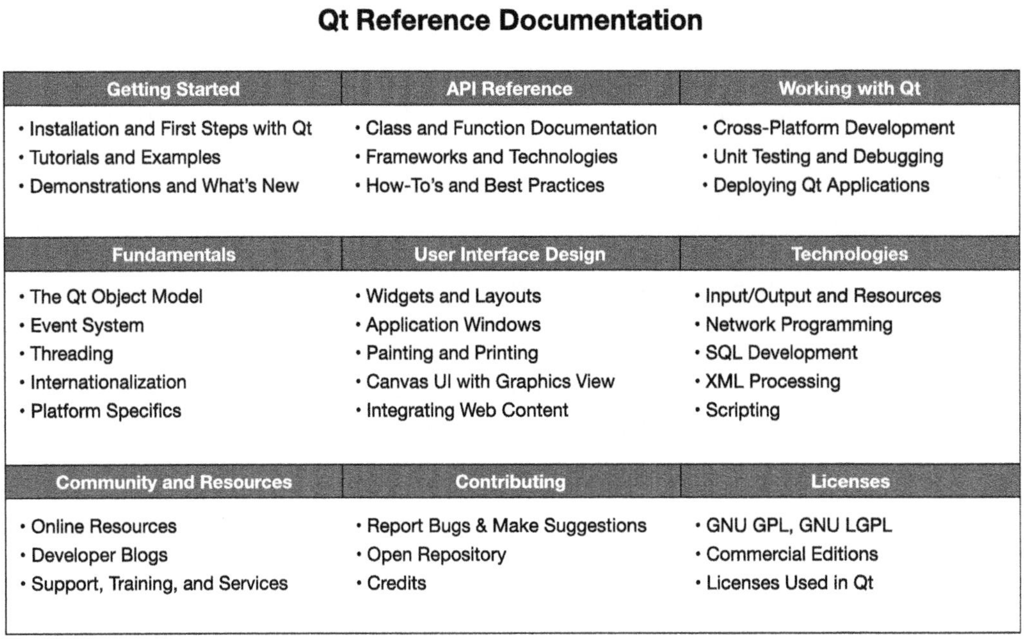

Qt Reference Documentation

Getting Started	API Reference	Working with Qt
• Installation and First Steps with Qt • Tutorials and Examples • Demonstrations and What's New	• Class and Function Documentation • Frameworks and Technologies • How-To's and Best Practices	• Cross-Platform Development • Unit Testing and Debugging • Deploying Qt Applications
Fundamentals	**User Interface Design**	**Technologies**
• The Qt Object Model • Event System • Threading • Internationalization • Platform Specifics	• Widgets and Layouts • Application Windows • Painting and Printing • Canvas UI with Graphics View • Integrating Web Content	• Input/Output and Resources • Network Programming • SQL Development • XML Processing • Scripting
Community and Resources	**Contributing**	**Licenses**
• Online Resources • Developer Blogs • Support, Training, and Services	• Report Bugs & Make Suggestions • Open Repository • Credits	• GNU GPL, GNU LGPL • Commercial Editions • Licenses Used in Qt

FIGURE 11.1 Diagram of the Qt reference documentation categories. For the latest version, see https://doc.qt.io/. The Qt product is copyright (c) 2023 The Qt Company.

> *TIP: Use an automated documentation tool that extracts API documentation from your header comments.*

It's perhaps obvious, but any user documentation comments should appear in your public headers, not `.cpp` files, so that they're more readily discoverable by your users. This is particularly important for closed source APIs, in which users don't have access to your `.cpp` files.

These source code comments are often maintained by developers. However, the quality of your API documentation can be greatly enhanced if you have professional technical writers perform a pass over the comments for style, grammar, and consistency checks. Also, as a fresh set of eyes, these writers can often provide good feedback and suggest additional information to highlight or improve examples to include. Some technical writers may not feel comfortable changing source code directly, but in those cases they can easily mark up a hard copy of the documentation and let the developer make the actual code changes.

There are several tools that let you create API documentation from comments in your C++ source code. These can normally generate output in various formats, such as

HTML, PDF, and LaTeX. The following list provides links to several of these tools. You can also find a good comparison of documentation generators on Wikipedia:

- **AutoDuck**: http://helpmaster.info/hlp-developmentaids-autoduck.htm
- **CcDoc**: http://ccdoc.sourceforge.net/
- **CppDoc**: https://gitlab.com/bproto/cppdoc
- **Doc-O-Matic**: http://www.doc-o-matic.com/
- **Doc++**: http://docpp.sourceforge.net/
- **Doxygen**: http://www.doxygen.org/
- **GenHelp**: http://www.frasersoft.net/
- **HeaderDoc**: https://en.wikipedia.org/wiki/HeaderDoc
- **ROBODoc**: https://rfsber.home.xs4all.nl/Robo/

Overview documentation

In addition to autogenerated API documentation, you should have manually written prose that provides higher-level information about the API. This will normally include an overview of what the API does and why the user should care about it. In a large organization, this task is normally performed by a technical writer. It may even be localized into several different languages. The overview documentation will often cover these topics:

- A high-level conceptual view of the API: what problem is being solved and how the API works. Diagrams are great if they are appropriate.
- The key concepts, features, and terminology.
- System requirements for using the API.
- How to download, install, and configure the software.
- How to provide feedback or report bugs.
- A statement on the life cycle stage of the API (e.g., prerelease, maintenance, stability, or deprecated; see the Versioning chapter).

There can also be opportunities to write custom scripts that make calls to your API or parse data files (or even parts of source code files) so that they can insert richer information into your overview documentation. For example, imagine an image processing API that provides several image filtering objects, each holding a name, a description, and a set of parameters. You could write a script to extract each filter object from the API and generate a simple markdown table in your overview documentation. That way, users have easy access to key information about the features of your API.

In the past, I've supported this kind of capability by writing documentation that embeds HTML comments to indicate the start and end of a section that will be updated by a script. (Most markdown systems can parse HTML comments but won't show them in the rendering of the documentation.) Then I've written a script that loads the markdown file, finds the special start and end comments, and inserts the autogenerated content between them. For example, the file may originally look like:

```
| ID        | Name        | Description |
| -------- | --------- | ----------- |
<!--- START GENERATED CONTENT --->
// entire content here is replaced each run of the script
<!--- END GENERATED CONTENT --->
```

and after running the script, the file may look like:

```
| ID        | Name        | Description |
| -------- | --------- | ----------- |
<!--- START GENERATED CONTENT --->
Median Filter | Removes noise by blending pixel brightness | window_size |
Gaussian Blur | Blurs an image with a Gaussian function | x_size, y_size |
Pixelize | Turns an image into large square pixels | block_size |
Sharpen | Sharpens edges without increasing noise | radius, amount |
<!--- END GENERATED CONTENT --->
```

This way, you can generate rich overview documentation that provides your clients with useful information about your API. Because these sections of your documentation are autogenerated, all you need to do is rerun your script and your documentation will instantly be up-to-date with your code.

Examples and tutorials

Examples are really, really important. An API overview is often too high-level to allow users to glean how to use your API, and even though you have documentation for every class and method in your API, this doesn't immediately tell the user how to use the API to perform a task. Adding even a few small examples can greatly enhance the utility of your documentation. Remember that an API is software for other developers. They just want to know how they can use your interface to get their job done. This can be part of a Getting Started section in your documentation and may include any of these:

- **Simple and short examples.** These should be minimal and easy to digest snippets of code that demonstrate the API's key functionality. They're not normally code that can be compiled, but instead focus on your API calls without all the boiler-plate code that goes around it.
- **Working demos.** These are complete real-world examples that show how the API can be used to perform a more complex and practical task. These should be easy

to reuse so that your users have a working starting point for their own projects. You should provide the source code for these with your API.
- **Tutorials and walkthroughs.** A tutorial illustrates the steps that you go through to solve a problem, rather than simply presenting the end result. This can be a useful way to build up a complex example and to address specific features of the API as you gradually add more calls to the worked example.
- **Sequence diagrams.** A UML sequence diagram can provide more context on how to perform certain tasks using your API. They can also be relatively easy to generate using freely available diagramming and charting tools. I provided an example of a sequence diagram back in Fig. 2.3.
- **User contributions.** Your users can be a great source of examples, too. Encourage your users to send you example code that can be added to your collection of demos, perhaps under a specific contrib directory so that it's clear that these are not supported by you.
- **FAQs.** A set of answers to frequently asked questions can be a very helpful addition to your documentation set. It lets your users quickly and easily discover whether the API suits their needs, how to avoid common pitfalls, or how to solve typical problems.

The act of writing documentation forces you to think from the user's perspective. As such, it can help you to identify shortcomings in the API or areas that could be simplified. It's therefore good practice to write the skeleton for the high-level documentation and some initial example code early on, to force you to think more deeply about the overall design and the use cases of your library.

Automation can also be useful here. For example, your sample code can be taken from your corpus of unit tests. That way, you guarantee that your code samples always compile and run successfully. You can then have a script that copies the relevant test code from your repository and embeds it into your documentation. This can use the same technique as earlier with special start/end comments in the markdown files to indicate where to insert the code. Again, this provides the benefit that your documentation is always up-to-date with your latest working code.

Release notes

Each release after the first one should include release notes. These tell your users what has changed since the last release. Release notes are normally terse documents that can include:

- An overview of the release, including a description of what's new or what the focus was for the release (e.g., bug fixes only).
- A link to where the release can be found.
- Identification of any source or binary incompatibilities from the previous release.
- A list of bugs fixed in the release.

- A list of features that were deprecated or removed.
- Migration tips for any changes in the API, such as how to use any upgrade scripts that are provided with the release.
- Any known issues, either introduced in this release or remaining from previous versions.
- Troubleshooting tips to work around known issues.
- Information on how users can send feedback and bug reports.

License information

You should always specify the license under which you're distributing your API. This lets your clients know what rights you are granting them and what their obligations are. Generally, you will use a license from one of these two categories:

1. **Proprietary**: Where the publisher retains ownership of a piece of software and may prohibit certain uses of it. Often these software products are closed source and require a licensing fee (although as a counterexample, the Qt commercial license includes the source code). Proprietary licenses may restrict certain activities such as reverse engineering the software, the number of users or machines, the number of developers, or concurrent use of the software by multiple users.
2. **Free and Open Software**: Software that can be used, studied, and modified without restriction. It can also be copied and redistributed in either modified or unmodified form with no or minimal restrictions. The term free refers to the usage rights of the software, not necessarily its price. Software that conforms to this category is referred to as free and open source software, commonly abbreviated to FOSS or FLOSS (free, libre, open source software).

There are two major bodies that approve FLOSS licenses. The Free Software Foundation (FSF), founded by Richard Stallman in 1985, approves licenses that comply with The Free Software Definition, and the Open Source Initiative (OSI), founded by Bruce Perens and Eric S. Raymond in 1998, approves licenses that comply with their Open Source Definition.

TIP: Specify the license terms for your API prominently.

There are also two principal types of FLOSS licenses:

1. **Copyleft**: Offers the right to distribute modified and unmodified copies of a piece of software and requires that any such derived works must be released under the same terms. There is a further subcategory of weak copyleft, which is often applied

to software libraries to allow clients to distribute code that links to that library without requiring that product to be distributed under the library's copyleft license.

2. **Permissive**: Offers the right to distribute modified and unmodified copies of a piece of software and allows the derived work to be distributed under terms that are more restrictive than those in the original license. This means you can provide an open source library, but your clients are not required to make all distributed derivates be available as open source.

Table 11.1 lists some common FLOSS software licenses and a brief overview of their impact on your users. All of these licenses are approved by both the FSF and OSI. This list is obviously incomplete and meant only as a rough guide. If you have access to a general counsel, you should consult with that person on the best license choice for your product. For more details, refer to http://www.fsf.org/ and http://www.opensource.org/.

This concludes our discussion of the various types of documentation that you might provide for your API. Fig. 11.2 presents another example of the documentation overview

Table 11.1 Common open source software licenses.

License name	Brief description
No license	If you do not specify a license, your users have no legal right to use your application programming interface (API) unless they directly ask for your permission (because you are the copyright holder).
GNU GPL license	The GNU general public license (GPL) is a copyleft license, which means that any derived works must also be distributed as GPL. An open source GPL library therefore cannot be used in a proprietary product. The Linux kernel and the GIMP image processing tool are released under the GPL. The Qt library was originally released under either a commercial or GPL license.
GNU LGPL license	The GNU Lesser GPL (LGPL) is a weak copyleft license that allows an open source API to be binary linked to proprietary code. The derived work can be distributed under certain specific conditions, such as providing the source code of the modified or unmodified LGPL library, among others constraints. GTK+ is licensed under the LGPL. Nokia added the LGPL license to Qt as of version 4.5.
BSD license	The BSD license is a simple permissive license that includes a legal disclaimer of liability by a named owner/organization. Normally a modified version of BSD is used, without the advertising clause. Proprietary code that links against BSD code can be freely distributed. Google released its Chrome browser under the BSD license. The Zend framework and libtiff library also uses BSD.
MIT/X11 license	This is a simple permissive license in the same vein as the BSD license. Proprietary code that links against MIT-licensed code can be freely distributed. MIT-licensed software includes Expat, Mono, Ruby on Rails, Lua 5.0 onward, and the X11 Window system.
Mozilla public license	This is a weak copyleft license that allows your open source library to be used to build proprietary software. Any code modifications must be redistributed under the MPL license. The Mozilla software products Firefox and Thunderbird are made available under the MPL.
Apache license	This is another permissive license that allows proprietary software to be distributed that is built upon Apache-licensed code. The Apache Web server obviously uses the Apache license. Google also uses it for many of its products, such as Android and the Google Web Toolkit.

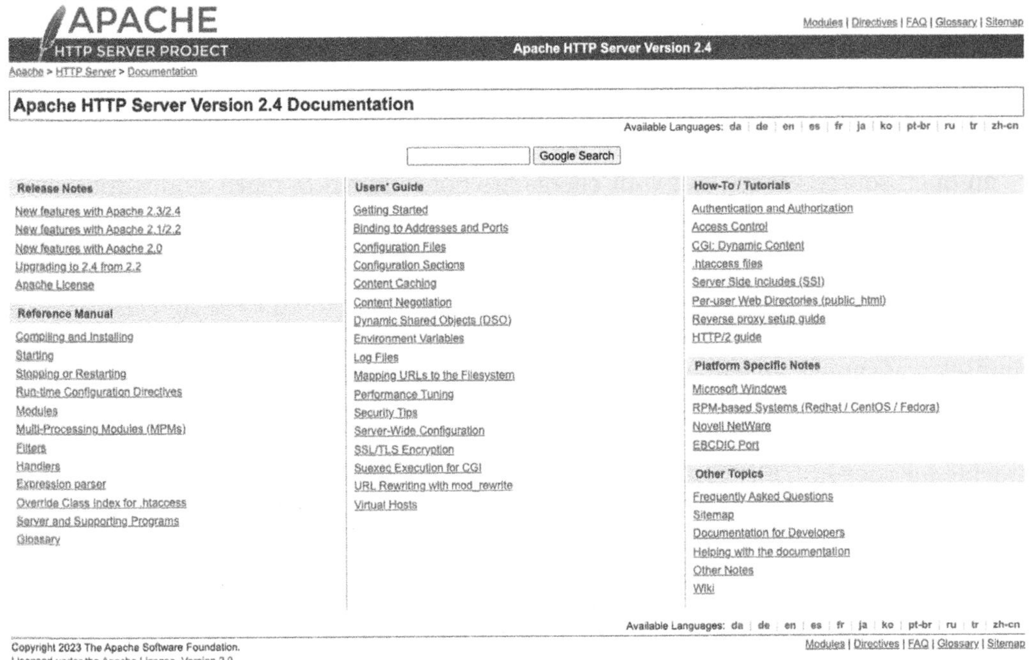

FIGURE 11.2 A screenshot of the documentation Web page for the Apache HTTP server, http://httpd.apache.org/docs/. Copyright (C) 2023 The Apache software foundation (ASF). *Reproduced with permission from the ASF.*

of a well-respected software project: the Apache HTTP Server. This documentation is also localized into several different languages.

Documentation usability

There are several research groups that investigate API documentation usability (Jeong et al., 2009; Robillard, 2009; Stylos and Myers, 2008). This work involves performing usability studies to see how well users can navigate through documentation and perform focused tasks. The aim is to be able to inform API designers about better ways to document their interfaces. This list summarizes some of these findings:

- **Index page.** Provide an overall index page that serves as a jumping-off point into the individual documentation elements. This gives users a conceptual map of the entire documentation set. Additionally, each starting point should provide some indication of what it will cover and what class of users it targets (developers, managers, legal, etc.)
- **Consistent look and feel.** API documentation will normally be composed of different elements, some autogenerated, some written by hand, and some contributed by users. You should use a consistent and unique style for all of these pages

so that users are always aware when they are browsing your documentation content, or if they have navigated to another website.

- **Code examples.** There can often be a lot of documentation for users to read. Consider the massive size of the Microsoft Developer Network library. Providing example code snippets and working demos can help users more quickly find and assimilate the information they need to use the API in their own code.
- **Diagrams**: Clear and concise diagrams that illustrate the high-level concepts can be helpful, particularly for users who just want to scan the documentation quickly. You should use familiar diagram formats where possible, such as UML or entity relationship diagrams.
- **Search**: A good search facility is important to let users find the information they need as fast as possible. All parts of the documentation should be searchable, including the autogenerated API specification as well as any manually written text.
- **Breadcrumbs**: Use navigation aids that let users keep track of their location within the documentation hierarchy and easily move back up the hierarchy. The term breadcrumbs is used to describe the common technique of displaying the current location as a series of pages with a separator symbol (e.g., "index > overview > concepts"). In addition, it can be useful to let users easily backtrack to various high-level pages.
- **Terminology**: Crucial terminology should be defined and used consistently. However, you should avoid specialized or esoteric terminology where it's not necessary because this can confuse and frustrate users.

Related to API usability is the property of ease of learning: a difficult-to-use API will likely also be difficult to learn. To this end, Martin Robillard investigated the question of what makes an API challenging to learn. He found that one of the largest obstacles to learning an API is the supporting documentation and resources. For example, he lists these hindrances to API learning (Robillard, 2009):

- **Lack of code examples**: Insufficient or inadequate examples are provided.
- **Incomplete content**: Documentation is missing or inadequately presented.
- **Lack of task focus**: No details are offered on how to accomplish specific tasks.
- **No design rationale**: Insufficient or inadequate details are provided on the high-level architecture and design rationale.
- **Inaccessible data formats**: Documentation isn't available in the desired format.

In addition to these points, a lot of research has focused on the fact that developers are often reluctant to read API documentation carefully or thoroughly (Zhong et al., 2009). This suggests that providing more documentation sometimes can be detrimental because your users may miss the really important points and caveats because they are buried too deeply. The use of higher-level tutorials and example code can help to address this problem. Also, cross-references between related classes and methods can lead the user to discover features about which they were unaware. Some researchers have also suggested using word clouds or other variable-font techniques to highlight important or common classes (Stylos et al., 2009).

Inclusive language

There has been a conscious move in recent years across the software engineering field to try to use more respectful and inclusive language. One high-profile example is when GitHub changed its default branch name for newly created repositories from master to main, to avoid racially charged associations.

Your software and its documentation will likely be used by a broad range of users from a diverse set of backgrounds and cultures. You may therefore want to consider the terminology and phrasing you use to ensure that they convey the concepts you intended in a positive and welcoming manner. It's important to appreciate that it's not your intent that matters, but the potential impact of your words on the users of your software or documentation.

One simple way to achieve this is to prefer more neutral terms where possible. This table lists common software terms that may be considered insensitive today, and some suggestions for more inclusive alternatives.

Term	Alternative
Black box/white box	Closed box/open box
	Functional/internal
	Behavioral/API
Black hat/white hat	Malicious/approved
	Unethical/ethical
Black list/white list	Deny list/allow list
	Deny list/permit list
	Reject list/accept list
	Unapproved list/approved list
Dummy	Boilerplate
	Placeholder
	Sample
	Test
Female/male	Socket/connector
	Port/plug
Man-in-the-middle	Intermediary attack
	Machine-in-the-middle
Master	Main
	Default
	Primary
	Root
Master/slave	Primary/secondary
	Primary/replica
	Main/secondary
	Host/client
	Active/standby
Scrum master	Scrum coach
	Scrum advocate
	Scrum facilitator
Webmaster	Website administrator

Many organizations have writing style guides that provide recommendations on how to write inclusively and respectfully. A common theme across many of these documents is to imagine how your content might be interpreted from different perspectives. For example, will your word choices be understood by all of your users, and do those words carry any harmful or negative associations? The following list offers some areas to consider when naming concepts in your software and when writing the associated documentation:

- **Color:** Avoid using color to convey positive or negative qualities, such as using white as a label for good and black for bad. Also avoid words such as nonwhite that imply white is the default.
- **Race:** Avoid using words that reinforce racial or ethnic stereotypes. More generally, simply avoid referring to race unless it's relevant to your software project. You may also want to prefer terms such as built-in instead of native.
- **Gender:** When the context doesn't require a specific gender, you can use gender-neutral pronouns such as they, their, or them, or use the plural form (e.g., "Users can enter their passwords here").
- **Age:** Try to avoid language that biases against certain age groups. For example, instead of saying "grandfathered" you could try using "legacy," "exempt," or "preexisting."
- **Names:** When using names to describe personas or theoretical users of your software, think about using diverse names such as Julia, Toby, Anaïs, Manuel, Hae-Won, Aki, Lakshmi, Sinéad, or Xiu.
- **Ability:** Try to avoid terms that could refer to physical abilities or mental health. For example, you might prefer degraded or immobilized instead of crippled, or consistency check instead of sanity check.

Language offers us many rich and expressive ways to convey meaning. It should be easy to find accurate and descriptive terms that are also welcoming and positive for our users. If you want to learn more about inclusive writing, these websites offer some great resources on the topic:

- Inclusive Naming Initiative: https://inclusivenaming.org/
- GSA/18F Inclusive Language guide: https://content-guide.18f.gov/our-style/inclusive-language/
- Conscious Style Guide: https://consciousstyleguide.com/
- The Diversity Style Guide: https://www.diversitystyleguide.com/
- Disability Language Style Guide: https://ncdj.org/style-guide/

Using Doxygen

Doxygen is a utility to generate API documentation automatically in various formats based upon the comments that you write in your source code. It has support for many languages, including C, C++, Objective-C, Java, Python, Fortran, C#, and PHP, and it can generate output in several formats, including HTML, LaTeX, PDF, XML, RTF, and Unix man pages, among others.

Doxygen is open source (released under the GNU GPL) and binaries are provided for several platforms, including Windows, Linux, and Mac. It's been developed since 1997 and is now a mature and powerful system. Because it's a popular tool and is used on many projects, I've dedicated the rest of this chapter to covering the basics of how you can use Doxygen in your own projects. However, similar guidelines apply to any of the other tools that I listed earlier in the Automated API documentation section.

The Doxygen website is http://www.doxygen.org/.

The configuration file

Doxygen is highly configurable, with over 200 options in recent versions. All of these options can be specified via the Doxygen configuration file. This is an ASCII text file with a simple `key = value` format. You can generate a default configuration file by running Doxygen with the `-g` command line argument.

You will then want to edit this configuration file to specify some details about your source code and to change the default behavior in certain cases. Some of the entries I find that I change a lot are:

```
PROJECT_NAME = <name of your project>
FULL_PATH_NAMES = NO
TAB_SIZE = 4
FILE_PATTERNS = *.h *.hpp *.dox
RECURSIVE = YES
HTML_OUTPUT = apidocs
GENERATE_LATEX = NO
```

With this initial setup performed, you can simply run Doxygen in your source directory and it will create an `apidocs` subdirectory with your API documentation (if you adopted the previous `HTML_OUTPUT` setting). In the following sections I'll describe how you can add comments to your code that Doxygen will pick up and add to this generated documentation.

Comment style and commands

You must use a special comment style to signal to Doxygen that you wish to add the comment text to the API documentation. There are various comment styles that Doxygen supports, including:

```
/**
 * ... text ...
 */

/*!
 * ... text ...
 */

///
/// ... text ...
///

//!
//! ... text ...
//!
```

Which style you adopt is a matter of personal taste: they all behave the same. I will adopt the triple slash style (///) for the rest of this chapter.

Within a comment, there are several commands you can specify to provide specific information to Doxygen. This information will often be formatted specially in the resulting documentation. The next list summarizes some of the most useful commands (refer to the Doxygen manual for the complete list):

- `\file` [<filename>]
- `\class` <class name> [<header-file>] [<header-name>]
- `\brief` <short summary>
- `\author` <list of authors>
- `\date` <date description>
- `\param` <parameter name> <description>
- `\param[in]` <input parameter name> <description>
- `\param[out]` <output parameter name> <description>
- `\param[in,out]` <input/output parameter name> <description>
- `\return` <description of the return result>
- `\code` <block of code> `\endcode`
- `\verbatim` <verbatim text block> `\endverbatim`
- `\exception` <exception-object> <description>
- `\deprecated` <explanation and alternatives>
- `\attention` <message that needs attention>
- `\warning` <warning message>
- `\since` <API version or date when the entity was added>

- `\version` <version string>
- `\bug` <description of bug>
- `\see` <cross-references to other methods or classes>

In addition to these commands, Doxygen supports various formatting commands to change the style of the next word. These include `\b` (bold), `\c` (typewriter font), and `\e` (italics). You can also use `\n` to force a new line, `\\` to enter a backslash character, and `\@` to enter the at sign.

API comments

Doxygen allows you to specify overview documentation for your entire API using the `\mainpage` comment. This lets you provide a high-level description of the API as well as a breakdown of the major classes. This description will appear on the front page of the documentation that Doxygen produces. It's common to store these comments in a separate file, such as `overview.dox` (this requires updating the `FILE_PATTERNS` entry of the Doxygen configuration file to include `*.dox`).

The text in this overview documentation may be long enough justify breaking it into separate sections. In that case, you can use the `\section` and `\subsection` commands to introduce this structure. You can even create separate pages to contain more detailed descriptions for certain parts of your API. This can be done with the `\page` command.

You may also find it useful to define groups of behavior for your API so that you can break up the various classes in your API into different categories. For example, you could create groups for classes or files pertaining to file handling, container classes, logging, versioning, and so forth. This is done by declaring a group with `\defgroup` and then using the `\ingroup` to add any specific element to that group.

Putting these features together, the next comment provides overview documentation for an entire API, which is broken down into three sections and cross-references two additional pages for more detailed descriptions. The pages include a link to show all of the API elements that have been tagged as part of a given group:

The resulting Doxygen HTML output for this comment (using the default style sheet) is shown in Fig. 11.3.

```
///
/// \mainpage API Documentation
///
/// \section sec_Contents Contents
///
/// \li \ref sec_Overview
/// \li \ref sec_Detail
/// \li \ref sec_SeeAlso
///
/// \section sec_Overview Overview
///
/// Your overview text here.
///
/// \section sec_Detail Detailed Description
///
/// Your more detailed description here.
///
/// \section sec_SeeAlso See Also
///
/// \li \ref page_Logging
/// \li \ref page_Versioning
///
///
/// \page page_Logging The Logging System
///
/// Overview of logging functionality
///
/// \link group_Logging View All Logging Classes \endlink
///
///
/// \page page_Versioning API Versioning
///
/// Overview of API Versioning
///
/// \link group_Versioning View All Versioning Classes \endlink
///


/// \defgroup group_Logging Diagnostic logging features
/// See \ref page_Logging for a detailed description.

/// \defgroup group_Versioning Versioning System
/// See \ref page_Versioning for a detailed description.
```

File comments

You can place a comment at the top of each header file to provide documentation for the entire module. Here's a sample template for these per-file comments:

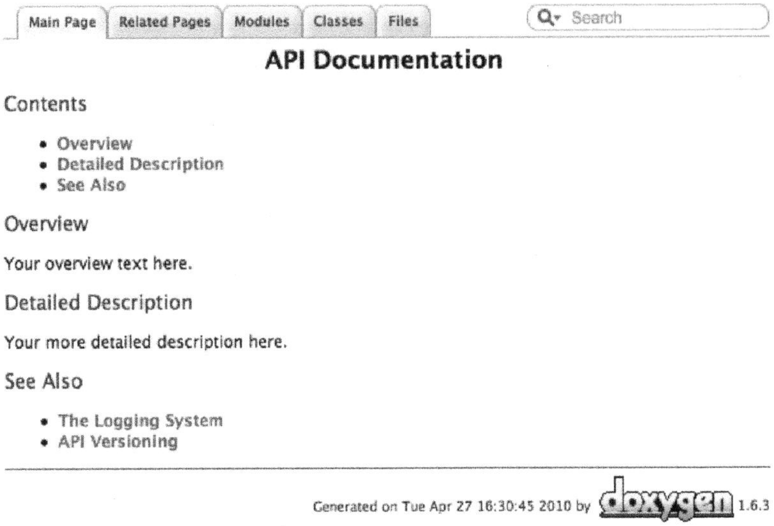

FIGURE 11.3 The Doxygen HTML output for the `\mainpage` example. API, *application programming interface.*

```
///
/// \file <filename>
///
/// \brief <brief description>
///
/// \author <list of author names>
/// \date <date description>
/// \since <API version when added>
///
/// <description of module>
///
/// <license and copyright information>
///
```

If this file contains functionality that you want to add to a group that you've defined, then you can also add the `\ingroup` command to this comment.

Class comments

Each class in your header can also have a comment to describe the overall purpose of the class. In addition to the sample template provided next, you may consider including the

`\ingroup` command if the class belongs to a group you've defined, `\deprecated` if the class has been deprecated, or `\code … \endcode` if you want to provide some example code:

```
///
/// \class <class name> [header-file] [header-name]
///
/// \brief <brief description>
///
/// <detailed description>
///
/// \author <list of author names>
/// \date <date description>
/// \since <API version when added>
///
```

Method comments

You can provide documentation for individual methods, detailing the name and a description for each parameter (and optionally if they are in, out, or in/out parameters) as well as a description of the return value. As with the class sample template, you may also consider adding `\ingroup` or `\deprecated`, as appropriate:

```
///
/// \brief <brief description>
///
/// <detailed description>
///
/// \param[in] <input parameter name> <description>
/// \param[out] <output parameter name> <description>
/// \return <description of the return value>
/// \since <API version when added>
/// \see <methods to list in the see also section>
/// \note <optional note about the method>
///
```

If you have methods in a class that fall into one or more logical groupings, you can specify this to Doxygen so that it will group the related methods together under a named subsection. This can be used to provide a more appropriate ordering of the class members, instead of Doxygen's default behavior. The next code snippet demonstrates the specification of two such member groups:

```
class Test
{
public:
    /// \name <group1-name>
    //@{
    void Method1InGroup1();
    void Method2InGroup1();
    //@}

    /// \name <group2-name>
    //@{
    void Method1InGroup2();
    void Method2InGroup2();
    //@}
};
```

Enum comments

Doxygen also lets you provide comments for enums, including documentation for individual values in the enum. The latter can be done using Doxygen's '<' comment syntax, which attaches the documentation to the previous element, instead of the next element:

```
///
/// \brief <brief description>
///
/// <detailed description>
///
enum MyEnum {
    ENUM_1,      ///< description of enum value ENUM_1
    ENUM_2,      ///< description of enum value ENUM_2
    ENUM_3       ///< description of enum value ENUM_3
}
```

Sample header with documentation

Putting all of this together, here's an entire header file with Doxygen style comments to describe the file, class, and each method. This example is provided in the supporting source code for the book, along with the generated HTML output that Doxygen produced for this file:

```
///
/// \file    version.h
///
/// \brief   Access the API's version information.
///
/// \author  Martin Reddy
/// \date    2010-07-07
/// \since   1.0
/// \ingroup group_Versioning
///
/// Copyright (c) 2010, Martin Reddy. All rights reserved.
///

#ifndef VERSION_H
#define VERSION_H

#include <string>

///
/// \class Version version.h API/version.h
///
/// \brief Access the version information for the API
///
/// For example, you can get the current version number as
/// a string using \c GetVersion, or you can get the separate
/// major, minor, and patch integer values by calling
/// \c GetMajor, \c GetMinor, or \c GetPatch, respectively.
///
/// This class also provides some basic version comparison
/// functionality and lets you determine if certain named
/// features are present in your current build.
///
/// \author Martin Reddy
/// \date   2010-07-07
/// \since  1.0
///
class Version
{
public:
    /// \name Version Numbers
    //@{
    ///
    /// \brief Return the API major version number.
    /// \return The major version number as an integer.
    /// \since 1.0
    ///
    static int GetMajor();

    ///
    /// \brief Return the API minor version number.
    /// \return The minor version number as an integer.
    /// \since 1.0
    ///
    static int GetMinor();

    ///
    /// \brief Return the API patch version number.
```

```
    /// \return The patch version number as an integer.
    /// \since 1.0
    ///
    static int GetPatch();

    ///
    /// \brief Return the API full version number.
    /// \return The version string, e.g., "1.0.1".
    /// \since 1.0
    ///
    static std::string GetVersion();
    //@}

    /// \name Version Number Math
    //@{
    ///
    /// \brief Compare the current version number against a specific
    ///        version.
    ///
    /// This method let's you check to see if the current version
    /// is greater than or equal to the specified version. This may
    /// be useful to perform operations that require a minimum
    /// version number.
    ///
    /// \param[in] major The major version number to compare against
    /// \param[in] minor The minor version number to compare against
    /// \param[in] patch The patch version number to compare against
    /// \return Returns true if specified version >= current version
    /// \since 1.0
    ///
    static bool IsAtLeast(int major, int minor, int patch);
    //@}

    /// \name Feature Tags
    //@{
    ///
    /// \brief Test whether a feature is implemented by this API.
    ///
    /// New features that change the implementation of API methods
    /// are specified as "feature tags." This method lets you
    /// query the API to find out if a given feature is available.
    ///
    /// \param[in] name The feature tag name, e.g., "LOCKING"
    /// \return Returns true if the named feature is available.
    /// \since 1.0
    ///
    static bool HasFeature(const std::string &name);
    //@}
};

#endif
```

12

Testing

Every developer, no matter how experienced and meticulous, will introduce bugs into the software they write. This is simply inevitable as an application programming interface (API) grows in size and complexity. The purpose of testing is to locate these defects as early as possible so they can be addressed before they affect your clients.

Modern software development relies heavily on the use of third-party APIs. As your own APIs become more ubiquitous, failures and defects in your code will have the potential to affect many clients and their applications.

As I noted earlier, your clients may eventually seek alternative solutions if the code you deliver is buggy or unpredictable, or crashes regularly. Conscientious testing is therefore a critical part of your API development process because it can be used to increase the reliability and stability of your product. This will ultimately contribute to the success of your API in the marketplace.

> *TIP: Writing automated tests is the most important thing you can do to ensure that you don't break your users' programs.*

In this chapter, I'll cover various types of automated testing that you can employ, such as unit testing, integration testing, and performance testing. I will also look at how to write good tests, as well as the equally important factor of how to design and implement APIs that are more amenable to testing. Finally, I'll complement this discussion by surveying some of the testing tools you can adopt for your project and look at how you can write tests using various popular automated testing frameworks. Along the way, I'll discuss process issues such as collaborating with a quality assurance (QA) team, using quality metrics, and how to integrate testing into your build process.

Reasons to write tests

It's a common fallacy that engineers don't like to write tests. From my experience, every good developer understands the need for testing, and most have encountered cases in which testing has caught a bug in their code. At the end of the day, software engineers are craftspeople who take pride in their work and will probably find it demoralizing to be expected to produce low-quality results. However, if the management for a project does not explicitly incorporate the need for testing into the schedule, then engineers are not given the resources and support they need to develop these tests. In general, a software

API Design for C++. https://doi.org/10.1016/B978-0-443-22219-1.00021-0

project can be date-driven, quality-driven, or feature-driven. You can pick two of these, but not all three. For example, in the case of the waterfall development process, feature creep and unforeseen problems can easily eliminate any time at the end of a project that was reserved for testing. Consequently, an engineer who attempts to spend time writing automated tests can appear to be less productive within a date- or feature-driven process. Instead, if engineers are empowered to focus on quality, I believe that they will relish the opportunity to write tests for their code.

I experienced this firsthand at Pixar, where we decided to introduce a new policy that engineers had to write unit tests for their code, and furthermore, that all non-GUI code had to achieve 100% code coverage. That is, every line of non-GUI code had to be exercised from test code. Rather than incite a mass rebellion, we found that developers thought this was a good use of their time. The key enabling factor was that we added time to the end of each iteration when all developers could focus on writing tests for their code. Even after 2 years of following this policy, there was universal agreement that the benefits of writing tests outweighed the costs, and that maintaining a 100% coverage target was still appropriate.

In case you still need some incentive, here are some reasons why you should employ testing for your own projects:

- **Increased confidence.** Having an extensive suite of automated tests can give you the confidence to make changes to the behavior of an API with the knowledge that you're not breaking functionality. Said differently, testing can reduce your fear of implementing a change. It's common to find legacy systems in which engineers are uneasy changing certain parts of the code because those parts are so complex and opaque that changes to its behavior could have unforeseen consequences (Feathers, 2004). Furthermore, the code in question may have been written by an engineer who is no longer with the organization, so there is no one who knows where the bodies are buried in the code.
- **Ensuring backward compatibility.** It's important to know that you haven't introduced changes in a new version of the API that break backward compatibility for code written against an older version of the API, or for data files generated by that older API. The use of automated tests can be used to capture the workflows and behavior from earlier versions so that these are always exercised in the latest version of the library.
- **Saving costs.** It's a well-known fact that fixing defects later in the development cycle is more expensive than fixing them earlier. This is because the defect becomes more deeply embedded in the code, and exorcising it may also involve updating data files. For example, Steve McConnell gives evidence that fixing a bug after release can be 10–25 times more expensive than during development (McConnell, 2004). Developing a suite of automated tests lets you discover defects earlier so that they can be fixed earlier, and hence is more economically overall.
- **Codify uses cases.** The use cases for an API represent supported workflows that your users should be able to accomplish. Developing tests for these use cases before you implement your API can let you know when you've achieved the

required functionality. These same tests can then be used continuously to catch any regressions in these important and key workflows.

- **Compliance assurance**. Software for use in certain safety- or security-critical applications may have to pass regulatory tests, such as Federal Aviation Administration certification. Also, some organizations may verify that your software conforms to their standard before allowing it to be branded as such. For example, the Open Geospatial Consortium (OGC) has a compliance testing program for software that is to be branded as Certified OGC Compliant. Automated tests can be used to ensure that you continue to conform to these regulatory and standards requirements.

These points can be summarized by stating that automated testing can help you to determine whether you're building the right thing (referred to as validation), and if you're building it right (called verification).

There can be a downside to writing many tests. As the size of your test suite grows, the maintenance for these tests grows commensurately. This can result in situations where a good code change that takes a couple of minutes to make might also break hundreds of tests and require many hours of effort also to update the test suite. This is a bad situation to get into because it disincentivizes an engineer from making a good fix only because of the overhead of updating the tests. I'll discuss ways to avoid this situation in the following sections. However, if the fix in question changes the public API, then the extra barrier may be a good thing because it forces the engineer to consider the potential impact on backward compatibility for existing clients.

SIDEBAR: The Cost of Untested Code

There are many examples of catastrophic software failures that have happened over the years. Numerous websites maintain catalogs of these bugs. In many of these cases, more thorough testing would have averted disaster. For example:

In May 1996, a software bug caused the bank accounts of 823 customers of a major US bank to be credited with $924,844,208.32 each. The American Bankers Association described this as the largest error in US banking history. It happened because the bank added new message codes to its ATM transaction software but failed to test them on all ATM protocols.

As an example of the need for backward compatibility testing, the Tokyo stock exchange halted trading for most of a day during Nov. 2005 owing to a software glitch after a systems upgrade.

Failing to meet client requirements can have a large impact on your business. For example, in Nov. 2007, a regional government in the United States brought a multimillion-dollar lawsuit against a software services vendor because the criminal justice information software they delivered "minimized quality" and did not meet requirements.

Finally, as an example of the importance of compliance testing, in Jan. 2009, regulators banned a large US health insurance company from selling Medicare programs because of computer errors that posed "a serious threat to the health and safety of Medicare beneficiaries."

Types of API testing

Testing an API is different from testing an end user or GUI application. However, there are still various techniques that are applicable to both cases. For example, a common categorization of software testing methodologies is:

1. **Open box testing**: Tests are developed with knowledge of the source code implementation and are normally automated using a programming language. This is sometimes called white box or clear box testing, implying that the tester can view the inner workings of the system.
2. **Closed box testing**: Tests are developed without knowledge of the underlying implementation and are instead based upon specifications such as functional requirements and use cases. This is sometimes called black box testing, meaning that the inner workings are hidden from the tester.

These terms can sometimes imply the use or nonuse of automation. For example, because open box testing generally involves writing code, it can be easy to automate these tests and run them automatically as part of a build process, whereas closed box testing is often associated with manual testing, in which a person walks through a series of manual steps to exercise a program physically from the point of view of a user. This doesn't always have to be the case, though: closed box testing could be automated, too. From an API testing perspective, you can imagine some automated tests being written with knowledge of the implementation and others being written only from the point of view of your clients' expectations with no knowledge of how the features were implemented. There are, however, several types of automated software testing techniques that are not applicable to API testing.

For example, the term system testing refers to testing that's performed on a complete integrated system. This is normally assumed to be an actual application that can be run by a user. Although it's conceivable to consider a large API as a complete integrated system, I will instead subscribe to the view that an API is a building block or component that's used to build entire systems. As such, I will not consider system testing to be part of the tasks involved in testing an API.

Furthermore, the area of automated GUI testing is generally inappropriate for APIs, i.e. the task of writing automated scripts that run an end user application and simulate user interactions, such as clicking on buttons or typing text. The exception to this would be if you're writing a GUI toolkit that creates these button and text entry widgets. However, in this case, you (and your clients) would be well-served by creating a custom testing tool that can navigate and interact with your widget hierarchy for the purposes of automated testing. For example, Froglogic provides an automated GUI testing tool for Qt applications called Squish.

Consequently, the primary functional testing strategies I'll concentrate on here are unit testing and integration testing. Unit testing verifies that the software does what the programmer expects, whereas integration testing satisfies clients that the software addresses their needs. You can also write tests to verify the nonfunctional requirements of your API. The subsequent list provides a selection of some of the most common nonfunctional testing strategies (I'll cover the topic of performance testing later in this chapter to provide an example of nonfunctional testing):

- **Performance testing**: Verifies that the functionality of your API meets minimum speed or memory usage requirements.
- **Load testing**: Puts demand, or stress, on a system and measures its ability to handle this load. This often refers to testing with many simultaneous users or performing many API requests per second. This is sometimes also called stress testing.
- **Scalability testing**: Ensures that the system can handle large and complex production data inputs instead of only simple test datasets. This is sometimes also called capacity or volume testing.
- **Soak testing**: Attempts to run the software continuously over an extended period to satisfy clients that it is robust and can handle sustained use (e.g., that there are no major memory leaks).
- **Security testing**: Ensures that any security requirements of your code are met, such as the confidentiality, authentication, authorization, integrity, and availability of sensitive information.
- **Concurrency testing**: Verifies the multithreaded behavior of your code to ensure that it behaves correctly and does not deadlock.

> *TIP: API testing should involve a combination of unit and integration testing. No-functional techniques can also be applied as appropriate, such as performance, concurrency, and security testing.*

Unit testing

A unit test is used to verify a single minimal unit of source code, such as an individual method or class. The purpose of unit testing is to take the smallest testable parts of an API and verify that they function correctly in isolation.

These kinds of tests tend to run very fast and often take the form of a sequence of assertions that return true or false, in which any false result will fail the test. Often these tests are colocated with the code that they test (such as in a `tests` subdirectory) and they can be compiled and run at the same point that the code itself is compiled. Unit tests tend to be written by developers using knowledge of the implementation; as such, they are an open box testing technique.

> *TIP: Unit testing is an open box testing technique to verify the behavior of functions and classes in isolation.*

To give a concrete example of a unit test, let's consider a function you want to test that converts a string to a double:

```
bool StringToDouble(const std::string &str, double &result);
```

This function accepts a string parameter and returns a Boolean to indicate whether the conversion was successful. If successful, the double value is written to the `result` reference parameter. Given this function, the next unit test performs a series of checks to ensure that it works as expected:

```
void TestStringToDouble()
{
    double value;

    Assert(StringToDouble("1", value), "+ve test");
    AssertEqual(value, 1.0, "'1' == 1.0");

    Assert(StringToDouble("-1", value), "-ve test");
    AssertEqual(value, -1.0, "'-1' == -1.0");

    Assert(StringToDouble("0.0", value), "zero");
    AssertEqual(value, 0.0, "'0.0' == 0.0");

    Assert(StringToDouble("-0", value), "minus zero");
    AssertEqual(value, -0.0, "'-0' == -0.0");
    AssertEqual(value, 0.0, "'-0' == 0.0");

    Assert(StringToDouble("3.14159265", value), "pi");
    AssertEqual(value, 3.14159265, "pi value");

    Assert(StringToDouble("2e10", value), "large scientific");
    AssertEqual(value, 2e10, "");

    Assert(StringToDouble("+4.3e-10", value), "small scientific");
    AssertEqual(value, 4.3e-10, "");

    AssertFalse(StringToDouble("", value), "empty");
    AssertFalse(StringToDouble("   ", value), "whitespace");
    AssertFalse(StringToDouble("+-1", value), "+-");
    AssertFalse(StringToDouble("1.1.0", value), "multiple points");
    AssertFalse(StringToDouble("text", value), "not a number");

    std::cout << "SUCCESS!" << std::endl;
}
```

Note the use of various helper functions to test the result of each operation: `Assert()`, `AssertFalse()`, and `AssertEqual()`. These are common functions in unit test frameworks that follow the JUnit style, although sometimes other, similar function names or capitalizations are used. If any of these JUnit-style assertions fail, then the entire test will fail, normally with a descriptive error message that pinpoints the failure.

SIDEBAR: JUnit

JUnit is a unit testing framework originally developed for the Java programming language by Kent Beck and Erich Gamma. It's designed around two key design patterns: Command and Composite.

Each test case is a command object that defines one or more `test*()` *methods, such as* `testMyObject()`*, as well as optional* `setUp()` *and* `tearDown()` *methods. Multiple test cases can be collated within a test suite. The test suite is a composite of test cases that automatically calls all of the* `test*()` *methods.*

JUnit also provides a number of methods to support assertion-based testing, such as `assertEquals()`*,* `assertNull()`*,* `assertTrue()`*, and* `assertSame()`*. If any of these is passed an expression that evaluates to false, then the test case will be marked as failed.*

Since the initial popularity of JUnit, the framework has been ported to many other languages and has become known by the more general moniker of xUnit. For example, there is PyUnit for Python, CUnit for C, and CppUnit for C++, among many other implementations.

That example is intentionally simple. However, in real software the method or object under test often depends upon other objects in the system or upon external resources such as files on disk, records in a database, or software on a remote server. This leads to two different views of unit testing:

1. **Fixture setup**. The classic approach to unit testing is to initialize a consistent environment, or fixture, before each unit test is run: for example, to ensure that dependent objects and singletons are initialized, to copy a specific set of files to a known location, or to load a database with a prepared set of initial data. This is often done in a `setUp()` function associated with each test, to differentiate the test setup steps from the actual test operations. A related `tearDown()` function is often used to clean up the environment once the test finishes. One of the benefits of this approach is that the same fixture can often be reused for many tests.
2. **Stub/mock objects**. In this approach, the code under test is isolated from the rest of the system by creating stub or mock objects that stand in for any dependencies outside the unit (Mackinnon et al., 2001). For example, if a unit test needs to communicate with a database, a stub database object can be created that accepts the subset of queries that the unit will generate and then return canned data in response, without making an actual connection to the database. The result is a completely isolated test that will not be affected by database problems, network issues, or file system permissions. The downside, however, is that the creation of these stub objects can be tedious and often cannot be reused by other unit tests. On the other hand, mock objects tend to be more flexible and can be customized for individual tests. I'll discuss each of these options in more detail later in the chapter.

TIP: If your code depends upon an unreliable resource, such as a database, file system, or network, consider using stub or mock objects to produce more robust unit tests.

Integration testing

In contrast to unit testing, integration testing is concerned with the interaction of several components in cooperation. Ideally, the individual components have already been unit tested. This form of testing can also be called component testing.

Integration tests are still necessary even if you have a high degree of unit test coverage, because testing individual units of code in isolation does not guarantee that they can be used together easily and efficiently, or that they meet your functional requirements and use cases. For example, the interface of one component may be incompatible with another component, or information required by another component may not be appropriately exposed for another component to use. The goal of integration testing is therefore to ensure that all of the components of your API work well together and are consistent, and that they enable users to perform the tasks they need.

Integration tests are normally developed against the specification of the API, such as the automatically generated API documentation, and therefore should not require an understanding of the internal implementation details. That is, they are written from the perspective of your clients. As such, integration testing is a closed box testing technique.

TIP: Integration testing is a closed box testing technique to verify the interaction of several components together.

You can often use the same tools to implement integration tests that you use to write unit tests. However, integration tests usually involve more complex ways to verify that a sequence of operations was successful. For example, a test may generate an output file that must be compared against a golden or baseline version that's stored with the test in your revision control system. This requires an efficient workflow to update the baseline version in cases where the failure is expected, such as the conscious addition of new elements in the data file or changes to the behavior of a function to fix a bug.

A good integration testing framework will therefore include dedicated comparison functions (or diff commands) for each file type that the API can produce. For example, an API may have an ASCII configuration file, in which an integration test failure should be triggered only if the value or number of settings changes, but not if the order of the settings in the file changes or if different whitespace characters are used to separate settings. As another example, an API may produce an image as its result. You therefore need a way to compare the output image against a baseline version of that image. For example, the R&D team at PDI/Dreamworks developed a perceptual image difference utility to verify that the rendered images for their film assets are visibly the same after a change to their animation system. This perceptually based comparison allows for minor imperceptible differences in the actual pixel values to avoid unnecessary failures (Yee and Newman, 2004).

This last example brings up the point that integration testing may also be data driven. That is, a single test program can be called many times with different input data. For

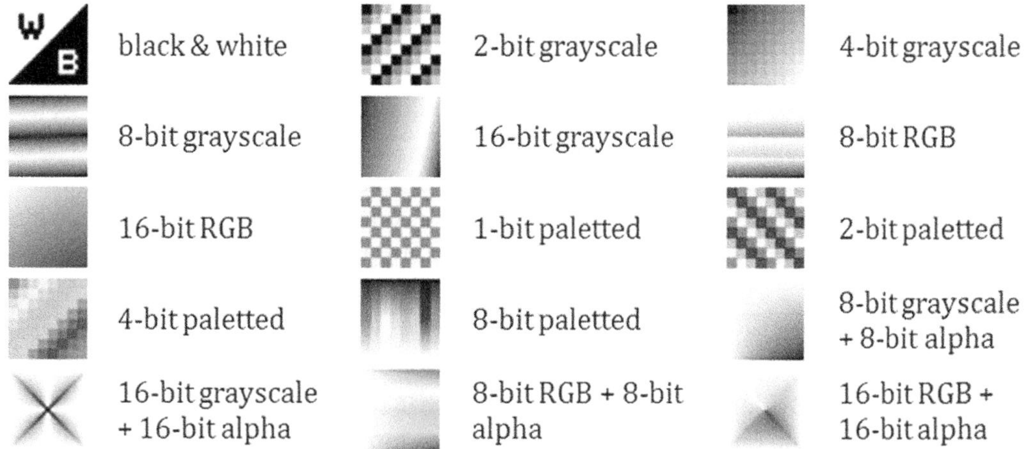

FIGURE 12.1 A subset of Willem van Schaik's PNG image test suite, called PngSuite. See http://www.schaik.com/ for details.

example, a C++ parser may be verified with a single integration test that reads a `.cpp` source file and outputs its derivation or abstract syntax tree. That test can then be called many times with different C++ source programs and its output compared against a correct baseline version in each case. Similarly, the libpng library has a `pngtest.c` program that reads an image and then writes it out again. This test is then run in a data-driven fashion using Willem van Schaik's suite of representative PNG images called PngSuite. Fig. 12.1 shows a few of the images in PngSuite. This integration test ensures that new changes to libpng don't break its ability to read and write various combinations of the PNG file format, including basic chunk handling, compression, interlacing, alpha transparency, filtering, gamma, and image comments, among other attributes.

Sometimes your API may be supported by one or more command line utilities to allow the functionality of your API to be accessed from the terminal or shell scripts. For example, the libjpeg library ships with various utility programs such as `cjpeg`, `djpeg`, and `jpegtran`. You should write automated tests for these tools as well because they are part of your SDK, too, and it can often be easy to break these tools without knowing it. You should therefore have tests in which you execute these tools with various command line arguments and then validate their outputs. Ideally, you would validate the success of these tests based on the results they produce, such as whether a new file was generated correctly on disk. However sometimes you may need to parse the stdout or stderr of the command to achieve what you need. Just be aware in that case you're building more brittle tests in which an engineer could easily edit an error message and break your tests.

Integration testing of APIs can be performed by developers, but in larger organizations it can also be a task that your QA team performs. In fact, a QA engineer will probably refer to this activity as API testing, which is a term that often implies ownership by QA. I've avoided using the specific term API testing here simply because this entire chapter is about testing APIs. Moreover, some of the examples I gave earlier, such as automated testing of command line utilities, could be done by QA engineers who don't know C++ or who focus more on writing shell scripts than compiled code.

Given that integration tests have a different focus than unit tests, may be maintained by a different team, and normally must be run after the build has completed successfully, these kinds of tests are commonly located in a different directory than unit tests. For example, they may live in a sibling directory to the top-level source directory, rather than being stored next to the actual code inside the source directory. This strategy also reflects the closed box nature of integration tests compared with open box unit tests.

Performance testing

Typically, your users will demand a reasonable level of performance from your API. For instance, if you've written a library that provides real-time collision detection between 3D objects for a game engine, your implementation must run fast enough during each frame that it doesn't slow your clients' games. You could therefore write a performance test for your collision detection code in which the test will fail if it exceeds a predefined performance threshold.

As a further example, when Apple was developing their Safari Web browser, page rendering speed was of paramount concern. They therefore added performance tests and defined acceptable speed thresholds for each test. They then put a process in place in which a code checkin would be rejected if it caused a performance test to exceed its threshold. The engineer would have to optimize the code (or somebody else's code if their code was already optimal) before it could be checked in.

TIP: Performance testing of your key use cases helps you avoid introducing speed or memory regressions unknowingly.

A related issue is that of stress testing, in which you verify that your implementation can scale to the real-world demands of your users: for example, a website that can handle many simultaneous users, or a particle system that can handle thousands or millions of particles. These are classed as nonfunctional tests because they don't test the correctness of a specific feature of your API, but instead are concerned with its operational behavior in the user's environment. That is, they test the nonfunctional requirements of your API.

The benefit of writing automated performance tests is that you can make sure any new changes don't adversely affect performance. For example, a senior engineer with whom I worked once refactored a data loading API to use an `std::string` object instead of a `char` buffer to store successive characters read from a data file. When the change was released, users found that the system took over 10 times longer to load their data files. It turns out that the `std::string::append()` method was reallocating the string each time, growing it by a single byte on each call and hence causing massive amounts of memory allocations to happen. This was ultimately fixed by using an `std::vector<char>` because the `append()` method for that container behaved more optimally. A performance test that monitored the time to load large data files could have discovered this regression before it was released to clients.

TIP: If performance is important for your API, consider writing performance tests for your key uses cases to avoid unwittingly introducing performance regressions.

However, performance tests tend to be much more difficult to write and maintain than unit or integration tests. One reason is that performance test results are real (floating-point) numbers that can vary from run to run. They're not discrete true or false values. It's therefore advisable to specify a tolerance for each test to deal with the variability of each test run. For example, you might specify 10 ms as the threshold for your collision detection algorithm but allow for a 15% fluctuation before marking the test as failed. Another technique is to fail the test only after several consecutive data points exceed the threshold, to factor out anomalous spikes in performance.

Also, it's best to have dedicated hardware for running your performance tests, so that other processes running on the machine don't interfere with the test results. Even with a dedicated machine, you may need to investigate turning off certain system background processes so that they don't affect your timings. This reveals another reason why performance tests are difficult to maintain: they are machine specific. This implies that you need to store different threshold values for each class of machine on which you run the tests.

A further complication of performance testing is the problem of information overload. You may end up with hundreds or even thousands of combinations of each performance test on different hardware, each producing a multitude of data points throughout a single day. As a result, you will want to store all of your performance results in a database. Also, if you don't have automatic measures to highlight tests that have exceeded their performance threshold, then you may never notice regressions. On the other hand, with so many tests, you will likely be inundated with false positives and spend most of your time updating baseline values. At this point, you may have more success considering the issue to be a data mining problem. In other words, collect as much data as possible and then have regular database searches that pick the top 5 or 10 most egregious changes in performance and flag those for investigation by a human.

Mozilla offers a great example of extensive performance testing. They've implemented a system in which performance tests are run for multiple products across a range of hardware. The results can be browsed with an interactive website that displays graphs for one or more performance tests at the same time. Fig. 12.2 shows an example of Facebook page load times for different browsers. (One thing to look out for when reading performance graphs is whether the y-axis starts at zero. If the results are scaled vertically to fit the screen, then what looks like a large degree of fluctuation could in reality be a tiny overall percentage change.)

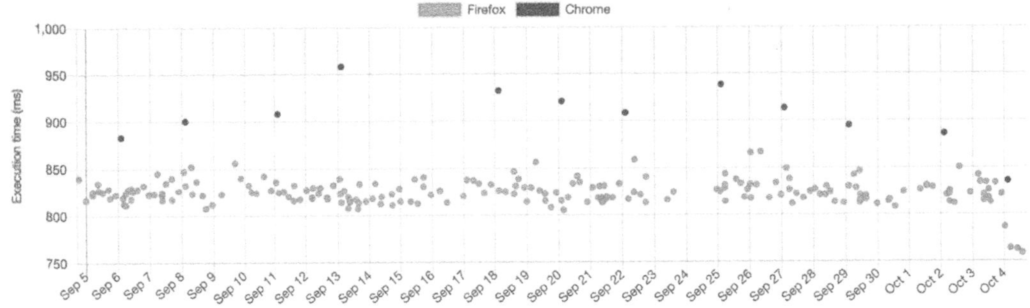

FIGURE 12.2 Firefox performance dashboard from https://arewefastyet.com/.

Writing good tests

Now that I've covered the basic types of API testing, I'll concentrate on how to write these automated tests. I will cover the qualities that make a good test and also present standard techniques for writing efficient and thorough tests. I'll also discuss how testing can be shared effectively with a QA team.

Qualities of a good test

Before I discuss the details of writing an automated test, I'll present a few high-level attributes of a good test. These are general qualities that you should always bear in mind when building your test suite. The overall message, however, is that you should treat test code with the same exacting standards that you use in your main API code. If you develop tests that exhibit these qualities, then you should end up with an easy to maintain and robust test suite that provides you with a valuable safety net for your API development:

- **Fast.** Your suite of tests should run quickly so that you get rapid feedback on test failures. Unit tests should always be fast, on the order of fractions of a second per test. Integration tests that perform actual user workflows, or data-driven integration tests that are run on many input files may take longer to execute. However, there are several ways to deal with this, such as favoring the creation of many unit tests but a few targeted integration tests. Also, you can have different categories of tests: the fast (or checkin or continuous) tests that are run during every build cycle, versus the slow (or complete or acceptance) tests that are run only occasionally, such as before a release.

- **Stable**. Tests should be repeatable, independent, and consistent: every time you run a specific version of a test you should get the same result. If a test starts failing erroneously or erratically, then your faith in the validity of that test's results will be diminished. You may even be tempted to turn the test off temporarily, which of course defeats the purpose of having the test. Using mock objects, in which all dependencies of a unit test are simulated, is one way to produce tests that are independent and stable to environmental conditions. It's also the only practical way to test date- or time-dependent code.
- **Portable**. If your API is implemented on multiple platforms, your tests should work across the same range of platforms. One of the most common areas of difference for test code running on different platforms is floating point comparisons. Rounding errors, architecture differences, and compiler differences can cause mathematical operations to produce slightly different results on different platforms. Floating point comparisons should therefore allow for a small error, or epsilon, rather than being compared exactly. This epsilon should be relative to the magnitude of the numbers involved and the precision of the floating-point type used. For instance, single-precision floats can represent only six to seven digits of precision. Therefore, an epsilon of 0.000001 may be appropriate when comparing numbers such as 1.234567, but an epsilon of 0.1 would be more appropriate when comparing numbers such as 123456.7.
- **High coding standards**. Test code should follow the same coding standards as the rest of your API: you should not slip standards just because the code will not be run directly by your users. Tests should be well-documented so that it's clear what's being tested and what a failure would imply. If you enforce code reviews for your API code, you should do the same for test code. In some places where I've worked, engineers prefer to start a code review by looking at unit test changes first. Similarly, you should not abandon your good engineering instincts simply because you are writing a test. If there's a case for factoring out common test code into a reusable test library, then you should do this. As the size of your test suite grows, you could end up with hundreds or thousands of tests. The need for robust and maintainable test code is therefore just as imperative as for your main API code.
- **Reproducible failure**. If a test fails, it should be easy to reproduce the failure. This means logging as much information as possible about the failure, pinpointing the actual point of failure as accurately as possible and making it easy for a developer to run the failing test in a debugger. Some systems employ randomized testing (called ad hoc testing), in which the test space is so large that random samples are chosen. In these cases, you should ensure that it's easy to reproduce the specific conditions that caused the failure, because simply rerunning the test will pick another random sample and may pass.

What to test

Finally, we get to the part about actually writing tests. The way you write a unit test is different from the way you write an integration test. This is because unit tests can have knowledge about the internal structure of the code, such as loops and conditions. However, in both cases the aim is to exercise the capabilities of the API methodically. To this end, there is a range of standard QA techniques that you can employ to test your API. A few of the most pertinent ones are:

- **Condition testing**. When writing unit tests, you should use your knowledge of the code under test to exercise all combinations of any `if/else`, `for`, `while`, and `switch` expressions within the unit. This ensures that all possible paths through the code have been tested. (I will discuss the details of statement coverage vs. decision coverage later in the chapter when I look at code coverage tools.)
- **Equivalence classes**. An equivalence class is a set of test inputs that all have the same expected behavior. The technique of equivalence class partitioning therefore attempts to find test inputs that exercise difference classes of behavior. For example, consider a square root function that is documented to accept values in the range 0–65,535. In this case there are three equivalence classes: the negative numbers, the valid range of numbers, and numbers greater than 65,535. You should therefore test this function with values from each of these three equivalence classes (e.g., −10, 100, 100,000).
- **Boundary conditions**. Most errors occur around the boundary of expected values. How many times have you inadvertently written code with an off-by-one error? Boundary condition analysis focuses test cases around these boundary values. For example, if you're testing a routine that inserts an element into a linked list of length n, you should test inserting at position 0, 1, $n-1$, and n.
- **Parameter testing**. A test for a given API call should vary all parameters to the function to verify the full range of functionality. For example, the `stdio.h` function `fopen()` accepts a second argument to specify the file mode. This can take the values "r," "w," and "a," in addition to optional "+" and "b" characters in each case. A thorough test for this function should therefore test all 12 combinations of the mode parameter to verify the full breadth of behavior.
- **Return value assertion**. This form of testing ensures that a function returns correct results for different combinations of its input parameters. These results could be the return value of the function, but they could additionally include output parameters that are passed as pointers or references. For instance, a simple integer multiplication function,

```
int Multiply(int x, int y)
```

could be tested by supplying a range of (x, y) inputs and checking the results against a table of known correct values.

- **Getter/setter pairs**. The use of getter/setter methods is extremely common in C++ APIs, and of course I've advocated that you should always prefer the use of these functions over directly exposing member variables in a class. You should therefore test that calling the getter before calling the setter returns an appropriate default result, and that calling the getter after the setter will return the appropriate value, such as:

```
AssertEqual(obj.GetValue(), 0, "test default");
obj.SetValue(42);
AssertEqual(obj.GetValue(), 42, "test set then get");
```

- **Operation order**. Varying the sequence of operations to perform the same test (where this is possible) can help to uncover order of execution assumptions and nonorthogonal behavior: that is, if the API calls have undocumented side effects that are being relied upon to achieve certain workflows.
- **Regression testing**. Backward compatibility with earlier versions of the API should be maintained whenever possible. It's therefore extremely valuable to have tests that verify this goal. For example, a test could try reading data files that were generated by older versions of the API to ensure that the latest version can still ingest them correctly. It's important that these data files are never updated to newer formats when the API is modified. That is, you will end up with live data files, which are up-to-date for the current version, and legacy data files, which verify the backward compatibility of the API.
- **Negative testing**. This testing technique constructs or forces error conditions to see how the code reacts to unexpected situations. For example, if an API call attempts to read a file on disk, a negative test might try deleting that file or making it unreadable, to see how the API reacts when it's unable to read the contents of the file. Another example of negative testing is supplying invalid data for an API call. For example, a credit card payment system that accepts credit card numbers should be tested with invalid credit card numbers (negative testing) as well as valid numbers (positive testing).
- **Buffer overruns**. A buffer overrun, or overflow, is when memory is written past the end of an allocated buffer. This causes unallocated memory to be modified, often resulting in data corruption and ultimately a crash. Data corruption errors can be difficult to track down because the crash may occur sometime after the actual buffer overrun event. It's therefore good practice to check that an API does not write to memory beyond the size of a buffer. This buffer could be an internal private member of a class or it could be a parameter that you pass into an API call. For example, the `string.h` function `strncpy()` copies at most n characters from one string to another. This could be tested by supplying source strings that are equal to

and longer than n characters, then verifying that no more than n characters (including the null terminator, '\0') are written to the destination buffer.
- **Memory ownership**. Memory errors are a common cause of crashes in C++ programs. Any API calls that return dynamically allocated memory should document whether the API owns the memory or if the client is responsible for freeing it. These specifications should be tested to ensure they are correct. For example, if the client is responsible for freeing the memory, a test could request the dynamic object twice and assert that the two pointers are different. A further test could free the memory and then rerequest the object from the API multiple times to ensure that no memory corruption or crashes occur.
- **Null input**. Another common source of crashes in C++ is passing a `nullptr` to a function that then immediately attempts to dereference the pointer without checking for null. You should therefore test all functions that accept a pointer parameter to ensure that they behave gracefully when passed a `nullptr`.

Focusing the testing effort

In all likelihood, it will be infeasible to test every possible code path in your API. You will therefore be faced with a decision about which subset of the overall functionality to test. To help you focus your testing effort, this list enumerates seven ways to determine the biggest bang for your testing buck:

1. Focus on tests that exercise the primary use cases or workflows of the API.
2. Focus on tests that cover multiple features or offer the widest code coverage.
3. Focus on the code that is the most complex and hence the highest risk.
4. Focus on parts of the design that are poorly defined.
5. Focus on features with the highest performance or security concerns.
6. Focus on testing problems that would cause the worst impact on clients.
7. Focus early testing efforts on features that can be completed early in the development cycle.

Working with quality assurance

If you're fortunate enough to have a good QA team to support your testing efforts, then they can share responsibility for writing automated tests. For example, it's standard practice for developers to write and own unit tests and for QA to write and own integration tests.

Different software development models produce different interactions with QA. For example, a traditional waterfall method, in which testing is performed as a final step before release, means that QA is often treated as a distinct group whose goal of quality is often negatively affected by delays during the development process. By contrast, more agile development processes such as Scrum favor embedding QA as part of the development process and including testing responsibilities within each short sprint or iteration. In either case, the benefit of working with QA engineers is that they become your first users. As such, they can help to ensure that the functional and business requirements of your API are met.

I noted earlier that API testing generally requires writing code, because an API is software that's used to build end user applications. This implies that your QA engineers must be able to write code to work on integration testing effectively. Related to this, Microsoft has traditionally used two broad terms to categorize QA engineers:

1. **Software test engineer (STE)**: has limited programming experience and may not even need a strong computer science background. An STE essentially performs manual closed box testing.
2. **Software design engineer in test (SDET)**: can write code and is therefore capable of performing open box testing, writing tools, and producing automated tests.

In terms of API testing, you will therefore more likely prefer a QA engineer who is an SDET rather than an STE. However, even most SDETs will not be able to program in C++, although most will be able to write code in a scripting language. Providing script bindings for your API can therefore offer greater opportunity for your QA team to contribute automated integration tests (see Chapter 14 on Scripting for details on adding scripting support). Another technique is to write programs that enable data-driven testing. The earlier reference to `pngtest.c` is an example of this: a single program written by a developer that can be used by QA engineers to produce a slew of data-driven integration tests. As noted earlier, your SDK may include one or more command-line utilities, which could be tested using a scripting language such as Python instead of C++.

Writing testable code

Testing an API shouldn't be something that you leave until the end of the process. There are decisions you make while you're designing and implementing an API that can improve your ability to write robust and extensive automated tests. In other words, you should consider how a class will be tested early on during its development. In the following sections, I will cover various techniques for writing software that's more amenable to automated unit and integration testing.

Test-driven development

Test-driven development (TDD), or test-first programming, involves writing automated tests to verify desired functionality before the code that implements this functionality is written. These tests will, of course, fail initially. The goal is then to write minimal code quickly to make these tests pass. Then finally, the code is refactored to optimize or clean up the implementation as necessary (Beck, 2002).

An important aspect of TDD is that changes are made incrementally, in small steps. You write a short test, then write enough code to make that test pass, then repeat. After every small change, you recompile your code and rerun the tests. Working in these small steps means that if a test starts to fail, then in all probability this will be caused by the code you wrote since the last test run. Let's look at an example to demonstrate this. I'll start with a small test to verify the behavior of a `MovieRating` class (Astels, 2003):

```
void TestNoRatings()
{
    MovieRating *nemo = new MovieRating("Finding Nemo");
    AssertEqual(nemo->GetRatingCount(), 0, "no ratings");
}
```

Given this initial test code, you now write the simplest possible code to make the test pass. Here's an example that satisfies this objective (I will inline the implementation for the API methods in these examples to make it clearer how the code under test evolves):

```
class MovieRating
{
public:
    MovieRating(const std::string &name) {}
    int GetRatingCount() const { return 0; }
};
```

This API clearly doesn't do a lot, but it does allow the previous test to pass. So now you can move on and add some more test code:

```
void TestAverageRating
{
    MovieRating *nemo = new MovieRating("Finding Nemo");
    nemo->AddRating(4.0f);
    nemo->AddRating(5.0f);
    AssertEqual(nemo->GetAverageRating(), 4.5f, "nemo avg rating");
}
```

Now it's time to write the minimal code to make this test pass:

```
class MovieRating
{
public:
    MovieRating(const std::string &name) {}
    int GetRatingCount() const { return 0; }
    void AddRating(float r) {}
    float GetAverageRating() const { return 4.5f; }
};
```

Writing another test will force us to make the implementation more general:

```
void TestAverageRatingAndCount
{
    MovieRating *cars = new MovieRating("Cars");
    cars->AddRating(3.0f);
    cars->AddRating(4.0f);
    cars->AddRating(5.0f);
    AssertEqual(cars->GetRatingCount(), 3, "three ratings");
    AssertEqual(cars->GetAverageRating(), 4.0f, "cars avg rating");
}
```

Now you should extend the implementation to return the number of ratings added and the average of those ratings. The minimal way to do this would be to record the current sum of all ratings and the number of ratings added. For example:

```
class MovieRating
{
public:
    MovieRating(const std::string &name) :
        mNumRatings(0),
        mRatingsSum(0.0f)
    {
    }

    int GetRatingCount() const
    {
        return mNumRatings;
    }

    void AddRating(float r)
    {
        mRatingsSum += r;
        mNumRatings++;
    }

    float GetAverageRating() const
    {
        return (mRatingsSum / mNumRatings);
    }

private:
    int mNumRatings;
    float mRatingsSum;
};
```

Obviously you can continue this strategy by adding further tests to verify that calling `GetAverageRating()` with zero ratings does not crash and to check that adding out-of-range rating values is treated appropriately, but I think you get the general principle.

One of the main benefits of TDD is that it forces you to think about your API before you start writing any code. You also must think about how the API will be used (i.e., you put yourself in the shoes of your clients). Another effect of TDD is that you implement only what your tests need. In other words, your tests determine the code you need to write (Astels, 2003). This can help you avoid premature optimization and keeps you focused on the overall behavior.

> *TIP: TDD means that you write unit tests first, then write the code to make the tests pass. This keeps you focused on the key use cases for your API.*

TDD doesn't have to be confined to the initial development of your API. It can also be helpful during maintenance mode. For example, when a bug is discovered in your API, you should first write a test for the correct behavior. This test will, of course, fail at first. You can then work on implementing the fix for the bug. You will know when the bug is fixed because your test will change to a pass state. Once the bug is fixed, you then have the added benefit of an ongoing regression test that will ensure the same bug is not introduced again in the future.

Stub and mock objects

One popular technique to make your unit tests more stable and resilient to failures is to create test objects that can stand in for real objects in the system. This lets you substitute an unpredictable resource with a lightweight controllable replacement for the purpose of testing. Examples of unpredictable resources include the file system, external databases, and networks. The stand-in object can also be used to test error conditions that are difficult to simulate in the real system, as well as events that are triggered at a certain time or are based upon a random number generator.

These stand-in objects will obviously present the same interface as the real objects they simulate. However, there are several different ways to implement these objects. This list presents some of the options and introduces the generally accepted terminology for each case:

- **Fake Object**: An object that has functional behavior but uses a simpler implementation to aid testing: for example, an in-memory file system that simulates interactions with the local disk.
- **Stub Object**: An object that returns prepared or canned responses: for example, a `ReadFileAsString()` stub might simply return a hardcoded string as the file contents, rather than reading the contents of the named file on disk.
- **Mock Object**: An instrumented object that has a preprogrammed behavior and that performs verification on the calling sequence of its methods: for example, a mock object (or simply a mock) can specify that a `GetValue()` function will return 10 the first two times it's called, then 20 after that. It can also verify that the function was called, say, only three times, or at least five times, or that the functions in the class were called in a given order.

The difference between a stub and a mock is often poorly understood, so let's demonstrate this with an example using the children's card game, War. This is a simple game in which a deck of cards is divided equally between two players. Each player reveals the top card and the player with the highest card takes both cards. If the cards have equal value, each player lays three cards face down and the fourth face up. The highest value card wins all of the cards on the table. A player wins the game by collecting all of the cards.

I'll model this game with three classes:

1. **Card**: represents a single card with the ability to compare its value against another card.
2. **Deck**: holds a deck of cards with functions to shuffle and deal cards.
3. **WarGame**: manages the game logic, with functions to play out the entire game and return the winner of the game.

During actual game play, the `Deck` object will return a random card. However, for the purposes of testing, you could create a stub deck that returns cards in a predefined order. If the `WarGame` object accepts the deck to use as a parameter to its constructor, you can easily test the logic of `WarGame` by passing it a `StubDeck` that defines a specific and repeatable sequence of cards.

This `StubDeck` would inherit from the real Deck class, which means that you must design `Deck` to be a base class (i.e., make the destructor virtual as well as any methods that need to be overridden for testing purposes). Here's an example declaration for the `Deck` class:

```
class Deck
{
public:
    Deck();
    virtual ~Deck();
    virtual void Shuffle();
    virtual int RemainingCards();
    virtual Card DealCard();
};
```

Our `StubDeck` class can therefore inherit from `Deck` and override the `Shuffle()` method to do nothing, because you don't want to randomize the card order. Then the constructor of `StubDeck` could create a specific order of cards. However, this means that the stub class is hardcoded to a single card order. A more general solution would be to extend the class with an `AddCard()` method. Then you can write multiple tests using `StubDeck` and simply call `AddCard()` a number of times to prepare it with a specific order of cards before passing it to `WarGame`. One way to do this would be to add a protected `AddCard()` method to the base Deck class (because it modifies private state) and then expose this as public in the `StubDeck` class. Then you can write:

```
#include "wargame.h"

void TestWarGame()
{
    StubDeck deck;
    deck.AddCard("9C");
    deck.AddCard("2H");
    deck.AddCard("JS");
    ...

    WarGame game(deck);
    game.Play();
    AssertEqual(game.GetWinner(), WarGame::PLAYER_ONE);
}
```

So, that's what a stub object would look like (in fact, this could even be considered a fake object, too, because it offers complete functionality but without the element of randomness). Let's now look at what testing with a mock object looks like.

One of the main differences between mock and stub objects is that mocks insist on behavior verification. That is, a mock object is instrumented to record all function calls for an object and it will verify behavior such as the number of times a function was called, the parameters that were passed to the function, or the order in which several functions were called. Writing code to perform this instrumentation by hand can be tedious and error prone. It's therefore best to rely upon a mock testing framework to automate this work for you. I'll use the Google Mock framework here (https://github.com/google/googletest) to illustrate how mocks can be used to test our `WarGame` class. The first thing you'll want to do is define the mock using the handy macros that Google Mock provides:

```
#include "wargame.h"
#include <gmock/gmock.h>
#include <gtest/gtest.h>

using namespace testing;

class MockDeck : public Deck
{
public:
    MOCK_METHOD0(Shuffle, void());
    MOCK_METHOD0(RemainingCards, int());
    MOCK_METHOD0(DealCard, Card());
};
```

The `MOCK_METHOD0` macro is used to instrument functions with zero arguments, which is the case for all methods in the `Deck` base class. If instead you have a method with one argument, then you would use `MOCK_METHOD1`, and so on. Now, let's write a unit test that uses this mock. Because I'm using Google Mock to create our mock, I'll also use Google Test as the testing framework. This looks like:

```
TEST(WarGame, Test1)
{
    MockDeck deck;

    EXPECT_CALL(deck, Shuffle())
        .Times(AtLeast(1));

    EXPECT_CALL(deck, DealCard())
        .Times(52)
        .WillOnce(Return(Card("JS")))
        .WillOnce(Return(Card("2H")))
        .WillOnce(Return(Card("9C")))
        ...
        ;

    WarGame game(deck);
    game.Play();
    ASSERT_EQ(game.GetWinner(), WarGame::PLAYER_ONE);
}
```

The clever bits are those two `EXPECT_CALL()` lines. The first one states that the `Shuffle()` method of our mock object should get called at least once, and the second one states that the `DealCard()` method should get called exactly 52 times, and that the first call will return `Card("JS")`, the second call will return `Card("2H")`, and so on. This approach means that you don't need to expose an `AddCard()` method for your mock object. The mock object will implicitly verify all of the expectations as part of its destructor and will fail the test if any of these are not met.

TIP: Both stub and mock objects can return canned responses, but mock objects also perform call behavior verification.

In terms of how this affects the design of your APIs, one implication is that you may wish to consider a model in which access to unpredictable resources is embodied within a base class that you pass into your worker classes, such as in the previous case in which you pass the `Deck` object into the `WarGame` object. This allows you to substitute a stub or mock version in your test code using inheritance. This is essentially the dependency injection pattern, in which dependent objects are passed into a class rather than that class being directly responsible for creating and storing those objects.

However, sometimes it's simply not practical to encapsulate and pass in all of the external dependencies for a class. In these cases, you can still use stub or mock objects, but rather than using inheritance to replace functionality, you can inject them physically at link time. In this case, you name the stub/mock class the same as the class you wish to replace. Then your test program links against the test code and not the code with the real implementation. Using our `ReadFileAsString()` example from earlier, you could create an alternate version of this function that returns canned data, then link the object`.o` file with this stub into our test program in place of the object file that holds the real implementation.

This approach can be powerful, although it necessitates you to create your own abstractions for accessing the file system, network, and so on. If your code directly calls `fopen()` from the standard library, then you can't replace this with a stub at link time unless you also provide stubs for all other standard library functions that your code calls.

Testing private code

The emphasis of this book has been on developing well-designed APIs that offer a logical abstraction while hiding implementation details. However, this can also make it difficult to write thorough unit tests. There will be times when you need to write a unit test that accesses private members of a class to achieve full code coverage. Given a class called `MyClass`, this can be done in several ways, including:

1. **Public member function**: Declaring a public `MyClass::SelfTest()` method for your test code to call.
2. **Friend function**: Creating a `MyClassSelfTest()` free function and declaring it as friend function in `MyClass`.
3. **Friend class**: Declaring a private `MyClass::SelfTest()` method and creating a friend `TestRunner` object that can run the testing function.

I detailed several reasons to avoid friends in Chapter 6 on C++ Usage, although in this case the use of a friend function or class can help you avoid exposing internal functions that are not meant to be used by your clients. For example, the friend function can be made relatively safe if the `MyClassSelfTest()` function is defined in the same library as the `MyClass` implementation, thus preventing clients from redefining the function in their own code. The Google Test framework provides a `FRIEND_TEST()` macro to support this kind of friend function testing.

I'll start by covering the public `SelfTest()` method, and then I'll provide a solution using a friend class. In both cases, I'll work with a simple bounding box class that includes a self-test method.

```
// bbox.h
class BBox
{
public:
    BBox();
    BBox(const Point &a, const Point &b);

    Point GetMin() const;
    Point GetMax() const;

    bool Contains(const Point &point) const;
    bool Intersects(const BBox &bbox) const;
    double CenterDistance(const BBox &bbox) const;

    void SelfTest() const;

private:
    Point CalcMin(const Point &a, const Point &b);
    Point CalcMax(const Point &a, const Point &b);
    Point GetCenter();

    Point mMin, mMax;
};
```

The `SelfTest()` public method can be called directly from a unit test to perform extra validation of the various private methods. This is convenient for testing, although there are some undesirable qualities of this approach: namely, you have to pollute your public API with a method that your clients should not call, and you may add extra bloat to your library by embedding the test code inside the `BBox` implementation.

In the first case, there are ways that you can discourage clients from using this function. One trivial way to do this would be simply to add a comment that the method is not for public use. Taking this one step further, you could remove the method from any API documentation you produce, so that users never see a reference to it (unless they look directly at your headers, of course). As I discussed in the Versioning chapter, you can achieve this with the Doxygen tool by surrounding the function declaration with the `\cond` and `\endcond` commands:

```
/// \cond TestFunctions
void SelfTest() const;
/// \endcond
```

As for the concern that the self-test function may add bloat to your code, there are a couple of ways to deal with this, if you feel it's necessary. One way would be to implement the `SelfTest()` method in your unit test code, not in the main API code (e.g., in `test_bbox.cpp` not `bbox.cpp`). Just because you declare a method in your `.h` file doesn't mean that you must define it. However, this opens up a similar security hole to using friends. That is, your clients could define the `SelfTest()` method in their own code as a way to modify the internal state of the object. Although the interface of this function restricts what they can do, because they cannot pass in any arguments or receive any results, they can still use global variables to circumvent this.

An alternative would be to compile the test code conditionally. For example,

```
// bbox.cpp
...
void SelfTest() const
{
#ifdef TEST
    // lots of test code
#else
    std::cout << "Self-test code not compiled in." << std::endl;
#endif
}
```

The downside of this approach is that you have to build two versions of your API: one with the self-test code compiled in (compiled with `-DTEST` or `/DTEST`) and one without the self-test code. If the extra build is a problem, you could compile the self-test code into debug versions of your library but remove it from release builds.

If, however, you really don't want to expose your `SelfTest()` function publicly, then you can use friends to help. For example, we can introduce a `TestRunner` class that's responsible for running the `SelfTest()` function, in which the object under test can make its testing function private and declare `TestRunner` as a friend. We can make this more reusable by introducing a `Testable` abstract interface from which objects can inherit if they want to provide a `SelfTest()` function for the `TestRunner` to run. For example:

```
class Testable
{
public:
    virtual ~Testable() = default;

private:
    virtual void SelfTest() const = 0;
    friend class TestRunner;
};

class TestRunner
{
public:
    void RunTests(const Testable &test) const
    {
        test.SelfTest();
    }
};

class BBox : public Testable
{
    ...

private:
    void SelfTest() const { ... }
};
```

Now you can easily add a self-testing function to any object by inheriting from the `Testable` interface and defining your private `SelfTest()` function, all without exposing your testing code to your clients. For example:

```
BBox bbox;
bbox.SelfTest();                // error: private member
TestRunner::RunTests(bbox);     // compiles and runs successfully
```

TIP: Use a `SelfTest()` member function to test private members of a class.

If you wish to provide a self-test function for a C API, then this is a much simpler proposition. For example, you could define an external linkage `SelfTest()` function in the `.c` file, that is, a nonstatic function that is decorated with `__declspec(dllexport)` on Windows, but provide no prototype for the function in the `.h` file. You then declare the function prototype in your test code so that you can call the function as part of your unit test. In this way, the function does not appear in your header file or any API documentation. In fact, the only way clients could discover the call is if they do a dump of all of the public symbols in your shared library.

Using assertions

An assertion is a way to verify assumptions that your code makes. You do this by encoding the assumption in a call to an assert function or macro. If the value of the expression evaluates to true, then all is well and nothing happens. However, if the expression evaluates to false, then the assumption you made in the code is invalid and your program will abort with an appropriate error (McConnell, 2004).

Assertions are essentially a way for you to include extra sanity tests for your program state directly in the code. As such, these are invaluable complementary aids to help testing and debugging.

Although you are free to write your own assertion routines, the C standard library includes an `assert()` macro in the `assert.h` header (also available in C++ as the `cassert` header). The next example uses this macro to show how you could document and enforce the assumption that a pointer you are about to dereference is nonnull:

```
#include <cassert>
#include <iostream>

MyClass::MyClass() :
    mStrPtr(new std::string("Hello"))
{
}

MyClass::PrintString()
{
    // mStrPtr should have been allocated in the constructor
    assert(mStrPtr != nullptr);
    std::cout << *mStrPtr << std::endl;
}
```

It is common practice to turn off all `assert()` calls for production code, so that an end user application doesn't abort needlessly when the user is running it. This is often done by making assert calls do nothing when they're compiled in release mode versus debug mode. (Thus you should never put code that must always be executed into an assertion.) Here's an example of a simple assertion definition that is active only in debug builds:

```
#ifdef DEBUG
#include <assert.h>
#else
#define assert(func)
#endif
```

You can also compile with the `NDEBUG` define to disable assertions in `assert.h`.

Assertions should be used to document conditions that you as the developer believe should never occur. They're not appropriate for run-time error conditions that might legitimately occur. If you can recover gracefully from an error, then you should always prefer that course of action rather than causing the client's program to crash. For example, if you have an API call that accepts a pointer from the client, you should never assume that it is nonnull. Instead, you should check for `nullptr` and return gracefully if that's the case, potentially emitting an appropriate error message. You should not use an assertion for this case. However, if your API enforces the condition that one of your private member variables is always nonnull, then it would be a programming error for it ever to be null. This is an appropriate situation for an assertion. Use assertions to check for programming errors; use normal error checking to test for user errors and attempt to recover gracefully in that situation.

TIP: Use assertions to document and verify programming errors that should never occur.

Assertions are commonly used in commercial products to diagnose errors. For instance, several years ago it was reported that the Microsoft Office suite is covered by over 250,000 assertions (Hoare, 2003). These are often used in conjunction with other automated testing techniques, such as running a large suite of unit and integration test cases on debug code with the assertions turned on. This test run will fail if any test code hits an assertion that fails, allowing a developer to follow up and investigate the reason for the failure before it leads to a crash in client code.

C++11 introduced support for compile-time assertions with `static_assert()`. If the constant expression for this assertion resolves to false at compile time, then the compiler displays the provided error message and fails; otherwise, the statement has no effect. For example,

```
// compile-time assertion in C++11
static_assert(sizeof(void *) == 32,
              "This code only works on 32-bit platforms.");
```

Contract programming

Bertrand Meyer coined and trademarked the term design by contract to prescribe the obligations between an interface and its clients (Meyer, 1987). For function calls, this means specifying the preconditions that a client must meet before calling the function, and the postconditions that the function guarantees on exit. For classes, this means defining the invariants that it maintains before and after a public method call (Hoare, 1969; Pugh, 2006).

In the previous Documentation chapter, I showed you how to communicate these conditions and constraints to your users via your API documentation. Here I'll illustrate how you can also implement them in code using assertion-style checks. For instance, continuing with the `SquareRoot()` function I introduced earlier, this code shows how to implement tests for its precondition and postcondition:

```
double SquareRoot(double value)
{
    // assert the function's precondition
    require(value >= 0);

    double result = 0.0;
    ...  // calculate the square root

    // assert the function's postcondition
    ensure(fabs((result * result) - value) < 0.001);
    return result;
}
```

The `require()` and `ensure()` calls in this example can be implemented in a similar fashion to the `assert()` macro I described in the previous section (i.e., they do nothing if the condition evaluates to true, otherwise they abort or throw an exception). Just as in the use of assertions, it's common to disable these calls for release builds to avoid their overhead in a production environment and to avoid aborting your clients' programs. In other words, you could simply define these functions as:

```
// check that a precondition has been met
#define require(cond) assert(cond)

// check that a postcondition is valid
#define ensure(cond) assert(cond)
```

Furthermore, you may implement a private method for your classes to test its invariants (i.e., that it's in a valid state). You can then call this method from inside your functions to ensure that the object is in a valid state when the function begins and ends. If you use a consistent name for this method (which you could enforce through the use of an abstract base class), then you could augment your `require()` and `ensure()` macros with a `check_invariants()` macro as:

```
#ifdef DEBUG
// turn on contract checks in a debug build
#define require(cond) assert(cond)
#define ensure(cond) assert(cond)
#define check_invariants(obj) assert(obj && obj->IsValid());
#else
// turn off contract checks in a non-debug build
#define require(cond)
#define ensure(cond)
#define check_invariants(obj)
#endif
```

Putting all of this together, here is a further example of contract programming for a string append method:

```
void String::append(const std::string &str)
{
    // no preconditions - references are always non-null

    // ensure the consistency of this string and the input string
    check_invariants(this);
    check_invariants(&str);

    // perform the actual string append operation
    size_t pre_size = size();
    ...

    // verify the postcondition
    ensure(size() == pre_size + str.size());

    // and ensure that this string is still self consistent
    check_invariants(this);
}
```

When Meyer originally conceived contract programming, he added explicit support for this technique in his Eiffel language. He also used an assertion model to implement this support, as I have done here. However, in Eiffel these assertions would get automatically extracted into the documentation for the class. C++ does not have this innate capability, so you must manually ensure that the assertions in your implementation match the documentation for your interface.

Nevertheless, one of the benefits of employing this kind of contract programming is that errors get flagged much closer to the actual source of the problem. This can make a huge difference when trying to debug a complex program, because often the source of an error and the point where it causes a problem are far apart. This is, of course, a general benefit of using assertions.

TIP: Enforcing an interface's contract implies the systematic use of assertions, such as `require()`, `ensure()`, *and* `check_invariants()`.

One particularly important piece of advice to remember when employing this programming style is to test against the interface, not the implementation. That is, your precondition and postcondition checks should make sense at the abstraction level of your API. They should not depend upon the specifics of your implementation; otherwise you will find that you have to change the contract whenever you change the implementation.

TIP: Perform contract checks against the interface, not the implementation.

There are also some design patterns that you can take advantage of to enforce preconditions and postconditions for your API. One such approach is to use the Thread-Safe Interface design pattern, which I covered in the chapter on Concurrency. Using this pattern, a public function can be implemented that performs the necessary preconditions and postconditions and then calls out to a separate protected virtual method to perform the actual work. Clients can then override the virtual method to perform different work, but the public API call will still enforce the same preconditions and postconditions for the client's derived class. For example:

```
class MyClass
{
public:
    virtual ~MyClass() = default;

    void Foo()
    {
        // check preconditions here, e.g., thread safety
        DoFoo();
        // check postconditions and invariants here
    }

protected:
    virtual void DoFoo() {}  // could also be pure virtual
};

class MyDerivedClass : public MyClass
{
protected:
    void DoFoo() override
    {
        // DoFoo functionality
    }
};
```

When thinking about the problem of reliably performing some logic before and after a block of code, you may be tempted to think about applying RAII principles (i.e., create an object in which the precondition is defined in the constructor and the postcondition in the destructor). However, destructors cannot return a result and should not trigger an exception, so RAII may not be a good tool to use in this case.

Finally, because this chapter is focused on testing, it's worth noting that most unit testing frameworks allow you to specify `setup()` and `teardown()` functions that will be called before and after every unit test within a class. This provides a great way to enforce preconditions and postconditions within your test code. I provide an example of this later in the Automated Testing Tools section.

Record and playback functionality

One feature that can be invaluable for testing (and many other tasks) is the ability to record the sequence of calls made to an API and then play them back again at will. Record and playback tools are common in the arena of application or GUI testing, in which user interactions such as button presses and keystrokes are captured and then played back to repeat the user's actions. However, the same principles can be applied to API testing. This involves instrumenting every function call in your API to be able to log its name, parameters, and return value. Then a playback module can be written that accepts this log, or journal, file and calls each function in sequence. This module can then check that the latest return values match the previously recorded responses.

Ordinarily this functionality will be turned off by default, so that the overhead of creating the journal file does not affect the performance of the API. However, it can be switched on in a production environment to capture actual end user activity. These journal files can then be added to your test suite as data-driven integration tests or they can be played back in a debugger to help isolate problems. You can even use them to refine the behavior of your API based upon real-world use information, such as detecting common invalid inputs and adding better error handling for these cases. Your clients could even expose this functionality in their applications to allow their end users to record their actions and play them back themselves, (i.e., to automate repetitive tasks in the application). This is often called a macro capability in end user applications.

There are several different ways that you could instrument your API in this fashion. One of the cleaner ways to do this is to introduce a proxy API that essentially forwards straight through to your main API, but which also manages all of the function call logging. In this way, you don't need to pollute your actual API calls with these details, and you always have the option of shipping a vanilla API without any logging functionality. This is demonstrated in this simple example:

```
bool PlaySound(const std::string &filename)
{
    LOG_FUNCTION("PlaySound");
    LOG_PARAMETER(filename);

    bool result = detail::PlaySound(filename);

    LOG_RESULT(result);
    return result;
}
```

Of course, if you already have a wrapper API such as a script binding or a convenience API, then you can simply reuse that interface layer.

Gerard Meszaros notes that on their face, record and playback techniques may appear to be counter to agile methodologies such as test-first development. However, he points out that it's possible to use record and playback in conjunction with test-first methodologies if the journal is stored in a human-readable file format such as XML (Meszaros, 2003). When this is the case, the record and playback infrastructure can be built early on and then tests can be written as data files rather than in code. This has the additional benefit that more junior QA engineers could also contribute data-driven integration tests to the test suite.

Adding robust record and playback functionality to your API can be a significant undertaking, but the costs are normally worth it when you consider the benefits of faster test automation and the ability to let your clients easily capture reproduction cases for bug reports.

Supporting internationalization

Internationalization (i18n) is the process of enabling a software product to support different languages and regional variations. The related term localization (l10n) refers to the activity of using the underlying internationalization support to provide translations of application text into a specific language and to define the locale settings for a specific region, such as the date format or currency symbol.

Internationalization testing can be used to ensure that a product fully supports a given locale or language. This tends to be an activity limited to end user application testing: that is, testing that an application's menus and messages appear in the user's preferred language. However, design decisions that you make during the development of your API can have an impact on how easily your clients can provide localization support in their applications.

For example, you may prefer to return integer error codes rather than error messages in a single language. If you do return error messages, then it would be helpful to define all potential error messages in an appropriate header file that your clients can access, so that they can be appropriately localized. Also, you should avoid returning dates or formatted numbers as strings, because these are interpreted differently across locales. For example, 100,000.00 is a valid number in the United States and the United Kingdom, but in France the same number would be formatted as 100 000,00 or 100.000,00.

There are several libraries that provide internationalization and localization functionality. You could use one of these libraries to return localized strings to your clients and let them specify the preferred locale for the strings that your API returns. These libraries are often easy to use. For example, the GNU gettext library provides a `gettext()` function to look up the translation for a string and return it in the language for the current locale (assuming that a translation has been provided). Often, this `gettext()` function is aliased to _, so that you can write simple code like:

```
std::cout << _("Please enter your username:");
```

Similarly, the Qt library provides excellent internationalization and localization features. All `QObject` subclasses that use the `Q_OBJECT` macro have a `tr()` member function that behaves similarly to GNU's `gettext()` function, such as:

```
button = new QPushButton(tr("Quit"), this);
```

Automated testing tools

In this section, I'll look at some tools that you can use to support your automated testing efforts. I will divide these into four broad categories:

1. **Test harnesses**. Software libraries and programs that make it easier to maintain, run, and report results for automated tests.
2. **Code coverage**. Tools that instrument your code to track the actual statements or branches that your tests executed.
3. **Bug tracking**. A database-driven application that allows defect reports and feature requests to be submitted, prioritized, assigned, and resolved for your software.
4. **Continuous build systems**. A system that rebuilds your software and reruns your automated tests whenever a new change is added.

Test harnesses

There are many unit test frameworks available for C and C++. Most of these follow a design similar to the classic JUnit framework and provide support for features such as assertion-based testing, fixture setup, grouping of fixtures for multiple tests, and mock objects. In addition to being able to define a single test, a good test framework should provide a way to run an entire suite of tests at once and report the total number of failures.

I'll not attempt to describe all available test harnesses here; a Web search on C++ test frameworks will turn up many tools for you to investigate if that is your desire. However, I will provide details for a few of the more popular or interesting frameworks.

- **CppUnit** (http://cppunit.sourceforge.net/): A port of JUnit to C++ originally created by Michael Feathers. This framework supports various helper macros to simplify the declaration of tests, capturing exceptions and a range of output formats including an XML format and a compiler-like output to ease integration with an IDE. CppUnit also provides a few different test runners, including Qt- and MFC-based GUI runners. Version 1 of CppUnit has reached a stable state, and future development is being directed toward CppUnit 2. Michael Feathers has also created an extremely lightweight alternative version of CppUnit called CppUnitLite. Here is a sample test case written using CppUnit, based upon an example from the CppUnit cookbook:

```
class ComplexNumberTest : public CppUnit::TestFixture
{
public:
    void setUp()
    {
        m_10_1 = new Complex(10, 1);
        m_1_1 = new Complex(1, 1);
        m_11_2 = new Complex(11, 2);
    }

    void tearDown()
    {
        delete m_10_1;
        delete m_1_1;
        delete m_11_2;
    }

    void testEquality()
    {
        CPPUNIT_ASSERT(*m_10_1 == *m_10_1);
        CPPUNIT_ASSERT(*m_10_1 != *m_11_2);
    }

    void testAddition()
    {
        CPPUNIT_ASSERT(*m_10_1 + *m_1_1 == *m_11_2);
    }

private:
    Complex *m_10_1;
    Complex *m_1_1;
    Complex *m_11_2;
};
```

- **Boost Test** (http://www.boost.org/): Boost includes a Test library for writing test programs, organizing tests into simple test cases and test suites, and controlling their run-time execution. A core value of this library is portability. As such, it uses a conservative subset of C++ features and minimizes dependencies on other APIs. This has allowed the library to be used for porting and testing of other Boost libraries. Boost Test provides an execution monitor that can catch exceptions in test code, as well as a program execution monitor that can check for exceptions and nonzero return codes from an end user application. The next example, derived from the Boost Test manual, demonstrates how to write a simple unit test using this library:

```
#define BOOST_TEST_MODULE MyTest
#include <boost/test/unit_test.hpp>

int add(int i, int j)
{
    return i + j;
}

BOOST_AUTO_TEST_CASE(my_test)
{
    // #1 continues on error
    BOOST_CHECK(add(2, 2) == 4);

    // #2 throws an exception on error
    BOOST_REQUIRE(add(2, 2) == 4);

    // #3 continues on error
    if (add(2, 2) != 4) {
        BOOST_ERROR("Ouch...");
    }

    // #4 throws an exception on error
    if (add(2, 2) != 4) {
        BOOST_FAIL("Ouch...");
    }

    // #5 throws an exception on error
    if (add(2, 2) != 4) {
        throw "Ouch...";
    }

    // #6 continues on error
    BOOST_CHECK_MESSAGE(add(2, 2) == 4,
                        "add() result: " << add(2, 2));

    // #7 continues on error
    BOOST_CHECK_EQUAL(add(2, 2), 4);
}
```

- **Google Test** (https://github.com/google/googletest): The Google C++ Testing Framework provides a JUnit-style unit test framework for C++. It's a cross-platform system that supports automatic test discovery (i.e., you don't have to enumerate all of the tests in your test suite manually) and a rich set of assertions, including fatal assertions (the `ASSERT_*` macros), nonfatal assertions (the `EXPECT_*` macros), and so-called death tests (checks that a program terminates expectedly). Google Test also provides various options for running tests and offers textual and XML report generation. As I mentioned earlier, Google also provides a mock object testing framework, Google Mock, which integrates well with Google Test. The next code demonstrates the creation of a suite of unit tests using Google Test:

```
#include <gtest/gtest.h>

bool IsPrime(int n);

TEST(IsPrimeTest, NegativeNumbers)
{
    EXPECT_FALSE(IsPrime(-1));
    EXPECT_FALSE(IsPrime(-100));
    EXPECT_FALSE(IsPrime(INT_MIN));
}

TEST(IsPrimeTest, TrivialCases)
{
    EXPECT_FALSE(IsPrime(0));
    EXPECT_FALSE(IsPrime(1));
    EXPECT_TRUE(IsPrime(2));
    EXPECT_TRUE(IsPrime(3));
}

TEST(IsPrimeTest, PositiveNumbers)
{
    EXPECT_FALSE(IsPrime(4));
    EXPECT_TRUE(IsPrime(5));
    EXPECT_FALSE(IsPrime(9));
    EXPECT_TRUE(IsPrime(17));
}

int main(int argc, char **argv)
{
    ::testing::InitGoogleTest(&argc, argv);
    return RUN_ALL_TESTS();
}
```

- **Template Unit Test** (http://tut-framework.sourceforge.net/): The Template Unit Test (TUT) Framework is a small, portable C++ unit test framework. It consists only of header files, so there's no library to link against or deploy. Tests are organized into named test groups and the framework supports automatic discovery of all tests that you define. Several test reporters are provided, including basic console output and a CppUnit-style reporter. It's also possible to write your own reporters using TUT's extensible reporter interface. Here is a simple canonical unit test written using the TUT framework:

```
#include <tut/tut.hpp>

namespace {
tut::factory tf("basic test");
}

namespace tut {

struct basic
{
};
typedef test_group<basic> factory;
typedef factory::object object;

template <> template <>
void object::test<1>()
{
    ensure_equals("2+2", 2 + 2, 4);
}

template<> template<>
void object::test<2>()
{
    ensure_equals("2*-2", 2 * -2, -4);
}

} // namespace tut
```

Code coverage

Code coverage tools let you discover precisely which statements of your code are exercised by your tests (i.e., these tools can be used to focus your testing activities on the parts of your code base that are not already covered by tests). There are different degrees of code coverage that can be measured. I will define each of these with reference to this simple code example:

```
void TestFunction(int a, int b)
{
    if (a == 1) { a++; }          // Line 1
    int c = a * b;                // Line 2
    if (a > 10 && b != 0) {       // Line 3
        c *= 2;                   // Line 4
    }
    return a * c;                 // Line 5
}
```

- **Function coverage.** In this coarsest level of code coverage, only function calls are tracked. In our previous example, function coverage will record only whether

`TestFunction()` was called at least once. The flow of control within a function has no effect on function code coverage results.

- **Line coverage**. This form of code coverage tests whether each line of code that contains an executable statement was reached. One limitation of this metric can be seen on Line 1 of our previous code. Line coverage will consider Line 1 to be 100% exercised even if the `a++` statement is not executed; it matters only if the flow of control hit this line. Obviously, you can get round this limitation by putting the `if` condition and the `a++` statement on separate lines.
- **Statement coverage**. This metric measures whether the flow of control reached every executable statement at least once. The primary limitation of this form of coverage is that it doesn't consider the different code paths that can result from the expressions in control structures such as `if`, `for`, `while`, or `switch` statements. For example, in our previous code, statement coverage will tell us if the condition on Line 3 evaluated to true, causing Line 4 to be executed. However, it will not tell us if that condition evaluated to false, because there is no executable code associated with that result.
- **Basic block coverage**. A basic block is a sequence of statements that cannot be branched into or out of. That is, if the first statement is executed, then all of the remaining statements in the block will also be executed. Essentially, a basic block ends on a branch, function call, throw, or return. This can be thought of as a special case of statement coverage, with the same benefits and limitations.
- **Decision coverage**. This code coverage metric measures whether the overall result of the expression in each control structure evaluated to both true and false. This addresses the major deficiency of statement coverage, because you will know if the condition in Line 3 evaluated to false. This is also called branch coverage.
- **Condition coverage**. Condition coverage determines whether each Boolean subexpression in a control structure has evaluated to both true and false. In our previous example, this means that Line 3 must be hit with $a > 10$, $a \leq 10$, $b != 0$, and $b == 0$. This does not necessarily imply decision coverage, because each of these events could occur in such an order that the overall result of the if statement always evaluates to false.

There are various programs that let you measure the code coverage of C++ code. Each of these supports different combinations of the previous metrics, normally by instrumenting the code that your compiler generates. Most of these tools are commercial offerings, although there are some free and open source options, too.

One feature in particular that can be useful is the ability to exclude certain lines of code from the analysis, which is often done by adding special comments to those lines of code. This can be used to turn off coverage for lines that legitimately can never be hit, such as methods in a base class that are always overridden, although in these cases it's important for the coverage tool to raise an error if that excluded code is ever exposed in the future.

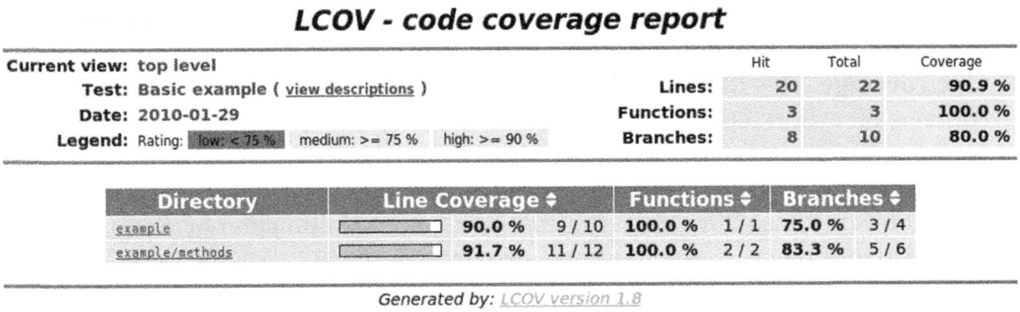

LCOV - code coverage report

			Hit	Total	Coverage
Current view:	**top level**				
Test:	**Basic example** (view descriptions)	**Lines:**	20	22	90.9 %
Date:	**2010-01-29**	**Functions:**	3	3	100.0 %
Legend:	Rating: low: < 75 % medium: >= 75 % high: >= 90 %	**Branches:**	8	10	80.0 %

Directory	Line Coverage ⇕			Functions ⇕		Branches ⇕	
example		90.0 %	9 / 10	100.0 %	1 / 1	75.0 %	3 / 4
example/methods		91.7 %	11 / 12	100.0 %	2 / 2	83.3 %	5 / 6

Generated by: LCOV version 1.8

FIGURE 12.3 Example HTML gcov code coverage report generated by lcov.

Another issue to bear in mind is that you should normally perform code coverage analysis on a build that has been compiled without optimizations, because compilers can reorder or eliminate individual lines of code during optimization.

This list provides a brief survey of some of the more prominent code coverage analysis tools:

- **Gcov** (http://gcc.gnu.org/onlinedocs/gcc/Gcov.html). This test coverage program is part of the open-source GNU GCC compiler collection. It operates on code generated by `g++` using the `-fprofile-arcs` and `-ftest-coverage` options. Gcov provides function, line, and branch code coverage. It outputs its report in a textual format; however, the accompanying lcov script can be used to output results as an HTML report (Fig. 12.3).
- **Intel Code Coverage Tool** (http://www.intel.com/). This tool is included with Intel compilers and runs on instrumentation files produced by those compilers. It provides function and basic block coverage and can restrict analysis to only modules of interest. It also supports differential coverage (i.e., comparing the output of one run against another run). The Code-Coverage Tool runs on Intel processors under Windows or Linux.
- **Bullseye Coverage** (http://www.bullseye.com/). This coverage tool, from Bullseye Testing Technology, provides function as well as condition/decision coverage, to give you a range of coverage precision. It offers features such as covering system-level and kernel mode code, merging results from distributed testing, and integration with Microsoft Visual Studio. It also gives you the ability to exclude certain portions of your code from analysis. Bullseye is a mature product that has support for a wide range of platforms and compilers.
- **Rational PureCoverage** (http://www.rational.com/). This code coverage analysis tool is sold as part of the PurifyPlus package from UNICOM Systems. It can report coverage at the executable, library, file, function, block, and line levels. PureCoverage can accumulate coverage over multiple runs and merge data from different programs that share the same source code. It offers both graphical and textual output to let you explore its results.

Once you have a code coverage build in place and can refer to regular coverage reports for your API, you can start instituting code coverage targets for your code. For example, you might specify that all code must achieve a particular threshold, such as 75%, 90%, or 100% coverage. The target you select will depend greatly upon the coverage metric that you adopt: attaining 100% function coverage should be relatively easy, whereas 100% condition/decision coverage will be far more difficult. From experience, a high and respectable degree of code coverage would be 100% function, 90% line, or 75% condition coverage.

It's also worth specifically addressing the issue of code coverage and legacy code. Sometimes your API must depend upon a large amount of old code that has no test coverage at all. Michael Feathers defines legacy code as code without tests (Feathers, 2004). In these cases, it may be impractical to enforce the same code coverage targets for the legacy code that you impose for new code. However, you can at least put some basic tests in place and then enforce that no checkin should lower the current coverage level. This effectively means that any new changes to the legacy code should be accompanied with tests. Enforcing this requirement for each checkin can sometimes be difficult (because you have to build the software with the change and run all tests to know whether it should be accepted), so another reasonable way to make this work is to record the legacy code coverage for the previous version of the library and ensure that the coverage for the new version equals or exceeds this threshold at the time of release. This approach offers a pragmatic way to deal with legacy code and lets you gradually increase the code coverage to acceptable levels over time.

In essence, different modules or libraries in your API may have different code coverage targets. In the past, I've made this clear by updating the code coverage report to display the target for each module and use a color scheme to indicate whether the targets have been met in each case. You can then quickly glance down the report to know whether your testing levels are on target.

Bug tracking

A bug tracking system is an application that lets you keep track of bugs (and often suggestions) for your software project in a database. An efficient bug tracking system that can be mapped well to your development and quality processes is an invaluable tool. Conversely, a poor bug tracking system that is difficult to use and does not fully reveal the state of your software can be an albatross around the neck of your project.

Most bug tracking systems support the triage of incoming bugs: that is, setting the priority (and perhaps severity) of a bug and assigning it to a particular developer. It's also standard practice to be able to define filters for the bugs in the system so that targeted bug lists can be created, such as a list of open crashing bugs or a list of bugs assigned to a particular developer. Related to this, some bug tracking systems will also provide report generation functions, often with the ability to display graphs and pie charts. This can be indispensable for generating quality metrics about your software, which together with code coverage results can be used to direct further testing efforts more efficiently. Another important feature is the ability to integrate your revision control system with your bug tracking system (e.g., so that bugs can be automatically marked as fixed or closed when you push a code change that fixes them).

It's also worth noting what a bug tracking system is not. For one, it is not a trouble ticket or issue tracking system. These are customer support systems that are used to receive feedback from users, many of which may not be related to software problems. Valid software issues that are discovered by customer support will then be entered into the bug tracking system and assigned to an engineer to work on. A bug tracking system is also generally not a task or project management tool: that is, a system that lets you track tasks and plan work. However, some bug tracking system vendors do provide complementary products that use the underlying infrastructure to provide a project management tool as well.

There are dozens of bug tracking systems on the market, and the best one for your project will depend upon several factors:

- You may prefer an open source solution so that you have the option to customize it if necessary. For example, many large open source projects use Bugzilla, including Mozilla, Gnome, Apache, and the Linux Kernel (see https://www.bugzilla.org/).
- There are also many commercial packages that provide excellent and flexible bug tracking capabilities that come with support and optionally secure hosting. Atlassian's JIRA is one such popular solution that provides an extremely customizable and robust bug tracking system. Atlassian also provides the related GreenHopper project management system for agile development projects, which lets you manage your user story backlog, task breakdowns, and sprint/iteration planning (see https://www.atlassian.com/).
- Alternatively, you may decide to go with a general project hosting solution that provides revision control features, disk storage quotas, discussion forums, and an integrated bug tracking system. GitHub is one popular option in this category (see https://github.com/).

Continuous build system

A continuous build system is an automated process that rebuilds your software as soon as a new change is checked into your revision control system. This is also commonly known as continuous integration (CI) and continuous delivery (CD). A continuous build system should be one of the first pieces of technology you put in place for a large project with multiple engineers, independent of any testing needs. It lets you know the current state of the build and identifies build failures as soon as they happen. It is also invaluable for cross-platform development, because even the most well-tested change for one platform can break the build for a different platform. There are several continuous build options on the market, including:

- **Jenkins** (https://www.jenkins.io/): An open source automation server for building, testing, and deploying software projects.
- **TeamCity** (https://www.jetbrains.com/teamcity/): A distributed build management and CI server from JetBrains.
- **CircleCI** (https://circleci.com/): A CI/CD platform that can be used to implement DevOps practices.
- **Travis CI** (https://travis-ci.com/): A hosted CI service that can build and test software on popular source code hosting services such as GitHub.

However, our focus here is on automated testing. The benefit of a continuous build system for testing is that it also provides a mechanism for running your automated tests regularly and hence discovering test breakages quickly. That is, the result of the automated build can be in one of four states: in progress, pass, build failure, or test failure.

As your test suite grows, so will the time it takes to run all of your tests. It's important that you receive fast turnaround on builds, so if your test run starts taking several hours to complete, then you should investigate some test optimization efforts. One way to do this is to segment your tests into different categories and run only the fast category of tests as part of the continuous build. Another solution is to have multiple automated builds: one that only builds the software and another that builds the software and then runs all tests. This gives engineers the ability to receive feedback quickly about build breakages while ensuring that a full test suite is run as often as possible.

13 Objective-C and Swift

Up to this chapter, I've focused on general aspects of application programming interface (API) design that could be applicable to any C++ project. Having covered the standard API design pipeline, the remaining chapters in this book deal with more specialized topics, such as integrating with other programming languages and extensibility via plugins. Although not all APIs need to be concerned with these topics, I felt that a comprehensive book on C++ API design should include coverage of these advanced topics. In this chapter, I will focus on interoperability with the Objective-C and Swift programming languages.

The Objective-C language was originally developed by Brad Cox and Tom Love in the 1980s. The language was adopted by NeXT, Inc., the company founded by Steve Jobs after he left Apple in 1985. NeXT eventually acquired the trademark to Objective-C before it was itself acquired by Apple in 1996. Since that time, Objective-C has become the standard programming language for developing iOS and macOS applications, at least until the release of Swift in 2014.

Objective-C is essentially an alternate approach to adding object-oriented features to the C programming language, influenced by the message-passing features of the Smalltalk language. However, there's a variant of the language called Objective-C++ that combines the syntax of C++ and Objective-C. As such, these two languages are closely linked, and given the prominence of the language as a vehicle for producing apps for Apple's App Store, I think there's value in looking at how a C++ API can interoperate with Objective-C.

I'll begin by briefly comparing the language features of C++ and Objective-C. This is not meant to be an exhaustive review, but it is sufficient for us to be able to discuss the API design implications. For a more complete introduction to the language, see (Kochan, 2013). After setting this context, I'll look at how Objective-C code can be hidden behind a pure C++ API. I'll then look at how a C++ API can be exposed to Objective-C. And finally, I'll look at how a C++ API can be exposed through Objective-C to Swift, the newest programming language for developing applications on Apple's ecosystem of platforms.

Interface design in C++ and Objective-C

Because C++ and Objective-C are both founded on C, they share a lot of common concepts, such as header files, preprocessor directives, typedefs, and free function calls. Header files in Objective-C follow the same C convention of using an `.h` suffix, whereas the implementation files use an `.m` suffix for Objective-C files and `.mm` for Objective-C++ files.

API Design for C++. https://doi.org/10.1016/B978-0-443-22219-1.00013-1

You're already familiar with C++'s object-oriented concepts at this point, but for the purposes of comparison, let's consider a simple class that holds the components of a person's name along with a function to return a formatted version of that name, as part of a larger address book library:

```
#include <string>

namespace address {

class Person
{
public:
    void SetGivenName(const std::string &name);
    std::string GetGivenName() const;

    void SetFamilyName(const std::string &name);
    std::string GetFamilyName() const;

    void SetNickname(const std::string &name);
    std::string GetNickname() const;

    std::string GetFullName() const;

private:
    class Impl;
    Impl *mImpl;
};

} // namespace address
```

An equivalent API in Objective-C might look like:

```
#import <Foundation/Foundation.h>

@interface ADRPerson : NSObject

@property(strong, nonatomic) NSString *givenName;
@property(strong, nonatomic) NSString *familyName;
@property(strong, nonatomic) NSString *nickname;

- (NSString *)getFullName;

@end
```

You'll notice a few differences about how interfaces are defined in Objective-C versus C++, such as:

1. Classes are declared between the `@interface` and `@end` keywords in header files (and optionally also in the `.m` files for private extensions to the interface, as we'll discuss later). Also note that Objective-C classes typically inherit from the `NSObject` class or one of its subclasses.
2. Objective-C offers the concept of a property to provide access to a piece of data. The compiler can automatically generate getter and setter functions for these

properties, unlike C++, in which you must manually write these. We could have also defined specific `getGivenName` and `setGivenName` functions, but it's more idiomatic to use properties for data members in Objective-C.

3. Objective-C doesn't have a concept of namespaces. As such, you must ensure that all of your public symbols have a unique prefix; otherwise they could conflict with another system or third-party library. The convention in Objective-C is to use two or three capital letters, such as CA for Core Animation, UI for UIKit, and NS for core Foundation classes. (The NS prefix comes from the name of the NeXTSTEP operating system.)
4. All objects in Objective-C are created and passed around as pointers (or const pointers), such as `NSString *`. Objective-C objects are always allocated on the heap, never statically on the stack. There is also no concept of references in the Objective-C language.
5. Although Objective-C supports the standard C `#include` preprocessor directive, it also adds a `#import` option. This behaves like `#include` with the exception that it won't include the same file twice. This means that you don't need to add include guards at the top of your Objective-C header files.

Data hiding in Objective-C

You may have noticed in the earlier example that the C++ code used the pimpl idiom to hide its data members, but this was not necessary for the Objective-C version. It's worth taking a moment to look at how we can achieve data hiding in Objective-C, because the language offers features different from C++ in this regard.

First, the use of properties in Objective-C can simplify access to data members because you can declare the properties in your header file, but the actual data members are defined within the `.m` file. This effectively achieves what we're trying to do with the pimpl idiom for C++ code: move the data member definitions out of the header files and into the implementation. For example, continuing our `ADRPerson` example from earlier, here is what the associated `.m` file might look like:

```
#import "adrperson.h"

@implementation ADRPerson {
    NSString *givenName;
    NSString *familyName;
    NSString *nickname;
}

- (NSString *)getFullName
{
    // function implementation
}

@end
```

It's also possible to put these data member definitions in the header file inside the `@interface` declaration, and you can also use Objective-C's access level keywords such as

@private to hide it from the public API (i.e., you can make the class look similar in form to the C++ version). However, the ability to hide these definitions within the implementation file is a far superior approach and should be preferred.

Another aspect of Objective-C that lets you hide information from your public API is based on a feature called categories. A category can be used to add new methods to any class, even if you don't have access to its source code, using this syntax:

```
@interface Person (CategoryName)
- (NSString *)getConversationalName;
@end
```

You can define multiple categories to extend any given object, in which each category is given a unique name inside parentheses, such as CategoryName in the previous example. One of the restrictions of categories is that you can't add new properties to the class, only functions (i.e., you can't change the size of the object after its initial declaration).

Although categories can be a powerful feature that let you add your own specializations to system and third-party objects, there is a variant of this feature called class extensions that is more relevant to our data hiding focus here. Class extensions can only be added to classes where you have the source code, but as a result they let you add new properties to a class in addition to new functions. The syntax is similar to declaring a category, except that you don't provide a name. For this reason, class extensions are sometimes also called anonymous categories.

For example, let's look at our adrperson.m implementation file again and include a class extension that adds a new data member and function to the ADRPerson class, completely hidden from the header file:

```
#import "adrperson.h"

@interface ADRPerson () {
    NSString *cachedFullName;
}

- generateFullName;

@end

@implementation ADRPerson {
    NSString *givenName;
    NSString *familyName;
    NSString *nickname;
}

- (NSString *)getFullName
{
    // function implementation
}

@end
```

In this example, we've used a class extension to add a new `cachedFullName` data member to the class as well as a new `generateFullName` function. Both are completely hidden from the public API declarations in the header file.

One limitation of the previous example is that the hidden data and function are visible only within the one `.m` file where they are declared. However, if you need to refer to these hidden members in another `.m` file you can do so by moving the declarations into a private header, such as an `adrperson_private.h` that can be included by your other source files but not distributed with the public headers for your API. For example, the private header might look like:

```
#import "adrperson.h"

@interface ADRPerson () {
    NSString *cachedFullName;
}

- generateFullName;

@end
```

And then the `.m` file would look like:

```
#import "adrperson_private.h"

@implementation ADRPerson {
    NSString *givenName;
    NSString *familyName;
    NSString *nickname;
}

- (NSString *)getFullName
{
    // function implementation
}

@end
```

Objective-C behind a C++ API

Now that I've introduced some of the basic concepts of Objective-C, let's discuss the various ways that C++ and Objective-C can be comingled.

One simple way is to maintain a pure C++ API and to leverage Objective-C in your implementation code. This is possible because both languages use header files and an Objective-C++ compiler can compile a header that contains C++ syntax.

There are several reasons why you might want to do this. One could be that you want your library to take advantage of certain functionality that's included with Objective-C's

Foundation library. Another reason might be to integrate with the native logging and performance measurement features for Apple platforms, such as `NSLog`, `os_log`, and `os_signpost`.

Let's look at an example. The following header file is pure C++ and provides an object for calculating and holding a globally unique identifier (GUID) such as `3a446875-61a5-42af-509c-ca0067405be4`. We'll keep this example simple by using a private data member in the header, although of course you could use the pimpl idiom to improve data hiding:

```
#include <string>

class GUIDHolder
{
public:
    GUIDHolder();

    // return the current GUID string
    std::string Get() const;

    // generate a new GUID string
    void Reset();

private:
    std::string mGUID;
};
```

We may decide that we want to use Objective-C's `NSUUID` object to generate the GUID string. In that case, we could create an `.mm` file (instead of a `.cpp`) with the implementation for this class and compile it with an Objective-C++ compiler. The `.mm` file might look like:

```
#include "guidholder.h"
#import <Foundation/Foundation.h>

using std::string;

GUIDHolder::GUIDHolder()
{
    Reset();
}

void GUIDHolder::Reset()
{
    NSString *uuid = [[NSUUID UUID] UUIDString];
    NSLog(@"Generated new GUID: %@", uuid);
    mGUID = (uuid) ? [uuid UTF8String] : "";
}

string
GUIDHolder::Get() const
{
    return mGUID;
}
```

Most of the code in this file is standard C++, except for the `#import` statement and the contents of the `Reset()` function. The latter code uses the `NSUUID` object from the Objective-C Foundation library and converts the `NSString *` representation to an `std::string`. That way, no knowledge of the fact that we've used Objective-C is leaked into the header file. However, your clients would now need to link their projects against the Objective-C runtime and the Foundation framework to resolve these symbols.

The following command line shows how to compile and link a simple `main.mm` Objective-C++ program. The `-lobjc` flag to pull in the Objective-C runtime may be optional with some compilers:

```
gcc main.mm -lobjc -framework Foundation -o myapp
```

Let's say that you wanted to maintain the flexibility to have your project be compiled by a regular C++ compiler for clients who don't have access to an Objective-C++ build environment. We can achieve this by checking the `__OBJC__` compiler define that most Objective-C compilers set (gcc and clang both set this define):

```
#include "guidholder.h"
#ifdef __OBJC__
#import <Foundation/Foundation.h>
#endif

using std::string;

GUIDHolder::GUIDHolder()
{
    Reset();
}

void GUIDHolder::Reset()
{
#ifdef __OBJC__
    NSString *uuid = [[NSUUID UUID] UUIDString];
    NSLog(@"Generated new GUID: %@", uuid);
    mGUID = (uuid) ? [uuid UTF8String] : "";
#else
    mGUID = SomeOtherGUIDGenerationFunction();
#endif
}

string GUIDHolder::Get() const
{
    return mGUID;
}
```

This comingling of Objective-C and C++ is possible because Objective-C++ is essentially just a union of the two syntaxes in the same file. However, they are still essentially two distinct languages, and so there is a limit to how much you can mix the features from each language. The following list offers some guidance on what you can and cannot do with Objective-C++:

- Memory management is handled natively for each language. For example, C++ uses its `new` and `delete` syntax and Objective-C uses `retain` and `release`, or more likely these days Objective-C will use Automatic Reference Counting to eliminate the need to manage retain counts manually.
- A C++ class cannot derive from an Objective-C one, and vice versa. Also C++ supports multiple inheritance whereas Objective-C uses single inheritance, and whereas there are similar member access level concepts, they use different syntax (e.g., `private` in C++ versus `@private` in Objective-C).
- You can pass C++ objects as Objective-C function arguments, and vice versa. For example, you can pass `std::string` to an Objective-C function and you can pass `NSString *` to a C++ function. Objective-C objects are always created on the heap and passed around as pointers.
- C++ objects can hold Objective-C objects because they are essentially just pointers. But Objective-C can hold C++ objects as data members only if they have a default constructor and no virtual functions.
- C++ uses the type `bool` with values `true` and `false` whereas Objective-C uses `BOOL` with values `YES` and `NO`. Similarly, C++ uses `nullptr` whereas Objective-C uses `nil`. These are each defined slightly differently, so it's good practice to pair the right type with the right language.
- C++ namespaces can't be intermingled with Objective-C (i.e., Objective-C declarations cannot appear inside a C++ namespace or include the declaration of a namespace).

C++ behind an Objective-C API

The previous section considered how to use Objective-C while hiding it behind a C++ API. In this section, I'll look at how you can leverage a C++ API behind an Objective-C one. You may want to do this, for example, if you want to offer a more standard API experience for building applications on Apple platforms.

Generally, using this approach you will want to use only Objective-C in your header files and not intermingle C++ syntax. That way you're not forcing your Objective-C clients to also understand C++ code. Essentially, the use of Objective-C++ lets you offer a plain Objective-C API while mixing Objective-C and C++ in your implementation (`.mm`) files.

Let's take our address book example from earlier. The pure Objective-C version of the API looked like:

```
#import <Foundation/Foundation.h>

@interface ADRPerson : NSObject

@property(strong, nonatomic) NSString *givenName;
@property(strong, nonatomic) NSString *familyName;
@property(strong, nonatomic) NSString *nickname;

- (NSString *)getFullName;

@end
```

Now, let's say that we want to implement this Objective-C API using our C++ library. In this case, the Objective-C interface will just be a thin layer on top of the underlying C++ API. So the implementation will store an instance of the C++ object and all Objective-C getters and setters will call through to their counterparts in the C++ interface:

```
#import "adrperson.h"
#include "person.h"

@interface ADRPerson () {
    address::Person mPerson;
}
@end

@implementation ADRPerson {
}

- (NSString *)givenName
{
    auto cppStr = mPerson.GetGivenName();
    return [NSString stringWithUTF8String:cppStr.c_str()];
}

- (void)setGivenName:(NSString *)name
{
    auto cppStr = (name) ? [name UTF8String] : "";
    mPerson.SetGivenName(cppStr);
}

- (NSString *)familyName
{
    auto cppStr = mPerson.GetFamilyName();
    return [NSString stringWithUTF8String:cppStr.c_str()];
}

- (void)setFamilyName:(NSString *)name
{
    auto cppStr = (name) ? [name UTF8String] : "";
    mPerson.SetFamilyName(cppStr);
}

- (NSString *)nickName
{
    auto cppStr = mPerson.GetNickname();
    return [NSString stringWithUTF8String:cppStr.c_str()];
}

- (NSString *)getFullName
{
    auto cppStr = mPerson.GetFullName();
    return [NSString stringWithUTF8String:cppStr.c_str()];
}

@end
```

We explicitly check for `nil` `NSString` pointers in the setter functions; otherwise our code will crash when trying to convert the `NSString *` to a `std::string`. `[nil UTF8String]` is not a problem because Objective-C specifically allows passing a message to a `nil` object, returning a `nil` result.

In essence, the Objective-C layer is a facade pattern, in which we're hiding one API behind another. In this case, we're also using a language barrier to hide our implementation API details further: if Objective-C developers want to delve into our implementation details, they will have to understand C++ code as well. Moreover we're hiding the existence of the `address::Person` C++ object from the Objective-C API by using a class extension.

You may have noticed in the previous example that I didn't create an `init` function (equivalent to a C++ constructor) to initialize our `address::Person` C++ object. Instead, I relied on the behavior of the Objective-C compiler to initialize all member variables automatically to `nil` or 0, or in the case of a C++ object, to call its default constructor.

However, we can extend this example to demonstrate using an explicit `init` and `dealloc` function, to allocate and deallocate any necessary resources. In this example, I'll switch to allocating the `address::Person` object on the heap to show how to delete the C++ object when the Objective-C object is destroyed:

```
#import "adrperson.h"
#include "person.h"

@interface ADRPerson () {
    address::Person *mPerson;
}
@end

@implementation ADRPerson {
}

- (id)init
{
    if (self = [super init]) {
        mPerson = new address::Person();
    }
    return self;
}

- (void)dealloc
{
    delete mPerson;
    [super dealloc];
}

// remaining code is unchanged
```

C++ behind a Swift API

Apple introduced the Swift programming language in 2014 as a replacement for Objective-C. Since that time, Swift has risen in popularity as the way to write code for platforms such as iOS and macOS. Interoperability with Objective-C was one of the core design principles of Swift. As such, this also provides a path for interoperability between Swift and C++.

In fact, if you already have an Objective-C layer for your C++ API, then you're most of the way toward also supporting Swift. For example, taking the Objective-C address book example from the previous section, you can access this directly from within Swift simply by importing the library and accessing the object, such as:

```
import adrperson

let person = ADRPerson()
person.givenName = "John"
print("Person = \(person.getFullName)")
```

However, there are several things you can do to augment your Objective-C APIs to make them more idiomatic when accessed from Swift. I'll discuss a number of these tactics in the remainder of this chapter.

Swift introduced the concept of optionals, similar to `std::optional` in C++17, in which a variable can hold a value or no value (i.e., `nil`). Any variable can be defined as an optional by adding a "?" suffix to the type name. For example, `String?` specifies a variable that can hold a string value or `nil`. When defining the various data types in your Objective-C interface, you need to declare whether they are optional. This is because all objects in Objective-C are pointers, but the compiler doesn't have enough information to know whether the Swift form of the object can hold `nil` (e.g., should `NSString *` in Objective-C map to `String` or `String?` In Swift?).

The concept of optionals in Swift maps to that of nullability in Objective-C (i.e., the use of the `_Nullable` and `_Nonnull` annotations for data type declarations), as shown here:

```
- (ADRPerson * _Nullable)findPerson:(NSString *_Nonnull)name;
```

In this example, the `name` parameter is defined as `_Nonnull`, which will map to a nonoptional `String` in Swift, whereas the return type is defined as `_Nullable`, which maps to an optional `ADRPerson?` in Swift. At a semantic level, this is stating that we require a string to work with, but we may not find an `ADRPerson` object that matches that name so the result could be `nil`.

Rather than annotate every data type in your Objective-C API, it's easier to declare a default for the entire header file and then provide only annotations for cases that differ from that default. For example, you can use the `NS_ASSUME_NONNULL_BEGIN` and `NS_ASSUME_NONNULL_END` macros to state that all Objective-C objects between this pair should be assumed to be nonnull:

```
#import <Foundation/Foundation.h>

NS_ASSUME_NONNULL_BEGIN

@interface ADRPerson : NSObject

@property(strong, nonatomic) NSString *givenName;
@property(strong, nonatomic) NSString *familyName;
@property(strong, nonatomic) NSString *nickname;

- (NSString *)getFullName;

@end

NS_ASSUME_NONNULL_END
```

In this case, all of the `NSString *` Objective-C objects will map to nonoptional `String` types in Swift. We could then use the `_Nullable` keyword to annotate specific declarations where we do want to support the use of an optional.

Another feature of Swift is support for named parameters. That is, it's possible to provide a name for each parameter of a Swift function and for the caller to specify that name when the function is called. For example, given this Swift function signature:

```
func find(given: String, family: String) -> ADRPerson?
```

you would write code to call this function as:

```
let person = find(given: "Alice", family: "Jones")
```

Objective-C uses a scheme known as interleaved parameters, in which the name of the function is interleaved with the arguments. So the previous Swift function might be represented in Objective-C syntax as:

```
- (ADRPerson *)findGiven:(NSString *)given family:(NSString *)family;
```

Because of these different parameter schemes, it's often necessary to provide additional context for how to map the interleaved names in Objective-C to named parameters in Swift. This can be done by using the NS_SWIFT_NAME macro at the end of the Objective-C function declaration. Here are some examples:

```
- (ADRPerson *)findGiven:(NSString *)given family:(NSString *)family
NS_SWIFT_NAME(find(given:family:));

- (NSString *)getFullName
NS_SWIFT_NAME(fullName());
```

In the first example, I've converted the interleaved function and parameter name of findGiven in the Objective-C case instead to use a function name of just find with an initial named parameter of given in the Swift case. In the second example, I've renamed the Objective-C function getFullName to be just fullName in Swift, to be more idiomatic.

The final topic I'll cover in this section is controlling how your Objective-C types are bridged to Swift types. By default, an NSString * in Objective-C will be bridged to a String in Swift, and an id will be bridged to Any. However, the various container classes provided by Objective-C can hold any objects that are derived from NSObject. So, unlike in C++ where you would declare the type of the elements in a vector, such as std::vector<std::string>, in Objective-C you would only declare that you're using an NSArray *.

What this means for bridging to Swift types is that the type of the container elements cannot be inferred by default, so they will be assumed to be Any or AnyHashable. Table 13.1 shows the default mappings of Objective-C types to Swift (the mappings were different before Swift 3, but I'll not worry about that here because Swift 3 came out in 2016).

It's possible to provide annotations for your Objective-C data types to indicate what type of object they hold. This information will then be used to bridge to a more specific type in Swift. This is done by providing typing information in angled brackets, as shown in Table 13.2.

Putting everything together in this section, you've seen how you can make a C++ API available in Swift by going through an Objective-C API, and how you can refine the Swift representation of that API by controlling the use of optionals, the naming of functions and parameters, and the bridging of Objective-C types to Swift. Given all of this, our initial ADRPerson object can now be expressed as follows to provide idiomatic interfaces in both Objective-C and Swift:

Table 13.1 A partial mapping of Objective-C types to Swift types.

Objective-C	Swift
Id	Any
NSArray *	[Any]
NSDictionary *	[AnyHashable: Any]
NSSet *	Set<AnyHashable>

Table 13.2 Example mappings of Objective-C container classes to Swift.

Objective-C	Swift
`NSArray<NSString *> *`	`[String]`
`NSDictionary<NSString *, id> *`	`[String: Any]`
`NSSet<NSString *> *`	`Set<String>`

```
#import <Foundation/Foundation.h>

NS_ASSUME_NONNULL_BEGIN

@interface ADRPerson : NSObject

@property(strong, nonatomic) NSString *givenName;
@property(strong, nonatomic) NSString *familyName;
@property(strong, nonatomic) NSString *nickname;

- (NSString *)getFullName;
NS_SWIFT_NAME(fullName());

@end

NS_ASSUME_NONNULL_END
```

14

Scripting

Continuing the focus from the previous chapter on interoperability with other programming languages, this chapter deals with the topic of scripting (i.e., allowing your C++ application programming interface (API) to be accessed from a scripting language such as Python, Ruby, Lua, Tcl, or Perl). I'll explain why you might want to do this and what some of the issues are that you need to be aware of, and then review some main technologies that let you create bindings for these languages.

To make this chapter more practical and instructive, I will take a detailed look at two script binding technologies and show how these can be used to create bindings for two scripting languages. Specifically, I will provide an in-depth treatment for how to create Python bindings for your C++ API using Boost Python, followed by a thorough analysis of how to create Ruby bindings using the Simplified Wrapper and Interface Generator (SWIG). I have chosen to focus on Python and Ruby because these are two of the most popular scripting languages in use, and in terms of binding technologies, Boost Python and SWIG are both freely available open source solutions that provide extensive control over the resulting bindings.

Adding script bindings

Extending versus embedding

A script binding provides a way to access a C++ API from a scripting language. This normally involves creating wrapper code for the C++ classes and functions that allow them to be imported into the scripting language using its native module loading features, such as the `import` keyword in Python, `require` in Ruby, or `use` in Perl.

There are two main strategies for integrating C++ with a scripting language:

1. **Extending the language**. In this model, the script binding is provided as a module that supplements the functionality of the scripting language. That is, users who write code with the scripting language can use your module in their own programs. Your module will look just like any other module for that language. For example, the expat and md5 modules in the Python standard library are implemented in C, not Python.
2. **Embedding within an application**. In this case, an end user C++ application embeds a scripting language inside it. Script bindings are then used to let end users write scripts for that specific application that call down into the core functionality of the program. Examples of this include the Autodesk Maya 3D modeling system, which offers Python and Maya Embedded Language scripting, and the Adobe

API Design for C++. https://doi.org/10.1016/B978-0-443-22219-1.00012-X

Director multimedia authoring platform, which embeds the Lingo scripting language.

Whichever strategy applies to your situation, the procedure for defining and building the script bindings is the same in each case. The only thing that really changes is who owns the C++ `main()` function.

Advantages of scripting

The decision to provide access to native code APIs from within a scripting language offers many advantages. These advantages can apply directly to you, if you provide a supported script binding for your C++ API, or to your clients, who may create their own script bindings on top of your C++-only API. I enumerate a few of these benefits here:

- **Cross-platform**. Scripting languages are interpreted, so they execute plain ASCII source code or platform-independent byte code. They will also normally provide their own modules to interface with platform-specific features, such as the file system. Writing code for a scripting language should therefore work on multiple platforms without modification. This can also be considered a disadvantage for proprietary projects because scripting code will normally have to be distributed in source form.
- **Faster development**. If you make a change to a C++ program, you must compile and link your code again. For large systems, this can be a time-consuming operation and can hinder engineers' productivity as they wait to be able to test their change. In a scripting language, you simply edit the source code and then run it; there's no compile and link stage. This allows you to prototype and test new changes quickly, often resulting in greater engineer efficiency and project velocity.
- **Write less code**. A given problem can normally be solved with less code in a higher-level scripting language versus C++. Scripting languages don't require explicit memory management, they tend to have a much larger standard library available to them than the C++ Standard Library, and they often take care of complex concepts such as reference counting behind the scenes. For example, the following single line of Ruby code will take a string and return an alphabetized list of all of the unique letters in that string. This would take a lot more code to implement in C++:

  ```
  "Hello World".downcase.split("").uniq.sort.join
  => " dehlorw"
  ```

- **Script-based applications**. The traditional view of scripting languages is that you use them for small command-line tasks, but you must write any large applications in a language such as C++ for maximum efficiency. However, an alternative view is that you can write the core performance-critical routines in C++, create script bindings for them, then write your application in a scripting language. In Model-View-Controller parlance, the Model and potentially also the View are written in C++ whereas the Controller is implemented using a scripting language. The key insight is that you don't need a superfast compiled language to manage user input that happens at a low rate of hertz.

At Pixar, we rewrote our in-house animation toolset as a Python-based main application that calls down into an extensive set of efficient Model and View C++ APIs. This gave us all of the advantages that I list here, such as removing the compile-link phase for many application logic changes while delivering an interactive 3D animation system for our artists.

- **Support for expert users**. Adding a scripting language to an end user application can allow advanced users to customize the functionality of the application by writing macros to perform repetitive tasks or tasks that are not exposed through the GUI. This can be done without sacrificing the usability of the software for novice users, who will interface with the application solely through the GUI.
- **Extensibility**. In addition to giving expert users access to under-the-covers functionality, a scripting interface can be used to let them add entirely new functionality to the application through plugin interfaces. Thus the developer of the application is no longer responsible for solving every user's problem. Instead, users have the power to solve their own problems. For example, the Firefox Web browser allows new extensions to be created using JavaScript as its embedded scripting language.
- **Scripting for testability**. One extremely valuable side effect of being able to write code in a scripting language is that you can write automated tests using that language. This is an advantage because you can enable your quality assurance (QA) team to write automated tests, too. Often (though not exclusively), QA engineers will not write C++ code. However, there are many skilled QA engineers who can write scripting language code. Involving your QA team in writing automated tests can let them contribute at a lower level and have greater insight into potential issues that should be covered by testing efforts.
- **Expressiveness**. The field of linguistics defines the principle of linguistic relativity (also known as the Sapir-Whorf hypothesis) as the idea that people's thoughts and behavior are influenced by their language. When applied to the field of computer science, this can mean that the expressiveness, flexibility, and ease of use of a programming language could affect the kinds of solutions that you can envision. That's because you don't have to be distracted by low-level issues such as memory management or statically typed data representations. This is obviously a more qualitative and subjective point than the previous technical arguments, but worth thinking about.

Language compatibility issues

One important issue to be aware of when exposing a C++ API in a scripting language is that the patterns and idioms of C++ will not map directly to those of the scripting language. As such, a direct translation of the C++ API into the scripting language may produce a script module that doesn't feel natural or idiomatic in that language. For example:

- **Naming conventions.** C++ functions are often written using either upper or lower camel case, such as `GetName()` or `getName()`. However, the Python convention (defined in PEP 8) is to use snake case for method names, such as `get_name()`. Similarly, Ruby specifies that method names should use snake case, too.
- **Getters and setters.** In this book, I have advocated that you should never directly expose data members in your classes, and that you should instead always provide getter/setter methods to access those members. However, many script languages allow you to use the syntax for accessing a member variable while forcing the access to go through getter and setter methods. In fact, in Ruby this is the only way that you can access member variables from outside a class. The result is that instead of C++ style code such as:

```
object.SetName("Hello");
std::string name = object.GetName();
```

you can simply write this, which still involves using underlying getter and setter methods:

```
object.name = "Hello"
name = object.name
```

- **Iterators.** Most scripting languages support the general concept of iterators to navigate through the elements in a sequence. However, the implementation of this concept will not normally harmonize well with the C++ Standard Library implementation. For example, C++ has five categories of iterators (forward, bidirectional, random access, input, and output), whereas Python has a single iterator category (forward). Making a C++ object iterable in a scripting language therefore requires specific attention to adapt it to the semantics of that language, such as adding an `__iter__()` method in the case of Python.
- **Operators.** You already know that C++ supports several operators, such as `operator+`, `operator+=`, and `operator[]`. Often these can be translated directly into the equivalent syntax of the scripting language, such as exposing C++'s stream `operator<<` as the `to_s()` method in Ruby (which returns a string representation of the object). However, the target language may support additional operators that are not supported by C++, such as Ruby's power operator (`**`) and its operator to return the quotient and modules of a division (`divmod`).

- **Containers**. The C++ Standard Library provides container classes such as `std::vector`, `std::set`, and `std::map`. These are statically typed class templates that can contain only objects of the same type. By comparison, many scripting languages are dynamically typed and support containers with elements of different types. It's a lot more common to use these flexible types to pass data around in scripting languages. For example, a C++ method that accepts several nonconst reference arguments might be better represented in a scripting language by a method that returns a tuple. For instance:

```
float width, height;
GetDimensions(&width, &height);      // C++

width, height = get_dimensions();    # Python
```

All of this means that creating a good script binding is often a process that requires a degree of manually tuning. Technologies that attempt to create bindings fully automatically will normally produce APIs that don't feel natural in the scripting language. For example, the PyObjC utility provides a bridge for Objective-C objects in Python, but it can result in cumbersome constructs in Python, such as methods called `setValue_()`. In contrast, a technology that lets you manually craft the way that functions are exposed in script will let you produce a higher-quality result.

Crossing the language barrier

The language barrier refers to the boundary where C++ meets the scripting language. The script bindings for an object will take care of forwarding method calls in the scripting language down into the relevant C++ code. However, having C++ code call up into the scripting language will not normally happen by default. This is because a C++ API that has not been specifically designed to interoperate with a scripting language will not know that it's running within a script environment.

For example, consider a C++ class with a virtual method that gets overridden in Python. The C++ code has no idea that Python has overridden one of its virtual methods. This makes sense because the C++ vtable is created statically at compile time and cannot adapt to Python's dynamic ability to add methods at run time. Some binding technologies provide extra functionality to make this cross-language polymorphism work. I will discuss how this is done for Boost Python and SWIG later in the chapter.

Another issue to be aware of is whether the C++ code uses an internal event or notification system. If this is the case, some extra mechanism will need to be put in place to forward any C++ triggered events across the language boundary into script code. For example, Qt and Boost offer a signal/slot system in which C++ code can register to receive notifications when another C++ object changes state. However, allowing scripts to receive these events will require you to write explicit code that can intercept the C++ events and send them over the boundary to the script object.

Finally, exceptions are another case in which C++ code may need to communicate with script code. For example, uncaught C++ exceptions must be caught at the language barrier and then be translated into the native exception type of the script language.

Script binding technologies

There are various technologies you can use to generate the wrappers that allow a scripting language to call down into your C++ code. Each offers its own specific advantages and disadvantages. Some are language-neutral technologies that support many scripting languages (such as COM or CORBA), some are specific to C/C++ but provide support for creating bindings to many languages (such as SWIG), some provide C++ bindings for a single language (such as Boost Python), whereas others focus on C++ bindings for a specific API (such as the Pivy Python bindings for the Open Inventor C++ Toolkit).

I will list several of these technologies here, and then in the remainder of the chapter I'll focus on two in more detail. I have chosen to focus on portable yet C++-specific solutions, rather than considering the more general and heavyweight interprocess communication models such as COM or CORBA. To provide greater utility, I will look at one binding technology that lets you define the script binding programmatically (Boost Python) and another that uses an interface definition file to generate code for the binding (SWIG).

Any script binding technology is essentially founded upon the Adapter design pattern. That is, it provides a one-to-one mapping of one API to another API while translating data types into their most appropriate native form and perhaps using more idiomatic naming conventions. Recognizing this fact means that you should also be aware of the standard issues that face API wrapping design patterns such as Proxy and Adapter. Of principal concern is the need to keep the two APIs synchronized over time. As you will see, both Boost Python and SWIG require you to keep redundant files in sync as you evolve the C++ API, such as extra C++ files in the case of Boost and separate interface files in the case of SWIG. This often turns out to be the largest maintenance cost when supporting a scripting API.

Boost Python

Boost Python (also written as `boost::python` or Boost.Python) is a C++ library that lets C++ APIs interoperate with Python. It's part of the excellent Boost libraries, available from https://www.boost.org/. With Boost Python, you can create bindings programmatically in C++ code and link the bindings against the Python and Boost Python libraries. This produces a dynamic library that can be imported directly into Python.

Boost Python includes support for these capabilities and features in terms of wrapping C++ APIs:

- C++ references and pointers.

- Translation of C++ exceptions to Python.
- C++ default arguments and Python keyword arguments.
- Manipulating Python objects in C++.
- Exporting C++ iterators as Python iterators.
- Python documentation strings.
- Globally registered type coercions.

Simplified wrapper and interface generator

SWIG is an open source utility that can be used to create bindings for C or C++ interfaces in a variety of high-level languages. The supported languages include scripting languages such as Perl, PHP, Python, Tcl, and Ruby, as well as non-scripting languages such as C#, Common Lisp, Java, Lua, Modula-3, OCAML, Octave, and R.

The central design concept of SWIG is the interface file, normally given an `.i` file extension. This file is used to specify the generic bindings for a given module using C/C++ as the syntax to define the bindings. The general format of an SWIG interface file is as follows.

```
// example.i
%module <module-name>
%{
// declarations needed to compile the generated C++ binding code
%}
// declarations for the classes, functions, etc. to be wrapped
```

The SWIG program can then read this interface file and generate bindings for a specific language. These bindings are then compiled to a shared library that can be loaded by the scripting language. For more information about SWIG, see https://www.swig.org/.

Python-SIP

SIP is a tool that lets you create C and C++ bindings for Python. It was originally created for the PyQt package, which provides Python bindings for the Qt tool kit. As such, Python-SIP has specific support for the Qt signal/slot mechanism. However, the tool can also be used to create bindings for any C++ API.

SIP works in a fashion similar to that of SWIG, although it doesn't support the range of languages that SWIG does. SIP supports much of the C/C++ syntax for its interface specification files and uses a syntax for its commands similar to that of SWIG (i.e., tokens that start with a % symbol), although it supports a different set and style of commands to customize the binding. Here's an example of a simple Python—SIP interface specification file:

```
// Define the SIP wrapper for an example library.

// define the Python module name and generation number
%Module example 0

class Example
{

// include example.h in the wrapper that SIP generates
%TypeHeaderCode
#include <example.h>
%End

public:
    Example(const char *name);

    char *GetName() const;
};
```

Component object model automation

Component Object Model (COM) is a binary interface standard that allows objects to interact with each other via interprocess communication. COM objects specify well-defined interfaces that enable software components to be reused and linked together to build end user applications. The technology was developed by Microsoft in 1993 and is still used today, predominantly on the Windows platform, although Microsoft now encourages the use of .NET and SOAP.

COM encompasses a large suite of technologies, but the part on which I will focus here is COM Automation, also known as OLE Automation, or simply Automation. This involves Automation objects (also known as ActiveX objects) being accessed from scripting languages to perform repetitive tasks or to control an application from script. Many target languages are supported, such as Visual Basic, JScript, Perl, Python, Ruby, and the range of Microsoft.NET languages.

A COM object is identified by a Universally Unique ID (UUID) and exposes its functionality via interfaces that are also identified by UUIDs. All COM objects support the `IUnknown` interface methods of `AddRef()`, `Release()`, and `QueryInterface()`. COM Automation objects additionally implement the `IDispatch` interface, which includes the `Invoke()` method to trigger a named function in the object.

The object model for the interface being exposed is described using an interface description language (IDL). IDL is a language-neutral description of a software component's interface, normally stored in a file with an `.idl` extension. This IDL description can then be translated into various forms using the `MIDL.EXE` compiler on Windows. The generated files include the proxy DLL code for the COM object and a type library that describes the object model. The next sample shows an example of the Microsoft IDL syntax:

```
// Example.idl
import "mydefs.h","unknown.idl";
[
    object, uuid(d1420a03-d0ec-11b1-c04f-008c3ac31d2f),
]
interface ISomething : IUnknown
{
    HRESULT MethodA([in] short Param1, [out] BKFST *pParam2);
    HRESULT MethodB([in, out] BKFST *pParam1);
};

[
    object, uuid(1e1423d1-ba0c-d110-043a-00cf8cc31d2f),
    pointer_default(unique)
]
interface ISomethingElse : IUnknown
{
    HRESULT MethodC([in] long Max,
                    [in, max_is(Max)] Param1[],
                    [out] long *pSize,
                    [out, size_is( , *pSize)] BKFST **ppParam2);
};
```

There's also a framework called Cross-Platform COM, or XPCOM. This is an open source project developed by Mozilla and used in many of their applications including the Firefox browser. XPCOM follows a design similar to that of COM, although their components are not compatible or interchangeable.

Common object request broker architecture

The Common Object Request Broker Architecture (CORBA) is an industry standard that enables software components to communicate with each other independently of their location and vendor. In this regard, it is similar to COM: both technologies solve the problem of communication between objects from different sources and both use a language-neutral IDL format to describe each object's interface.

CORBA is cross-platform with several open source implementations and provides strong support for Unix platforms. It was defined by the Object Manage Group in 1991 (the same group that manages the UML modeling language). CORBA offers a wide range of language bindings, including Python, Perl, Ruby, Smalltalk, JavaScript, Tcl, and the CORBA Scripting Language (IDLscript). It also supports interfaces with multiple inheritance, versus COM's single inheritance.

In terms of scripting, CORBA doesn't require a specific automation interface, as does COM. All CORBA objects are scriptable by default via the Dynamic Invocation Interface, which lets scripting languages determine the object's interface dynamically. As an example of accessing CORBA objects from a scripting language, here is a simple IDL description and how it maps to the Ruby language:

```
// example.idl
interface myInterface
{
    struct MyStruct { short value; };
};

// example.rb
require 'myInteface'
s = MyInterface::MyStruct.new
s.value = 42
```

Adding Python bindings with Boost Python

The rest of this chapter is dedicated to giving you a concrete understanding of how to create script bindings for your C++ API. I will begin by showing you how to create Python bindings using the Boost Python libraries.

Python is an open source dynamically typed language that was designed by Guido van Rossum and first appeared in 1991. Python is strongly typed and features automatic memory management and reference counting. It comes with a large and extensive standard library, including modules such as os, sys, re, difflib, codecs, datetime, math, gzip, csv, socket, json, and xml, among many others. One of the more unusual aspects of Python is that indentation is used to define scope, as opposed to curly braces in C and C++. The original CPython implementation of Python is the most common, but there are other major implementations, such as Jython (written in Java) and Iron Python (targeting the .NET framework). For more details on the Python language, refer to https://www.python.org/.

As I've noted, Boost Python is used to define Python bindings programmatically, which can then be compiled to a dynamic library that can be directly loaded by Python. Fig. 14.1 illustrates this basic workflow.

Building Boost Python

Many Boost packages are implemented solely as headers, using templates and inline functions, so you only need to make sure that you add the Boost directory to your compiler's include search path. However, using Boost Python requires you to build and link against the boost_python library, so you need to know how to build Boost.

The recommended way to build Boost libraries is to use the bjam utility, a descendant of the Perform Jam build system. First you will need to download bjam. Prebuild executables are available for most platforms from https://www.boost.org/.

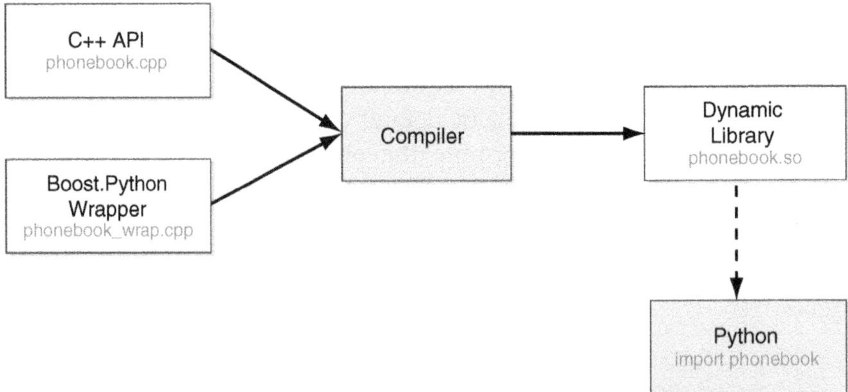

FIGURE 14.1 The workflow for creating Python bindings of a C++ application programming interface (API) using Boost Python. *White boxes* represent files and *shaded boxes* represent commands.

Building the Boost libraries on a Unix variant such as Linux or Mac involves these steps:

```
% cd <boost-root-directory>
% ./bootstrap.sh --prefix=<install-dir>
% ./bjam toolset=<toolset> install
```

The <toolset> string is used to define the compiler under which you wish to build, such as "gcc," "darwin," "msvc," or "intel."

If you have multiple versions of python installed on your machine, you can specify which version to use via bjam's configuration file, a file called `user-config.bjam` that you should create in your home directory. You can find more details about configuring bjam in the Boost.Build manual, but essentially you will want to add something like these entries to your `user-config.bjam` file:

```
using python
     : 2.6                              # version
     : /usr/local/bin/python2.6         # executable Path
     : /usr/local/include/python2.6     # include path
     : /usr/local/lib/python2.6         # lib path
```

On Windows, you can perform similar steps from the command prompt. Just run `bootstrap.bat` instead of `boostrap.sh`.

Wrapping a C++ API with Boost Python

Let's start by presenting a simple C++ API, which I will then expose to Python. I'll use the example of a phone book that lets you store phone numbers for multiple contacts. This gives us a manageable yet nontrivial example to build upon throughout the chapter. Here's the public definition of our phone book API:

```
// phonebook.h
#include <string>

class Person
{
public:
    Person();
    explicit Person(const std::string &name);

    void SetName(const std::string &name);
    std::string GetName() const;

    void SetHomeNumber(const std::string &number);
    std::string GetHomeNumber() const;
};

class PhoneBook
{
public:
    int GetSize() const;

    void AddPerson(Person *p);
    void RemovePerson(const std::string &name);
    Person *FindPerson(const std::string &name);
};
```

This will let us demonstrate several capabilities, such as wrapping multiple classes, the use of C++ Standard Library containers, and the use of multiple constructors. I will also take the opportunity to demonstrate the creation of Python properties in addition to the direct mapping of C++ member functions to Python methods.

The `Person` class is essentially just a data container: it contains only getter and setter methods that access underlying data members. These are good candidates for translating to Python properties. A property in Python behaves like a normal object but uses getter and setter methods to manage access to that object (as well as a deleter method for destroying the object). This makes for more intuitive access to class members that you want to behave like a simple data member, while letting you provide logic that controls getting and setting the value.

Now that I've presented our C++ API, let's look at how you can specify Python bindings for it using `boost::python`. You will normally create a separate `.cpp` file to specify the bindings for a given module, in which it's conventional to use the same base filename as the module with a `_wrap` suffix appended; I will use `phonebook_wrap.cpp` for our example. This wrap file is where you specify the classes you want to expose and the methods you want to be available on those classes. This file presents the `boost::python` code necessary to wrap our `phonebook.h` API:

```
// phonebook_wrap.cpp
#include "phonebook.h"

#include <boost/python.hpp>

using namespace boost::python;

BOOST_PYTHON_MODULE(phonebook)
{
    class_<Person>("Person")
        .add_property("name", &Person::GetName, &Person::SetName)
        .add_property("home_number", &Person::GetHomeNumber,
                      &Person::SetHomeNumber);

    class_<PhoneBook>("PhoneBook")
        .def("size", &PhoneBook::GetSize)
        .def("add_person", &PhoneBook::AddPerson)
        .def("remove_person", &PhoneBook::RemovePerson)
        .def("find_person", &PhoneBook::FindPerson,
            return_value_policy<reference_existing_object>());
}
```

Note that for the `Person` class, I defined two properties, called `name` and `home_number`, and I provided the C++ getter and setter functions to control access to those properties (if I provided only a getter method, then the property would be read only). For the `PhoneBook` class, I defined standard methods, called `size()`, `add_person()`, `remove_person()`, and `find_person()`, respectively. I also had to specify explicitly how I want the pointer return value of `find_person()` to behave.

You can then compile the code for `phonebook.cpp` and `phonebook_wrap.cpp` to a dynamic library. This will involve compiling against the headers for Python and `boost::python` as well as linking against the libraries for both. The result should be a `phonebook.so` library on Mac and Linux, or `phonebook.dll` on Windows. (Python doesn't recognize the `.dylib` extension on the Mac.) For example, on Linux:

```
g++ -c phonebook.cpp
g++ -c phonebook_wrap.cpp -I<boost_includes> -I<python_include>
g++ -shared -o phonebook.so phonebook.o phonebook_wrap.o \
    -lboost_python -lpython
```

TIP: Make sure that you compile your script bindings with the same version of Python against which your Boost Python library was built.

At this point, you can directly load the dynamic library into Python using its `import` keyword. Here's some sample Python code that loads our C++ library and demonstrates the creation of `Person` and `Phonebook` objects. Note the property syntax for accessing the Person object, such as `p.name`, versus the method call syntax for the `PhoneBook` members, such as `book.add_person()`:

```
#!/usr/bin/python

import phonebook

# create the phonebook
book = phonebook.PhoneBook()

# add one contact
p = phonebook.Person()
p.name = 'Martin'
p.home_number = '(123) 456-7890'
book.add_person(p)

# add another contact
p = phonebook.Person()
p.name = 'Genevieve'
p.home_number = '(123) 456-7890'
book.add_person(p)

# display number of contacts added (2)
print "No. of contacts =", book.size()
```

Constructors

In our previous `phonebook_wrap.cpp` file, I didn't specify the constructors for the `Person` or `PhoneBook` classes explicitly. In this case, Boost Python will expose the default constructor for each class. That's why I was able to write:

```
book = phonebook.PhoneBook()
```

However, in the C++ API, the `Person` class has two constructors, a default constructor and a second constructor that accepts a string parameter:

```
class Person
{
public:
    Person();
    explicit Person(const std::string &name);
    ....
};
```

You can tell Boost Python to expose both of these constructors by updating the wrapping code for Person to specify a single constructor in the class definition and then list further constructors using the `.def()` syntax:

```
class_<Person>("Person", init<>())
    .def(init<std::string>())
    .add_property("name", &Person::GetName, &Person::SetName)
    .add_property("home_number", &Person::GetHomeNumber,
                  &Person::SetHomeNumber);
```

Now you can create `Person` objects from Python using either constructor:

```
#!/usr/bin/python

import phonebook

p = phonebook.Person()
p = phonebook.Person('Martin')
```

Extending the Python API

It's also possible to add new methods to the Python API that don't exist in the C++ API. This is most commonly used to define some of the standard Python object methods, such as `__str__()` to return a human-readable version of the object or `__eq__()` to test for equality.

In the next example, I have updated the `phonebook_wrap.cpp` file to include a static free function that prints out the values of a `Person` object. I then use this function to define the `Person.__str__()` method in Python:

```
// phonebook_wrap.cpp
#include "phonebook.h"

#include <boost/python.hpp>

#include <iostream>
#include <sstream>

using namespace boost::python;

static std::string PrintPerson(const Person &p)
{
    std::ostringstream stream;
    stream << p.GetName() << ": " << p.GetHomeNumber();
    return stream.str();
}

BOOST_PYTHON_MODULE(phonebook)
{
    class_<Person>("Person", init<>())
        .def(init<std::string>())
        .add_property("name", &Person::GetName, &Person::SetName)
        .add_property("home_number", &Person::GetHomeNumber,
                      &Person::SetHomeNumber)
        .def("__str__", &PrintPerson);

    ....
}
```

This demonstrates the general ability to add new methods to a class. However, in this particular case, Boost Python provides an alternative way to specify the `__str__()` function in a more idiomatic fashion. You could define `operator<<` for `Person` and tell Boost to use this operator for the `__str__()` method. For example:

```
#include "phonebook.h"

#include <boost/python.hpp>

#include <iostream>

using namespace boost::python;

std::ostream &operator<<(std::ostream &os, const Person &p)
{
    os << p.GetName() << ": " << p.GetHomeNumber();
    return os;
}

BOOST_PYTHON_MODULE(phonebook)
{
    class_<Person>("Person", init<>())
        .def(init<std::string>())
        .add_property("name", &Person::GetName, &Person::SetName)
        .add_property("home_number", &Person::GetHomeNumber,
                      &Person::SetHomeNumber)
        .def(self_ns::str(self));
}
```

With this definition for Person.__str__() you can now write code such as this (entered at the interactive Python interpreter prompt, >>>):

```
>>> import phonebook
>>> p = phonebook.Person('Martin')
>>> print p
Martin:
>>> p.home_number = '(123) 456-7890'
>>> print p
Martin: (123) 456-7890
```

While I am talking about extending the Python API, I will note that the dynamic nature of Python means that you can add new methods to a class at run time. This is not a Boost Python feature, but a core capability of the Python language itself. For example, you could define the __str__() method at the Python level as:

```
#!/usr/bin/python

import phonebook

def person_str(self):
    return "Name: %s\nHome: %s" % (self.name, self.home_number)

# override the __str__ method for the Person class
phonebook.Person.__str__ = person_str

p = phonebook.Person()
p.name = 'Martin'
p.home_number = '(123) 456-7890'

print p
```

This will output out the next text to the shell:

```
Name: Martin
Home: (123) 456-7890
```

Inheritance in C++

Both C++ and Python support multiple inheritance, and Boost Python makes it easy to expose all of the base classes of any C++ class. I'll show how this is done by turning the Person class into a base class (i.e., provide a virtual destructor) and adding a derived class called PersonWithCell, which adds the ability to specify a cell phone number. This is not a particularly good design choice, but it serves our purposes for this example:

```
// phonebook.h
#include <string>

class Person
{
public:
    Person();
    explicit Person(const std::string &name);
    virtual ~Person();

    void SetName(const std::string &name);
    std::string GetName() const;

    void SetHomeNumber(const std::string &number);
    std::string GetHomeNumber() const;
};

class PersonWithCell : public Person
{
public:
    PersonWithCell();
    explicit PersonWithCell(const std::string &name);

    void SetCellNumber(const std::string &number);
    std::string GetCellNumber() const;
};
```

You can then represent this inheritance hierarchy in Python by updating the wrap file as:

```
BOOST_PYTHON_MODULE(phonebook)
{
    class_<Person>("Person", init<>())
        .def(init<std::string>())
        .add_property("name", &Person::GetName, &Person::SetName)
        .add_property("home_number", &Person::GetHomeNumber,
                      &Person::SetHomeNumber);

    class_<PersonWithCell, bases<Person>>("PersonWithCell")
        .def(init<std::string>())
        .add_property("cell_number", &PersonWithCell::GetCellNumber,
                      &PersonWithCell::SetCellNumber);
    ...
}
```

Now you can create `PersonWithCell` objects from Python as:

```
#!/usr/bin/python

import phonebook

book = phonebook.PhoneBook()

p = phonebook.Person()
p.name = 'Martin'
p.home_number = '(123) 456-7890'
book.add_person(p)

p = phonebook.PersonWithCell()
p.name = 'Genevieve'
p.home_number = '(123) 456-7890'
p.cell_number = '(213) 097-2134'
book.add_person(p)
```

Cross-language polymorphism

You can create classes in Python that derive from C++ classes that you've exposed with Boost Python. For example, the next Python program shows how you could create the `PersonWithCell` class directly in Python and still be able to add instances of this class to `PhoneBook`:

```
#!/usr/bin/python

import phonebook

book = phonebook.PhoneBook()

class PyPersonWithCell(phonebook.Person):
    def get_cell_number(self):
        return self.cell
    def set_cell_number(self, n):
        self.cell = n
    cell_number = property(get_cell_number, set_cell_number)

p = PyPersonWithCell()
p.name = 'Martin'
p.home_number = '(123) 456-7890'
p.cell_number = '(213) 097-2134'
book.add_person(p)
```

Of course, the `cell_number` property on `PyPersonWithCell` will be callable only from Python. C++ will have no idea that a new method has been dynamically added to an inherited class.

Moreover, even C++ virtual functions that are overridden in Python will not be callable from C++ by default. However, Boost Python provides a way to do this if cross-language polymorphism is important for your API. This is done by defining a wrapper class that multiply inherits from the C++ class being bound as well as Boost Python's wrapper class template. This wrapper class can then check to see whether an override has been defined in Python for a given virtual function and then call that method if it's defined. For example, given a C++ class called `Base` with a virtual method, you can create the wrapper class as:

```
class Base
{
public:
    virtual ~Base();
    virtual int f();
};

class BaseWrap : Base, wrapper<Base>
{
public:
    int f()
    {
        // check for an override in Python
        if (override f = this->get_override("f")) {
            return f();
        }

        // or call the C++ implementation
        return Base::f();
    }

    int default_f()
    {
        return this->Base::f();
    }
};
```

Then you can expose the `Base` class as:

```
class_<BaseWrap, boost::noncopyable>("Base")
    .def("f", &Base::f, &BaseWrap::default_f);
```

Supporting iterators

Boost Python also lets you create Python iterators based upon C++ Standard Library iterator interfaces that you define in your C++ API. This lets you create objects in Python that behave more "Pythonically" in terms of iterating through the elements in a container. For example, you can add `begin()` and `end()` methods to the `PhoneBook` class that provide access to Standard Library iterators for traversing through all of the contacts in the phone book:

```
class PhoneBook
{
public:
    int GetSize() const;

    void AddPerson(Person *p);
    void RemovePerson(const std::string &name);
    Person *FindPerson(const std::string &name);

    typedef std::vector<Person *> PersonList;
    PersonList::iterator begin();
    PersonList::iterator end();
};
```

With these additional methods, you can extend the wrapping for the `PhoneBook` class to specify the `__iter__()` method, which is the Python way for an object to return an iterator:

```
BOOST_PYTHON_MODULE(phonebook)
{
    ...
    class_<PhoneBook>("PhoneBook")
        .def("size", &PhoneBook::GetSize)
        .def("add_person", &PhoneBook::AddPerson)
        .def("remove_person", &PhoneBook::RemovePerson)
        .def("find_person", &PhoneBook::FindPerson,
             return_value_policy<reference_existing_object>())
        .def("__iter__", range(&PhoneBook::begin, &PhoneBook::end));
}
```

Now, you can write Python code that iterates through all contacts in a PhoneBook object as:

```
#!/usr/bin/python

import phonebook

book = phonebook.PhoneBook()
book.add_person(phonebook.Person())
book.add_person(phonebook.Person())

for person in book:
    print person
```

Putting it all together

Combining all of the features that I've introduced in the preceding sections, here's the final definition of the `phonebook.h` header and the `phonebook_wrap.cpp` `boost::python` wrapper:

```
// phonebook.h
#include <string>
#include <vector>

class Person
{
public:
    Person();
    explicit Person(const std::string &name);
    virtual ~Person();

    void SetName(const std::string &name);
    std::string GetName() const;

    void SetHomeNumber(const std::string &number);
    std::string GetHomeNumber() const;
};

class PersonWithCell : public Person
{
public:
    PersonWithCell();
    explicit PersonWithCell(const std::string &name);

    void SetCellNumber(const std::string &number);
    std::string GetCellNumber() const;
};

class PhoneBook
{
public:
    int GetSize() const;

    void AddPerson(Person *p);
    void RemovePerson(const std::string &name);
    Person *FindPerson(const std::string &name);

    typedef std::vector<Person *> PersonList;
    PersonList::iterator begin() { return mList.begin(); }
    PersonList::iterator end() { return mList.end(); }
};


// phonebook_wrap.cpp
#include "phonebook.h"

#include <boost/python.hpp>

#include <iostream>
#include <sstream>

using namespace boost::python;

std::ostream &operator<<(std::ostream &os, const Person &p)
{
    os << p.GetName() << ": " << p.GetHomeNumber();
    return os;
}

static std::string PrintPersonWithCell(const PersonWithCell *p)
{
    std::ostringstream stream;
```

```
    stream << "Name: " << p->GetName() << ", Home: ";
    stream << p->GetHomeNumber() << ", Cell: ";
    stream << p->GetCellNumber();
    return stream.str();
}

BOOST_PYTHON_MODULE(phonebook)
{
    class_<Person>("Person", init<>())
        .def(init<std::string>())
        .add_property("name", &Person::GetName, &Person::SetName)
        .add_property("home_number", &Person::GetHomeNumber,
                      &Person::SetHomeNumber)
        .def(self_ns::str(self));

    class_<PersonWithCell,
           bases<Person>>("PersonWithCell")
        .def(init<std::string>())
        .add_property("cell_number", &PersonWithCell::GetCellNumber,
                      &PersonWithCell::SetCellNumber)
        .def("__str__", &PrintPersonWithCell);

    class_<PhoneBook>("PhoneBook")
        .def("size", &PhoneBook::GetSize)
        .def("add_person", &PhoneBook::AddPerson)
        .def("remove_person", &PhoneBook::RemovePerson)
        .def("find_person", &PhoneBook::FindPerson,
             return_value_policy<reference_existing_object>())
        .def("__iter__", range(&PhoneBook::begin, &PhoneBook::end));
}
```

Adding Ruby bindings with SWIG

In the following sections, I'll look at another example of creating script bindings for C++ APIs. In this case I'll use the SWIG and I'll use this utility to create bindings for the Ruby language.

Ruby is an open source dynamically typed scripting language that was released by Yukihiro "Matz" Matsumoto in 1995. Ruby was influenced by languages such as Perl and Smalltalk with an emphasis on ease of use. In Ruby, everything is an object, even types that C++ treats separately as built-in primitives such as `int`, `float`, and `bool`. Ruby is an extremely popular scripting language and is often cited as being more popular than Python in Japan, where it was originally developed. For more information on the Ruby language, see https://www.ruby-lang.org/.

SWIG works by reading the binding definition within an interface file and generating C++ code to specify the bindings. This generated code can then be compiled to a dynamic library that can be directly loaded by Ruby. Fig. 14.2 illustrates this basic workflow. SWIG supports many scripting languages. I will use it to create Ruby bindings, but it could just as easily be used to create Python bindings, Perl bindings, or bindings for several other languages.

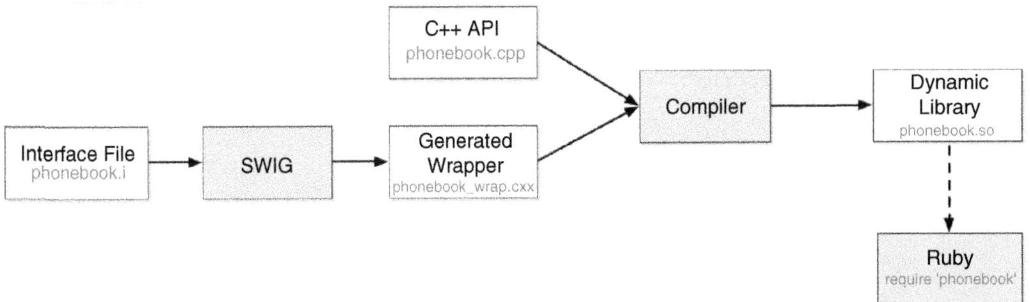

FIGURE 14.2 The workflow for creating Ruby bindings of a C++ application programming interface (API) using Simplified Wrapper and Interface Generator (SWIG). *White boxes* represent files and *shaded boxes* represent commands.

Wrapping a C++ API with SWIG

I'll start with the same phone book API from the Python example and then show how to create Ruby bindings for this interface using SWIG. The phone book C++ header looks like:

```
// phonebook.h
#include <string>

class Person
{
public:
    Person();
    explicit Person(const std::string &name);

    void SetName(const std::string &name);
    std::string GetName() const;

    void SetHomeNumber(const std::string &number);
    std::string GetHomeNumber() const;
};

class PhoneBook
{
public:
    bool IsEmpty() const;
    int GetSize() const;

    void AddPerson(Person *p);
    void RemovePerson(const std::string &name);
    Person *FindPerson(const std::string &name);
};
```

Let's look at a basic SWIG interface file to specify how you want to expose this C++ API to Ruby:

```
// phonebook.i
%module phonebook
%{

// we need the API header to compile the bindings
#include "phonebook.h"

%}

// pull in the SWIG Standard Library wrappings (note the '%')
%include "std_string.i"
%include "std_vector.i"

class Person
{
public:
    Person();
    explicit Person(const std::string &name);

    void SetName(const std::string &name);
    std::string GetName() const;

    void SetHomeNumber(const std::string &number);
    std::string GetHomeNumber() const;
};

class PhoneBook
{
public:
    bool IsEmpty() const;
    int GetSize() const;

    void AddPerson(Person *p);
    void RemovePerson(const std::string &name);
    Person *FindPerson(const std::string &name);
};
```

You can see that the interface file looks similar to the `phonebook.h` header file. In fact, SWIG can parse most C++ syntax directly. If your C++ header is simple, you can even use SWIG's `%include` directive to tell it to read the C++ header file directly. I've chosen not to do this so that you have direct control over what you do and don't expose to Ruby.

Now that you have an initial interface file, you can ask SWIG to read this file and generate Ruby bindings for all of the specified C++ classes and methods. This will create a `phonebook_wrap.cxx` file, which you can compile together with the C++ code to create a dynamic library. For example, the steps on Linux are:

```
swig -c++ -ruby phonebook.i    # creates phonebook_wrap.cxx
g++ -c phonebook_wrap.cxx -I<ruby-include-path>
g++ -c phonebook.cpp
g++ -shared -o phonebook.so phonebook_wrap.o phonebook.o \
    -L<ruby-lib-path> -lruby
```

Tuning the Ruby API

This first attempt at the Ruby binding is rudimentary. There are several issues that you'll want to address to make the API feel more natural to Ruby programmers. First, the naming convention for Ruby methods is to use snake case instead of camel case: that is, `add_person()` instead of `AddPerson()`. SWIG supports this by letting you rename symbols in the scripting API using its `%rename` command. For example, you can add these lines to the interface file to tell SWIG to rename the methods of the `PhoneBook` class:

```
%rename("size") PhoneBook::GetSize;
%rename("add_person") PhoneBook::AddPerson;
%rename("remove_person") PhoneBook::RemovePerson;
%rename("find_person") PhoneBook::FindPerson;
```

Recent versions of SWIG support an `-autorename` command line option to perform this function renaming automatically. It's expected that this option will eventually be turned on by default.

Second, Ruby has a concept similar to Python's properties to provide convenient access to data members. In fact, elegantly, all instance variables in Ruby are private and must therefore be accessed using via getter/setter methods. The `%rename` syntax can be used to accomplish this ability, too. For example:

```
%rename("name") Person::GetName;
%rename("name=") Person::SetName;
%rename("home_number") Person::GetHomeNumber;
%rename("home_number=") Person::SetHomeNumber;
```

Finally, you may have noticed that I added an extra `IsEmpty()` method to the `PhoneBook` C++ class. This method returns true if no contacts have been added to the phone book. I've added this because it lets us demonstrate how to expose a C++ member function as a Ruby query method. This is a method that returns a Boolean return value, and by convention it ends with a question mark. I would therefore like the `IsEmpty()` C++ function to appear as `empty?` in Ruby. This can be done using SWIG's `%predicate` or `% rename` directives:

```
%rename("empty?") PhoneBook::IsEmpty;
```

With these amendments to our interface file, our Ruby API is starting to feel more idiomatic. If you rerun SWIG on the interface file and rebuild the phone book dynamic library, you can import it directly into Ruby and write code such as:

```
#!/usr/bin/ruby

require 'phonebook'

book = Phonebook::PhoneBook.new

p = Phonebook::Person.new
p.name = 'Martin'
p.home_number = '(123) 456-7890'
book.add_person(p)

p = Phonebook::Person.new
p.name = 'Genevieve'
p.home_number = '(123) 456-7890'
book.add_person(p)

puts "No. of contacts = #{book.size}"
```

Note the use of the `p.name` getter and `p.name=` setter, as well as the snake case `add_person()` method name.

Constructors

Our `Person` class has two constructors: a default constructor that takes no parameters and a nondefault constructor that takes an `std::string` name. Using SWIG, you simply have to include those constructor declarations in the interface file and it will automatically create the relevant constructors in Ruby. That is, given the previous interface file, you can already do:

```
#!/usr/bin/ruby

require 'phonebook'

p = Phonebook::Person.new
p = Phonebook::Person.new('Genevieve')
```

In general, method overloading is not quite as flexible in Ruby as it is in C++. For example, SWIG will not be able to disambiguate between overloaded functions that map to the same types in Ruby: for example, a constructor that takes a `short` and another that takes an `int`, or a constructor that takes a pointer to an object and another that takes a reference to the same type. SWIG provides a way to deal with this by letting you ignore a given overloaded method (using `%ignore`) or renaming one of the methods (using `%rename`).

Extending the Ruby API

SWIG lets you extend the functionality of your C++ API (e.g., to add new methods to a class that will appear only in the Ruby API). This is done using SWIG's `%extend` directive. I will demonstrate this by adding a `to_s()` method to the Ruby version of our `Person` class. This is a standard Ruby method that is used to return a human-readable representation of an object, equivalent to Python's `__str__()` method:

```
// phonebook.i

%module phonebook
%{

#include "phonebook.h"
#include <sstream>
#include <iostream>

%}

...

class Person
{
public:
    Person();
    explicit Person(const std::string &name);

    void SetName(const std::string &name);
    std::string GetName() const;

    void SetHomeNumber(const std::string &number);
    std::string GetHomeNumber() const;

    %extend {
        std::string to_s()
        {
            std::ostringstream stream;
            stream << self->GetName() << ": ";
            stream << self->GetHomeNumber();
            return stream.str();
        }
    }
};

...
```

Using this new definition for our `Person` binding, you can write this Ruby code:

```
#!/usr/bin/ruby

require 'phonebook'

p = Phonebook::Person.new
p.name = 'Martin'
p.home_number = '(123) 456-7890'

puts p
```

The `puts p` line will print out the `Person` object using our `to_s()` method. In this case, this results in the following output:

```
Martin: (123) 456-7890
```

Inheritance in C++

As with the previous constructor case, there's nothing special that you need to do to represent inheritance using SWIG. You just declare the class in the interface file using the standard C++ syntax. For example, you can add this `PersonWithCell` class to our API:

```
// phonebook.h
#include <string>

class Person
{
public:
    Person();
    explicit Person(const std::string &name);
    virtual ~Person();

    void SetName(const std::string &name);
    std::string GetName() const;

    void SetHomeNumber(const std::string &number);
    std::string GetHomeNumber() const;
};

class PersonWithCell : public Person
{
public:
    PersonWithCell();
    explicit PersonWithCell(const std::string &name);

    void SetCellNumber(const std::string &number);
    std::string GetCellNumber() const;
};

...
```

Then you can update the SWIG interface file as:

```
// phonebook.i

...

%rename("name") Person::GetName;
%rename("name=") Person::SetName;

%rename("home_number") Person::GetHomeNumber;
%rename("home_number=") Person::SetHomeNumber;

class Person
{
public:
    Person();
    explicit Person(const std::string &name);
    virtual ~Person();

    void SetName(const std::string &name);
    std::string GetName() const;

    void SetHomeNumber(const std::string &number);
    std::string GetHomeNumber() const;

    ...
};

%rename("cell_number") PersonWithCell::GetCellNumber;
%rename("cell_number=") PersonWithCell::SetCellNumber;

class PersonWithCell : public Person
{
public:
    PersonWithCell();
    explicit PersonWithCell(const std::string &name);

    void SetCellNumber(const std::string &number);
    std::string GetCellNumber() const;
};

...
```

You can then access this derived C++ class from Ruby as:

```
#!/usr/bin/ruby

require 'phonebook'

p = Phonebook::Person.new
p.name = 'Martin'
p.home_number = '(123) 456-7890'

p = Phonebook::PersonWithCell.new
p.name = 'Genevieve'
p.home_number = '(123) 456-7890'
p.cell_number = '(213) 097-2134'
```

Ruby supports only single inheritance, with support for additional mixin classes. C++ supports multiple inheritance, of course. Therefore, by default SWIG will consider only the first base class listed in the derived class: member functions in any other base classes will not be inherited. However, versions of SWIG support an optional `-minherit` command line option that will attempt to simulate multiple inheritance using Ruby mixins (although in this case a class no longer has a true base class in Ruby).

Cross-language polymorphism

By default, if you override a virtual function in Ruby, you'll be unable to call the Ruby method from C++. However, SWIG gives you a way to enable this kind of cross-language polymorphism via its "directors" feature. When you enable directors for a class, SWIG generates a new wrapper class that derives from the C++ class as well as SWIG's Director class. The director class stores a pointer to the underlying Ruby object and works out whether a function call should be directed to an overridden Ruby method or the default C++ implementation. This is analogous to the way that Boost Python supports cross-language polymorphism. However, SWIG creates the wrapper class for you behind the scenes: all you have to do is specify for which classes you want to create directors and enable the directors feature in your `%module` directive. For example, this update to our interface file will turn on cross-language polymorphism for all of our classes:

```
%module(directors="1") phonebook
%{

#include "phonebook.h"
#include <sstream>
#include <iostream>

%}

%feature("director");

...
```

Putting it all together

I have evolved our simple example through several iterations to add each incremental enhancement. I'll finish this section by presenting the entire C++ header and SWIG interface file for your reference. First, here's the C++ API:

```
// phonebook.h
#include <string>

class Person
{
public:
    Person();
    explicit Person(const std::string &name);
    virtual ~Person();

    void SetName(const std::string &name);
    std::string GetName() const;

    void SetHomeNumber(const std::string &number);
    std::string GetHomeNumber() const;
};

class PersonWithCell : public Person
{
public:
    PersonWithCell();
    explicit PersonWithCell(const std::string &name);

    void SetCellNumber(const std::string &number);
    std::string GetCellNumber() const;
};

class PhoneBook
{
public:
    bool IsEmpty() const;
    int GetSize() const;

    void AddPerson(Person *p);
    void RemovePerson(const std::string &name);
    Person *FindPerson(const std::string &name);
};
```

and here's the final SWIG interface (`.i`) file:

```
%module(directors="1") phonebook
%{

#include "phonebook.h"
#include <sstream>
#include <iostream>

%}

%feature("director");

%include "std_string.i"
%include "std_vector.i"

%rename("name") Person::GetName;
%rename("name=") Person::SetName;

%rename("home_number") Person::GetHomeNumber;
%rename("home_number=") Person::SetHomeNumber;

class Person
{
public:
    Person();
    explicit Person(const std::string &name);
    virtual ~Person();

    void SetName(const std::string &name);
    std::string GetName() const;

    void SetHomeNumber(const std::string &number);
    std::string GetHomeNumber() const;

    %extend {
        std::string to_s()
        {
            std::ostringstream stream;
            stream << self->GetName() << ": ";
            stream << self->GetHomeNumber();
            return stream.str();
        }
    }
};

%rename("cell_number") PersonWithCell::GetCellNumber;
%rename("cell_number=") PersonWithCell::SetCellNumber;

class PersonWithCell : public Person
{
public:
    PersonWithCell();
    explicit PersonWithCell(const std::string &name);

    void SetCellNumber(const std::string &number);
    std::string GetCellNumber() const;
};

%rename("empty?") PhoneBook::IsEmpty;
%rename("size") PhoneBook::GetSize;
%rename("add_person") PhoneBook::AddPerson;
%rename("remove_person") PhoneBook::RemovePerson;
%rename("find_person") PhoneBook::FindPerson;

class PhoneBook
```

```
{
public:
    bool IsEmpty() const;
    int GetSize() const;

    void AddPerson(Person *p);
    void RemovePerson(const std::string &name);
    Person *FindPerson(const std::string &name);
};
```

15

Extensibility

In this final chapter, I'll discuss the topic of application programming interface (API) extensibility. By this, I mean the ability for your clients to modify the behavior of your interface without requiring you to evolve the API for their specific needs. This can be a critical factor in your ability to maintain a clean and focused interface while delivering a flexible system that lets your users solve problems that you never anticipated. This concept is expressed by the open/closed principle, which we discussed in Chapter 4 on Design: that an API should be open for extension but closed for modification (Meyer, 1997).

To offer a real-world example, the Marionette animation system at Pixar supported key-frame animation with a range of possible interpolation schemes between animation keys, such as Bézier, Catmull-Rom, linear, and step interpolation. However, during the development of *The Incredibles and Cars*, it became necessary to allow our production users to devise and iterate on more sophisticated interpolation routines. Rather than continually update the core animation system every time our users needed to change their custom interpolation algorithm, we devised a plugin system that allowed production users to create dynamic libraries that could be discovered at run time and would then be added to the set of built-in interpolation routines. This proved an effective way to resolve the production-specific needs while maintaining a general-purpose film-making system.

This chapter is dedicated to various techniques that allow you to achieve the same level of flexibility in your own APIs. I will spend most of the chapter detailing how to create industrial-strength cross-platform plugin architectures for your C and C++ APIs, but I will also cover other extensibility techniques using inheritance and templates.

Extending via plugins

In the most common scenario, a plugin is a dynamic library that's discovered and loaded at run time, as opposed to a dynamic library against which an application is linked at build time. Plugins can therefore be written by your users, using a well-defined Plugin API that you provide. This allows them to extend the functionality of your API in designated ways. Fig. 15.1 illustrates this concept, in which the white boxes represent artifacts that your users produce.

However, static library plugins are also possible, such as for embedded systems in which all plugins are statically linked into the application at compile time. This is useful to ensure that a plugin can be found at run time and that it's been built under the same environment as the main executable. However, I'll focus on the dynamic library model in this chapter, because this poses the most design challenges and gives users the ability to add new plugins to the system at run time.

API Design for C++. https://doi.org/10.1016/B978-0-443-22219-1.00004-0

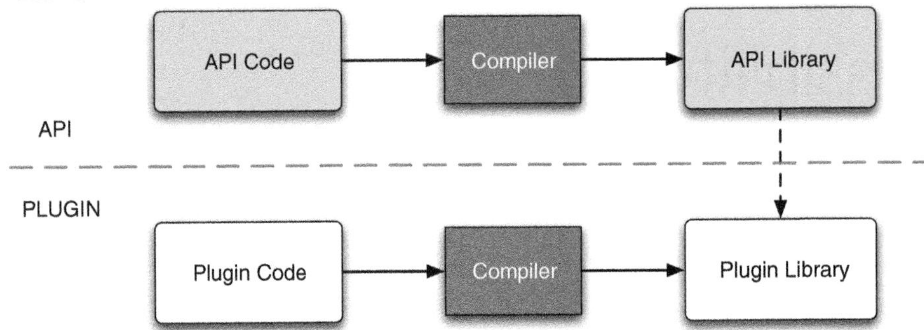

FIGURE 15.1 A plugin library is a dynamic library that can be compiled separately from a Core application programming interface (API) and explicitly loaded by the API on demand.

Plugin model overview

There are many examples of commercial software packages that allow their core functionality to be extended by C/C++ plugins. For example, the Apache Web server supports C-based "modules," Adobe Photoshop supports a range of plugin types to manipulate images, and Web browsers such as Firefox, Chrome, and Opera support the Netscape Plugin API (NPAPI) for the creation of browser plugins such as the PDF Reader plugin. The Qt toolkit can also be extended via the `QPluginLoader` class. (A server-based Plugin API such as Apache's module interface is sometimes referred to as a Server API.)

Some of the benefits of adopting a plugin model in your API are:

- **Greater versatility**. Your API can be used to solve a greater range of problems without requiring you to implement solutions for all of them.
- **Community catalyst**. By giving your users the ability to solve their own problems within the framework of your API, you can spark a community of user-contributed additions to your base design.
- **Smaller updates**. Functionality that exists as a plugin can be easily updated independently of the application simply by dropping in a new version of the plugin. This can often be a much smaller update than distributing a new version of the entire application.
- **Future proofing**. Your API may reach a level stability where you feel that no more updates are necessary. However, further evolution of the functionality of your API can continue through the development of plugins, allowing the API to maintain its

usefulness and relevance for a greater period. For example, the NPAPI has changed little in recent years, but it remains a popular method to write plugins for many Web browsers.

- **Isolating risk**. Plugins can be beneficial for in-house development, too, by letting engineers change functionality without destabilizing the core of your system. This is, of course, because by their nature, plugins are loosely coupled from the core system.

As I hinted earlier, a plugin system doesn't have to be used only by your clients. You can develop parts of your Core API implementation as plugins, too. In fact, this is a good practice because it ensures that you fully exercise your plugin architecture and that you live in the same world as your users ("eat your own dog food"). For example, the GNU Image Manipulation Program (GIMP) ships many of its built-in image processing functions as plugins using its GIMP Plugin API.

SIDEBAR: Netscape Plugins

The Netscape Plugin API provides a cross-platform Plugin API to embed custom functionality inside various Web browsers. The interface grew out of work from Adobe Systems to integrate a PDF viewer into early versions of the Netscape browser. This plain C Plugin API is still used today to embed native code extensions inside Web browsers such as Mozilla Firefox, Apple Safari, and Google Chrome. For example, Shockwave Flash, Apple QuickTime, and Microsoft Silverlight are implemented as browser plugins (try typing "about:plugins" into Firefox to see a list of installed plugins).

The NPAPI gives native code these capabilities inside a Web browser:

- *Register new MIME types.*
- *Draw into a region of the browser window.*
- *Receive mouse and keyboard events.*
- *Send and receive data over HTTP.*
- *Add hyperlinks or hotspots that link to new URLs.*
- *Communicate with the Document Object Model.*

While I was at SRI International in the 1990s, we developed a Web-based 3D terrain visualization system called TerraVision (Leclerc and Lau, 1994). This started life as a desktop application that required a 4-processor SGI Onyx RealityEngine^2. However, as commodity graphics hardware advanced, we were eventually able to make it run as a plugin inside Netscape Navigator and Microsoft Explorer on a standard PC.

Continued

SIDEBAR: *Netscape Plugins*—cont'd

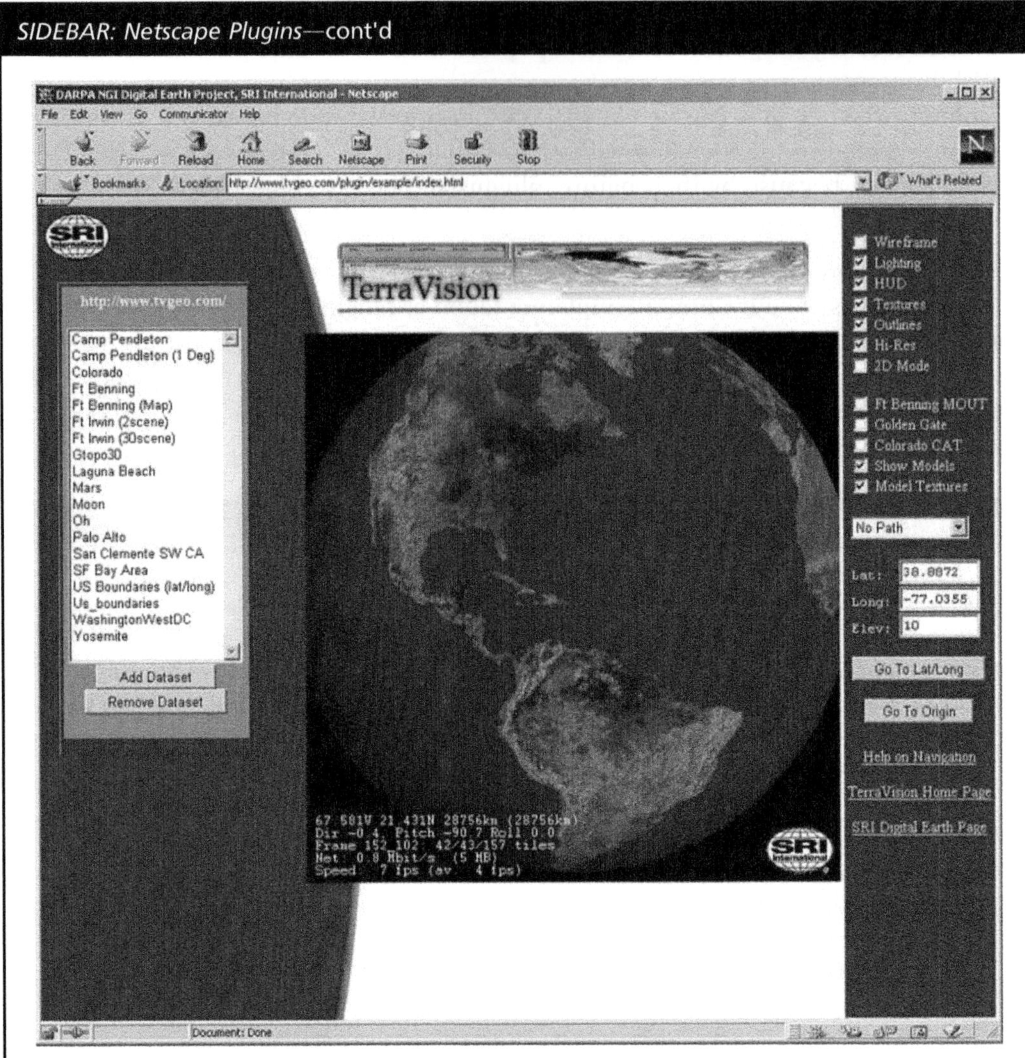

TerraVision was a complex multithreaded application, so to make this work in a non–thread-safe browser environment we had to run it in a separate process. Our Netscape plugin would create a new process for TerraVision to run in and pass it a window handle to draw into. All communication between the plugin and the TerraVision processes happened via pipe I/O.

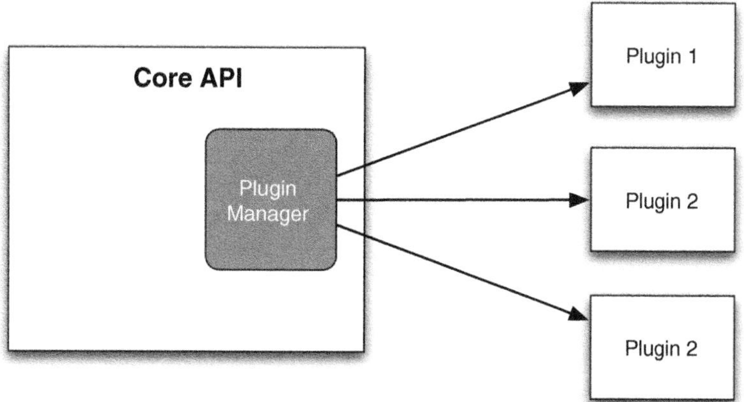

FIGURE 15.2 The Plugin manager lives in the core application programming interface (API). It discovers and loads plugins that have been built against the Plugin API.

Plugin system design issues

There are many ways to design a plugin system. The best solution for your current project may not be the best choice for your next project. I will therefore start by trying to tease out some of the high-level issues that you should be aware of when devising a plugin system.

At the same time, a few general concepts are applicable to all plugin systems. For example, when supporting dynamic library plugins, you'll always need a mechanism to load a dynamic library and access symbols in that file. Also, when creating any plugin system, there are two major features that you must design (Fig. 15.2):

1. **The Plugin API**: This is the API that your users must compile and link against to create a plugin. I differentiate this from your Core API, which is the larger code base into which you are adding the plugin system.
2. **The Plugin Manager**: This is an object (often a singleton) in the Core API code that manages the life cycle of all plugins, such as loading, registration, and unloading. This object can also be called the Plugin registry.

With these general concepts in hand, let's look at some of the design decisions that will affect the precise plugin architecture that you should build for your API:

- **C versus C++**. As I discussed earlier in the Versioning chapter, the C++ specification does not define a specific application binary interface (ABI). Therefore, different compilers, and even different versions of the same compiler, can produce code that is binary incompatible. The implication for a plugin system is that plugins developed by clients using a compiler with a different ABI may not be loadable. In contrast, the ABI for plain C code is well-defined and will work across platforms and compilers.

- **Versioning**. You will want some way to know whether a plugin was built with an incompatible version of your API. Determining what constitutes an incompatible API can be difficult to surmise automatically, so it's often left to the plugin writer to specify. For example, Firefox's extensions API lets you specify a minimum and maximum version range with which the extension is known to work (with a system to update an extension's max version easily for occasions when an incompatible API is released). It's also useful to know which version of the API a plugin was compiled against. This could be embedded in the plugin automatically or again, it could be left to the plugin writer to specify. For example, Google's Android API lets you specify an `android:targetSdkVersion` in addition to `android:minSdkVersion` and `android:maxSdkVersion`.
- **Internal versus external metadata**. Metadata, such as a human-readable name and version information, can either be defined within the plugin code itself or specified in a simple external file format. The benefit of using an external metadata file is that you don't have to load all plugins to know the set of all available objects. For example, you may want to present a list of all plugins to the user and then load only the plugins they chose to use. The downside, however, is that you can't simply drop a new plugin into a directory and have it be loaded automatically. You must also include a per-plugin metadata file, or update a global metadata file for all plugins, depending upon the approach you adopt.
- **Generic versus specialized Plugin Manager**: One approach to implementing the Plugin Manager is to make it low-level and generic (i.e., it simply loads plugins and accesses symbols in those plugins). However, doing so can mean that the Plugin Manager does not know about the existence of concrete types in your API. As a result, it will probably have to return objects as `void*` pointers and you must cast those to concrete types before using them. Alternatively, a Plugin Manager that can, at a minimum, forward declare the types for any objects in a plugin can produce a more type-safe solution, although as a result it can't be implemented independently of your API. A middle ground is to introduce a dynamic run time typing system into your API, in which the Plugin Manager can return references in terms of a generic type that can be registered later by your API.
- **Security**: You must decide how much you'll trust user plugins. Plugins are arbitrary compiled code that you allow to run in your process. A plugin could therefore potentially do anything, from accessing data that it shouldn't, to deleting files on the end user's hard drive, and to crashing the entire application. If you need to protect against such malicious plugins, then you may consider creating a socket-based solution, in which plugins run in a separate process and communicate with the Core API through an IPC channel. Alternatively, you could implement bindings for a language that supports sandboxing of user scripts, such as JavaScript or TypeScript, and require all plugins to be written in that scripting language.

- **Static versus dynamic libraries**: As I've mentioned, it's possible to define plugins as static libraries, meaning that they must be compiled into the application program. The more common solution for consumer applications is to use dynamic libraries, so that users can write their own plugins and extend the application at run time. A constraint for writing static plugins is that you must ensure that no two plugins define the same symbols: that is, the initialization function for each plugin must be named uniquely, such as `<PluginName>_PluginInit()`, whereas in the case of dynamic library plugins you can use the same initialization function name for every plugin, such as `PluginInit()`.

Implementing plugins in C++

I've identified that supporting C++ plugins can be difficult because of cross-platform and cross-compiler ABI problems. However, this is a book about C++ API design, so let's take a few more moments to present some solutions that let you use C++ plugins more robustly.

First off, if you're happy requiring that plugin developers use the identical version of the same compiler that you use for building your API, then you should have nothing to worry about. If that's not the case, one solution is to use a binding technology for your plugins: for example, an IPC solution such as COM, XPC, or gRPC, or creating script bindings for your API and letting users write extensions with a cross-platform scripting language such as Python or Ruby (as I covered in the previous chapter).

If you absolutely need to use C++ plugins for maximum performance or feel that creating a COM or script binding is too heavyweight for your needs, there are still ways that you can use C++ more safely in plugins. This list offers several best practices, many of which are implemented by the open source DynObj library available on http://www.codeproject.com/:

- **Use abstract base classes**: Implementing virtual methods of an abstract base class can insulate a plugin from ABI problems because a virtual method call is usually represented as an index into a class's vtable. Theoretically, the vtable format can differ between compilers, but in practice this tends not to happen. (However, different compilers may order overloaded virtual methods differently, so it's best to avoid these.) All of the methods in the interface need to be pure virtual, although inlined methods can be used safely, too, because the code will get embedded directly into the plugin.
- **Use C linkage for free functions**: All global functions in your Plugin API should use C linkage to avoid C++ ABI issues (i.e., they should be declared with `extern "C"`). Similarly, function callbacks that a plugin passes to the Core API should also use C linkage for maximum portability.
- **Avoid the C++ Standard Library and exceptions**: Different implementations of Standard Library classes such as `std::string` and `std::vector` may not be ABI compatible. It's therefore best to avoid these containers in any function calls

between the Core API and Plugin API. Similarly, the ABI for exceptions tends to be unstable across compilers, and so these should also be avoided in your Plugin API.

- **Don't mix allocators**: It's possible for plugins to be linked against a memory allocator different from your API. For example, on Windows it's common for debug builds to use a different allocator from release builds. The implication for the design of our plugin system is that either the plugin must allocate and free all of its objects or the plugin should pass control to the Core API to create and destroy all objects. However, your Core API should never free objects that were allocated by a plugin, and vice versa.

Putting all of this information together, I'll now develop a flexible and robust cross-platform C++ plugin system. The plugin system will allow new C++ classes to be registered with the Core API by providing one or more factory methods. I'll continue our extensible factory example from Chapter 3 (Patterns) and augment it to allow new `IRenderer` classes to be registered from plugins, in which these plugins are dynamically loaded at run time rather than being compiled into the Core API. Furthermore, the plugin architecture will support different approaches to storing plugin metadata, either within an accompanying external file or within the plugins themselves.

The plugin API

The Plugin API is the interface that you provide to your users to create plugins. I'll call it `pluginapi.h` in our example here. This header file will contain functionality that allows plugins to communicate with the Core API.

When the Core API loads a plugin, it needs to know which functions to call or which symbols to access to let the plugin do its work. This means that you should define specifically named entry points in the plugin that your users must provide. There are several ways in which you can do this. For example, when writing a GIMP plugin, you must define a variable called `PLUG_IN_INFO` that lists the various callbacks defined in the plugin:

```
#include <libgimp/gimp.h>

GimpPlugInInfo PLUG_IN_INFO = {
    NULL,    /* called when GIMP starts */
    NULL,    /* called when GIMP exits */
    query,   /* procedure registration and arguments definition */
    run,     /* perform the plugin's operation */
};
```

Netscape Plugins use a similar, although slightly more flexible technique. In this case, plugin writers define an `NP_GetEntryPoints()` function and fill in the appropriate fields of the `NPPluginFuncs` structure that the browser passes in during plugin registration. The NPPluginFuncs structure includes size and version fields to handle future expansion.

Another solution is to have specifically named functions that the Core API can call if they are exported by the plugin. I'll adopt this approach for our example because it's simple and scalable (e.g., it doesn't rely on a fixed size array or structure).

The two most basic callbacks that a plugin should provide are an initialization function and a cleanup function. As I noted earlier, these functions should be declared with C linkage to avoid name mangling differences between compilers. If you want to develop a cross-platform plugin system, you'll also have to deal with correctly using __declspec(dllexport) and __declspec(dllimport) decorators on Windows. Rather than require our plugin developers to know all of these details, I'll provide some macros to simplify everything. (I stated earlier that you should avoid preprocessor macros for declaring things such as API constants, but they are perfectly valid to affect compile-time configurations such as this.)

Also, I've decided that our plugin should be allowed to register new IRenderer derived classes, so I'll provide a Plugin API call to let plugins do just that. Here's a first draft of our Plugin API:

```
// pluginapi.h
#include "defines.h"
#include "renderer.h"

#define CORE_FUNC extern "C" CORE_API
#define PLUGIN_FUNC extern "C" PLUGIN_API

#define PLUGIN_INIT() PLUGIN_FUNC int PluginInit()
#define PLUGIN_FREE() PLUGIN_FUNC int PluginFree()

using RendererInitFunc = IRenderer *(*) ();
using RendererFreeFunc = void (*)(IRenderer *);

CORE_FUNC void RegisterRenderer(const char *type,
                                RendererInitFunc init_cb,
                                RendererFreeFunc free_cb);
CORE_FUNC void UnregisterRenderer(const char *type);
```

This header provides macros to define the initialization and cleanup functions for a plugin: PLUGIN_INIT() and PLUGIN_FREE(), respectively. I also provide the PLUGIN_FUNC() macro to let plugins export functions for the Core API to call, as well as the CORE_FUNC() macro that exports Core API functions for plugins to call. Finally I provide a function, RegisterRenderer(), which allows plugins to register new IRenderer classes with the Core API. A plugin must provide both an init function and a free function for their new IRenderer classes to ensure that allocations and frees happen within the plugin (to address the point that you should not mix memory allocators).

You may also note our use of the CORE_API and PLUGIN_API defines. These let us specify the correct DLL export/import decorators under Windows. CORE_API is used to decorate functions that are part of the Core API, and PLUGIN_API is used for functions that will be defined in plugins. The definition of these macros is contained in the defines.h header and looks like:

```
// defines.h
#ifdef _WIN32
#ifdef BUILDING_CORE
#define CORE_API    __declspec(dllexport)
#define PLUGIN_API  __declspec(dllimport)
#else
#define CORE_API    __declspec(dllimport)
#define PLUGIN_API  __declspec(dllexport)
#endif
#else
#define CORE_API
#define PLUGIN_API
#endif
```

You must build your Core API with the BUILDING_CORE define set for these macros to work correctly (e.g., add /DBUILDING_CORE to the compile line on Windows). This define is not needed when compiling plugins.

Finally, for completeness, here are the contents of the renderer.h file, which is included by pluginapi.h:

```
// renderer.h
#include <string>

class IRenderer
{
public:
    virtual ~IRenderer() {}
    virtual bool LoadScene(const char *filename) = 0;
    virtual void SetViewportSize(int w, int h) = 0;
    virtual void SetCameraPos(double x, double y, double z) = 0;
    virtual void SetLookAt(double x, double y, double z) = 0;
    virtual void Render() = 0;
};
```

This is essentially the same definition that I presented in the Patterns chapter, except that I have changed the LoadScene() method to accept a const char * parameter instead of an std::string (to address our concerns about binary compatibility of C++ Standard Library classes between compilers).

An example plugin

Now that I've developed a rudimentary Plugin API, let's examine what a plugin built against this API might look like. The basic parts that you need to include are:

1. The new IRenderer class.

2. Callbacks to create and destroy this class.
3. A plugin initialization routine that registers the create/destroy callbacks with the Core API.

Here's the code for such a plugin. This plugin defines and registers a new renderer called "opengl." This is defined in a new `OpenGLRenderer` class that derives from our `IRenderer` abstract base class:

```
// plugin1.cpp
#include "pluginapi.h"

#include <iostream>

class OpenGLRenderer : public IRenderer
{
public:
    ~OpenGLRenderer() {}
    bool LoadScene(const char *filename) { return true; }
    void SetViewportSize(int w, int h) {}
    void SetCameraPos(double x, double y, double z) {}
    void SetLookAt(double x, double y, double z) {}
    void Render() { std::cout << "OpenGL Render" << std::endl; }
};

PLUGIN_FUNC IRenderer *CreateRenderer()
{
    return new OpenGLRenderer;
}

PLUGIN_FUNC void DestroyRenderer(IRenderer *r)
{
    delete r;
}

PLUGIN_INIT()
{
    RegisterRenderer("opengl", CreateRenderer, DestroyRenderer);
    return 0;
}

PLUGIN_FREE()
{
    UnregisterRenderer("opengl");
    return 0;
}
```

In this example, I've defined a `PLUGIN_INIT()` function, which will be run whenever the plugin is loaded. This registers our `OpenGLRenderer` factory function, `CreateRenderer()`, and the associated destruction function, `DestroyRenderer()`. These are both defined using

PLUGIN_FUNC to ensure that they're correctly exported with C linkage. The PLUGIN_FREE() function will then unregister the renderer when the plugin is unloaded.

The RegisterRenderer() function essentially just calls the RendererFactory ::RegisterRenderer() method that I presented in the Patterns chapter (with the addition of being able to pass a destruction callback as well as the CreateCallback). There are a couple of reasons why I added an explicit registration function to the Plugin API rather than letting plugins register themselves directly with the RendererFactory. One reason is simply to give us a layer of abstraction so that you could change RendererFactory in the future without breaking existing plugins. Another reason is to avoid plugins calling methods that use C++ Standard Library strings: RegisterRenderer uses a const char * to specify the renderer name.

The Plugin manager

Now that you have a Plugin API and you can build plugins against this API, you need to be able to load and register those plugins into the Core API. This is the role of the Plugin Manager. Specifically, the Plugin Manager needs to handle these tasks:

- Load metadata for all plugins. These metadata can be either stored in a separate file (such as an XML or JSON file) or embedded within the plugins themselves. In the latter case, the Plugin Manager will need to load all available plugins to collate the metadata for all plugins. These metadata let you present users with a list of available plugins among which they may choose.
- Load a dynamic library into memory, provide access to the symbols in that library, and unload the library if necessary. This involves using dlopen(), dlclose(), and dlsym() on UNIX platforms (including macOS) and LoadLibrary(), FreeLibrary(), and GetProcAddress() on Windows. I provide details about these calls in Appendix A (Libraries).
- Call the plugin's initialization routine when the plugin is loaded and call the cleanup routine when the plugin is unloaded. These are the functions that are defined by PLUGIN_INIT() and PLUGIN_FREE() within the plugin.

The Plugin Manager provides a single point of access to all plugins in the system; as such it's often implemented as a singleton. In terms of design, the Plugin Manager can be thought of as a collection of plugin instances, in which each plugin instance represents a single plugin and offers functionality to load and unload that plugin. Here's an example implementation for a Plugin Manager:

```
// pluginmanager.cpp
#include "defines.h"

#include <string>
#include <vector>

class CORE_API PluginInstance
{
public:
    explicit PluginInstance(const std::string &name);
    ~PluginInstance();

    bool Load();
    bool Unload();
    bool IsLoaded();

    std::string GetFileName();
    std::string GetDisplayName();

private:
    class Impl;
    Impl *mImpl;
};

class CORE_API PluginManager
{
public:
    static PluginManager &GetInstance();

    bool LoadAll();
    bool Load(const std::string &name);

    bool UnloadAll();
    bool Unload(const std::string &name);

    std::vector<PluginInstance *> GetAllPlugins();

private:
    PluginManager();
    ~PluginManager();
    PluginManager(const PluginManager &);
    const PluginManager &operator=(const PluginManager &);

    std::vector<PluginInstance *> mPlugins;
};
```

This design decouples the ability to access the metadata for all plugins from the need to load those plugins. That is, if metadata such as the plugin's display name are stored in an external file, you can call `PluginManager::GetAllPlugins()` without loading the actual plugins. On the other hand, if the metadata are stored in the plugins, then `GetAllPlugins()` can simply call `LoadAll()` first. The following example presents a sample external metadata file based upon an XML syntax:

```
<?xml version="1.0" encoding='UTF-8'?>
<plugins>
    <plugin filename="oglplugin">
        <name>OpenGL Renderer</name>
    </plugin>
    <plugin filename="dxplugin">
        <name>DirectX Renderer</name>
    </plugin>
    <plugin filename="mesaplugin">
        <name>Mesa Renderer</name>
    </plugin>
</plugins>
```

Irrespective of the approach to store plugin metadata within an external file or embedded within each plugin, this code outputs the display name for all available plugins:

```
std::vector<PluginInstance *> plugins =
    PluginManager::GetInstance().GetAllPlugins();

std::vector<PluginInstance *>::iterator it;
for (it = plugins.begin(); it != plugins.end(); ++it) {
    PluginInstance *pi = *it;
    std::cout << "Plugin: " << pi->GetDisplayName() << std::endl;
}
```

A related issue is that of plugin discovery. The previous API doesn't restrict the ability for the implementation of `PluginManager::Load()` to search multiple directories to discover all plugins. The name passed to this `Load()` method can be a base plugin name without any path or file extension, such as "glplugin." The `Load()` method can then search various directories and look for files with extensions that may be platform specific, such as `libglplugin.dylib` on macOS or `glplugin.dll` on Windows. Of course, you can always introduce your own plugin filename extension. For example, Adobe Illustrator uses the `.aip` extension for its plugins, and Microsoft Excel uses the `.xll` extension.

The following Core API initialization code registers a single built-in renderer and then loads all plugins, allowing additional renderers to be added to the system at run time:

```
class MesaRenderer : public IRenderer
{
public:
    bool LoadScene(const char *filename) { return true; }
    void SetViewportSize(int w, int h) {}
    void SetCameraPos(double x, double y, double z) {}
    void SetLookAt(double x, double y, double z) {}
    void Render() { std::cout << "Mesa Render" << std::endl; }
    static IRenderer *Create() { return new MesaRenderer; }
};

...

// create a built-in software renderer
RendererFactory::RegisterRenderer("mesa", MesaRenderer::Create);

// discover and load all plugins
PluginManager::GetInstance().LoadAll();
```

Plugin versioning

As a final note, I will expand on the topic of plugin versioning. As with API versioning, you'll want to make sure that the release of your first plugin system includes a versioning system. You could either coopt the version number of your Core API or you could introduce a specific Plugin API version number. I suggest the latter, because the Plugin API is a separate interface from the Core API and the two may change at different rates. For example, Google's Android API uses the notion of API Level (Table 15.1). This is a single integer that increases monotonically with each new version of the Android API.

One of the most important pieces of information you'll want to access is the Plugin API version against which a given plugin was built. This can let you determine whether a plugin is incompatible with the current release and therefore shouldn't be registered, such as if the plugin was built with a later version of the API or an incompatible older API. Given the importance of this information, it's advisable to embed this information automatically in every plugin. This ensures that the correct version is always compiled

Table 15.1 The Android application programming interface (API) Level for each version of the Android platform.

Platform version	API Level
Android 2.1	7
Android 2.0.1	6
Android 2.0	5
Android 1.6	4
Android 1.5	3
Android 1.1	2
Android 1.0	1

into the plugin every time it's successfully rebuilt. With the Plugin API I've proposed, you could include this information in the `PLUGIN_INIT()` macro, because users must call this for the plugin to do anything. For example:

```
// pluginapi.h

...

#define PLUGIN_API_VERSION 1

#define PLUGIN_INIT()                                   \
    const int PluginVersion = PLUGIN_API_VERSION; \
    PLUGIN_FUNC int PluginInit()

...
```

In addition, users can optionally specify a minimum and maximum version of the API with which the plugin will work. The minimum version number will be more commonly specified. For example, if the plugin uses a new feature that was added to the API in a specific release, that release should be specified as the minimum version. Specifying a maximum version number is useful only after a new version of the API has been released and the plugin writer finds that it breaks the plugin. Normally, the maximum version will be unset, because plugin writers should assume that future API releases will be backward compatible.

This min/max version number could be specified in an external metadata format, such as:

```
<?xml version="1.0" encoding='UTF-8'?>
<plugins>
    <plugin filename="plugin1">
        <name>OpenGL Renderer</name>
        <minversion>2</minversion>
    </plugin>
    ...
</plugins>
```

Alternatively, you can extend the Plugin API with additional calls to let plugins specify this information in code:

```
...

#define PLUGIN_MIN_VERSION(version) \
    PLUGIN_API int PluginMinVersion = version

#define PLUGIN_MAX_VERSION(version) \
    PLUGIN_API int PluginMaxVersion = version

...
```

Extending via inheritance

The focus of this chapter thus far has been supporting API extensibility at run time via plugins. However, there are other ways in which your clients can extend the functionality of your API for their own purposes. The primary object-oriented mechanism for extending a class is inheritance. This can be used to let your users define new classes that build upon and modify the functionality of existing classes in your API.

Adding functionality

Jonathan Wood has a video on Microsoft's Visual C++ Developer Center in which he demonstrates extending MFC's `CString` class via inheritance to create a `CPathString` class that adds some path manipulation functions to the basic string class. The resulting class looks something like:

```
class CPathString : public CString
{
public:
    CPathString();
    ~CPathString();

    CString GetFileName();
    CString GetFileExtension();
    CString GetFileBase();
    CString GetFilePath();
};
```

This is a simple example of extending an existing class in which only new methods are added to the base class.

An important point to reiterate is that this can be done safely only if the class was designed to be inherited from. The primary indicator for this is whether the class has a virtual destructor. In that example, MFC's `CString` class does not have a virtual destructor. This means that there are cases when the destructor for `CPathString` will not be called, such as:

```
CString *str = new CPathString;
delete str;
```

In this case, this is not an issue because all of the new `CPathString` methods are stateless; that is, they do not allocate any memory that must be freed by the `CPathString` destructor. However, this does highlight the issue that if you expect your users to inherit from any of your classes, you should declare the destructor for those classes to be virtual.

> *TIP: Declaring a virtual destructor for a class is a signal to your users that you have designed it to be inherited from.*

Modifying functionality

In addition to adding new member functions to a class, we know that C++ allows you to define functions in a derived class that override existing functions in a base class if they've been marked as virtual in the base class. This can provide even more avenues for your users to customize the behavior of a class if you give them the hooks to do so.

For example, all UI widgets in the Qt library provide these virtual methods:

```
virtual QSize minimumSizeHint() const;
virtual QSize sizeHint() const;
```

This allows users of the Qt library to create derived classes of these widgets and change their appearance. To illustrate this, the following class inherits from the standard Qt button widget and overrides the `sizeHint()` method to specify the preferred size of the button. The result of this can be seen in Fig. 15.3:

```
class MySquareButton : public QPushButton
{
public:
    QSize sizeHint() const
    {
        return QSize(100, 100);
    }
};
```

This code works because `sizeHint()` is a known method of every widget and is called by the layout classes to determine the widget's preferred size. That is, the creators of the

FIGURE 15.3 A standard Qt `QPushButton` widget (*left*), and a derived class of `QPushButton` that overrides the `sizeHint()` virtual method (*right*).

Qt library explicitly designed this point of customization into the tool kit and allowed users to modify it in their own derived classes by deliberately declaring the method to be virtual.

Simply changing the size of the widget by calling the `resize()` method in the `MySquareButton` constructor is not the same thing. The effect of this would be to set the size of the button forcibly. However, the point of `sizeHint()` is to provide an indication of the preferred size to the UI layout engine (i.e., to other classes in the API), so that it can override this size when necessary to satisfy other widget size constraints.

This could be implemented without a virtual `sizeHint()` method. For example, non-virtual `setSizeHint()` and `getSizeHint()` methods could be added to the widget base class. However, this would require the base class to store the hint information as a data member in the object and hence increase the size of every object that inherits from it. In contrast, the use of the `sizeHint()` virtual method supports the ability for a class to simply calculate the preferred size on each invocation without the need to store the size within the object instance.

In the chapter on Performance, I cautioned you to add virtual methods to a class only when you need them. That advice is still valid. In the previous example, the designers of the Qt API added these virtual methods to their API carefully and consciously, to produce a flexible way for their users to extend the base functionality of their classes.

Inheritance and the standard library

Programmers who are new to C++ and the C++ Standard Library often try to subclass containers, such as:

```
#include <string>

class MyString : public std::string
{
public
    ...
};
```

However, as I've noted, you should only attempt to derive from a class that defines a virtual destructor. The Standard Library container classes don't provide virtual destructors, and in fact they have no virtual methods for you to override at all. This is a clear indication that these classes were not meant to be inherited from. Attempting to do so could introduce subtle and difficult-to-debug resource leaks into your code and your clients' code. The general rule is therefore that you should never inherit from C++ Standard Library container classes.

As an alternative, you could use composition to add functionality to a Standard. Library container in a safe manner. That is, use an `std::string` as a private member and provide accessor methods that thinly wrap the underlying `std::string` methods. Then you can add your own methods to this class. For example:

```
class MyString
{
public:
    MyString() :
        mStr("")
    {}
    explicit MyString(const char *str) :
        mStr(str)
    {}

    bool empty() const { return mStr.empty(); }
    void clear() { mStr.clear(); }
    void push_back(char c) { mStr.push_back(c); }
    ...

private:
    std::string mStr;
};
```

However, the Standard Library does provide a few classes that were designed for inheritance. One of the most obvious of these is `std::exception`. This is the base class for all Standard Library exceptions, including: `bad_alloc`, `bad_cast`, `bad_exception`, `bad_typeid`, `lock_error`, `logic_error`, and `runtime_error`. You can define your own exceptions that derive from `std::exception` simply:

```
class DivByZeroException : public std::exception
{
public:
    const char *what() const
    {
        return "Division by zero attempted";
    }
};
```

Another part of the Standard Library that supports extension through inheritance is the iostream library. This is actually a powerful, well-designed, and extensible API that provides various stream abstractions. A stream can be thought of simply as a sequence of bytes waiting to be processed, such as the standard `cin` input stream and `cout` output stream. You can write custom stream classes by deriving from a particular stream class or from the `streambuf` base class. For example, you could create custom stream classes to send and receive HTTP data to/from a Web server.

There is also the Boost Iostreams library, which makes it easier to work with Standard Library streams and stream buffers, and provides a framework for defining filters on streams and buffers. The library comes with a collection of handy filters including regular expression filtering and data compression schemes such as zlib, gzip, and bzip2.

Inheritance and enums

There are times when your users may want to extend an enum that you define in one of your base classes, such as to add further enumerators for new features that they have added in their derived classes. This can be done easily in C++:

```
class Base
{
public:
    enum SHAPE {
        CIRCLE,
        SQUARE,
        TRIANGLE,
        SHAPE_END = TRIANGLE
    };
};

class Derived : public Base
{
public:
    enum SHAPE {
        OVAL = SHAPE_END + 1,
        RECTANGLE,
        HEXAGON
    };
};
```

Here is an example that demonstrates the use of this extended enum:

```
std::cout << "CIRCLE="    << Derived::CIRCLE    << std::endl;
std::cout << "SQUARE="    << Derived::SQUARE    << std::endl;
std::cout << "TRIANGLE="  << Derived::TRIANGLE  << std::endl;
std::cout << "OVAL="      << Derived::OVAL      << std::endl;
std::cout << "RECTANGLE=" << Derived::RECTANGLE << std::endl;
std::cout << "HEXAGON="   << Derived::HEXAGON   << std::endl;
```

The important part that makes this work robustly is that the `Base` class defined `SHAPE_END`, so that the `Derived` class could add new values after the last value defined by `Base::SHAPE`. Therefore this is a good practice for you to adopt when defining enums in classes that you expect to be subclassed by your clients. Without this, your clients could pick an arbitrarily large integer to start numbering their enumerators (e.g., `OVAL=100`), although this is a less elegant solution.

> *TIP: For enums in a base class, add an enumerator for the last value in the enum, such as <enum-name>_END.*

You can also do this with C++11's `enum class`, although an `enum class` is not implicitly convertible to its underlying type, so you would have to add an explicit cast, such as `OVAL = static_cast<int>(Base::SHAPE::SHAPE_END) + 1`.

The Visitor pattern

We presented various generic design patterns back in Chapter 3 (Patterns). However, we deferred discussion of the Visitor design pattern until now because it's specifically targeted at API extensibility. The core goal of the Visitor pattern is to allow clients to traverse all objects in a data structure and perform a given operation on each of those objects. This pattern is essentially a way to simulate adding new virtual methods to an existing class. It therefore provides a useful pattern for your clients to extend the functionality of your API. For example, a Visitor pattern could be used to let clients provide a set of methods that operate on every node in a scene graph hierarchy or to traverse the derivation tree of a programming language parser and output a human-readable form of the program.

Let's develop a Visitor example to illustrate how this pattern works. I'll use the example of a scene graph hierarchy that describes a 3D scene, such as that used by Open Inventor, OpenSceneGraph, or the Virtual Reality Modeling Language. To keep the example simple, our scene graph will contain only three different node types: `Shape`, `Transform`, and `Light`. Fig. 15.4 shows an example hierarchy using these node types.

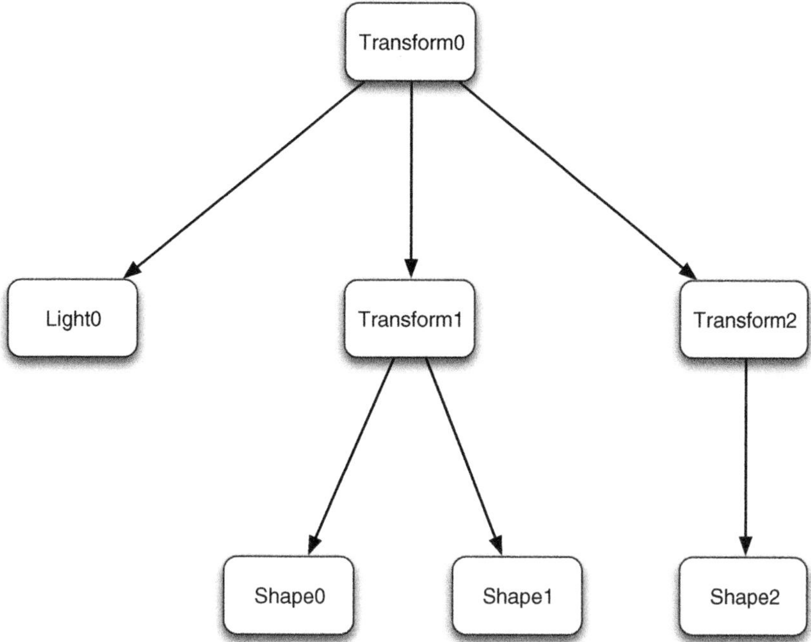

FIGURE 15.4 Example scene graph hierarchy showing nodes of different types.

I'll begin by defining our abstract Visitor interface. Clients can create concrete subclasses of this interface to add custom operations to the scene graph. It essentially declares a `Visit()` method for each node type in the scene graph:

```
// nodevisitor.h

class ShapeNode;
class TransformNode;
class LightNode;

class INodeVisitor
{
public:
    virtual void Visit(ShapeNode &node) = 0;
    virtual void Visit(TransformNode &node) = 0;
    virtual void Visit(LightNode &node) = 0;
};
```

Now let's look at our scene graph API. This provides the declarations for each of our node types, as well as a skeleton `SceneGraph` class. Each node type derives from a base node type, called `BaseNode`:

```
// scenegraph.h
#include <string>

class INodeVisitor;

class BaseNode
{
public:
    BaseNode(const std::string &name);

    virtual void Accept(INodeVisitor &visitor) = 0;

private:
    std::string mName;
};

class ShapeNode : public BaseNode
{
public:
    ShapeNode(const std::string &name);

    void Accept(INodeVisitor &visitor);
    int GetPolygonCount() const;
};

class TransformNode : public BaseNode
{
public:
    TransformNode(const std::string &name);

    void Accept(INodeVisitor &visitor);
};

class LightNode : public BaseNode
{
public:
    LightNode(const std::string &name);

    void Accept(INodeVisitor &visitor);
};


class SceneGraph
{
public:
    SceneGraph();
    ~SceneGraph();

    void Traverse(INodeVisitor &visitor);

private:
    class Impl;
    Impl *mImpl;
};
```

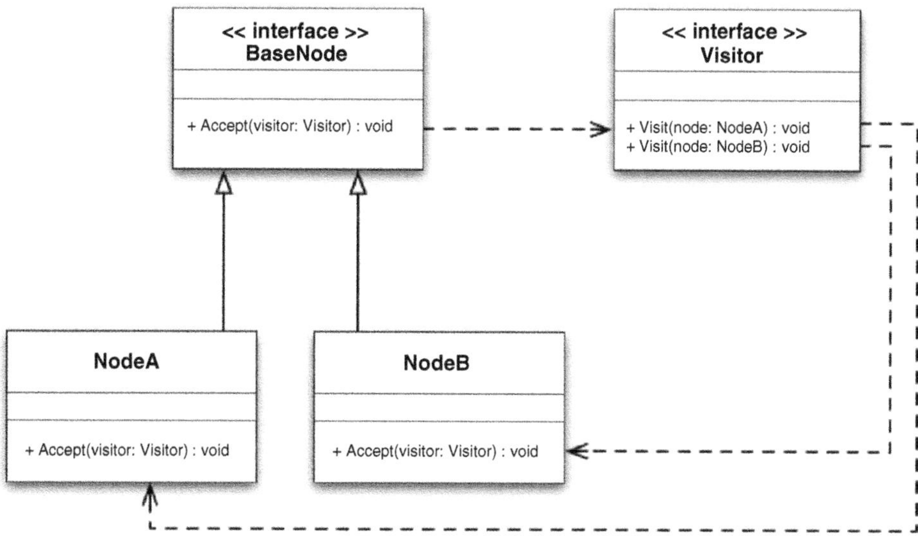

FIGURE 15.5 UML diagram of the visitor design pattern.

Each node type declares an `Accept()` method, taking a Visitor object as its parameter. This method is used to call the appropriate `Visit()` method in the Visitor class. This can be thought of as a way to have a single virtual method in each node that can then call any user-supplied virtual method. Fig. 15.5 portrays a unified modeling language (UML) diagram that shows this Visitor pattern:

```
// scenegraph.cpp
void ShapeNode::Accept(INodeVisitor &visitor)
{
    visitor.Visit(*this);
}

void TransformNode::Accept(INodeVisitor &visitor)
{
    visitor.Visit(*this);
}

void LightNode::Accept(INodeVisitor &visitor)
{
    visitor.Visit(*this);
}
```

Building upon this infrastructure, the `SceneGraph::Traverse()` method can be implemented by navigating the scene graph hierarchy and then calling the `Accept()` method for every node in the graph. Your clients can then define custom Visitor classes to perform arbitrary operations on the scene graph. This is done without exposing details about how the scene graph is implemented. For example, the next code demonstrates how a client could write a Visitor to count the number of each node type in the scene graph and sum the polygon count for all shape nodes in the scene graph:

```
class MyVisitor : public INodeVisitor
{
public:
    int mNumShapes;
    int mNumPolygons;
    int mNumTransforms;
    int mNumLights;

    MyVisitor() :
        mNumShapes(0),
        mNumPolygons(0),
        mNumTransforms(0),
        mNumLights(0)
    {
    }

    void Visit(ShapeNode &node)
    {
        mNumPolygons += node.GetPolygonCount();
        ++mNumShapes;
    }

    void Visit(TransformNode &node)
    {
        ++mNumTransforms;
    }

    void Visit(LightNode &node)
    {
        ++mNumLights;
    }
};

...

MyVisitor visitor;
scenegraph.Traverse(visitor);

std::cout << "Shapes:     " << visitor.mNumShapes << std::endl;
std::cout << "Polygons:   " << visitor.mNumPolygons << std::endl;
std::cout << "Transforms:" << visitor.mNumTransforms << std::endl;
std::cout << "Lights:     " << visitor.mNumLights << std::endl;
```

This example demonstrates many of the benefits of the Visitor pattern. The most relevant benefit to the topic of extensibility is that clients can effectively plug in their own methods into your class hierarchy. Other benefits include the colocation of all code that performs a single coherent operation. For example, all of the code that implements the node counting functionality in that example is contained with the single `MyVisitor` class rather than being distributed across all individual node classes. A further benefit is that the state required to count the various nodes and the number of polygons (`mNumShapes`, `mNumPolygons`, `mNumTransforms`, and `mNumLights`) is isolated in the `MyVisitor` class rather than being stored directly in the SceneGraph object, and hence increasing the size of it.

However, there are some significant downsides to the Visitor pattern, too. The flexibility of being able to add new methods to a related set of classes comes at the cost of making it more difficult to add new related classes. In visitor.h, the visitor interface must know about every class that can be visited (i.e., all of our node types). Therefore, adding a new node type to our scene graph will require the Visitor interface also to be updated. As a result, the Visitor pattern is most appropriately used in cases where the class hierarchy is stable (Alexandrescu, 2001).

To address this problem, let's consider adding a new node type, called CameraNode, to our scene graph. The naive way to do this would be to add another Visit() pure virtual method to the INodeVisitor interface that accepts a CameraNode reference. However, we know that adding a pure virtual method to an interface is a bad thing to do in terms of API backward compatibility, because it will break all existing client code. Instead, there are a couple of alternative ways to solve this problem:

1. Thinking ahead, you could release the first version of INodeVisitor with a Visit() pure virtual method for BaseNode. This will effectively become a catch-all method that will be called if a node type is encountered for which there is not an explicit Visit() method. The inelegant consequence of this is that users must use a sequence of dynamic_cast calls inside this catch-all method to work out which node type has been passed in. Adopting this solution would change the visitor interface as follows:

```
// nodevisitor.h
class ShapeNode;
class TransformNode;
class LightNode;
class BaseNode;

class INodeVisitor
{
public:
    virtual void Visit(ShapeNode &node) = 0;
    virtual void Visit(TransformNode &node) = 0;
    virtual void Visit(LightNode &node) = 0;
    virtual void Visit(BaseNode &node) = 0;  // catch-all
};
```

2. A better solution is to add a new Visit() virtual method for the new node type, instead of a pure virtual method. That is, you provide an empty implementation for the new method so that existing code will continue to compile, while allowing

users to implement a type-safe `Visit()` method for the new node type where appropriate. This would change the `INodeVisitor` interface thus:

```
// nodevisitor.h
class ShapeNode;
class TransformNode;
class LightNode;
class CameraNode;

class INodeVisitor
{
public:
    virtual void Visit(ShapeNode &node) = 0;
    virtual void Visit(TransformNode &node) = 0;
    virtual void Visit(LightNode &node) = 0;
    virtual void Visit(CameraNode &node);
};
```

Extending via templates

C++ is often referred to as a multiparadigm language because it supports different styles of programming, such as procedural, object-oriented, and generic programming. Inheritance is the primary way to extend classes using object-oriented concepts. However, when programming with templates, the default way to extend an interface is to specialize a template with concrete types.

For example, the C++ Standard Library provides various container classes, such as `std::vector` and `std::set`. You can use these container classes to create data structures that contain arbitrary data types, such as `std::vector<MyCustomClass *>`, which the creators of the Standard Library had no way of knowing about when they designed the library.

Similarly, the Standard Library provides the ability to create reference-counted pointers that can hold any pointer type, without having to resort to using `void *`. This provides a powerful and generic facility that can be customized by clients to create type-safe shared pointers to any object, such as `std::shared_ptr<MyCustomClass>`.

Templates therefore offer an excellent way for you to write extensible code that can be applied to many different types, including types that your clients define in their own code. In the next couple of sections, I will present the concept of policy-based templates to help you maximize the flexibility of your class templates. I will also investigate a curiously common template pattern that provides static polymorphism as an alternative to the dynamic polymorphism of object-oriented programming.

Policy-based templates

Andrei Alexandrescu popularized the use of policy-based templates in his book *Modern C++ Design* in 2001. The term refers to the approach of building complex behaviors out of smaller classes, called policy classes, each of which define the interface for a single aspect of the overall component. This concept is implemented using class templates that accept several template parameters (often template template parameters), instantiated with classes that conform to the interface for each policy. By plugging in different policy classes, you can produce an exponentially large number of concrete classes.

For example, Alexandrescu presents this design for a smart pointer class template that accepts several policy classes to customize its behavior (Alexandrescu, 2001):

```
template <
    typename T,
    template <class> class OwnershipPolicy = RefCounted,
    class ConversionPolicy = DisallowConversion,
    template <class> class CheckingPolicy = AssertCheck,
    template <class> class StoragePolicy = DefaultSPStorage>
class SmartPtr;
```

The type that `SmartPtr` points toward is represented by the template parameter `T`. The remaining parameters specify various policies, or behaviors, for the smart pointer. These can be instantiated with classes that conform to a defined interface for each parameter and provide alternative implementations for the smart pointer. Here's an overview of each parameter's purpose:

- **OwnershipPolicy**: Specifies the ownership model for the smart pointer. Predefined policy classes include `RefCounted`, `DeepCopy`, and `NoCopy`.
- **ConversionPolicy**: Determines whether implicit conversion to the type of the object being pointed to is allowed. The two available classes are `AllowConversion` and `DisallowConversion`.
- **CheckingPolicy**: Specifies the error checking strategy. The predefined policy classes for this parameter include `AssertCheck`, `RejectNull`, and `NoCheck`.
- **StoragePolicy**: Defines how the object being pointed to is stored and accessed, including `DefaultSPStorage`, `ArrayStorage`, and `HeapStorage`.

Policy-based design recognizes that there is a multiplicity of solutions for every problem in computer science. The use of these generic components means that clients can choose among literally thousands of solutions simply by supplying different combinations of policy classes at compile time.

The first step in creating your own policy-based templates is to decompose a class into orthogonal parts. Anything that can be done in multiple ways is a candidate for factoring out as a policy. Policies that depend upon each other are also candidates for further decomposition or redesign. There is, of course, nothing new here. The essence of good software engineering is being able to recognize the more general and flexible abstraction for a particular problem.

Taken to the extreme, a host class (as policy-based templates are often called) becomes a shell that simply assembles a collection of policies to produce aggregate behavior. However, Alexandrescu states that you should try to keep the number of policy classes small for any given host class, because it becomes awkward to work with more than four to six template parameters. This correlates well with the cognitive limit of our working memory, which is believed to be 7 ± 2 (Miller, 1956).

It is also useful to provide typedefs for specific combinations of policy classes that you use for a given task. For example, if your API passes around smart pointers that uses nondefault policies, it would be tedious to have to specify all of those parameters all of the time, and changing those policies in the future would require your clients to update all of their code accordingly. Instead, you can introduce a typedef for the specific pointer type, such as:

```
using ShapePtr = SmartPtr<Shape, RefCounted,
                          AllowConversion, NoChecked>;
```

The curiously recurring template pattern

In this final section on extensibility via templates, I will present an interesting C++ idiom that was first observed by James Coplien in early template code (Coplien, 1995) and which may prove useful in your own API designs. The Curiously Recurring Template Pattern (CRTP) involves a template class that inherits from a base class using itself as a template parameter. Said differently (perhaps to make that last sentence clearer), it's when a base class is templated on the type of its derived class. This provides the fascinating quality that the base class can access the namespace of its derived class. The general form of this pattern is:

```
template <typename T>
class Base;
class Derived : public Base<Derived>;
```

The CRTP is essentially just a way to provide compile-time polymorphism. That is, it allows you to inherit an interface from a base class, but to avoid the overhead of virtual method calls at run time. In this way, it can be thought of as a "mixin" class (i.e., an interface class with implemented methods).

As a practical example of this pattern, the CRTP can be used to track statistics for each specialization of a template. For example, you can use it track a count of all existing objects of a given type, or the total amount of memory occupied by all existing objects of a given type. I will demonstrate the latter.

This class provides the base class declaration for our memory tracker interface:

```
// curious.h
template <typename TrackedType>
class MemoryTracker
{
public:
    // return memory used by existing objects:
    static size_t BytesUsed();

protected:
    MemoryTracker();
    MemoryTracker(MemoryTracker<TrackedType> const &);
    ~MemoryTracker();

private:
    size_t ObjectSize();
    static size_t mBytes;  // byte count of existing objects
};

#include "curious_priv.h"
```

For completeness, I also provide the associated definitions for this base class:

```
template <typename TrackedType>
size_t MemoryTracker<TrackedType>::BytesUsed()
{
    return MemoryTracker<TrackedType>::mBytes;
}

template <typename TrackedType>
MemoryTracker<TrackedType>::MemoryTracker()
{
    MemoryTracker<TrackedType>::mBytes += ObjectSize();
}

template <typename TrackedType>
MemoryTracker<TrackedType>::MemoryTracker(MemoryTracker<TrackedType>
const &)
{
    MemoryTracker<TrackedType>::mBytes += ObjectSize();
}

template <typename TrackedType>
MemoryTracker<TrackedType>::~MemoryTracker()
{
    MemoryTracker<TrackedType>::mBytes -= ObjectSize();
}

template <typename TrackedType>
inline size_t MemoryTracker<TrackedType>::ObjectSize()
{
    // [*] access details of the derived class
    return sizeof(*static_cast<TrackedType *>(this));
}

// initialize counter with zero
template <typename TrackedType>
size_t MemoryTracker<TrackedType>::mBytes = 0;
```

The clever part is the line directly after comment marked [*]. Here, the base class is accessing details of the derived class, in this case the size of the derived class. However, in a different example, it could just as easily call a method in the derived class.

Now you can derive from this `MemoryTracker` class, using the CRTP, to keep track of all memory currently consumed by a certain class. This can even be used to track memory use of individual template specializations, as the next example shows. All of these derived classes will essentially inherit the `BytesUsed()` method from our previous base class, but significantly, the method will be bound at compile time, not run time:

```
template <typename T>
class MyClass1 : public MemoryTracker<MyClass1<T>>
{
public:
    int mValue;   // sizeof(MyClass1) == sizeof(int)
};

class MyClass2 : public MemoryTracker<MyClass2>
{
public:
    int mValue;   // sizeof(MyClass2) == sizeof(int)
};

...

MyClass1<char> c1, c2;
MyClass1<wchar_t> w1;
MyClass2 i1, i2, i3;
std::cout << MyClass1<char>::BytesUsed() << std::endl;
std::cout << w1.BytesUsed() << std::endl;
std::cout << MyClass2::BytesUsed() << std::endl;
```

This code will print out the values 8, 4, and 12, assuming a 32-bit system in which `sizeof(MyClass1) == sizeof(MyClass2) == 4` bytes. That is, there are two instances of `MyClass1<char>` (8 bytes), one instance of `MyClass1<wchar_t>` (4 bytes), and three instances of `MyClass2` (12 bytes).

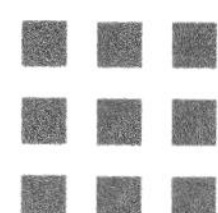

Appendix A: Libraries

A library lets you package the compiled code and data that implement your application programming interface (API) so that your clients can embed these into their own applications. Libraries are the instruments of modularization. This appendix will cover the different types of libraries that you can use and how you can create them on various platforms. It also covers physical aspects of API design, namely exposing the public symbols of your API in the symbol export table of its library file.

The characteristics, usage, and supporting tools for libraries are inherently platform-specific. How you work with a Dynamic Link Library (DLL) on Windows is different from how you work with a Dynamic Shared Objects (DSO) on Unix. I have therefore decided to organize the bulk of the content in this appendix by platform, specifically Windows, Linux, and macOS. This also has the benefit of not distracting you with platform-specific details that you don't care about for your current project.

Static versus dynamic libraries

There are two main forms of libraries that you can create. The decision about which one you employ can have a significant impact on your clients' applications in terms of tangible factors such as load time, executable size, and robustness to different versions of your API. These two basic types are static libraries and dynamic (or shared) libraries. I will describe each in detail over the following sections.

Static libraries

A static library contains object code that's linked with an end-user application and then becomes part of that executable. Fig. A.1 illustrates this concept. A static library is sometimes called an archive because it's essentially just a package of compiled object files. These

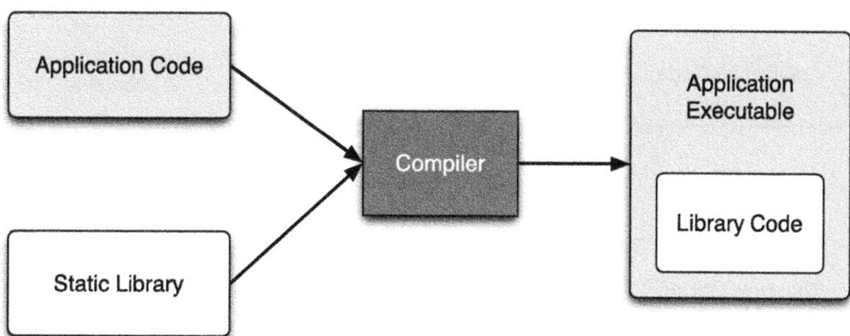

FIGURE A.1 Linking a static library into an application causes the library code to be embedded in the resulting executable.

libraries normally have a file extension of `.a` on Unix and macOS machines or `.lib` on Windows (e.g., `libjpeg.a` or `jpeg.lib`).

Some implications of distributing your API's implementation as a static library are:

- A static library is needed only to link an application. It is not needed to run that application because the library code is essentially embedded inside the application. As a result, your clients can distribute their applications without additional run-time dependencies.
- If your clients wish to link your library into multiple executables, each one will embed a copy of your code. If your library is 10 MB in size and your client wishes to link this into five separate programs, then you could be adding up to 50 MB to the total size of the product. Note that only the object files in the static library that are actually used are copied to the application. So in reality the total size of each application could be less than this worst case.
- Your clients can distribute their applications without concerns that it will find an incompatible library version on the end user's machine or a completely different library with the same name from another vendor. This avoids the library version and dependency issues that were often described as "DLL hell" on early Windows platforms.
- On the other hand, if your clients want to be able to hot patch their application (i.e., they want to update the version of your API used by their application), they must replace the entire executable to achieve this. If this is done as an Internet-based update, the end user may have to download a much larger update and hence wait longer for the update to complete.

Dynamic libraries

Dynamic libraries are files that are linked against at compile time to resolve undefined references and then distributed with the end user application so that the application can load the library code at run time (Fig. A.2). This normally requires the use of a dynamic linker on

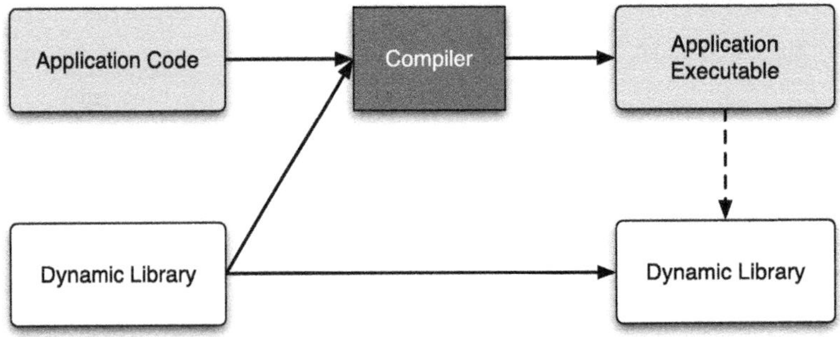

FIGURE A.2 A dynamic library is used to link an application and is then distributed with that application so that the library can be loaded at run time.

the end user's machine to find and load all dynamic library dependencies at run time, perform the necessary symbol relocations, and then pass control to the application. For example, the Linux dynamic linker is called `ld.so`, and on the Mac it is called `dyld`. Often, the dynamic linker supports several environment variables to modify or debug its behavior. For example, refer to `man dyld` on the Mac.

Dynamic libraries are sometimes called shared libraries because they can be shared by multiple programs. On Unix machines they can be called DSOs and on Windows they're referred to as DLLs. They have an `.so` file extension on Unix platforms, `.dll` on Windows, and `.dylib` on macOS (e.g., `libjpeg.so` or `jpeg.dll`).

Some implications of using dynamic libraries to distribute your API include:

- Your clients must distribute your dynamic library with their application (as well as any dynamic libraries on which your library depends) so that it can be discovered when the application is run.
- Your clients' applications will not run if the dynamic library can't be found: for example, if the library is deleted or moved to a directory that's not in the dynamic library search path. Furthermore, the application may not run if the dynamic library is upgraded to a newer version or overwritten with an older version.
- Using dynamic libraries can often be more efficient than static libraries in terms of disk space if more than one application needs to use the library. This is because the library code is stored in a single shared file and not duplicated inside each executable. Note that this is not a hard and fast rule, however. As I noted earlier, the executable only needs to include the object code from the static library that's actually used. So if each application uses only a small fraction of the total static library, the disk space efficiency can still rival that of a complete dynamic library.
- Dynamic libraries may also be more efficient in terms of memory. Most modern operating systems will attempt only to load the dynamic library code into memory once and share it across all applications that depend upon it. This may also lead to better cache utilization. By comparison, every application that's linked against a static library will load duplicate copies of the library code into memory.
- If your clients wish to hot patch their application with a new (backward-compatible) version of your shared library, they can simply drop in the replacement library file and all of their applications will use this new library without having to recompile or relink.

TIP: You should prefer to distribute your library as a dynamic library to give your users greater flexibility. If your library is sufficiently small and stable, you may additionally decide to provide a static library version.

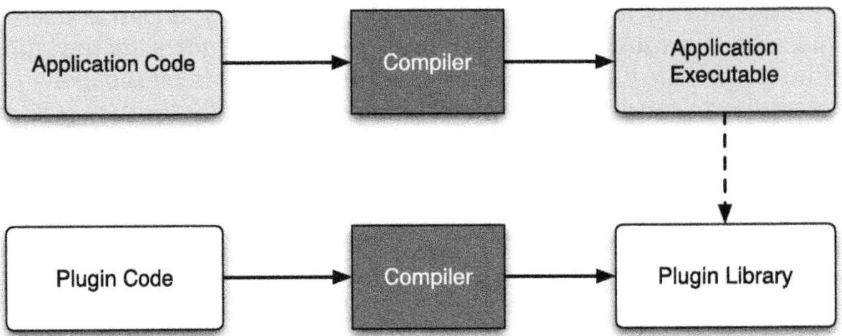

FIGURE A.3 A plugin library is a dynamic library that can be compiled separately from the application and explicitly loaded by the application on demand.

It's also important to understand the behavior of dynamic libraries that depend upon other dynamic libraries. If your library depends upon other dynamic libraries, then you must also ship those libraries with your software development kit (SDK), unless you can reasonably expect the end user's platform to have those libraries preinstalled. This is a transitive property: you must also ship all of the dependencies of your dependencies. Essentially, the entire chain of dynamic libraries must be available to the application at run time for it to be able to run. Of course, if your dynamic library links against a static library, then you don't need to provide that static library to your clients because the code from the static library is added directly to your dynamic library.

Dynamic libraries as plugins

Dynamic libraries are normally linked against an application and then distributed with that application so that the operating system can load the library when the application is launched. However, it's also possible for an application to load a dynamic library on demand, without the application having been compiled and linked against that library.

This can be used to create plugin interfaces, in which the application can load additional code at run time that extends the basic capabilities of the program. For example, most Web browsers support plugins to handle specific content types, such as viewing Apple QuickTime movies. This use of dynamic libraries is illustrated in Fig. A.3.

In terms of API development, this gives you the capability to create extensible APIs that allow your clients to drop in new functionality that your API will then load and execute. The Netscape Plugin API is an example of this: it's the API that you develop against to create a plugin (i.e., dynamic library) that browsers such as Firefox, Safari, Opera, and Chrome can then load and run. I discussed the use of plugins to create extensible APIs in Chapter 15.

Libraries on Windows

On Windows, static libraries are represented with `.lib` files, whereas dynamic libraries have `.dll` file extensions. Additionally, you must accompany each `.dll` file with an import library, or `.lib` file. The import library is used to resolve references to symbols exported in the DLL. For example, the Win32 User Interface API is implemented in `user32.dll`, with an accompanying `user32.lib`. Note that although they share the same `.lib` file extension, a static library and an import library are different file types. If you plan to distribute both static and dynamic library versions of your API, then you will need to avoid a filename collision by either naming the static library differently or placing each in a separate directory. For example:

static:	`foo_static.lib`	or	**static:**	`static/foo.lib`
dynamic:	`foo.dll`		**dynamic:**	`dynamic/foo.dll`
import:	`foo.lib`		**import:**	`dynamic/foo.lib`

On Windows, several other file formats are implemented as DLLs:

- ActiveX Controls files (`.ocx`)
- Device driver files (`.drv`)
- Control Panel files (`.cpl`)

Importing and exporting functions

As I discussed in Chapter 6, if you want a function to be callable from a DLL on Windows, you must explicitly mark its declaration with this keyword:

```
__declspec(dllexport)
```

For example,

```
__declspec(dllexport) void MyFunction();
class __declspec(dllexport) MyClass;
```

Conversely, if you want to use an exported DLL function in an application, then you must prefix the function prototype with this keyword:

```
__declspec(dllimport)
```

It's therefore common to employ preprocessor macros to use the export decoration when building an API and the import decoration when using the same API in an application. It's also important to note that these `__declspec` decorations may cause compile errors on non-Windows compilers, so you should use them only when compiling under Windows. The next preprocessor code provides a simple demonstration of this (see the Exporting Symbols Section in Chapter 6 for a more complete cross-platform example):

```
// On Windows, compile with /D "_EXPORTING" to build the DLL
#ifdef _WIN32
#ifdef _EXPORTING
#define DECLSPEC __declspec(dllexport)
#else
#define DECLSPEC __declspec(dllimport)
#endif
#else
#define DECLSPEC
#endif
```

You can then decorate the symbols you want to export from your DLL as:

```
DECLSPEC void MyFunction();
class DECLSPEC MyClass;
```

As an alternative to modifying your source code with these `__declspec` declarations, you can create a module definition `.def` file to specify the symbols to export. A minimal DEF file contains a `LIBRARY` statement to specify the name of the DLL with which the file is associated, and an `EXPORTS` statement followed by a list of symbol names to export:

```
// MyLIB.def
LIBRARY "MyLIB"
EXPORTS
    MyFunction1
    MyFunction2
```

The DEF file syntax also supports more powerful manipulations of your symbols, such as renaming symbols or using an ordinal number as the export name to help minimize the size of the DLL. The ordinal number represents the position of a symbol's address pointer in the DLL's export table. Using ordinal numbers for your DLL symbols can produce slightly faster and smaller libraries. However, from an API stability perspective this can be risky because seemingly innocuous changes to the DEF file can then change the exported symbols for your API. Therefore, I recommend using full symbol names rather than ordinal values when specifying your DLL exports.

The Dynamic Link Library entry point

DLLs can provide an optional entry point function to initialize data structures when a thread or process loads the DLL, or to clean up memory when the DLL is unloaded. This is managed by a function called `DllMain()` that you define and export within the DLL. If the entry point function returns `FALSE`, this is assumed to be a fatal error and the application will fail to start. This code provides a DLL entry point template:

```
BOOL APIENTRY DllMain(HANDLE dllHandle,
                      DWORD reason,
                      LPVOID lpReserved)
{
    switch (reason) {
    case DLL_PROCESS_ATTACHED:
        // A process is loading the DLL
        break;

    case DLL_PROCESS_DETACH:
        // A process unloads the DLL
        break;

    case DLL_THREAD_ATTACHED:
        // A process is creating a new thread
        break;

    case DLL_THREAD_DETACH:
        // A thread exits normally
        break;
    }

    return TRUE;
}
```

Creating libraries on Windows

The next steps describe how to create a static library on Windows. These steps are for Microsoft Visual Studio 2017, although the steps are similar for other versions of Visual Studio:

1. Select the menu **File > New > Project.**
2. Select the **Visual C++ > Win32** option and the **Win32 Project** icon.
3. The Win32 Application Wizard should appear.
4. Select the **Static library** option under Application type (see Fig. A.4).

You can then add new or existing source files to your project under the Source Files folder in the left-hand pane. Then, when you perform a build for your project, the result will be a static library `.lib` file.

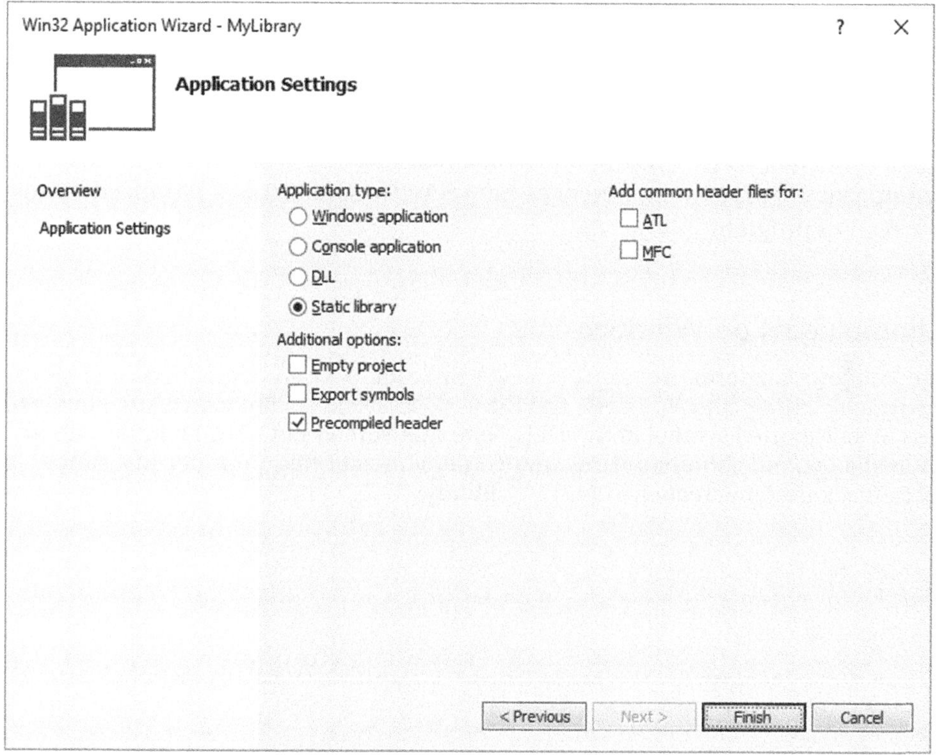

FIGURE A.4 Creating a new static library or Dynamic Link Library (DLL) in Visual Studio 2017.

The steps to create a DLL are very similar. The only difference is during Step 4, where you will select the **DLL** option instead of Static library. Then, when you build your project, Visual Studio will generate a `.dll` file and an associated `.lib` import file.

Useful Windows utilities

There are various programs that can help you manage DLLs on Windows and to investigate DLL problems. Many of these are command-line tools that you can run from the MS-DOS prompt. A few of these DLLs utilities are:

- **tasklist.exe**: This program can be used to find out which dynamic libraries a running Windows EXE file depends upon, for example:

```
tasklist /m /fi "IMAGENAME eq APPNAME.EXE"
```

- **depends.exe:** The dependency walker utility will recursively scan an executable to discover all of its dependent DLLs. It will check for missing DLLs, invalid DLLs, and circular dependencies, among other error conditions.
- **dlister.exe**: This utility provides a log of all the DLLs installed on your computer. This can be output as a text file or a database file.
- **dcomp.exe**: Displays the differences between two DLL listings produced by the `dlister.exe` program.

Loading plugins on Windows

On the Windows platform, the `LoadLibrary()` or `LoadLibraryEx()` functions can be used to load a dynamic library into a process, and the `GetProcAddress()` function is used to obtain the address of an exported symbol in the DLL. Note that you do not need an import library `.lib` file to load a dynamic library in this way. To demonstrate this, consider this simple plugin interface that's used to create a `plugin.dll` library:

```
#ifndef PLUGIN_H
#define PLUGIN_H

#include <string>

extern "C" __declspec(dllexport) void DoSomething(const char *name);

#endif
```

Then the following code snippet illustrates how to load this DLL on demand and call the `DoSomething()` method from that library:

```
// open the dynamic library
HINSTANCE handle = LoadLibrary("plugin.dll");
if (!handle) {
    std::cout << "Cannot load plugin!" << std::endl;
    exit(1);
}

// get the DoSomething() function from the plugin
FARPROC fptr = GetProcAddress(handle, "DoSomething");
if (fptr == (FARPROC) NULL) {
    std::cout << "Cannot find function in plugin: " << error;
    std::cout << std::endl;
    FreeLibrary(handle);
    exit(1);
}

// call the DoSomething() function
(*fptr)("Hello There!");

// close the shared library
FreeLibrary(handle);
```

Libraries on Linux

In the following sections, I'll provide an overview of creating and managing static and dynamic libraries on Linux. The emphasis here is to surface the important issues and techniques. However, for a deeper treatment, I recommend reading Ulrich Drepper's excellent article, "How to write shared libraries," available online at:

https://akkadia.org/drepper/dsohowto.pdf

Creating static libraries on Linux

On Unix, a static library is simply an archive of object (`.o`) files. You can use the Unix `ar` command to compile several object files into a static library. For example, the following commands demonstrate how to compile three `.cpp` files to `.o` files using the GNU C++ compiler and then creating a static library from those object files:

```
g++ -c file1.cpp
g++ -c file2.cpp
g++ -c file3.cpp
ar -crs libmyapi.a file1.o file2.o file3.o
```

The `-c` option to `g++` tells the compiler to produce an `.o` file from the input `.cpp` file. The options to `ar` are: `-c` creates an archive, `-r` inserts the supplied `.o` files into that archive, and `-s` creates an index for the archive (equivalent to the older convention of running `ranlib` on the resulting archive).

Your users can then link against your library using the -l option to ld or g++. This specifies the name of the library to link against. The -L linker option can also be used to specify the directory where your library can be found. For example:

```
g++ usercode.cpp -o userapp -L. -lmyapi
```

In this example, the end user application userapp is created by compiling usercode.cpp and linking against the libmyapi.a static library in the same directory.

The order of archives on this command line is significant. For each archive that the linker finds on the command line, it looks to see if that archive defines any symbols that were referenced from any object files that were specified earlier on the command line. If it does define any needed symbols, the object files with those symbols are copied into the executable. It's therefore best practice to specify libraries at the end of the command line (Mitchell et al., 2001).

While I'm discussing the creation of static libraries, it's worth noting the proper usage of the -static compiler option. This flag is used for the creation of executables, not libraries. It's therefore applicable to users of your API, but not to the building of your API itself. This flag instructs the compiler to prefer linking the static versions of all dependent libraries into the executable so that it depends upon no dynamic libraries at run time.

Creating dynamic libraries on Linux

Creating a dynamic library on Linux is very similar to creating a static library. Using the GNU C++ compiler, you can simply use the -shared linker option to generate an .so file instead of an executable.

On platforms where it's not the default behavior, you should also specify either the -fpic or -fPIC command line option to instruct the compiler to emit position-independent code (PIC). This is needed because the code in a shared library may be loaded into a different memory location for different executables. It's therefore important to generate PIC code for shared libraries so that that user code doesn't depend upon the absolute memory address of symbols. The following example illustrates how to compile three source files into a dynamic library:

```
g++ -c -fPIC file1.c
g++ -c -fPIC file2.c
g++ -c -fPIC file3.c
g++ -shared -o libmyapi.so -fPIC file1.o file2.o file3.o
```

Users can then link your dynamic library into their code using the same compile line shown earlier for the static library case: i.e.,

```
g++ usercode.cpp -o userapp -L. -lmyapi
```

If you have both a static library and a dynamic library with the same base name in the same directory (i.e., libmyapi.a and libmyapi.so), the linker will use the dynamic library. To favor the use of a static library over a dynamic library with the same base name, you can use the -static linker option, or you could place the static library in a different directory and ensure that this directory appears earlier in the library search path (using the -L linker option).

Note that in a dynamic library, all code is essentially flattened into a single object file. This contrasts with static libraries that are represented as a collection of object files that can be individually copied into an executable as needed (i.e., object files in a static archive that are not needed are not copied into the executable image). As a result, loading a dynamic library will involve loading all the code defined in that `.so` file (Mitchell et al., 2001).

By default, all symbols in a DSO are exported publicly (unless you specify the `-fvisibility=hidden` compiler option). However, the GNU C++ compiler supports the concept of export maps to explicitly define the set of symbols in a dynamic library that will be visible to client programs. This is a simple ASCII format where symbols can be listed individually or using glob-style expressions. For example, the following map file, `export.map`, specifies that only the `DoSomething()` function should be exported, and all other symbols should be hidden:

```
{
    global: DoSomething;
    local: *
};
```

This map file can then be passed to the compiler when building a dynamic library using the `-version-script` linker option, as in this example:

```
g++ -shared -o libmyapi.so -fPIC file1.o file2.o file3.o \
    -Wl,--version-script=export.map
```

Shared library entry points

It's possible to define functions that will be called automatically when your shared library is loaded or unloaded. This can be used to perform library initialization and cleanup operations without requiring your users to call explicit functions to perform this.

One way to do this is using static constructors and destructors. This will work for any compiler and any platform, although you should remember that the order of initialization of static constructors is not defined across translation unit boundaries (i.e., you should never depend upon static variables in other `.cpp` files being initialized). Bearing this caveat in mind, you could create a shared library entry point in one of your `.cpp` files as:

```
class APIInitMgr
{
public:
    APIInitMgr()
    {
        std::cout << "APIInitMgr Initialized." << std::endl;
    }

    ~APIInitMgr()
    {
        std::cout << "APIInitMgr Destroyed." << std::endl;
    }
};

static APIInitMgr sInitMgr;
```

There is an alternative, more elegant approach. However, it's specific to the GNU compiler. This involves using the constructor and destructor `__attribute__` decorations for functions. For example, the following code shows you how to define library initialization and cleanup routines and hide these within one of your `.cpp` files:

```
static void __attribute__((constructor)) APIInitialize()
{
    std::cout << "API Initialized." << std::endl;
}

static void __attribute__((destructor)) APICleanup()
{
    std::cout << "API Cleaned Up." << std::endl;
}
```

If you use this approach, you should be aware that your shared library must not be compiled with the GNU GCC arguments `-nostartfiles` or `-nostdlib`.

Useful Linux utilities

There are several standard Linux utilities that can help you work with static and shared libraries. Of note is the GNU `libtool` shell script. This command provides a consistent and portable interface for creating libraries on different Unix platforms. The `libtool` script can be used in various ways, but in its simplest form you can just give `libtool` a list of object files, specify either the `-static` or `-dynamic` option, and it will then create a static or dynamic library, respectively. For example,

```
libtool -static -o libmyapi.a file1.o file2.o file3.o
```

The `libtool` script can be very useful if you want your source code to compile easily on a range of Unix platforms without worrying about the idiosyncrasies of creating libraries on each platform.

Another useful command for working with libraries is `nm`, which can be used to display the symbol names in an object file or library. This is useful to find out whether a library defines or uses a given symbol. For example, this command line will output all global (external) symbols in a library:

```
nm -g libmyapi.a
```

This will produce output such as:

```
00000000 T _DoSomething
00000118 S _DoSomething.eh
         U __ZNSolsEPFRSoS_E
         U __ZNSt8ios_base4InitC1Ev
         U __ZNSt8ios_base4InitD1Ev
         U __ZSt4cout
...
```

The character in the second column specifies the symbol type, where "T" refers to a text section symbol that's defined in this library and "U" refers to a symbol that's referenced by the library but is not defined by it. An uppercase letter represents an external symbol, whereas a lowercase character represents an internal symbol. The string in the third column provides the mangled symbol name. This can be unmangled using the `c++filt` command. For example,

```
c++filt __ZNSt8ios_base4InitD1Ev
std::ios_base::Init::~Init()
```

Another useful command is `ldd`. This can be used to display the list of dynamic libraries on which an executable depends. This will display the full path that will be used for each library, so you can see which version of a dynamic library will be loaded and whether any dynamic libraries cannot be found by the operating system. For example, the following output is produced on Linux for a simple executable:

```
% ldd userapp
    linux-gate.so.1 =>  (0xb774a000)
    libstdc++.so.6 => /usr/lib/libstdc++.so.6 (0xb7645000)
    libm.so.6 => /lib/tls/i686/cmov/libm.so.6 (0xb761f000)
    libgcc_s.so.1 => /lib/libgcc_s.so.1 (0xb7600000)
    libc.so.6 => /lib/tls/i686/cmov/libc.so.6 (0xb74bb000)
    /lib/ld-linux.so.2 (0xb774b000)
```

An executable that's been linked with the `-static` option will not depend upon any dynamic libraries. Running `ldd` on such an executable produces this output on Linux:

```
% ldd userapp
    not a dynamic executable
```

Finally, if you have a static library, it's possible to convert it to a dynamic library. Recall that a static archive (`.a`) is just a packaging of object files (`.o`). You can therefore extract the individual object files using the `ar` command and then relink them as a dynamic library. For example,

```
ar -x libmyapi.a
g++ -shared -o libmyapi.so *.o
```

Loading plugins on Linux

On Linux platforms, you can use the `dlopen()` function call to load an `.so` file into the current process. Then you can use the `dlsym()` function to access symbols within that library. This lets you create plugin interfaces, as described earlier in the book. For example, consider the following very simple plugin interface:

```
#ifndef PLUGIN_H
#define PLUGIN_H

#include <string>

extern "C" void DoSomething(const char *name);

#endif
```

You can build a dynamic library for this API, such as `libplugin.so`. Then the following code demonstrates how to load this library and call the `DoSomething()` function within that `.so` file:

```
using FuncPtrT = void (*)(const char *);
const char *error;

// open the dynamic library
void *handle = dlopen("libplugin.so", RTLD_LOCAL | RTLD_LAZY);
if (!handle) {
    std::cout << "Cannot load plugin!" << std::endl;
    exit(1);
}
dlerror();

// get the DoSomething() function from the plugin
FuncPtrT fptr = (FuncPtrT) dlsym(handle, "DoSomething");
if ((error = dlerror())) {
    std::cout << "Cannot find function in plugin: " << error;
    std::cout << std::endl;
    dlclose(handle);
    exit(1);
}

// call the DoSomething() function
(*fptr)("Hello There!");

// close the shared library
dlclose(handle);
```

Finding dynamic libraries at run time

When you run an executable that depends upon a dynamic library, the system will search for this library in various standard locations, normally `/lib` and `/usr/lib`. If the `.so` file cannot be found in any of these locations the executable will fail to launch. Recall that the `ldd` command can be used to tell you if the system cannot find any dependent dynamic library. This is obviously a concern for creating executable programs that depend upon your API. There are three main options available to your clients to ensure that any executable they build using your API can find your library at run time:

1. The client of your API ensures that your library is installed in one of the standard library directories on the end user's machine, such as `/usr/lib`. This will require the end user to perform an installation process and to have root privileges to copy files into a system directory.
2. The `LD_LIBRARY_PATH` environment variable can be set to augment the default library search path with a colon-separated list of directories. Your clients could therefore distribute a shell script to run their application where that script sets the `LD_LIBRARY_PATH` variable to an appropriate directory where your dynamic library can be found.
3. Your clients can use the rpath (run path) linker option to compile the preferred path to search for dynamic libraries into their executable. For example, this

compile line will produce an executable that will cause the system to search in `/usr/local/lib` for any dynamic libraries:

```
g++ usercode.cpp -o userapp -L. -lmyapi -Wl,-rpath,/usr/local/lib
```

Libraries on macOS

The macOS operating system is built upon a version of BSD Unix called Darwin. As such, many of the details that I've presented earlier for Linux apply equally well to the Mac. However, there are a few differences that are worth highlighting between Darwin and other Unix platforms such as Linux. Note that I'll refer to macOS here for simplicity, but everything I present applies equally well to all other Apple operating systems, such as iOS, iPadOS, watchOS, tvOS, and visionOS.

Creating static libraries on macOS

Static libraries can be created on macOS in the same way as for Linux (i.e., using `ar` or `libtool`). However, there are some different behaviors when linking a static library into an application.

Apple discourages the use of the `-static` compiler option to generate executables with all library dependencies linked statically. This is because Apple wants to ensure that applications always pull in the latest system libraries that they distribute. In fact the `gcc` man page states that `-static` "will not work on macOS unless all libraries (including `libgcc.a`) have also been compiled with `-static`. Since neither a static version of `libSystem.dylib` nor `crt0.o` are provided, this option is not useful to most people."

Essentially, the `-static` option on the Mac is reserved for building the kernel, or for the very brave.

Related to this situation, by default the Mac linker will scan through all paths in the library search path looking for a dynamic library. If it fails, it will then scan the paths again looking for a static library. This means that you cannot use the trick of favoring a static library by placing it in a directory that appears earlier in the library search path. However, there is a linker option called `-search-paths-first` that will cause the linker to look in each search path for a dynamic library and then, if not found, to look for a static library in the same directory. This option makes the Mac linker behavior more like the Linux linker in this respect. Note, however, that there is no way to favor linking against a static library over a dynamic library on the Mac when both are located in the same directory.

Creating dynamic libraries on macOS

Dynamic libraries can be created on macOS in a way very similar to that in the Linux instructions given earlier. There's one important difference that you should use the `-dynamiclib` option to `g++` to create dynamic libraries on the Mac, instead of `-shared`:

```
g++ -c -fPIC file1.c
g++ -c -fPIC file2.c
g++ -c -fPIC file3.c
g++ -dynamiclib -o libmyapi.so -fPIC file1.o file2.o file3.o \
    -headerpad_max_install_names
```

Also, note the use of the `-headerpad_max_install_names` option. This flag is highly recommended when building dynamic libraries on the Mac, for reasons I'll explain in a moment.

It should also be noted that the `ldd` command is not available on macOS. Instead, you can use the `otool` command to list the collection of dynamic libraries that an executable depends upon, such as:

```
otool -L userapp
```

Frameworks on macOS

The macOS operating system also introduces the concept of frameworks to distribute all of the files necessary to compile and link against an API in a single package. A framework is simply a directory with a `.framework` extension that can contain various resources such as dynamic libraries, header files, and reference documentation. It's essentially a bundle, in the `NSBundle` sense. This bundling of all necessary development files in a single package can make it easier to install and uninstall a library. Also, a framework can contain multiple versions of a library in the same bundle to make it easier to maintain backward compatibility for older applications.

The following directory listing gives an example layout for a framework bundle, where the `->` symbol represents a symbol link:

```
MyAPI.framework/
    Headers -> Versions/Current/Headers
    MyAPI   -> Versions/Current/MyAPI
    Versions/
        A/
            Headers/
                MyHeader.h
            MyAPI
        B/
            Headers/
                MyHeader.h
            MyAPI
        Current -> B
```

Most of Apple's APIs are distributed as frameworks, including Cocoa, Foundation, and Core Services. You may therefore wish to distribute your API as a framework on the Mac to make it appear more like a system Mac library.

You can build your API as a framework using Apple's XCode development environment. This can be done by selecting the **File > New Project** menu and then selecting Framework in the left-hand panel (Fig. A.5). By default, you can choose to have your project setup to use either the Carbon or Cocoa frameworks. If you don't need either of these (i.e., you are writing a pure C++ library), you can simply remove these frameworks after XCode has created the project for you.

Clients can link against your framework by supplying the `-framework` option to `g++` or `ld`. They can also specify the `-F` option to specify the directory to find your framework bundle.

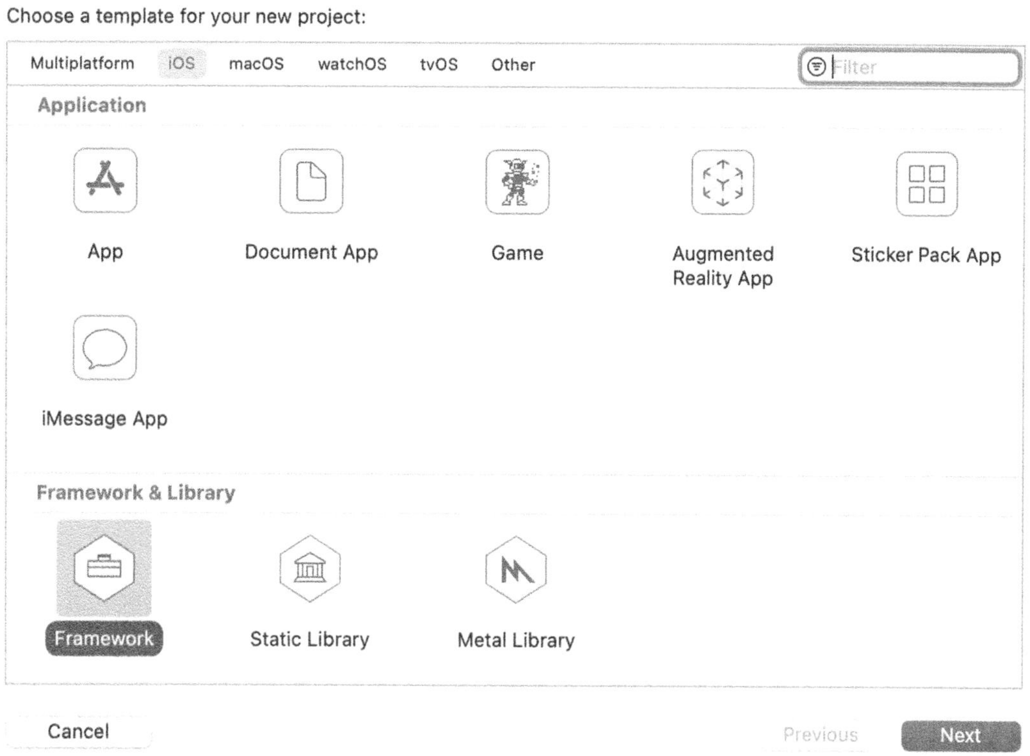

FIGURE A.5 Creating a framework project using Apple's XCode integrated development environment (IDE).

Text-based InstallAPI

When you install Xcode on your Mac you are installing a version of the clang compiler, but you are also installing one or more SDKs that let your code compile against the system libraries for different OS types (e.g., macOS, iOS, tvOS, watchOS) and different OS versions. You can find all these SDKs inside of the Xcode app bundle. For example,

```
Xcode.app/
  Contents/
    Developer/
      Platforms/
        MacOSX.platform/      <- platform
          Developer/
            SDKs/
              MacOSX13.sdk/   <- OS version
```

Originally, Apple included all of the actual Mach-O library binary files for each of these SDKs, such as the `.dylib` files for your code to link against. However, this produced very large SDK sizes, and those sizes would be compounded if you installed many SDKs for different platforms and OS versions. So Apple introduced the Text-based InstallAPI to address this issue.

When compiling a library with the "Supports Text-Based InstallAPI" Xcode setting turned on, Xcode will generate a `.tbd` file for your library. This is a plain text YAML file that contains all the information needed to link code against the library, without any of the actual executable code. This includes information such as the filesystem path where the library lives in an OS distribution as well as the names of all of the exported symbols in the library. These `.tbd` files can be vastly smaller than the actual library files themselves, often by two orders of magnitude, so Apple was able to reduce the size of their SDK distributions drastically.

This feature is not likely to be something that you'll use yourself. It's more useful within Apple for distributing OS SDKs. However, it's useful to know about these files in case you come across them in an SDK and wonder what they are, or in case you're pondering why you can't find any `.dylib` files in an SDK.

It's also worth noting that in recent versions of Apple operating systems, you will no longer find most system libraries in `/usr/lib`. For example, even though the `.tbd` file for `libxml2` says that the library lives at `/usr/lib/libxml2.2.dylib`, that file is no longer present as of macOS Big Sur. The reason is that the OS now ships with a built-in dynamic library cache of all system libraries. So checking for the library's existence on the file system will fail. However, if you attempt to `dlopen()` the library, then it will work.

Finding dynamic libraries at run time

The same Linux principles for finding dynamic libraries at run time apply to applications that are run under macOS, with a couple of small differences. The first is that the environment variable that is used to augment the library search path is called `DYLD_LIBRARY_PATH` (although `LD_LIBRARY_PATH` is now also supported by more recent versions of macOS).

Also, Apple added support for the `-rpath` linker option only in recent versions of macOS. Traditionally, you would instead use the notion of install names. An install name is a path that is burned into a Mach-O binary to specify the path to search for dependent dynamic libraries. This path can be specified relative to the executable program by starting the install name with the special string `@executable_path`.

You can specify an install name when you build your dynamic library, but your clients may also change this path using the `install_name_tool` utility. However, they cannot specify a path that is longer than the original path in the `.dylib`. This is why it is always advisable to build your dynamic libraries with the `-headerpad_max_install_names` option on the Mac: to give your clients the flexibility to change the library's install name to whatever they wish.

The following commands demonstrate how a client could change the install name for your library, and change the install name for their executable:

```
install_name_tool -id @executable_path/../Libs/libmyapi.dylib \
   libmyapi.dylib

install_name_tool -change libmyapi.dylib \
   @executable_path/../Libs/libmyapi.dylib \
   UserApp.app/Contents/MacOS/executable
```

References

Albrecht, T., 2009. Pitfalls of object oriented programming. In: Proceedings of Game Connect: Asia Pacific (GCAP) 2009 (Melbourne, Australia).

Alexander, T. (Ed.), 2003. Massively Multiplayer Game Development. Charles River Media. ISBN 1584502436.

Alexandrescu, A., 2001. Modern C++ Design: Generic Programming and Design Patterns Applied. Addison-Wesley Professional. ISBN 0201704315.

Arnold, K., 2005. Programmers are people, too. ACM Queue 3 (5). July 6 2005.

Astels, D., 2003. Test-Driven Development: A Practical Guide, second ed. Prentice Hall. ISBN 0131016490.

Astrachan, O.L., 2000. A Computer Science Tapestry: Exploring Computer Science with C++, second ed. McGraw–Hill. ISBN 0072465360.

Bass, L., Clements, P., Kazman, R., 2003. Software Architecture in Practice, second ed. Addison-Wesley Professional. ISBN 0321154959.

Beck, K., 2002. Test Driven Development: By Example. Addison-Wesley Professional. ISBN 0321146530.

Beck, K., 2007. Implementation Patterns. Addison-Wesley Professional. ISBN 0321413091.

Blanchette, J., 2008. The Little Manual of API Design. Trolltech. June 19 2008.

Bloch, J., 2008. Effective Java, second ed. Prentice Hall. ISBN 0321356683.

Booch, G., Maksimchuk, R.A., Engle, M.W., Young, B.J., Conallen, J., Houston, K.A., 2007. Object Oriented Analysis and Design with Applications, third ed. Addison-Wesley Professional. ISBN 020189551X.

Booch, G., Rumbaugh, J., Jacobson, I., 2005. Unified Modeling Language User Guide, second ed. Addison-Wesley Professional. ISBN 0321267974.

Bourque, P., Dupuis, R., Abran, A., Moore, J.W., Tripp, L. (Eds.), 2004. The Guide to the Software Engineering Body of Knowledge. IEEE Press, 2004 Version. ISBN 0769523307.

Brooks, F., 1995. The Mythical Man-Month: Essays on Software Engineering, second ed. Addison-Wesley Professional. ISBN 0201835959.

Clarke, S., 2004. Measuring API Usability. Dr. Dobbs Journal, pp. S6–S9. May 2004.

Cline, M.P., Lomow, G., Girou, M., 1998. C++ FAQs, second ed. Addison-Wesley Professional. ISBN 0201309831.

Cockburn, A., 2000. Writing Effective Use Cases. Addison-Wesley Professional. ISBN 0201702258.

Cohn, M., 2004. User Stories Applied: For Agile Software Development. Addison-Wesley Professional. ISBN 0321205685.

Conway, M.E., 1968. How do committees invent? Datamation 14 (5), 28–31.

Coplien, J.O., 1991. Advanced C++ Programming Styles and Idioms. Addison-Wesley Professional. ISBN 0201548550.

Coplien, J.O., 1995. Curiously recurring template patterns. C++ Report 7 (2), 24–27. ISSN 10406042.

Čukić, I., 2018. Functional Programming in C++. Manning Publications. ISBN 9781617293818.

Cunningham, W., 1992. The WyCash portfolio management system. March 26. In: OOPSLA '92 Experience Report, pp. 29–30 (Vancouver, Canada).

DeLoura, M. (Ed.), 2001. Game Programming Gems 2. Charles River Media. ISBN 1584500549.

Dewhurst, S.C., 2002. C++ Gotchas: Avoiding Common Problems in Coding and Design. Addison-Wesley Professional. ISBN 0321125185.

Evans, E., 2003. Domain-Driven Design: Tackling Complexity in the Heart of Software. Addison-Wesley Professional. ISBN 0321125215.

Faber, R., 2010. Architects as service providers. IEEE Software 27 (2), 33–40. March/April 2010.

Feathers, M.C., 2004. Working Effectively with Legacy Code. Prentice Hall. ISBN 0131177052.

Foote, B., Yoder, J., 1997. Big ball of mud. In: Fourth Conference on Pattern Languages of Programs (PLoP '97). September 1997, Monticello, Illinois.

Fowler, M., Beck, K., Brant, J., Opdyke, W., Roberts, D., 1999. Refactoring: Improving the Design of Existing Code. Addison-Wesley Professional. ISBN 0201485672.

Friedman, T.L., 2007. The World Is Flat 3.0: A Brief History of the Twenty-First Century. Picador. ISBN 0312425074.

Gamma, E., Helm, R., Johnson, R., Vlissides, J.M., 1994. Design Patterns: Elements of Reusable Object-Oriented Software. Addison-Wesley Professional. ISBN 0201633612.

Gamma, E., Helm, R., Johnson, R., O'Brien, L., 2009. Design Patterns 15 Years Later: An Interview with Erich Gamma, Richard Helm, and Ralph Johnson. InformIT. Oct 22.

Hanson, D., 1996. C Interfaces and Implementations: Techniques for Creating Reusable Software. Addison-Wesley Professional. ISBN 0201498413.

Headington, M.R., 1995. Removing implementation details from C++ class declarations. In: Proceedings of the 26th ACM SIGCSE Symposium on Computer Science Education, pp. 24–28. Nashville, TN.

Hennessy, J.L., Patterson, D.A., 2006. Computer Architecture: A Quantitative Approach, fourth ed. Morgan Kaufmann. ISBN 0123704901.

Henning, M., 2009. API design matters. Communications of the ACM 52 (5), 46–56.

Hoare, C.A.R., 1969. An axiomatic basis for computer programming. Communications of the ACM 12 (10), 576–580.

Hoare, C.A.R., 2003. Assertions: a personal perspective. IEEE Annals of the History of Computing 25 (2), 14–25.

Hofmeister, C., Nord, R., Soni, D., 2009. Applied Software Architecture. Addison-Wesley Professional. ISBN 0321643348.

Hunt, A., Thomas, D., 1999. The Pragmatic Programmer: From Journeyman to Master. Addison-Wesley Professional. ISBN 020161622.

ISO/IEC 14882:1998, 1998. Programming Languages—C++. International Organization for Standardization, Geneva, Switzerland. Revised by ISO/IEC 14882:2003.

ISO/IEC 14882:2003, 2003. Programming Languages—C++. International Organization for Standardization, Geneva, Switzerland.

ISO/IEC TR 19768:2007, 2007. Information Technology—Programming Languages—Technical Report on C++ Library Extensions. International Organization for Standardization, Geneva, Switzerland.

Jacobson, I., 1992. Object-Oriented Software Engineering: A Use Case Driven Approach. Addison-Wesley Professional. ISBN 0201544350.

Jeong, S., Xie, Y., Beaton, J., Myers, B., Stylos, J., Ehret, R., Karstens, J., Efeoglu, A., Busse, D., 2009. Improving documentation for eSOA APIs through user studies. In: Second International Symposium on End User Development (IS-EUD'2009). Springer-Verlag, Siegen, Germany, pp. 86–105. March 2–4, 2009.

Josuttis, N.M., 1999. The C++ Standard Library: A Tutorial and Reference. Addison-Wesley Professional. ISBN 0201379260.

Knuth, D., 1974. Structured programming with go to statements. ACM Computing Surveys 6 (4), 268.

Kochan, S., 2013. Programming in Objective-C. Addison-Wesley. ISBN 0321967607.

Korson, T.D., Vaishnavi, V.K., 1986. An empirical study of the effects of modularity on program modifiability. In: Papers Presented at the 1st Workshop on Empirical Studies of Programmers, pp. 168–186. Washington, DC.

Lakos, J., 1996. Large-Scale C++ Software Design. Addison-Wesley Professional. ISBN 0201633620.

Leclerc, Y.G., Lau, S.Q., 1994. TerraVision: a terrain visualization system. In: Technical Note 540, Artificial Intelligence Center. SRI International, Menlo Park, CA.

Lieberherr, K.J., Holland, I., 1989. Assuring good style for object-oriented programs. IEEE Software 6 (5), 38–48.

Liskov, B., Zilles, S., 1974. Programming with abstract data types. ACM SIGPLAN Notices 9 (4), 50–59.

McConnell, S., 2004. Code Complete: A Practical Handbook of Software Construction, second ed. Microsoft Press. ISBN 0735619670.

Mackinnon, T., Freeman, S., Craig, P., 2001. Endo-testing: unit testing with mock objects. In: Extreme Programming Examined. Addison Wesley Longman, pp. 287–301 ch. 17.

Martelli, A., 2011. API Design Anti-patterns. PyCon, Atlanta, 2011, Mar 9–17.

Martin, R.C., 2000. Design Principles and Design Patterns. Object Mentor article.

Martin, R.C., 2002. Agile Software Development: Principles, Patterns, and Practices. Prentice Hall. ISBN 0135974445.

Martin, R.C., 2008. Clean Code: A Handbook of Agile Software Craftsmanship. Pearson Education. ISBN 9780132350884.

Medvidovic, N., Taylor, R.N., 2000. A classification and comparison framework for software architecture description languages. IEEE Transactions on Software Engineering 26 (1), 70–93.

Mellor, S.J., Balcer, M.J., 2002. Executable UML: A Foundation for Model-Driven Architecture. Addison-Wesley Professional. ISBN 0201748045.

Meszaros, G., 2003. Agile regression testing using record and playback. Proceedings of OOPSLA 353–360. October 26–30, Anaheim, CA.

Meyer, B., 1987. Programming as contracting. In: Report TR-EI-12/CO. Interactive Software Engineering, Goleta, CA.

Meyer, B., 1997. Object-Oriented Software Construction, second ed. Prentice Hall. ISBN 0136291554.

Meyers, S., 1998. More Effective C++: 35 New Ways to Improve Your Programs and Designs. Addison-Wesley Professional. ISBN 020163371X.

Meyers, S., 2000. How non-member functions improve encapsulation. Dr. Dobb's Journal 18, 44–52. Feb 1, 2000.

Meyers, S., 2004. The most important design guideline? IEEE Software 21 (4). July/August 2004.

Meyers, S., 2005. Effective C++: 55 Specific Ways to Improve Your Programs and Designs, third ed. Addison-Wesley Professional. ISBN 0321334876.

Meyers, S., Alexandrescu, A., 2004. C++ and the Perils of Double-Checked Locking. Dr. Dobb's Journal. July 2004.

Miller, G.A., 1956. The magical number seven, plus or minus two: some limits on our capacity for processing information. Psychological Review 63 (2), 81–97.

Mitchell, M.L., Samuel, A., Oldham, J., 2001. Advanced Linux Programming. Sams. ISBN 0735710430.

Neighbors, J.M., 1980. Software construction using components. Ph.D. Dissertation. In: Technical Report UCI-ICS-TR-160. Department of Information and Computer Science, University of California, Irvine, 1980.

Parnas, D.L., 1972. On the criteria to be used in decomposing systems into modules. Communications of the ACM 15 (12), 1053–1058.

Parnas, D.L., 1979. Designing software for ease of extension and contraction. IEEE Transactions on Software Engineering 5 (2), 128–138. March 1979.

Parnas, D.L., 1994. Software aging. In: Proceedings of the 16th International Conference on Software Engineering, pp. 279–287 (Sorrento, Italy).

Pierce, B.C., 2002. Types and Programming Languages. The MIT Press. ISBN 0262162091.

Pugh, K., 2006. Interface Oriented Design. Pragmatic Bookshelf. ISBN 0976694050.

Rainsberger, J.B., 2001. Use Your Singletons Wisely. IBM Developer Works. Jul 1.

Raymond, E.S., 2003. The Art of UNIX Programming. Addison-Wesley Professional. ISBN 0131429019.

Rhoten, G., Reddy, M., 2020. Authoring grammatically correct conversational templates for siri. In: 44th Internationalization and Unicode Conference. October 14-16, 2020.

Robillard, M.P., 2009. What makes APIs hard to learn? answers from developers. IEEE Software 26 (6), 27–34.

Rooney, G., 2005. Preserving Backward Compatibility. O'Reilly OnLamp.com.

Schmidt, D.C., 1999. Strategized locking, thread-safe interface, and scoped locking: patterns and idioms for simplifying multi-threaded C++ components. C++ Report 11.

Shlaer, S., Mellor, S.J., 1988. Object-Oriented Systems Analysis: Modeling the World in Data. Prentice Hall. ISBN 013629023X.

Shreiner, D. (Ed.), 2004. OpenGL Reference Manual: The Official Reference Document to OpenGL, Version 1.4, fourth ed. Addison-Wesley Professional. ISBN 032117383X.

Snyder, A., 1986a. Encapsulation and inheritance in object-oriented programming languages. Proceedings of OOPSLA 38–45.

Snyder, A., 1986b. Encapsulation and inheritance in object-oriented programming languages. SIGPLAN Notices 21 (11). November 1986.

Spinellis, D., 2010. Code documentation. IEEE Software 27 (4), 18–19.

Stevens, W., Myers, G., Constantine, L., 1974. Structured design. IBM Systems Journal 13 (2), 115–139.

Stroustrup, B., 2000. The C++ Programming Language: Special Edition, third ed. Addison-Wesley Professional. ISBN 0201700735.

Stroustrup, B., 2001. Exception safety: concepts and techniques. In: Dony, C., Lindskov Knudsen, J., Romanovsky, A., Tripathi, A. (Eds.), Advances in Exception Handling Techniques. Springer Verlag Lecture Notes in Computer Science, LNCS-2022. ISSN 0302-9743. ISBN 3-540-41952-7. April 2001.

Stylos, J., Faulring, A., Yang, Z., Myers, B.A., 2009. Improving API documentation using API usage information. In: Proceedings of IEEE Symposium on Visual Languages and Human-Centric Computing (VL/HCC), pp. 119–126. September 20–24, Corvallis, Oregon.

Stylos, J., Graf, B., Busse, D., Ziegler, C., Ehret, R., Karstens, J., 2008. A case study of API redesign for improved usability. In: IEEE Symposium on Visual Languages and Human-Centric Computing (VL/HCC). Herrsching am Ammersee, Germany, pp. 189–192. September 15–19, 2008.

Stylos, J., Myers, B., 2008. The implications of method placement on API learnability. In: Proceedings of the 16th ACM SIGSOFT Symposium on Foundations of Software Engineering (FSE), pp. 105–112. November 9–14. Atlanta, GA.

Sutter, H., 1999. Exceptional C++: 47 Engineering Puzzles, Programming Problems, and Solutions. Addison-Wesley Professional. ISBN 0201615622.

Sutter, H., Alexandrescu, A., 2004. C++ Coding Standards: 101 Rules, Guidelines, and Best Practices. Addison-Wesley Professional. ISBN 0321113586.

Tulach, J., 2008. Practical API Design: Confessions of a Java Framework Architect. Apress. ISBN 1430209739.

Vandevoorde, D., Josuttis, N.M., 2002. C++ Templates: The Complete Guide. Addison-Wesley Professional. ISBN 0201734842.

Wiegers, K.W., 2003. Software Requirements, second ed. Microsoft Press. ISBN 0735618798.

Williams, A., 2012. C++ Concurrency in Action. Practical Multithreading, Manning. ISBN 1933988770.

Wolf, M., 2004. Why Globalization Works. Yale University Press. ISBN 0300102526.

Yee, H.Y., Newman, A., 2004. A perceptual metric for production testing. In: ACM SIGGRAPH 2004 Sketches. August 8–12, Los Angeles, CA.

Zhong, H., Zhang, L., Xie, T., Mei, H., 2009. Inferring resource specifications from natural language API documentation. In: Proceedings of 24th IEEE Conference on Automated Software Engineering (ASE), pp. 307–318. November 16–20. Auckland, New Zealand.

Index

'*Note:* Page numbers followed by "f" indicate figures "t" indicate tables and "b" indicate boxes.'

A

Abstract base classes (ABCs), 101
  backward compatibility, 401
  overview, 101–103
  plugin implementation, 541
Abstract interfaces, use of, 156
Abstractions
  application programming interface (API)
    data-driven, 179
    problem domain model, 25–27
  architecture design, 146–147
  class design, 154–167
  iterators, 210
Access control
  and #define, 240
  and friends, 242–243
  and getter/setter routines, 35–36
  proxy pattern, 110
  violation, 243
Access levels
  encapsulation, language variations, 33–35
  examples, 33f
  and logical hiding, 33–35
Active Server Pages, web development, 9b
Actors, use case template elements, 134b
Actual classes, 154
Ad hoc testing, 455
Adaptable systems, architectural patterns, 151
Adapter design pattern, 81–82
Adapter pattern, 111b
  definition, 81–82
  overview, 110–112
  Universal Modeling Language (UML) diagram, 111f
Advanced API versions, parallel products, 392b
Aggregation, Universal Modeling Language (UML) class diagram, 27b
Agile development, definition, 137–139
AHA, 45
AIX, memory-based analysis, 364
Algorithms
  application programming interface (API)
    language, 14
    orthogonality, 57
  behavioral patterns, 82t
  collision detection, 452
  functionality mismatch, 12
  memory optimization, 343–348
  memory-based analysis, 364–365
  and namespaces, 330
  parallel development example, 10–11
  and Standard Template Library (STL), 16b
  use cases, 110
Allocators
  mixing, 60
  and Pimpl, 81
  plugin implementation, 541
Alternate-functionality application programming interface (APIs), façade pattern, 115
Alternate function style, 275–276
AMD CodeXL, for performance analysis, 363
American National Standards Institute (ANSI) C application programming interface, 182
  benefits, 182–183
  features, 181–182
  FMOD case study, 186
  functions calls from C++, 185
  keywords, 181–182
  type checking, 182b

American National Standards Institute (ANSI) C application programming interface (*Continued*)
writing, 183–185
Amiga, header file, 337–338
Android platform, application programming interface (API) versions, 547t
Anonymous namespace, external linkage and, 245
Any, C++17 application programming interface features, 296–298
Apache HTTP Server, 430f
Apache License, 429t
Apache Portable Runtime, 429t
Apache's module interface, 534
Apache Web server, 534
API compatibility. *See* Source compatibility
Apple, 452, 497
Application binary interface (ABI), 252, 393–394, 537–538
backward compatibility, 272
compatibility, 396–399
Application programming interface (API) basics, 1, 2f
avoid, 11–13
in C++, elements, 3–4
code reuse, 8–10
contracts and contractors, 1–3
definition, 7–11
design, 4–7
examples, 13–17
file formats, 19–21
layers of, 13–16
libraries, frameworks, and software development kits, 17–19
network protocols, 19–21
parallel development, 10–11
purpose, 11–12
real-life example, 16–17
reasons to avoid, 11–13
robust code, 7–8
and web development, 9b
Application programming interface (API) design, 32, 123
architectural patterns, 151–153
communicating, 153–154
architecture design, 142–154
constraints, 144–146
developing, 143–144, 143f
identifying major abstractions, 146–147
inventing key objects, 148–151
for OpenSceneGraph application programming interface, 147f
case for good design, 123–129
class design, 154–167
class design options, 156–157
composition, 162–163
using inheritance, 157–159
law of Demeter (LoD), 165–166
Liskov substitution principle, 159–163
naming, 166–167
object-oriented concepts, 155–156
open/closed principle (OCP), 163–165
private inheritance, 161–162
SOLID principles, 157
development, 124f
elements, 139–142
function design, 167–178
error handling, 173–178, 174b
naming, 168–169
options, 167–168
parameters, 169–173, 171b–172b
functional requirements, 130–131
example, 131–132
gathering, 129–132
maintaining, 132
for long term, 128–129
technical debt
accruing, 124–126
paying back, 126–128
use cases, 134b
Agile development, 137–139
creating, 132–139
developing, 133
requirements, 137–139
templates, 134
writing good, 135–137
Application programming interface (API) design pattern, 81, 82t
factory methods, 101–107

abstract base classes (ABCs), 101–103
adapter pattern, 110–112
extensible factory example, 105–107
Façade pattern, 113–115
interfaces, 101–103
proxy pattern, 108–110
simple factory example, 103–105
wrapping, 107–115
observer pattern, 115–121
dependencies in, 117f
implementing, 117–120
Model–View–Controller (MVC), 116–117
push *vs.* pull observers, 120–121
Universal Modeling Language (UML) representation of, 118f
Pimpl idiom, 83–92, 83f, 83b
using, 83–87, 86b
advantages, 89–90
copy semantics, 87–88
disadvantages, 90–91
opaque pointers in C, 91–92
Pimpl and smart pointers, 88–89
singleton, 93–100, 93b
dependency injection *vs.*, 96–98
implementing in C++, 94–96
monostate *vs.*, 98–99
session state *vs.*, 100
singletons and thread safety, 96, 96b
Application programming interface (API) performance
constant declaration, 338–341
constexpr, consteval, and constinit keywords, 340–341
copy-on-write (COW), 351–356, 352f
#include minimization, 333–338
"Winnebago" headers, 333
forward declarations, 333–336
redundant #include guards, 336–338, 337t
initialization lists, 341–343
inline functions, 348–351
iterating over elements, 356–360
array references, 358–359
extern templates, 359–360
iterators, 356–357
random access, 357–358
memory optimization, 343–348
pass input arguments by const reference, 331–332
performance analysis, 360–365
memory-based analysis, 364–365
multithreading analysis, 365
time-based analysis, 361–364
Application programming interface (API) qualities (C++-specific)
avoid #define for constants, 239–241
avoid using friends, 242–243
coding conventions, 247–250
const correctness, 216–220, 216b
method, 216–218
parameter, 218, 218b
return value, 219–220
constructors and assignment, 211–216
explicit keyword, 215–216
exporting symbols, 244–247
function parameters, 236–239
default arguments, 238–239
pointer *vs.* reference parameters, 236–238
namespaces, 209–211
operator overloading, 227–236
adding operators to class, 231–233
conversion, 236
free operators *vs.* member operators, 229–231
overloadable, 228
syntax, 233, 234t–235t
templates, 220–227
explicit instantiation application programming interface design, 224–227
implicit instantiation application programming interface design, 223–224
terminology, 220–222

Application programming interface (API) qualities (general)
coupling, 65–79
callbacks, observers, and notifications, 74–79
intentional redundancy, 69–71
manager classes, 71–73, 73f
by name only, 67–68
reducing class, 68–69
easy to use, 50–65
consistent, 54–56
difficult to misuse, 51–54
discoverability, 50–51
orthogonal, 57–60
platform independent, 63–65
robust resource allocation, 60–63
implementation detail hiding, 31–42
classes, 40–42
logical hiding, 33–35
member variables, 35–38
method, 38–40
physical hiding, 31–33
minimally complete, 42–50
convenience application programming interfaces (API), 45–48
don't repeat yourself (DRY), 44–45
overpromise, 43–44
virtual functions judiciously, 48–50
problem domain model, 25–30
abstraction, 25–27
key objects model, 27–29
solve core problems, 29–30
stable, documented, and tested, 80
Application programming interface (API) reviews
overviews, 408–414
precommit, 412–414
prerelease, 410–412
purpose, 409–410
Application programming interface (API) styles
American National Standards Institute (ANSI) C, 182
application programming interface benefits, 182–183
features, 181–182
FMOD C application programming interface, 186
writing application programming interface in, 183–185
data-driven, 197–207
advantages, 199–201
data-driven web services, 203–204
disadvantages, 201
FMOD data-driven application programming interface, 202–203
idempotency, 204–207
supporting variant argument lists, 201–202
flat C, 180–186
functional, 193–197
advantages, 196
disadvantages, 196–197
example, 195–196
functional programming concepts, 193–195
object-oriented C++, 186–188
FMOD C++ application programming interface, 188
object-oriented application programming interfaces advantages, 187
object-oriented application programming interfaces disadvantages, 187–188
template-based, 189–193
advantages, 191–192
disadvantages, 192–193
example, 189–190
templates *vs.* macros, 190–191
Application programming interface (API), 25, 81, 123, 179, 209, 251, 329, 367, 383, 390–391, 415, 443, 487, 501, 533
in American National Standards Institute (ANSI) C, 183–185
comments, Doxygen, 436
life cycle, 392–393
testing, types of, 446–453
Architectural patterns, 151–153
communicating, 153–154

Architecture Description Languages, architecture design communication, 153
Architecture design
abstractions, 146–147
application programming interface (API) design, 140
communication, 153–154
constraints, 144–146
development, 143–144, 143f
key objects, 148–151
layers examples, 152f
menvshared, 152b
OpenSceneGraph example, 147f
overview, 142–154
patterns, 151
Second Life Viewer, 16–17, 17f
Architecture Review Board (ARB), 410
Array
American National Standards Institute (ANSI) C features, 181–182
and constructors, 184
deallocation, 60
and initialization lists, 341–343
and iterators, 356–357
plugin application programming interface (API), 533
random access, 357–358
reference parameters, 237–238
references, 358–359
variant argument lists, 201–202
Array subscript
binary operators, 458
syntax, 451
Assertions, 469–470
contract programming, 470–473
data-driven application programming interface (APIs), 179
implicit instantiation design, 221
JUnit, 448–449
return value tests, 456
test harnesses, 476
testable code writing, 459–475
unit testing, 449
Assignment operator, 156, 253
class design, 154–167
compiler-generated functions, 257
copy semantics, 87–88
copy-on-write (COW), 351–356
C++ usage, 211–216
C++-specific application programming interface (APIs)
compiler-generated functions, 257
defining, 213–215
initialization lists, 341–343
overloading, 227–236
and Pimpl, 81
and singletons, 93
Association, UML class diagram, 27b
Asynchronous tasks, 372
at() method, and random access, 357
Auto return type, C++14 application programming interface features, 284–285
Autodesk Maya 3D modeling system, 501–502
AutoDuck, documentation generation, 425
Automated application programming (API) interface documentation, 423–425
Automated documentation, creation, 424b
Automated GUI testing, overview, 446
Automated teller machine (ATM), 131–132
functional requirements, 131–132
use case example, 132–139
user stories, 137
Automated testing, 6, 443
application programming interface (API) design, 6
bug tracking, 483–484
code coverage, 480–483, 482f
continuous build system, 484–485
test harnesses, 476–480
tools, 470, 476–485
Automation, 44, 427
objects, 508

B

Backward compatibility, 394–395, 408–409
application programming interface (API) design, 5

Backward compatibility (*Continued*)
application programming interface (API) reviews
overviews, 408–414
precommit, 412–414
prerelease, 410–412
purpose, 409–410
binary compatibility, 398–399
default arguments, 238–239
functional compatibility, 395
inline namespaces for versioning, 407–408
maintenance, 401–408
deprecating functionality, 404–406
functionality addition, 401–402
functionality changing, 402–404
patch version, 384
precommit, 412–414
prerelease, 410–412
removing functionality, 406–407
requirement, 6
source compatibility, 395–396
and visitor pattern, 554–560
Base case, 281
Basic block coverage, testing tools, 481
Basic course
automated teller machine (ATM) use case example, 134b
use case template elements, 136b–137b
Behaviors
defining with documentation, 421–422
documentation, 423–430
key objects, 148
mock objects, 449
Best practices
application programming interface (API) design, 5
application programming interface (API) qualities, 80
coding conventions, 249
#define, 239–241
error reporting, 177
function parameters, 236–239
template-based APIs, 189
"Big Three" rule, 213
Binary compatibility, 396–399, 402
American National Standards Institute (ANSI) C, 183–185
binary compatible application programming interface (API) changes, 398–399
binary incompatible application programming interface (API) changes, 397–398
maintenance, 401–408
Pimpl idiom, 81
Binary instrumentation, time-based performance analysis, 361
Binary literals, C++14 application programming interface features, 288
Binary operators
overloadable operators, 228
symmetry, 230
Bit fields, memory optimization, 344–345
Bitwise operators, 305
Black box testing. *See* Closed box testing
Booleans, 52b
Boost, 367
Boost headers, 351
Boost Iostreams library, 553
Boost library, 189
and copy-on-write (COW), 351–356
and coupling, 66
error handling, 173–178
for extending, 501–502
implicit instantiation, 221
template-based APIs, 179
Boost Python, 506–507
application programming interface (API) extension, 514–516
bindings, 510–520
building, 510–511
constructors, 513–514
cross-language polymorphism, 518–519
features, 520
inheritance in C++, 516–517
iterators support, 519
wrapping C++ application programming interface (API) with, 511–513
Boost test, overview, 477
Boundary conditions, definition, 456

Branch coverage. *See* Decision coverage
Branching strategies, 388
Breadcrumbs, documentation usability, 431
Browser plugins, 534
BSD License, 429t
Buffer overruns, definition, 457–458
Bug tracking
  definition, 476
  testing tools, 483–484
Bugzilla, 484
Bullseye coverage, testing tools, 482
Business logic, data-driven application programming interface (APIs), 199–201
Business requirements, definition, 130
Byte type, C++17 application programming interface features, 305

C

C++, 25, 341, 560
  application programming interface (API), 501
    objective-C behind, 491–494
  Coding Conventions, 248
  implementing plugins in, 539–540
  implementing singleton in, 94–96
  inheritance in, 516–517
  language basics, 14, 251
  multithreading with, 367–369
  and objective-C, interface design in, 489–491
  behind objective-C application programming interface (API), 494–496
  plugin system design, 537–538
  resource allocation, 60
  Standard Library approach, 55, 356, 505, 560
  behind swift application programming interface (API), 497–500
C++03, 286, 343
  code, 252, 256
  standard, 96
C++0x, 252, 283
C++11 application programming interface (API) features, 252–283
  alternate function style, 275–276
  constant expressions, 277–279
  constructors, 252–256
  default and deleted functions, 256–258
  enum classes, 264–265
  initializer list constructors, 261
  inline namespaces, 270–272
  migrating to C++11, 282–283
  noexcept specifier, 269–270
  nullptr keyword, 279
  object construction, 258–260
  override and final specifiers, 266–269
  rule of five, 252–256
  smart pointers, 261–264
  tuples, 276–277
  type aliases with using, 272–273
  user-defined literals, 273–275
  variadic templates, 280–282
C++14 application programming interface (API) features, 283–289
  auto return type, 284–285
  binary literals and digit separators, 288
  const expression improvements, 287–288
  deprecated attribute, 285–286
  migrating to C++14, 288–289
  variable templates, 286–287
C++17 application programming interface (API) features, 289–307
  any, 296–298
  byte type, 305
  checking for header availability, 303–304
  fold expressions, 301–303
  inline variables, 289–291
  maybe_unused attribute, 305–307
  migrating to C++17, 307
  nested namespaces, 300–301
  optional, 293–296
  string views, 291–293
  variant, 298–300
C++1y, 283
C++20 application programming interface (API) features, 307–322
  abbreviated function templates, 318–319
  consteval specifier, 319–320
  constinit specifier, 320–321

C++20 application programming interface (API) features (*Continued*)
- constraints and concepts, 316–318
- migrating to C++20, 321–322
- modules, 308–311
- spaceship operator, 311–315

C++23 application programming interface (API) features, 322–327
- expected values, 323–324
- migrating to C++23, 327
- multidimensional subscript operator, 324–325
- preprocessor directives, 325–327

C++98 standard, 96, 291

C++ revisions
- C++11 application programming interface features, 252–283
  - alternate function style, 275–276
  - constant expressions, 277–279
  - constructors, 252–256
  - default and deleted functions, 256–258
  - enum classes, 264–265
  - initializer list constructors, 261
  - inline namespaces, 270–272
  - migrating to C++11, 282–283
  - noexcept specifier, 269–270
  - nullptr keyword, 279
  - object construction, 258–260
  - override and final specifiers, 266–269
  - rule of five, 252–256
  - smart pointers, 261–264
  - tuples, 276–277
  - type aliases with using, 272–273
  - user-defined literals, 273–275
  - variadic templates, 280–282
- C++14 application programming interface features, 283–289
  - auto return type, 284–285
  - binary literals and digit separators, 288
  - const expression improvements, 287–288
  - deprecated attribute, 285–286
  - migrating to C++14, 288–289
  - variable templates, 286–287
- C++17 application programming interface features, 289–307
  - any, 296–298
  - byte type, 305
  - checking for header availability, 303–304
  - fold expressions, 301–303
  - inline variables, 289–291
  - maybe_unused attribute, 305–307
  - migrating to C++17, 307
  - nested namespaces, 300–301
  - optional, 293–296
  - string views, 291–293
  - variant, 298–300
- C++20 application programming interface (API) features, 307–322
  - abbreviated function templates, 318–319
  - consteval specifier, 319–320
  - constinit specifier, 320–321
  - constraints and concepts, 316–318
  - migrating to C++20, 321–322
  - modules, 308–311
  - spaceship operator, 311–315
- C++23 application programming interface (API) features, 322–327
  - expected values, 323–324
  - migrating to C++23, 327
  - multidimensional subscript operator, 324–325
  - preprocessor directives, 325–327
- to use, 251–252

C# language, 25
- API qualities, 25
- Doxygen, 157
- formal design notations, 150–151
- robust resource allocation, 60

C++ usage
- avoid #define for constants, 239–241
- avoid using friends, 242–243
- coding conventions, 247–250
- const correctness, 216–220, 216b
  - method, 216–218
  - parameter, 218, 218b
  - return value, 219–220
- constructors and assignment, 211–216
- design pattern

abstract base classes (ABCs), 101–103
adapter pattern, 110–112
copy semantics, 87–88
extensible factory example, 105–107
façade pattern, 113–115
opaque pointers in C, 91–92
Pimpl and smart pointers, 88–89
proxy pattern, 108–110
simple factory example, 103–105
singleton *vs.* dependency injection, 96–98
singleton *vs.* monostate, 98–99
singletons and thread safety, 96, 96b
design
composition, 162–163
error handling, 173–178, 174b
function parameters, 169–173, 171b–172b
law of Demeter (LoD), 165–166
Liskov substitution principle, 159–163
private inheritance, 161–162
explicit keyword, 215–216
exporting symbols, 244–247
function parameters, 236–239
default arguments, 238–239
pointer *vs.* reference parameters, 236–238
namespaces, 209–211
operator overloading, 227–236
adding operators to class, 231–233
conversion, 236
free operators *vs.* member operators, 229–231
overloadable, 228
syntax, 233, 234t–235t
qualities
callbacks, 74–79
coupling by name, 67–68
difficult to misuse, 51–54
intentional redundancy, 69–71
member variables hiding, 35–38
method hiding, 38–40
notifications, 74–79
observers, 74–79
orthogonal, 57–60
physical hiding, 31–33
platform independent, 63–65
robust resource allocation, 60–63
styles
American National Standards Institute (ANSI) C application programming interface, 182
C functions from, 185
data-driven application programming interfaces, 197–207
FMOD C++ application programming interface, 188
functional application programming interfaces, 193–197
template-based application programming interfaces, 189–193
templates, 220–227
explicit instantiation application programming interface design, 224–227
implicit instantiation application programming interface design, 223–224
terminology, 220–222
versioning, 538
application binary interface (ABI) compatibility, 396–399
application programming interface (API) life cycle, 392–393
application programming interface (API) reviews, 408–414
backward compatibility maintenance, 401–408
backward compatibility, 394–395
binary compatibility, 396–399
compatibility levels, 393–401
deprecating functionality, 404–406
forward compatibility, 399–401
functional compatibility, 395
functionality addition, 401–402
functionality changing, 402–404
inline namespaces for versioning, 407–408
removing functionality, 406–407
software branching strategies, 388–392

C++ usage (*Continued*)
source compatibility, 395–396
Caching, 36
and getter/setter routines, 35–36
and performance, 358
Callbacks, 242
and architecture design, 140
and coupling, 74–79
and factory methods, 101–107
and plugin application programming interface (API), 533
Callgrind, for performance analysis, 363
Calling C functions from C++, 185
Calling convention, adapter pattern, 111
Case study
FMOD C application programming interface, 186
FMOD C++ application programming interface, 188
FMOD data-driven application programming interface, 202–203
Category, 490
CcDoc, documentation generation, 425
CDD Standard Library, 189
Chaining
function calls, 165
operators, 212
Change control process, application programming interface (API) design, 6
Change request process, application programming interface (API) design, 6
CheckingPolicy, 561
Chrome, 534
CircleCI, 485
Circular dependency, architecture design, 152f
Clang C++ compiler, 246
C language basics, 291
flat C application programming interfaces, 180
opaque pointers in, 91–92
plugin system design, 537–538
preprocessor, 325
programmers, 31
Class abstractions, 44
Class adapters, 112
Class comments, Doxygen, 438–439
Class concept
coding conventions, 247–250
definition, 155
Class design, 154–167
class design options, 156–157
composition, 162–163
using inheritance, 157–159
law of Demeter (LoD), 165–166
Liskov substitution principle, 159–163
naming, 166–167
object-oriented concepts, 155–156
open/closed principle (OCP), 163–165
options, 156–157
private inheritance, 161–162
SOLID principles, 157
Class diagrams
design notations, 150–151
implementation hiding, 7
Unified Modeling Language (UML), 26
Classes, coding conventions, 248
Class hierarchy
application programming interface (API) design, 140
example, 141f
Class invariant, contract programming, 418
Class uncopyable, 87
Classic synchronization bug, 62
Client applications, and application programming interface (API) performance, 349
Client protocol, 19–20
Clients, 237
C linkage
plugin API, 539
plugin implementation, 541
Closed box testing, 446
Cocoa (Apple)
event ordering, 75
and Model–View–Controller (MVC), 116
"Winnebago" headers, 333
Code bloat, 349

Code coverage
definition, 476
testing tools, 480–483
Code-driven application programming interfaces, 197
Code duplication, 8
Code examples, documentation usability, 431
Code readability, 196
Code redundancy, 70
Code reuse, 8–10
and application programming interface (API) usage, 7–8
scripting, 363
Code robustness, and application programming interface (API) usage, 7–8
Coding conventions, C++-specific application programming interface (APIs), 247–250
Coding standards, as test quality, 409
Coercions, 157
Cohesion, definition, 65–66
Comment style, coding conventions, 248
Commenting, 22–23
coding conventions, 247–250
defining behavior, 416–417
Doxygen
application programming interface (API) comments, 436
class comments, 438–439
enum comments, 440–442
file comments, 437–438
method comments, 439
style, 436
Commit semantics, 176
Common Lisp, 507
Common Object Request Broker Architecture (CORBA), 506, 509–510
Communication, 153–154
architecture design, 142–154
behavioral changes, 420–421
Community catalyst, plugin-based extensibility, 534
Compatibility levels, 393–401
application binary interface (ABI) compatibility, 396–399
backward compatibility, 394–395
binary compatibility, 396–399
forward compatibility, 399–401
functional compatibility, 395
source compatibility, 395–396
Compile errors
and copy semantics, 87–88
and easy to use, 45
and parameter lists, 171b–172b
and type checking, 182
Compiler-generated default constructor, 257
Compile-time constant index, 277
Compile-time speed
definition, 329
and explicit instantiation, 192–193
redundant #include guards, 336–338
template-based APIs, 179
Completion, application programming interface (API) life cycle, 393
Complex code, 262
Compliance testing, 445
Component object model (COM) automation, 506, 508–509, 539
Component testing. *See* Integration testing
Components, 1
Composition, 27b
and adapters, 112
class design, 162–163
coding conventions, 247–250
concept, 155
and inheritance, 155, 157–159
and intimacy, 66
and Liskov substitution principle (LSP), 157
and open/closed principle (OCP), 157
UML class diagram, 27b
use of, 156
Concepts, 316–318
Concurrency, 196, 367
accessing shared data, 373–378
initializing shared data, 375–377

Concurrency (*Continued*)
stateless application programming interfaces (API), 373–375
synchronized data access, 377–378
concurrent application programming interface (API) design, 378–381
concurrency best practices, 378–379
thread-safe interface pattern, 379–381
multithreading with C++, 367–369
terminology, 369–373
asynchronous tasks, 372
data races and race conditions, 369–370
parallelism, 372–373
reentrancy, 371–372
thread safety, 370–371
testing, 447
Condition coverage, 481
code coverage, 444
testing tools, 443
Condition testing, definition, 456
Configuration file
Boost Python, 501
Doxygen, 434
factory method example, 97
integration testing, 450–452
member variable hiding, 35–38
and parallel development, 10–11
Conscientious testing, 443
Consistency, 56
adapter pattern, 112
application programming interfaces (API) easy to use, 54–56
automated application programming interface (API) documentation, 434
documentation "look and feel", 430–431
random access functionality, 358
Constants (general)
constexpr, consteval, and constinit keywords, 340–341
declaration, 338–341
#define for, 239–241
expressions, 277–279
naming conventions, 248
Const correctness, 216b
definition, 216–220
method, 216–218
parameter, 218, 218b
return value, 219–220
Consteval keywords, 340–341
Consteval specifier, 319–320
Const expression improvements, C++14 application programming interface features, 287–288
Constexpr keywords, 340–341
Constinit keywords, 340–341
Constinit specifier, 320–321
Constraints, 316–318
Const references
forward declarations, 333–336
function design, 167–178
function parameters, 169–173
input arguments passing, 331–332
return value const correctness, 219–220
Constructors, 156, 253, 260
American National Standards Institute (ANSI) C application programming interface, 182
Boost Python bindings, 513–514
C++ usage, 211–216
in C++, 101
class design, 154–167
coding conventions, 247–250
copy semantics, 87–88
default arguments, 238–239
defining, 213–215
error handling, 173–178
explicit keyword, 215–216
factory methods, 101–107
function design, 167–178
function parameters, 169–173
implementation method hiding, 38–40
initializer list, 261
input argument passes, 331–332
operator symmetry, 230
Pimpl, 81
resource allocation, 60–63
ruby bindings with simplified wrapper and interface generator (SWIG), 525
and singletons, 81
stub objects, 449

virtual functions, 49
Const, use of, 156
Container classes
application programming interface (API) comments, 411
array references, 358–359
and consistency, 55
coupling, 66
dependency, 67
documentation, 423–430
forward compatibility, 399–401
and friends, 243
function naming, 168–169
language compatibility, 505
namespaces, 209–211
orthogonality, 57–60
performance testing, 452–453
robust code, 7–8
variant argument lists, 201–202
Continuous build system
definition, 476
testing tools, 484–485
Continuous delivery (CD), 484–485
Continuous integration (CI), 484–485
Contractor analogy, 1–3
Contract programming, 418
analogy, 1–3
programming, 470–473
Convenience application programming interfaces (API), 474
façade pattern, 115
usage, 45–48
Conversation Authoring Template (CAT), 198b
Conversion operators
class design, 156
and explicit keyword, 215–216
overloading, 227–236
ConversionPolicy, 561
Conway's law, 146
Copy constructor
argument passing, 239
C++ usage, 211–212
class design, 156
compiler-generated functions, 257
copy semantics, 87–88
error handling, 173–178
input argument passes, 331–332
Pimpl, 81
and singletons, 81
Copy semantics, Pimpl idiom, 87–88
Copyleft license, definition, 428–429
Copy–on–write (COW), 351–352
and application programming interface (API) performance, 351–356
example, 352f
Core Animation (CA), 489
Core application programming interface (API) initialization code, 546
plugin API communication, 533
Plugin Manager, 537f
plugin versioning, 547–548
Counter monitoring, time-based performance analysis, 363
Coupling, 65–79
application programming interface (API) qualities, 80
callbacks, 74–79
intentional redundancy, 69–71
manager classes, 71–73, 73f
by name only, 67–68
notifications, 74–79
observers, 74–79
and operator overloading, 227–236
Pimpl idiom, 90
reducing class, 68–69
Coverity, for performance analysis, 365
CppDoc, documentation generation, 425
CppUnit
and JUnit, 448–449
overview, 476
CppUnitLite, 476
CPython, 510
Creational design patterns, examples, 81
Cross-language polymorphism
Boost Python bindings, 518–519
Ruby bindings with simplified wrapper and interface generator (SWIG), 528
in script binding, 459

Cross-Platform component object model (XPCOM), 509
Cross-platform plugin system, 541
CUnit, 449b
Curiously Recurring Template Pattern (CRTP), 562–564
Custom interpolation algorithm, 533
Cyclic dependencies, 153

D
Data, 375–376
  data-driven dialogue engine, 198b
  data-driven web services, 203–204
  data-oriented features, 394–395
  hiding in Objective-C, 491–494
  races and race conditions, 369–370
Data-Driven application programming interfaces (API), 179, 197–207
  advantages, 199–201
  data-driven web services, 203–204
  disadvantages, 201
  FMOD data-driven application programming interface, 202–203
  idempotency, 204–207
  supporting variant argument lists, 201–202
Data formats, documentation usability, 431
Data types, 4
  adapter pattern, 110–112
  API orthogonality, 57
Dates, as version numbers, 385–386
Death tests, 478
Debt types, 125
Debugging
  API performance, 329–330
  and getter/setter routines, 36
Debug mode, proxy pattern, 110
Decision coverage, testing tools, 481
Declarations
  American National Standards Institute (ANSI) C application programming interface, 182
  automated API documentation, 434
  coding conventions, 247–250
  constants, 339
  constexpr keyword, 340–341
  constructors, 513–514
  constructors and assignment, 211–216
  copy-on-write (COW), 351–356
  copy semantics, 87–88
  coupling by name, 67–68
  definition, 31
  enum and, 264
  deprecation, 393
  and explicit keyword, 215–216
  external linkage, 468–469
  implementation detail hiding, 31–42
  of physical hiding, 31–33
  private inheritance, 161–162
  separation model, 224
  template terminology, 220–222
  template-based APIs, 179
  test harnesses, 476–480
  "Winnebago" headers, 333
Declarative style, 196
Default arguments, 181, 238–239
  American National Standards Institute (ANSI) C features, 181–182
  binary compatibility, 398–399
  Boost Python, 506–507
Default constructor
  Boost Python, 506–507
  C++ usage, 211
  copy semantics, 87–88
  initializer list, 261
  ruby bindings with simplified wrapper and interface generator (SWIG), 525
  singleton implementations, 81
Default functions, C++11 application programming interface features, 256–258
#define
  coding conventions, 247–250
  and constants, 239–241
  version API creation, 386
Definition
  application programming interfaces (APIs) and parallel branches, 390–391
  assertions, 469–470
  and composition, 162–163
  copy-on-write (COW), 351–356

defining behavior, 416–417
explicit instantiation design, 158
forward declarations, 333–336
implicit instantiation design, 221
orthogonal design, 59–60
physical hiding, 31–33
template, 220–222
template-based APIs, 179
"Winnebago" headers, 333
Delegating constructor, 259
Deleted functions, C++11 application programming interface features, 256–258
Dependency injection, 44, 96–98
and stub objects, 449
Dependency problems, 151
#include minimization
forward declarations, 333–336
"Winnebago" headers, 333
Deprecated attribute, C++14 application programming interface features, 285–286
Deprecation, 404
application programming interface (API) life cycle, 393
backward compatibility, 401
overview documentation, 425–426
release notes, 427–428
Description, 131
ATM use case example, 131–132
use case template elements, 127
Design by contract, 418
Design pattern, 81
Destruction model, 156
Destructors, 260
abstract base classes (ABCs), 101
American National Standards Institute (ANSI) C application programming interface, 182
in C++, 101
usage, 211
class design, 154–167
coding conventions, 247–250
copy semantics, 87–88
input argument passes, 331–332
resource allocation, 60–63
and singletons, 81
stub objects, 449
virtual functions, 49
Detailed design, application programming interface (API) design, 140
Devirtualization, 269
Diagrams, as documentation type, 425, 431
Digit separators, C++14 application programming interface features, 288
DirectX, 3D graphics application programming interface (APIs), 15
Discoverability
application programming interface (API), 50–51
automated API documentation, 434
function naming, 167
function parameters, 169–173
Distributed SCM systems, 389–390
Distributed systems, architectural patterns, 151
Doc++, documentation generation, 425
Doc-O-Matic, documentation generation, 425
Documentation, 4, 11–12, 421–422
automated application programming (API) interface documentation, 423–425
behavioral changes, 420–421
defining behavior, 416–417
Doxygen, 434–442
application programming interface (API) comments, 436
class comments, 438–439
comment style and commands, 435–436
configuration file, 434
enum comments, 440–442
file comments, 437–438
method comments, 439
examples and tutorials, 426–427
interface's contract, 418–420
lack of, 13
lead, prerelease API reviews, 410–411
license information, 428–430
overview, 425–426
reasons, 415–422

Documentation (*Continued*)
release notes, 427–428
types of, 423–430
usability, 430–433
inclusive language, 432–433
Domains
Evans, 149
key objects, 148–149
neighbors, 149
Shlaer–Mellor, 148–149
Don't repeat yourself (DRY), 44–45
Double Check Locking Pattern, 96
Double freeing, 60
Doxygen, 188, 415, 434–442
application programming interface (API) comments, 436
class comments, 438–439
comment style and commands, 435–436
configuration file, 434
documentation generation, 425
enum comments, 440–442
sample header with documentation, 440–442
file comments, 437–438
method comments, 439
tool, 407
DRD, multithreading analysis, 365
Dry run mode, proxy pattern, 110
DTrace, for performance analysis, 364
Dynamic binding, 155–156
Dynamic libraries
binary compatibility, 322
Boost Python bindings, 518–519
extensibility, 487
files, 4
plugin system design, 539
ruby bindings with simplified wrapper and interface generator (SWIG), 525
symbol exporting, 244

E

Easy to use, 50–65
consistency, 54–56
difficult to misuse, 51–54
discoverability, 50–51
orthogonality, 57–60
platform independence, 63–65
resource allocation, 60–63
Eiffel language, contract programming, 472
Embedding
forward declarations, 333–336
private code testing, 466–469
script bindings, 501–502
Encapsulation, 33–35
classes, 40–42
concept, 155
implementation method hiding, 38–40
in languages, 33b–34b
logical hiding, 33–35
return value const correctness, 219–220
symbol exporting, 244
Enum
Booleans, 52b
C++11 application programming interface features, 264–265
comments, 440–442
inheritance and, 553–554
Environmental factors, architecture design, 144
Equivalence classes, definition, 456
Error codes approach, 174
contract programming, 418
template-based API, 179
Error handling, 173–178, 174b
function design, 167–178
record/playback functionality, 474
Error reporting best practices, 177
Event-based application programming interfaces (APIs), 197–198
Event domains, key objects, 148
Event ordering, definition, 75
Evolution strategy, definition, 127
Examples, as documentation type, 426–427
Exceptions, 174
application programming interface (API) consistency, 55
binary compatibility, 322
Boost Python, 518–519
coding conventions, 247–250
specifier, 269–270

and Standard Template Library (STL), 16b
test harnesses, 476
Executable Universal Modeling Language, 150–151
Expected values, 323–324
Expert users, scripting advantages, 503
Explicit instantiation
application programming interface (API) design, 224–227
backward compatibility, 401
inlining, 168
template-based API, 179
templates, 221
Explicit keyword, 215–216
class design, 154–167
conversion operators, 236
C++ usage, 215–216
function design, 167–178
Explicit specialization, templates, 222
Explicit, use of, 156
Exported visibility, 244
Expressiveness, scripting advantages, 503
Extending, script bindings, 501–502
Extensibility
via inheritance, 549–560
and enums, 553–554
functionality addition, 549–550
modifying functionality, 550–551
and standard library, 551–553
visitor pattern, 554–560
via plugins, 533–548
example plugin, 542–544
implementing plugins in C++, 539–540
Netscape Plugin application programming interface (API), 535b–536b
overview, 534–536
plugin application programming interface (API), 540–542
plugin library, 534f
Plugin Manager, 544–546
system design issues, 537–539
versioning, 547–548
scripting advantages, 503
via templates, 560–564
curiously recurring template pattern (CRTP), 562–564
policy-based templates, 560–562
Extensions, 387–388
automated teller machine (ATM) use case example, 134b
Firefox web browser, 503
Netscape Plugin API, 534
use case template elements, 136b–137b
External linkage level, 244
External metadata, plugin system design, 538
Extern templates, 359–360
Extra computation, and getter/setter routines, 36
Extreme programming (XP), 137

F

Façade pattern
definition, 81–82, 113b
overview, 113–115
Universal Modeling Language (UML) diagram, 114f
Facebook, 9b
Factory methods, 81, 101–107
abstract base classes (ABCs), 101–103
adapter pattern, 110–112
extensible factory example, 105–107
façade pattern, 113–115
interfaces, 101–103
proxy pattern, 108–110
simple factory example, 103–105
wrapping, 107–115
Factory object, 105
Failure transparency, 176
Fake object, definition, 462
Faster compiles, Pimpl idiom, 90
Fast Fourier transform (FFT), 12
Feature tags, version application programming interface (API) creation, 387–388
Federal Aviation Administration certification, 445
File comments, Doxygen, 437–438

File formats
and application programming interface (APIs), 19–21
software branching strategies, 388–392
File handling, 4
Final specifiers, C++11 application programming interface features, 266–269
Finer access control, and getter/setter routines, 36
Firefox, 534
Fixture setup, unit testing, 449
Flat C application programming interface (API), 179–186
American National Standards Institute (ANSI) C, 182
application programming interface benefits, 182–183
features, 181–182
writing application programming interface in, 183–185
calling C functions from C++, 185
Flexibility
and coupling, 66
nonfunctional requirements, 130
FMOD application programming interface (API)
C case study, 186
C++ case study, 188
data-driven, 202–203
Fold expressions, C++17 application programming interface features, 301–303, 302t
Formal design notations, 150–151
Formatting, coding conventions, 248
Fortran, Doxygen, 434
Forward compatibility
overview, 399–401
patch version, 384
Forward declarations, 333–336
coding conventions, 247–250
coupling by name, 67–68
private inheritance, 157–158
Foundation classes, 489
Foundation Class library (Microsoft), and, 116
Foundation framework, 493
Fourier transform (FFT), 12
Fragility, technical debt, 125
Frameworks, 17–19
application/GUI, 15
Mac OS X, 17f
Free and Open Software (FOSS), 428
Free functions, 1
function design, 167–178
naming, 168–169
overloading, 227–236
plugin implementation, 541
Free, libre, open source software (FLOSS), 428
definition, 428–429
license types, 428
Free operators *vs.* member operators, 229–231
Free Software Foundation (FSF), 428
FreeBSD, 364
Frequently Asked Questions (FAQs), as documentation type, 427
Friend function
function design, 167–178
private code testing, 466
Friends
C++ usage, 242–243
class or function, 242
law of Demeter (LoD), 165–166
private code testing, 466
use of, 157
Friendship, and encapsulation, 33b–34b
Function calls
architectural patterns, 151–153
communicating, 153–154
architecture design, 142–154
constraints, 144–146
developing, 143–144, 143f
identifying major abstractions, 146–147
inventing key objects, 148–151
for OpenSceneGraph application programming interface, 147f
case for good design, 123–129

class design, 154–167
class design options, 156–157
composition, 162–163
using inheritance, 157–159
law of Demeter (LoD), 165–166
Liskov substitution principle, 159–163
naming, 166–167
object-oriented concepts, 155–156
open/closed principle (OCP), 163–165
private inheritance, 161–162
SOLID principles, 157
development, 124f
elements, 139–142
flat C application programming interfaces (APIs), 180
function design, 167–178
error handling, 173–178, 174b
naming, 168–169
options, 167–168
parameters, 169–173, 171b–172b
functional requirements, 130–131
example, 131–132
gathering, 129–132
maintaining, 132
for long term, 128–129
technical debt
accruing, 124–126
paying back, 126–128
use cases, 134b
Agile development, 137–139
creating, 132–139
developing, 133
requirements, 137–139
templates, 134
writing good, 135–137
Function coverage, 480–481
code coverage, 480–483, 482f
testing tools, 470, 476–485
Function design, 167–178
error handling, 173–178, 174b
naming, 168–169
options, 167–168
parameters, 169–173, 171b–172b
Function documentation, 418
Function parameters, 236–239
default arguments, 238–239
pack, 280
pointer *vs.* reference parameters, 236–238
Function refactoring, 44
Functional application programming interfaces (API), 179, 193–197
advantages, 196
disadvantages, 196–197
example, 195–196
functional programming concepts, 193–195
Functional compatibility
deprecation, 393
overview, 395
Functional programming concepts, 193–195
Functional requirements, 130
definition, 130–131
example, 131–132
gathering, 129–132
maintaining, 132
Functionality addition, via inheritance, 549–550
Functionality mismatch, 12
Functionality modification, via inheritance, 550–551
Future changes, proxy pattern, 110
Future proofing, plugin-based extensibility, 534–535

G

Gcov
example, 482f
testing tools, 482
General Public License (GPL), 12
Generalization, 27b
class hierarchy, 140
factory methods, 101
Unified Modeling Language (UML) class diagram, 26
Generic programming, 179
GenHelp, documentation generation, 425
Getter and setter methods, 349, 457
Boost Python, 501
class coupling, 68–69

Getter and setter methods (*Continued*)
- inlining, 168
- language compatibility, 504

Git, 389–390
GitHub, 413–414, 485
Global state, Singleton design pattern, 93
Globalization, 10
Globally unique identifier (GUID), 492
"glplugin", 546
Gnome, 484
GNOME Glib, flat C application programming interfaces, 180
GNU, 246
GNU General Public License (GPL), 12
- API and licence restrictions, 12
- Doxygen, 407
- FLOSS license, 429

GNU Image Manipulation Program (GIMP), 535–536
GNU LGPL license, 429t
Goal
- ATM use case example, 131–132
- use case template elements, 127

Good documentation, 6
Google Mock, testable code writing, 478
Google Test, overview, 478
GProf, for performance analysis, 363
GPS devices, 64
Graphical user interface (GUI), 4–5
- application programming interface (APIs), 15
- application programming interface (APIs) design, 4

Greater binary compatibility, Pimpl idiom, 90
gRPC, 539

H

Hardcoded assumptions, 8
Hardware compatibility, 197
Header availability, checking for, 303–304
Header files, 3, 18
- assertions, 469–470
- automated documentation, creation, 424b
- Boost, 367
- callbacks, 74–79
- class comments, 438–439
- coding conventions, 247–250
- inclusion model, 224
- inlining, 168
- Pimpl idiom, 83–92, 83f, 83b
- template-based APIs, 189
- winnebago headers, 333

Header units, 310–311
HeaderDoc, documentation generation, 425
Heavyweight interprocess communication models, 506
Helgrind, multithreading analysis, 365
"Hello World" program, 5
Hiding implementation details, 7
HP-UX, memory-based analysis, 364–365
Hypertext Mark-up Language (HTML), 424–425
- Doxygen sample header, 434
- gcov/lcov code example, 482f

I

Idempotency, data-driven application programming interfaces, 204–207
IDLscript, 509
Image application programming interface (API), 14
Image Libraries, flat C application programming interfaces, 181
Immobility, technical debt, 125
Implementation, 95, 489, 495
Implementation classes hiding, 40–42
Implementation detail hiding, 31–42
- classes, 40–42
- logical hiding, 33–35
- member variables, 35–38
- method, 38–40
- physical hiding, 31–33

Implementation method hiding, 38–40
Implicit instantiation
- application programming interface (API) design, 223–224
- templates, 221

Implicit sharing, 355

import keyword, Boost Python bindings, 513
#include minimization, 333–338
forward declarations, 333–336
redundant #include guards, 336–338, 337t
"Winnebago" headers, 333
Inclusion model, templates, 224
Inclusive language, 432–433
Independence, application programming interfaces (API) orthogonality, 59
Independent, Negotiable, Valuable, Estimable, Small, Testable (INVEST), 138–139
Index page, documentation usability, 430
Information hiding, 38
American National Standards Institute (ANSI) C application programming interface, 182
API wrapping, 506
constants, 339
copy-on-write (COW), 351–356
#define for constants, 239–241
factory methods, 81
Pimpl idiom, 83–92, 83f, 83b
virtual functions, 102
Information overload, 453
Inheritance, 157–159
in C++
Boost Python bindings, 516–517
ruby bindings with simplified wrapper and interface generator (SWIG), 526–528
concept, 155
extending via, 549–560
and enums, 553–554
functionality addition, 549–550
functionality modification, 550–551
and standard library, 551–553
visitor pattern, 554–560
private, 161–162
and standard library, 551–553
use of, 156
In-house instrumentation, time-based performance analysis, 361
Initialization
factory methods, 81
lists, 341–343
model, 156
reference parameters, 236–238
Initialization function, 376–377
Inline functions, 348–351, 361
Inline namespaces
C++11 application programming interface features, 270–272
for versioning, 407–408
Inline variables, C++17 application programming interface features, 289–291
Input argument pass, by const reference, 331–332
Instantiation, templates, 221
Integration testing, 443, 450–452
Intel Code-Coverage Tool, testing tools, 482
Intel Inspector, 365
Intel Parallel Studio, multithreading analysis, 365
Intel VTune, for performance analysis, 363, 365
Intentional debt, definition, 125
Interaction domains, key objects, 149
Interactive systems, architectural patterns, 151
Interface description language (IDL), 508–509
Interfaces, 4–5
contract documentation, 418–420
design in C++ and Objective-C, 489–491
factory methods, 101–103
Interleaved parameters, 498
Internal code, 13
Internal metadata, plugin system design, 538
Internal visitor, 242
Internationalization, testable code writing, 475
Interoperability, adapter pattern, 112
Intimacy, and coupling, 66
Invariant relationships, and getter/setter routines, 36

INVEST, definition, 138–139
iOS, 497
 SDK, 18–19
Iostream library, 552
Iron Python, 510
Iterators, 356–357
 language compatibility, 504
 support, 519

J

Java, 25, 102
 java-based NetBeans project, 413
 robust resource allocation, 60
Java Archive (JAR) file, 33b–34b
Java Swing, 116
JavaScript, 538
 and CORBA, 506
 extensibility, 503
JavaScript Object Notation (JSON) object, 202, 297
Jenkins, 485
JPEG File Interchange Format (JFIF), 19
 and APIs, 19
 header specification, 20t
 image APIs, 14
JScript, 508
JUnit testing, overview, 449b
Jython, 510

K

KDU JPEG-2000 decoder library, façade pattern, 115
Key objects
 architecture design, 140
 invention, 148–151
 model, 27–29
 object-oriented API advantages, 187
 Unified Modeling Language (UML) diagram example, 29f

L

Lambda calculus, 193
Lambert Conformal Conic, adapter pattern, 112
Language application programming interface (API), 14
Language barrier, scripting, 505–506
Language compatibility and scripting, 504–505
Late binding, 155–156
LaTeX, 424–425
Latter syntax, 35–36
Law of Demeter (LoD), 165–166, 166b
Lazy evaluation, and getter/setter routines, 36
Lazy initialization, 99b
Lazy instantiation
 proxy pattern, 110
 singleton thread safety, 96
 templates, 221
Lcov
 example, 482f
 testing tools, 482
Leaking memory, 262
Least knowledge principle, 165
Legacy code
 code coverage, 482
 façade pattern, 115
libjpeg library, image APIs, 14
libpng library
 adapter pattern, 110–112
 image APIs, 14
Libraries, 17–19
 adapter pattern, 112
 files, 4, 18
 size, 330
libtiff library, image APIs, 14
License information, as documentation type, 428–430
 common open source software licenses, 429t
License restrictions, API avoidance, 12
Licensing fee, 428
Life cycle of application programming interface (API), 392–393, 392f
Lifetime management, definition, 74
Linden Lab (LL), 210
Line coverage, testing tools, 481
Line protocol, 19–20

Linux, 511
  architecture constraints, 144–146
  Boost Python, 506–507
  caching, 36
  callgrind, for performance analysis, 363
  code coverage, 482
  gprof, 363
  Helgrind, multithreading analysis, 365
  Kernel application programming interfaces (API)
    flat C application programming interfaces, 180
  Intel VTune, 363
  memory-based analysis, 364–365
  multithreading analysis, 365
  numbering schemes, 385–386
  Open SpeedShop, 363
  OProfile, 363
  Sysprof, 363
Linux Kernel API
  American National Standards Institute (ANSI) C application programming interface, 182
  bug tracking, 413
  Second Life Viewer architecture, 17f
  version numbers, 386
Liskov substitution principle, 159–163
  class design, 154–167
  composition, 162–163
  private inheritance, 161–162
Load testing, definition, 447
Localization, testable code writing, 475
Logging domains, key objects, 149
Logical hiding. *See* Physical hiding
  definition, 33–35
  implementation details, 31
  Pimpl, 81
Loki library, 189
Longevity, API code robustness, 7
Loose coupling, 66, 196
Low coupling, 66
Low-level class, 79
Lua, 501
  MIT/X11 license, 429t
  and Simplified Wrapper and Interface Generator (SWIG), 501

M

Mac, 511
macOS, 497
  APIs in C++, 3
  architecture constraints, 144–146
  Boost Python, 506–507
  callgrind, for performance analysis, 363
  gprof, 363
  Helgrind, multithreading analysis, 365
  Second Life Viewer architecture, 17f
  Valgrind, for performance analysis, 364
  "Winnebago" headers, 333
Macros, 190–191
  assertions, 469–470
  backward compatibility, 394–395
  coding conventions, 247–250
  contract programming, 470–473
  and #define, 240
  internationalization support, 475
  private code, 466–469
  template-based APIs, 189
Maintenance, 11–12
  application programming interface (API) life cycle, 393
  application programming interface (API) reviews, 393
  invariant relationships, and getter/setter routines, 36
  overviews, 408–414
  precommit, 412–414
  prerelease, 410–412
  purpose, 409–410
Major version number, definition, 384
Manager classes
  and coupling, 71–73, 73f
  singleton, 81
  singleton *vs.* monostate, 98–99
Marionette animation system, 533
Mark Kilgard's OpenGL Utility Toolkit (GLUT), 47
Maya API, 358
Median value, 258

Member functions, 550
 backward compatibility, 394–395
 function design, 167–178
 naming, 168–169
 operator overloading, 227–236
 private code, 466–469
Member operators *vs.* free operators, 229–231
Member variables, 342
 hiding, 35–38
 initialization lists, 341–343
 memory optimization, 343–348
Memory-based analysis, 364–365
Memory bugs, 364
Memory errors, 458
Memory latency, 343–344
Memory leaks, 60
Memory management, API resource allocation, 494
Memory model semantics, API consistency, 368
Memory optimization, 343–348
 API performance, 329–330
 member variables, 343–348
 memory layout, 348f
Memory overhead, 344
Memory ownership, definition, 458
Memory performance, analysis tools, 364
Memory usage, 197
Menvshared, architectural patterns, 152b
Mercurial, 389–390
Message passing application programming interfaces, 197–198
Metadata, 538
 internal *vs.* external, 538
 Plugin Manager, 538
Method comments, Doxygen, 439
Microsoft C Runtime (MSVCRT), 252
Microsoft.NET languages
 and COM automation, 508
 web development, 9b
 Python bindings, 501
Microsoft style options, 249
Microsoft Visual Studio
 external linkage and, 245
 version numbers, 385–386
Microsoft's Win32 application programming interface, 13–14
Middleware services, 19–20
Minimal completeness, 42–50
 convenience application programming interfaces (API), 45–48
 don't repeat yourself (DRY), 44–45
 overpromise, 43–44
 virtual functions judiciously, 48–50
Minimalism, 50
Minor version number, definition, 384
Misuse considerations, 51–54
MIT/X11 license, 429t
Mock objects, 462
 testable code writing, 462–466
 unit testing, 449
Model code, 117
Model–View–Controller (MVC), 116, 116b, 151
 dependencies overview, 117f
 overview, 116–117
 script-based applications, 502–503
Modula-3, 507
Modularization, 7–8
 and API code reuse, 8–10
 and API code robustness, 7–8
Module implementation units, 308–309
Module interface units, 3
Modules, 308–311
 header units, 310–311
 named modules, 308–310
Mono, MIT/X11 license, 429t
Monostate pattern, 98–99, 99b
Move constructors, C++11 application programming interface features, 252–256
Mozilla, 453
 bug tracking, 476
 performance testing, 453
Mozilla Public License, definition, 429t
Multidimensional subscript operator, 324–325
Multithreading analysis, 365
Multithreading with C++, 367–369
Mutex locking, 367

N

Name
  ATM use case example, 131–132
  class design, 154–167
  coding conventions, 247–250
  use case template elements, 127
Named modules, 308–310
Named Parameter Idiom, 202–203
Namespaces, 188, 270, 300
  C++ usage, 209–211
  naming conventions, 248
Naming
  class design, 166–167
  collisions, enum and, 264
  conventions, 248
    language compatibility, 504
Natural language, key objects, 148
Negative testing, definition, 457
Nested namespaces, C++17 application programming interface features, 300–301
NetBeans bug tracking system, 413–414
NetBeans project, precommit API reviews, 413
Netscape Plugin API (NPAPI), 534, 535b–536b, 540–541
Netscape Portable Runtime
  C++ usage, 209–211
  flat C application programming interfaces, 181
Network protocols, 19–21
Next-generation project, 127
NeXTSTEP (NS) operating system, 489
No license, definition, 429t
Noexcept specifier, C++11 application programming interface features, 269–270
Non-default constructors
  explicit keyword, 215–216
  factory methods, 101–107
  function design, 167–178
Nonfunctional constraints, 157
Nonfunctional requirements
  class design, 166–167
  definition, 130
Nontransferability, technical debt, 125–126
Notes, use case template elements, 134b
Notifications
  and coupling, 74–79
  and getter/setter routines, 36
Null dereferencing, 60
Null input, 458
NULL pointer, 291
  behavioral changes, 420–421
  error handling, 173–178, 174b
  factory methods, 101–107
  functional compatibility, 395
  resource allocation, 60–63
Nullptr keyword, 279

O

Object adapters, 112
Object concept, definition, 155
Object construction, C++11 application programming interface features, 258–260
Object hierarchy
  application programming interface (API) design, 140
  input arguments, 331–332
Object modeling, 27, 157
  abstractions, 146–147
  architecture design, 140
  class naming, 166–167
Objective–C
  C++ application programming interface (API), 491–494
  C++ behind Objective-C application programming interface (API), 494–496
  data hiding in, 491–494
  interface design in C++ and, 489–491
  language, 487
Objective-C++, 487
ObjectName, 260
Object-oriented C++ application programming interfaces, 186–188
  binary compatibility, 396–399
  callbacks, 74–79
  FMOD C++ application programming interface, 188

Object-oriented C++ application programming interfaces (*Continued*)
  object-oriented application programming interfaces advantages, 187
  object-oriented application programming interfaces disadvantages, 187–188
  template-based APIs, 189
Object-Oriented CDD application programming interface (API), 179
Object-oriented concepts, 155–156
Object-oriented design
  class naming, 166–167
  composition *vs.* inheritance, 155
  concept definitions, 27
  law of Demeter (LoD), 165–166
  Universal Modeling Language (UML) diagram, 111f
Object–oriented programming (OOP), 3, 186–187
  contractor analogy, 3
  definition, 37
  inheritance, 380
Observers
  coupling, 74–79
  pattern, 82
  Universal Modeling Language (UML) diagram, 111f
OCAML, 507
Octave, 507
OGRE, 15
OLE Automation. *See* Component object model (COM) automation
Online 3D multiplayer game, 74
Opaque pointers
  in C, 91–92
  forward compatibility, 399–401
  method hiding, 38–40
  Pimpl idiom, 83–92
Open box testing, 446
Open/closed principle (OCP), class design, 163–165
OpenCV API, 52
Open Geospatial Consortium (OGC), 445
OpenGL
  convenience API, 45–48, 387–388
  façade pattern, 113–115
OpenGL Utility Library (GLU), 47
OpenGL Utility Toolkit (GLUT), 47
Open Inventor, 554
OpenJPEG library, façade pattern, 115
OpenSceneGraph
  example, 147f
  3D graphics APIs, 15
Open Source Initiative (OSI), 428
Open SpeedShop, for performance analysis, 363
Opera, 534
Operating system (OS), 13–16
Operation order, testing, 457
Operational factors, architecture design, 144
Operator overloading, 227–236
  adding operators to class, 231–233
  conversion, 236
  free operators *vs.* member operators, 229–231
  overloadable, 288
  syntax, 233, 234t–235t
Operators, 156
  language compatibility, 504
  symmetry, and overloading, 230
Operator syntax, overloading, 233
OProfile, for performance analysis, 363
Optimization, 96
Optional, C++17 application programming interface (API) features, 293–296
Optionals concept, 497
Organizational factor, 144
  architecture design, 140
  function design, 167–178
Orthodox canonical class form, definition, 213
Orthogonality
  application programming interfaces (API) easy to use, 57–60
  definition, 57
Output parameters, 237
Overflow, 457–458
Overloadable operators, categories, 228

Overloading, virtual functions, 49
Override specifiers, C++11 application programming interface features, 266–269
Overview documentation, characteristics, 425–426
OwnershipPolicy, 561

P

Package-private, definition, 33b–34b
Parallel branches, software branching strategies, 390–391
Parallel development, and API usage, 10–11
Parallelism, 367, 372–373
Parallel products, software branching strategies, 391–392
Parameter const correctness, 218
Parameter order, API consistency, 54
Parameter testing, definition, 456
Parasoft Insure++, for performance analysis, 364–365
ParaTools ThreadSpotter, 365
Partial specialization, templates, 222
Patch version number, definition, 384
PDF, 424–425
  Doxygen, 434
  Reader plugin, 534
Peer-to-peer applications and APIs, 19–20
Perforce, 389–390
Perforce TotalView MemoryScape, 364
Performance, 197
  analysis, 360–365
  testing, 443, 447, 452–453
Perl, 501
  and COM automation, 508
  and CORBA, 509
  Ruby creation, 521
  and SWIG, 521
Permissive license, definition, 429
Persona modeling, 139
PHP, 507
  array references, 358–359
  Doxygen, 434
  and SWIG, 521
  web development, 9b
Physical hiding, 31–33
  access levels for C++ classes, 33f
  declaration *vs.* definition, 31–33
  Pimpl idiom, 40
Pimpl idiom, 40, 81, 83–92, 83f, 83b, 399
  using, 83–87, 86b
  advantages, 89–90
  copy semantics, 87–88
  disadvantages, 90–91
  opaque pointers in C, 91–92
  Pimpl and smart pointers, 88–89
Pimpl pointers, 88–89, 89b
Pivy Python, 506
Pixar, 330
  API performance approach, 330
  API reviews, 411
  design team, 154b
  extensibility example, 533
  menvshared use, 152b
  script-based applications, 503
Planning, 11–12
Platform compatibility, 131
  nonfunctional requirements, 130
  and scripting, 459
Platform idioms, 56
  API consistency, 55
  Pimpl idiom, 40
Platform independence, 65
Playback functionality, testable code writing, 473–475
Plugin application programming interface (API), 533, 537
  definition, 537
  example, 544
  overview, 540–542
Plugin-based extensibility, 533–548
  example plugin, 542–544
  implementing plugins in C++, 539–540
  Netscape Plugin application programming interface (API), 535b–536b
  overview, 534–536
  Plugin application programming interface (API), 540–542
  plugin library, 534f
  Plugin Manager, 544–546

Plugin-based extensibility (*Continued*)
  system design issues, 537–539
  versioning, 547–548
Plugin Manager
  as Core API component, 537f
  definition, 537
  generic *vs.* specialized, 538
  overview, 544–546
Plugin registry, 537
Plugin system, 533, 535–536
Plugin versioning, 547–548
PngSuite
  definition, 450–451
  example, 451f
Pointer *vs.* reference parameters, 236–238
Policy-based design, 561
Policy-based templates, for extending, 560–562
Policy classes, 560–561
Policy decisions, software branching strategies, 388
Polymorphism
  Boost Python, 506–507
  cross-language, 505
  Curiously Recurring Template Pattern (CRTP), 562
  definition, 155–156
  template-based APIs, 189
Portability
  coding conventions, 249
  as test quality, 477
POSIX API, 13–14
Postconditions, 418
  ATM use case example, 131–132
  contract programming, 419b
  use case template elements, 127
Precommit application programming interface (API) reviews, 412–414
Precompiled headers, 333
Preconditions, 418
  contract programming, 419b
  use case template elements, 127
Preprocessor directives, 325–327
Preprocessor symbols, overloadable operators, 157
Prerelease
  application programming interface (API)
    life cycle, 393
    reviews, 410–412
Primary functional test strategies, 446–447
Primitiveness, convenience APIs, 45–46
Principle of linguistic relativity, 503
Private access level, 33
Private code
  inlining, 168
  testing, 466–469
Private inheritance
  adapter pattern, 111b
  class design, 161–162
Problem domain model, 25–30
  abstraction, 25–27
  architecture design, 140
  class naming, 166–167
  functional requirements, 129–132
  key objects model, 27–29
  platform independence, 65
  solve core problems, 29–30
Process, coding conventions, 249
Processor–memory performance gap, 343–344
Product owner, prerelease application programming interface (API) reviews, 410
Properties, key objects, 148
Proprietary license, definition, 428
Protected access level, 33
Prototypes, key objects, 148
Proxy application programming interface (API), record/playback functionality, 108
Proxy design pattern, 81–82
Proxy pattern, 108–110, 109b
pthreads, 367
Public access level, 33
Public headers, 18
  #define, 240
  automated documentation, 424b
  explicit instantiation, 221
  factory methods, 101–107
  forward declarations, 333–336

implicit instantiation application programming interface design, 223–224
inlining, 168
platform independence, 63–65
source compatibility, 395–396
template-based API, 179
Public member function, private code testing, 466
Public symbols, 210
namespaces, 209–211
private code testing, 466–469
"Winnebago" headers, 333
Pull observer pattern, 120–121
Pure functions, 194
Purpose application programming interface (API) reviews, 409–410
Push observer pattern, 120–121
Python, 25, 167, 510, 539
application programming interface (API) extension, 514–516
convention, 504
Python Boost Python, 506–507
application programming interface (API) extension, 514–516
bindings, 510–520
building, 510–511
constructors, 513–514
cross-language polymorphism, 518–519
features, 520
inheritance in C++, 516–517
iterators support, 519
wrapping C++ application programming interface (API) with, 511–513
Python-SIP, features, 507–508
PyUnit, 449b

Q

Qt API, 145–146, 173, 209
automated GUI testing, 446
backward compatibility, 401–408
copy-on-write (COW), 351–356
extensibility, 551
GUI APIs, 15
Python-SIP, features, 507–508
Qt library, 550
Quality, 11–12
Quality assurance (QA), 389
integration testing, 443
team, 443
record and playback functionality, 474
working with, 458–459
QuickTime (Apple), 535b–536b

R

Race conditions, 369–370
Random access, and iterators, 357–358
Randomized testing, 455
Rational PureCoverage, testing tools, 482
Record functionality, testable code writing, 473–475
Reduced coupling
operator overloading, 230
Pimpl idiom, 89–90
Reduced-functionality application programming interface (APIs), façade pattern, 115
Redundancy
application programming interfaces (API) orthogonality, 59
and coupling, 69–71
#include guards, 336–338
Reentrancy, 74, 371–372
Reference objects, 212
Reference symbols, coding conventions, 210–211
Reference *vs.* pointer parameters, 236–238
Regression testing, definition, 457
Relational operators, 231–232
Release notes, as documentation type, 427–428
Resource Acquisition Is Initialization (RAII), 62, 473
Resource allocation, 60–63
Resource sharing, proxy pattern, 110
Return value
assertion, 456
const correctness, 219–220
conversion operators, 236

Review process, software branching strategies, 390
Revision number. *See* Patch version number
Revolution strategy, definition, 127
Rigidity, technical debt, 125
R language, 507
ROBODoc, documentation generation, 425
Robust code, 7–8
Role domains, key objects, 148
Rollback semantics, 176
RTF, Doxygen, 434
Ruby, 521, 539
application programming interface (API)
extension, 525–526
tuning, 524–525
bindings with simplified wrapper and interface generator (SWIG), 521–529
constructors, 525
cross-language polymorphism, 528
inheritance in C++, 526–528
wrapping C++application programming interface (API), 522–523
Rule of five, 252–256
“Rule of Three”, 213, 254
Rule of Zero, 260
Run-time memory overhead, definition, 330
Run-time speed, definition, 330

S

Safari (Apple), 452
Sample header with documentation, 440–442
Sampling, time-based performance analysis, 362
Sapir-Whorf hypothesis, 503
Scalability, nonfunctional requirements, 131
Scalability testing, 447
Scaling process, 10
Scene graph, 554
array references, 358–359
hierarchy example, 555f
nodes example, 557f
OpenSceneGraph application programming interface, 147f
visitor pattern, 554–560
Scoped enums, 264
Scoped locking idiom, 377
Scoped pointers, 61
and Pimpl, 83–92
definition, 61
Scoping
#define and, 240
templates *vs.* macros, 265
Script-based applications, scripting advantages, 502–503
Script binding technologies, 474
Boost Python, 506–507
Common Object Request Broker Architecture (CORBA), 509–510
component object model (COM) automation, 508–509
Python-SIP, 507–508
simplified wrapper and interface generator (SWIG), 507
technologies, 506–510
Scripting
Boost Python bindings, 510–520
application programming interface (API) extension, 514–516
building, 510–511
constructors, 513–514
cross-language polymorphism, 518–519
features, 520
inheritance in C++, 516–517
iterators support, 519
wrapping C++ application programming interface (API) with, 511–513
language, 501, 507
ruby bindings with simplified wrapper and interface generator (SWIG), 521–529
application programming interface (API) tuning, 524–525
constructors, 525
cross-language polymorphism, 528
inheritance in C++, 526–528
Ruby application programming interface (API) extension, 525–526

wrapping C++ application programming interface (API), 522–523
script bindings addition, 501–506
advantages of scripting, 502–503
extending *vs.* embedding, 501–502
language barrier, 505–506
language compatibility issues, 504–505
for testability, 503
Scrum, 458–459
Search, documentation usability, 431
Second Life Viewer, The
architecture diagram, 361
façade pattern, 114f
monostate design pattern, 98
Security domains, key objects, 148
Security issues
architecture constraints, 144–146
nonfunctional requirements, 135
patch version, 384
plugin system design, 538
Security testing
compliance assurance, 445
definition, 447
private code, 466–469
Self-test method, example, 466
Separation model, templates, 224
Sequence diagrams, as documentation type, 427
Server protocol, 19–20
Server-based Plugin API, 534
Server-based solution, 389–390
Session state pattern, 100, 100b
Shared data
accessing, 373–378
initializing shared data, 375–377
stateless application programming interfaces (API), 373–375
synchronized data access, 377–378
initializing, 375–377
Shared pointers, 60
copy-on-write (COW), 351–356
and Pimpl, 81
Shark (Apple), for performance analysis, 364
Shlaer-Mellor method, 148–151
Short examples, as documentation type, 426
Short-term glue code, 125–126
Signals and slots, and coupling, 79
Simple code, 262
Simple examples, as documentation type, 426
Simplified Wrapper and Interface Generator (SWIG), 501, 507
for ruby bindings, 521–529
application programming interface (API) extension, 525–526
application programming interface (API) tuning, 524–525
constructors, 525
cross-language polymorphism, 528
inheritance in C++, 526–528
wrapping C++application programming interface (API) with, 522–523
wrapping C++application programming interface (API) with, 522–523
Single responsibility principle, Open/closed principle, Liskov substitution principle (LSP), Interface segregation principle, Dependency inversion principle (SOLID) principle, 157
Single test program, 450–451
Singleton design pattern, 81, 93–100, 93b
dependency injection *vs.*, 96–98
implementing in C++, 94–96
monostate *vs.*, 98–99
session state *vs.*, 100
and thread safety, 96, 96b
SIP, 507
Size, and coupling, 66
Size-based types, and memory optimization, 345–346
Smalltalk language
and CORBA, 506
and Ruby, 507
Smart pointers, 88–89, 89b
architecture constraints, 144–146
C++11 application programming interface features, 261–264
class design, 154–167

Smart pointers (*Continued*)
  implicit instantiation, 221
  Pimpl, 81
  policy-based templates, 560–562
Soak testing, 447
SOAP, and COM automation, 508
Software branching strategies, 388–392
  application programming interfaces (API) and parallel branches, 390–391
  branching strategies, 388
  file formats, 391–392
  parallel products, 391–392
  policies, 388–390
Software design engineer in test (SDET), 459
Software development kit (SDK), 4, 17–19
Software development models, 9, 458–459
Software library, 17–18
Software test engineer (STE), 459
Solaris, 364–365
  memory-based analysis, 364–365
  multithreading analysis, 365
Source code, lack of, 12–13
Source compatibility
  adding functionality, 549–550
  changing functionality, 402–404
  overview, 395–396
Source control management (SCM), software branching strategies, 389–390
Spaceship operator, 311–315
Specialization
  class hierarchy, 558–559
  Curiously Recurring Template Pattern (CRTP), 562
  explicit instantiation, 221
  lazy instantiation, 221
  templates, 222
Specialized Plugin Manager, plugin system design, 538
Square root function, 418–419
Squish (automated GUI testing tool), 446
Stability, 45–46
  application programming interface (API) life cycle, 392–393
  application programming interface (API) reviews, 408–414
  convenience application programming interfaces, 45–48
  data-driven APIs, 179
  version numbers, 383–388
Stack declaration, template-based API, 189–190
Stakeholders, 138
  ATM use case example, 131–132
  use case template elements, 127
Standard application, 4–5
Standard C Library, 236
  and API design, 4
  and consistency, 56
  conversion operators, 236
Standard design patterns, use of, 156
Standard Library, 560
  containers and algorithms, 369
  inheritance and, 551–553
Standard Template Library (STL), 16b
  API consistency, 56
  forward declarations, 333–336
  input arguments via const reference, 331–332
  iterators, 356–357
  language APIs, 14
  plugin implementation, 541
  template-based APIs, 189
Start-up time
  definition, 330
  performance testing, 452–453
Stateless application programming interfaces (API), 373–375
Stateless functions, 374
Statement coverage, testing tools, 481
Static declaration, external linkage and, 245
Static initialization, 320
Static libraries
  binary compatibility, 396–399
  files, 4
  plugin system design, 539
  plugins, 533
Static variable, 376
std::expected class template, 323

std::optional, 296
std::variant class template, 298
StoragePolicy, 561
Stream operators, 232
Stress testing. *See* Scalability testing
String views, C++17 application programming interface features, 291–293
Structs
  American National Standards Institute (ANSI) C application programming interface, 182
  coding conventions, 247–250
  definition, 180
  flat C application programming interfaces, 180
  implementation detail hiding, 31–42
  logical hiding, 33–35
  template-based APIs, 189
Structural design patterns
  API wrapping, 506
  examples, 108
Structural patterns, architectural patterns, 151
Stub objects, 462
  testable code writing, 462–466
  unit testing, 449
Subclassing, 213
  constructors and assignment, 211–216
  virtual functions, 48–50
Swap function, 257
Swift
  C++ behind Swift application programming interface (API), 497–500
  programming language, 497, 507
Symbian platform, 412–413
  precommit API reviews, 412
  public interface change request (CR) process, 413f
Symbol export, 244–247
Symbol names, 300
  binary compatibility, 396–399
  default arguments, 238–239
  and #define, 240
  flat C application programming interfaces, 180
  namespaces, 330
Symbol table, 241
  #define and, 240
  forward declarations, 333–336
Synchronization
  and getter/setter routines, 36
  mechanisms, 368
Synchronized data access, 377–378
Synopsys Coverity, 365
Sysprof, for performance analysis, 363
System testing, 446

T

Tangible domains, key objects, 148
Target constructor, 259
Task execution, 369
Task focus, documentation usability, 431
Tcl, 501
  and CORBA, 506
  and SWIG, 506
TeamCity, definition, 485
Technical debt
  accruing, 124–126
  paying back, 126–128
Technical lead, prerelease application programming interface (API) reviews, 410
Template arguments, 221
Template-Based application programming interface (API), 179
  advantages, 191–192
  disadvantages, 192–193
  example, 189–190
  overview, 189–193
  templates *vs.* macros, 190–191
Template parameters, 220, 280
Templates, 220–227, 349
  in C++03, 280
  explicit instantiation application programming interface design, 224–227
  extending via, 560–564

Templates (*Continued*)
curiously recurring template pattern (CRTP), 562–564
policy-based templates, 560–562
implicit instantiation application programming interface design, 223–224
terminology, 220–222
use of, 156
Template Unit Test (TUT), 479
Terminology, 220–222
concurrency, 369–373
asynchronous tasks, 372
data races and race conditions, 369–370
parallelism, 372–373
reentrancy, 371–372
thread safety, 370–371
documentation usability, 431
Testability, 196
Testable code, 459–475
assertions, 469–470
contract programming, 470–473
internationalization support, 475
private code, 466–469
record and playback functionality, 473–475
stub and mock objects, 462–466
test-driven development (TDD), 459–462
Test–driven development (TDD), 459–462
Test-first programming, 459
Test harnesses
Boost Test, 477
CppUnit, 476
definition, 476
Google Test, 478
Template Unit Test (TUT), 479
testing tools, 476–480
Testing, 11–12
integration testing, 450–452
lead, prerelease API reviews, 411
performance testing, 452–453
reasons, 443–445
tools
bug tracking, 483–484
code coverage, 480–483, 482f
continuous build system, 484–485
test harnesses, 476–480
types of application programming interface (API) testing, 446–453
unit testing, 447–449
untested code cost, 445b
writing, 454–459
focus, 458
qualities, 454–455
working with quality assurance (QA), 458–459
TeX document processing system, 385
singletons and, 96, 96b
Threading
errors, 368
model, 368–369
Threading Build Blocks, 210
Thread safety, 370–371
application programming interface (API), 370
interface design pattern, 379–381, 473
proxy pattern, 110
3D
feature films, 139
graphics application programming interface (API), 15
Three Laws of Interfaces, error handling, 173
TIFF files, 14
Time-based analysis, 361–364
TotalView MemoryScape, for performance analysis, 364
Traditional destructor, 254
Traditional function style, 276
Traditional waterfall method, 458–459
Transmission control protocol (TCP), 21
Travis CI, 485
Trigger, 417
ATM use case example, 131–132
use case template elements, 127
Trunk code line, software branching strategies, 388
Tuples, 276–277
Tutorials, as documentation type, 426–427
Type checking, 182

#define and, 240
templates *vs.* macros, 190–191
Type coercions, 157
Typedefs, 180
API design, 187
flat C application programming interfaces (APIs), 180
forward compatibility, 399–401
forward declarations, 333–336
naming conventions, 248
policy-based templates, 560–562
template-based APIs, 189
Type-safe fashion, 298–299
Type safety, enum and, 264
TypeScript, 538

U

Ubuntu, numbering schemes, 385–386
UIKit (UI), 489
Unary operators, overloadable operators, 228
Undefined size, enum and, 264
UNICOM Systems PurifyPlus, for performance analysis, 364
Unified Modeling Language (UML), 26
adapter pattern, 111f
application programming interface (API) abstraction, 26f
application programming interface (API) key objects, 27
class diagrams, 27b
Façade pattern, 86, 114f
key objects, 29f
manager classes, 71–72
observer pattern, 117f–118f
proxy patterns, 109f
sequence diagram to change telephone number type, 30f
singleton design pattern, 93f
types, 150f
visitor patterns, 555f
Unintentional debt, definition, 125
Union, 345
data-driven APIs, 179
memory optimization, 343–348
variant argument lists, 201–202
Unique pointers, 61
Unit testing, 443, 446–447
JUnit, 449b
overview, 447–449
test-driven development (TDD), 459
Universal Transverse Mercator, adapter pattern, 112
Universally unique identifier (UUID), 508
API key object modeling, 27
COM automation, 508
and coupling, 28
Universal Transverse Mercator, adapter pattern, 112
Unix, 511
CORBA, 506
Doxygen, 407
layers of, 13–16
memory optimization, 343–348
Plugin Manager, 544–546
TotalView MemoryScape, 364
Usability, 131
API reviews, 411
architecture constraints, 144–146
documentation, 412
nonfunctional requirements, 131
Use cases, 134b
agile development, 137–139
analysis, 136
creating, 132–139
developing, 133
requirements, 137–139
templates, 134
writing good, 135–137
User contributions, as documentation type, 427
User datagram protocol (UDP), 21
User-defined literals, 273–275

V

Valgrind, for performance analysis, 364
Validation, and getter/setter routines, 35–36
Value objects, 211–212

Variable templates, C++14 application programming interface features, 286–287
Variadic templates, 280–282
Variant, C++17 application programming interface features, 298–300
Variant argument lists, data-driven application programming interfaces, 201–202
Vectorized API, 329
Versioning, 538
  application programming interface (API) life cycle, 392–393
  application programming interface (API) reviews, 408–414
    precommit, 412–414
    prerelease, 410–412
    purpose, 409–410
  backward compatibility maintenance, 401–408
    deprecating functionality, 404–406
    functionality addition, 401–402
    functionality changing, 402–404
    inline namespaces for versioning, 407–408
    removing functionality, 406–407
  compatibility levels, 393–401
    application binary interface (ABI) compatibility, 396–399
    backward compatibility, 394–395
    binary compatibility, 396–399
    forward compatibility, 399–401
    functional compatibility, 395
    source compatibility, 395–396
  software branching strategies, 388–392
    application programming interfaces (API), 390–391
    branching strategies, 388
    file formats, 391–392
    parallel branches, 390–391
    parallel products, 391–392
    policies, 388–390
Version number, 383–388
  esoteric numbering schemes, 385–386
  progression, 384f
  significance, 383–385
  version application programming interface (API) creation, 386–388
Virtual constructor in C++, 101
Virtual destructor, 549
  coding conventions, 247–250
  compiler-generated functions, 257
  stub objects, 449
Virtual functions, 268
  backward compatibility, 401
  calls, 48
  memory optimization, 343–348
  and minimal completeness, 48–50
  observer pattern, 115–121
Virtual inheritance, diamond problem, 158–159
Virtual Reality Modeling Language, 554
Visibility
  and coupling, 66
  encapsulation, 33–35
  symbol export, 310
  UML class diagram, 29
Visitor pattern
  and inheritance, 554–560
  UML diagram, 29
Visual Basic, 508
Visualization ToolKit (VTK), 107
void keyword, definition, 160
void* pointers
  data-driven APIs, 179
  Plugin Manager, 538

W

Walkthroughs, as documentation type, 427
Waterfall development process, 443–444
Weak pointers, definition, 60
Web-based API, 329
Web browsers, 534
Web development, 9b
Web services, data-driven application programming interface (API), 203–204
White box testing. *See* Open box testing
Wiki pages, 153
Windows (Win32), 4, 403

architecture constraints, 144–146
backward compatibility, 401
flat C application programming interfaces, 180
Intel VTune, 363
Second Life Viewer, 16–17, 17f
"Winnebago" headers, 333
"Winnebago" headers, 333
Working demos, as documentation type, 426–427
World of Warcraft wiki, 423
Wrapper application programming interface (API), 108, 474
Wrapping patterns
with Boost Python, 511–513
overview, 107–115
with simplified wrapper and interface generator (SWIG), 522–523
wxWidgets library, GUI APIs, 15

X

XML, 391, 474
API examples, 16–17
data-driven web services, 203–204
code reuse, 8–10
Doxygen, 434–442
file formats, 391–392
Google Test reports, 478
Plugin Manager, 544–546
Python, 510
record and playback functionality, 473–475
software branching strategies, 388–392
test harnesses, 476
X/Motif, GUI APIs, 15
XPC, 539
xUnit, 449b

Z

Zero-argument constructor, 260

Printed and bound by CPI Group (UK) Ltd, Croydon, CR0 4YY

03/10/2024

01040327-0017